T0230792

OPERATIONS RESEARCH
Theory and Practice

Dr. N.V.S. Raju

BSP   **BS Publications**

CRC Press
Taylor & Francis Group
Boca Raton London New York

CRC Press is an imprint of the
Taylor & Francis Group, an **informa** business

CRC Press
Taylor & Francis Group
6000 Broken Sound Parkway NW, Suite 300
Boca Raton, FL 33487-2742

First issued in paperback 2023

© 2020 by N.V.S Raju and BSP Publications
CRC Press is an imprint of the Taylor & Francis Group, an informa business

No claim to original U.S. Government works

ISBN 13: 978-1-03-265420-1 (pbk)
ISBN 13: 978-0-367-36596-7 (hbk)
ISBN 13: 978-0-367-36607-0 (ebk)

DOI: 10.1201/9780367366070

Print edition not for sale in South Asia (India, Sri Lanka, Nepal, Bangladesh, Pakistan, Myanmar or Bhutan)

This book contains information obtained from authentic and highly regarded sources. Reasonable efforts have been made to publish reliable data and information, but the author and publisher cannot assume responsibility for the validity of all materials or the consequences of their use. The authors and publishers have attempted to trace the copyright holders of all material reproduced in this publication and apologize to copyright holders if permission to publish in this form has not been obtained. If any copyright material has not been acknowledged please write and let us know so we may rectify in any future reprint.

Except as permitted under U.S. Copyright Law, no part of this book may be reprinted, reproduced, transmitted, or utilized in any form by any electronic, mechanical, or other means, now known or hereafter invented, including photocopying, microfilming, and recording, or in any information storage or retrieval system, without written permission from the publishers.

Trademark notice: Product or corporate names may be trademarks or registered trademarks, and are used only for identification and explanation without intent to infringe.

Publisher's Note
The publisher has gone to great lengths to ensure the quality of this reprint but points out that some imperfections in the original copies may be apparent.

Library of Congress Cataloging-in-Publication Data
A catalog record has been requested

Visit the Taylor & Francis Web site at
http://www.taylorandfrancis.com

and the CRC Press Web site at
http://www.crcpress.com

Namo Mathrudevo bhava
Namo Pithrudevo bhava
Namo Achaaryadevo bhava
Namo Athidhidevo bhava
Ashees Shishyaah

Dedicated with Tributes

To

The Holy Feet of My Parents

Late Sri. N.V. Murali Manohar Raju & Late Smt. Ammaayi

Preface

Science is unbounded and no science was born on one day or developed overnight. It is the result of continuous development through centuries. Man has been trying to explore the best possible methods to perform his jobs to result in the most efficient and effective output within the available resources. One such attempt is made through this book Operations Research, to make our students understand and master the subject easily.

We know that management uses both art and science to make optimal decisions. These two concepts (art and science) are fully endorsed through Operations Research and thus can become a powerful tool in engineer's/manager's hands to aid his decision power. Not only engineers or managers, but also every human being requires the awareness of these Operations Research concepts. Some might have been even using unknowingly but can apply these more confidently if they are familiar with the concepts.

OBJECTIVE

My intention of writing this book is two-fold. The first objective is to dispel the phobia that Operations Research is all mathematics and that it requires a special pre-requisite of mathematical knowledge, which is not totally true. Perhaps, the orientation of teaching by mathematical lecturers might have given a scope to create such a feeling in the students. The approach in this book went on minimizing mathematical derivations, which are required neither for practical applications nor for university examinations. However, mathematical knowledge of school level is necessary and more than sufficient to read this book. Most of my students who do not have mathematical background approached me to explain the concepts with minimum or no use of mathematics. This influenced me to attempt to make their needs feasible in an interesting and lucid manner.

My second objective on writing this book is to tailor according to the growing needs of my students and staff of all colleges under Jawaharlal Nehru Technological University. Therefore a wide coverage of all the topics has been organized in five units (divided into 12 chapters in total). To maximize the utility of this book each chapter is exhaustively dealt with large number of ($\geq$ 300) solved examples with a wide variety to resolve the confusion in students. Not less than 300 practice problems along with hints and answers (almost all half solved) in addition to $\geq$ 300 solved illustrations (most of them are taken from JNTU exam papers) are shown with algorithmic approach.

A sub-objective of this book is to make this book a comprehensive volume to the faculty also. Hence, it is made in such a way that a new faculty member to this field can also learn and teach.

CONSTRAINTS

My students of B.E., B. Tech., and M. Tech. of JNTU and other universities approached me to provide an exhaustive but simplified, conceptual and handy book that must give greater than or equal to the best success in their university examination and as well that can be applied in their real life working conditions after their graduation. Best efforts have been put in to cater their needs, for which, most questions have been selected from their university examination papers and explained conceptually with examples taken from daily life with an engineering and practical orientation.

SURPLUS

Keeping all the available and requirement constraints in view, every chapter has been added with flow charts and figures in surplus to ease the understanding. Moreover, the second page of each chapter is designed in such a way to represent the entire chapter at a glance that can be very useful to students for last minute preparation of their examinations.

Further, I do not claim the originality of the concepts but for the presentation in a user-friendly form. While writing this book, I was benefited immensely by referring to many books and publications. I express my gratitude to all such authors, publishers, professors and institutions. If anybody has been missed inadvertently, I seek his or her pardon.

CONDITIONS

In spite of the utmost care taken to make this book equal to zero mistakes (error free), still some mistakes might have gone oversight. I seek your non-negative approval for the omission and commission that might have crept into this volume. Also any suggestions to improve this book are always unrestricted and gratefully acknowledged.

Dr. N.V.S. Raju
e-mail: srijanataraj@yahoo.com

Acknowledgement

At the outset, I express my deep sense of gratitude for encouragement and support extended by…

- **Dr. K. Narayana Rao,** Prof. (Retd.) of ME,OU, Former Member Secretary, AICTE, New Delhi.
- **Dr. G. Venugopal Reddy,** Prof. of CSE, Hon'ble Vice Chancellor, JNTUH, Hyderabad.
- **Dr. N.V.Ramana Rao, Rector,** Prof. of CE, JNTUH, Hyderabad
- **Dr. N. Yadaiah, Professor of EEE, Registrar, JNTUH**
- **Dr. A. Vinaya Babu,** Prof. of CSE, Former Director, Admissions, JNTUH, Hyderabad & Former Principal JNTUHCEH, Director, Indur Group of Institutions
- **Dr. C.B. Krishna Murthy,** Prof. (Retd.) of ME, JNTU (Former Principal for JPNEC, Mahboobnagar & NIET, Deshmukhi, Nalgonda).
- **Dr.** (Retd. Col.) **K. Prabhakar Rao,** Former Principal for NIET-Deshmukhi, VREC-Nizamabad & Rajamahendra Engg College.

I am grateful to my present colleagues, **Dr. K. Vasantha Kumar** APME, **Dr. Suresh Arjula** APME, **Mrs. M. Shailaja,** APME, JNTUCEJ for their moral support and motivation.

I thank my former colleagues **Dr. N. Venkateshwarlu** Assoc. Prof. in MED-SOET, IGNOU **Mr. Soma Sundaram Kalaga, Dr. B. Kranthi Kiran,** APCSE, JNTUHCEJ, **Dr. D.V. Ravi Shankar,** Prof. of ME & Principal, TKR Engg College, **Mr. A. Suresh Rao,** Prof. in CSE, **Mr. N.L. Narayana,** Prof. of ME, SVIT, and all my present & former colleagues and students.

My special thanks are due to my students and Lecturers (Ad-hoc) **Mr. S. Shiva Rama Krishna (SSRK), Mr. Shivraj Chawariya, Mr. M.Pradeep Kumar** (Presently APME, KITS-S), for helping me in preparing the manuscript.

While authoring this book, I had to go through numerous books on Operations Research and optimization techniques and met many professors/practitioners. I thank all those authors and who directly or indirectly helped me while authoring this book. If feel sorry if I missed any inadvertently.

I further acknowledge the computerization team of this book and active follow up and processing by the publishers **Sri Nikhil Shah** and **Sri Anil Shah**. The work of **Mr. M. Vasudeva Rao, Mr. Naresh, Ms. Sandhya** and **Mr. Hari Prasad** are praiseworthy during the publishing process.

Last but not the least, I thank my wife **Mrs. Prasanna** and children **Kum. Srija Nataraj, Kum. Himaja Padma Priya** and **Kum. Sanja Karunamayi** for their co-operation & patience and my Parents **Smt. & Sri (Late) N.V. Murali Manohar Raju** and Brothers **Dr. Venkata N. Raju** and **Mr. N.V. Srinivasa Raju** and sisters **Smt Hema latha, Smt. Prema latha** and **Smt. Suma latha** for their encouragement. Finally I thank entire **Kasu** and **Chennamadhavuni** families for showering their good wishes on me.

Dr. N.V.S. Raju

Contents

CHAPTER 2

Formulation of Linear Programming Problems

CHAPTER 3

Graphical Solution of LPP

CHAPTER 4

Simplex Method

CHAPTER 5

Transportation Problem

CHAPTER 6

Assignment Problem

CHAPTER 7

Inventory Management

CHAPTER 8

Queue Models (Waiting Lines)

CHAPTER 9

Sequencing

Chapter 10

Replacement Analysis

CHAPTER 11

Game Theory

CHAPTER 12

Dynamic Programming

1 Chapter

Operations Research – An Overview

CHAPTER AT A GLANCE

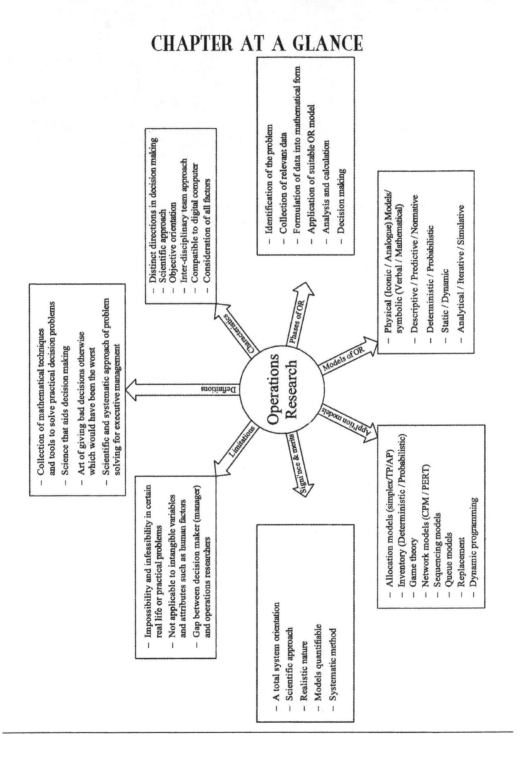

1.0 *Introduction*

"Science that aids decision making is Operations Research which is tested and implemented through appropriate modelling."

We take several decisions in our daily life. Most of these decisions are taken by common sense. But the decisions taken by mere common sense sometimes may mislead or confuse us. Also, such decisions may neither provide any evidential support nor stand on any scientific base. Therefore, it has become necessary for managers and engineers to believe in the science that provides the evidential support and scientific base.

Operations Research (OR) is one such science that provides better solutions to the managers, engineers and any practitioners with better solutions. This science came into existence during World War II. Though it was first employed for military operations, its applications are extended to any field on the earth in some form or other.

Thus OR is considered as the science that deals with decision making and this book deals with formulating, analysing, testing and application of various OR models.

1.1 *Origin and Development of OR Models*

The term, *"Operations Research"* was first coined by Mc Closky and Trefthen in 1940 in a small town, Bowdsey of United Kingdom.

The name operations research was given to this subject because it has started with the *research* of (military) *operations*. During world war - II, the military commands of UK and USA engaged several teams of scientists to discover tactical and strategic military operations. Their mission was to formulate specific proposals and to arrive to the decisions that can optimally utilize the scarce resources to acquire maximum possible level of effective results. In simple words, it was to uncover the methods that can yield greatest results with little efforts. Thus it has gained popularity and was called *"an art of winning the war without actually fighting it"*.

Following the end of the war, the success and encouraging results of British teams have attracted industrial managers to apply these methods to solve their complex problems. The first method in this direction was simplex method (LPP) developed in 1947 by G.B. Dantzig, USA. Since then several scientists have been developing this science in the interest of making operations to yield high profits or least costs.

Now, this science has become universally applicable to any area such as transportation, hospital management, agriculture, libraries, city planning, financial institutions, construction management and so forth. In India, many industries have been realising the advantages by implementing the OR models.

A few to quote in this regard are Delhi Cloth Mills, Indian Airlines, Indian Railways, Hindustan Liver Ltd., (HLL), Tata Iron & Steel Co., (TISCO), Fertilizer Corporation of India (FCI), Life Insurance Corporation (LIC) of India etc.

1.2 *Operations Research : Some Definitions*

Because of the wide scope of applications of operations research, giving a precise definition is difficult. However, a few definitions of OR are as under.

- *Operational research is the application of the methods of science to complex problems, in the direction and management of large systems of men, machines, materials and money in industry, business, government and defense. The distinctive approach is to develop a scientific model of the system incorporating measurements of factors such as change and risk, with which to predict and compare the outcome of alternative decisions, strategies or controls. The purpose is to help management in determining its policy and actions scientifically.*   Operations Research Society, UK

- *Operations Research is concerned with scientifically deciding how to best design and operate man - machine systems usually requiring the allocation of scarce resources.*
                                                                    Operations Research Society, America

Apart from being lengthy, the definition given by Operational Research Society of UK, has been criticised because it emphasizes complex problems and large systems, leaving the reader with the impression that it is a highly technical approach suitable only to large organisation. The definitions of OR society of America contains an important reference to the allocation of scarce resources. The key words used in the above definitions are *scientific approach, scarce resources, system and model.* The British definition contains reference to optimisation, while the American definition quietly slips in the word, best. A few other definition which are commonly used and widely acceptable are as follows :

- *Operations Research is the systematic application of quantitative methods, techniques and tools to the analysis of problems involving the operation of systems.*
                                                                    Daellenbach and George, 1978

- *Operations Research is essentially a collection of mathematical techniques and tools which in conjunction with a systems approach, are applied to solve practical decision problems of an economic or engineering nature.*   Daellenbach and George, 1978

These two definitions project another view of OR as the collection of models and methods which have grown up largely independent of one another.

- *Operations Research utilizes the planned approach (updated scientific method) and an interdisciplinary providing a quantitative basis for decision making and uncovering new problems of quantitative analysis.*   Thierauf and Klekamp, 1975

- *This new decision making field has been characterised by the use of scientific knowledge through inter disciplinary team effort for the purpose of determining the best utilisation of limited resources.*   H.A Taha, 1976

These two definitions refer to the inter disciplinary nature of OR. However, one of the best definitions, given by Churchman, Ackoff and Arnoff, is as follows :

- *Operations Research, in the most general sense, can be characterised as the application of scientific methods, techniques and tools, to problems involving the operations of a system so as to provide those in control of the operations with optimum solutions to the problems.*
  Churchman, Ackoff and Arnoff, 1957

- *Operation Research has been described as a method, an approach, a set of techniques, a team activity, a combination of many disciplines, an extension of particular disciplines (mathematics, engineering and economics), a new discipline, a vocation, even a religion. It is perhaps some of all these things.*
  S L Cook, 1977

- *OR is the art of giving bad answers to problems to which otherwise worse answers are given.*
  T L Saaty, 1958

- *Operations Research may be described as a scientific approach to decision making that involves the operations of organisational system.*   F S Hiller and G J Lieberman, 1980

- *Operations Research is a scientific method of providing executive departments with a quantitative basis for decisions under their control.*   P M Morse and G E Kimball

- *Operations Research is applied decision theory. It uses any scientific, mathematical, or logical means to attempt to cope with the problems that confront the executive, when he tries to achieve a through - going rationality in dealing with his decision problems.*
  D W Miller and M K Stan

- *Operations Research is a scientific approach to problems solving for executive management.*   H M Wagner

As the discipline of operations research grew, many other names such as Optimization Techniques, Operation Analysis, Systems Analysis, Decision Analysis, Management Science were given to it. However, each of these emphases the quantitative approach to the analysis and solution of management problems.

1.3 *Phases (or Steps) of Operation Research Method (Modelling)*

About four to five decades ago it would have been difficult to get a single operations researcher to describe a procedure for conducting OR project. But today, every engineer and manager are presenting the procedure in different ways as suits to their situations. However, the procedure for an OR study generally involves the following major phases :

Phase - I :   Formulating the problem

Phase - II :   Fitting a Suitable 'OR' model

Phase - III :   Analysis and Deriving the solutions from the model

Phase - IV :   Testing the Model and its solution

Phase - V :   Sensitivity analysis and Controlling the solution

Phase - VI :   Decision making and Implementing the solution

Phase - I : *Formulating the Problem :*

To find the solution of a problem, the problem must be formulated in the form of an appropriate model. This requires the following information.

1. Who has to take the decision?

2. What are the objectives?

3. What are the ranges of controllable variables?

4. What are the uncontrolled variables that may affect the possible solutions?

5. What are the restrictions (or) constraints on the variables?

6. What are other conditions or nature of variables?

The formulation should be considerably careful while executing this phase because a wrong formulation cannot give a right decision (solution), and even may be disastrous in some cases.

Phase - II : *Fitting a Suitable 'OR' Model :*

The next phase of the investigation is concerned with the searching of suitable OR model in an appropriate form which is convenient for analysis. It requires the identification of both static and dynamic structural elements. The OR model consists of the following three important basic factors.

1. Decision variables and OR parameters;

2. Constraints (or) restrictions;

3. Objective function.

Various models are discussed in the next sections (1.4 & 1.5)

Phase - III : *Analysis and Deriving the Solutions from the Model :*

The third phase involves in the computation of those values of decisions variables that maximise or minimise the objective function. Such solution is called an optimal solution which is always in the best interest of the problem under consideration.

Phase - IV : *Testing the Model and its Solution :*

After computing and deriving the solution from the model, it is once again tested as a whole for the errors if any. A model may be said to be valid if it can provide a reliable prediction of the system's performance. A good practitioner of Operations Research believes that his model be applicable for a longer time and thus updates the model time to time by considering the past, present and future specifications of the problem.

Phase - V : *Sensitivity* Analysis and Controlling the Solution :

The fifth phase establishes controls over the solution with desired degree of satisfaction. The model requires immediate modification as soon as the controlled variables (one or more) change significantly, otherwise the model goes out of control. This action is often referred to as sensitivity analysis. As the conditions are constantly changing in the world, the model and the solution may not remain valid for a long time. Therefore it leaves a lot of scope for further improvement along with controlling.

Phase - VI : *Decision Making and Implementing the Solution :*

Finally, the tested results of the model are implemented to work. This phase is primarily executed with the co-operation of operations Research experts and those who are responsible for managing and operating the systems. However, this job is more concerned with the production dept of executors of the model. Moreover, it essentially requires to take into confidence, the willingness and belief of the people who implement it or involve in it.

1.4 *Operation Research Models : Classification*

OR model is a representation of a real life situation. OR models can be studied by classifying in many ways. These are discussed here below :

1.4.1 *Classification Based on Structure*

1. **Physical** Models : **These models give a physical appearance of the real object either in reduced form or scaled up. These are further divided into two categories.**

   (a) *Iconic* Models : *These are representations in either idealised form or a scaled version. (i.e., enlarging or reducing in size) of real objects. e.g., Blue prints, Globe, Photographs, Drawings, Templates etc.*

   (b) *Analogue* Models : *These models represent a system by a set of properties different from that of the original system, and physically do not resemble. After attaining solution, it is re-interpreted in terms of original system. e.g., frequency curves, flow charts, organisation charts etc.*

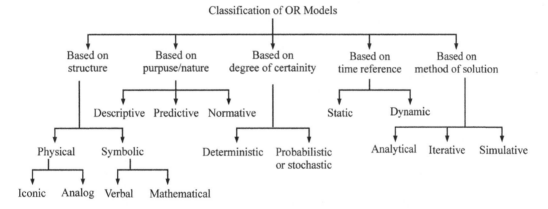

FIGURE 1.1 : CLASSIFICATION OF OR MODELS

2. **Symbolic** Models : **These models use symbols in the form of letters, mathematical operators or any other symbols to represent the properties of the system. These are often described in two types.**

   (a) *Verbal* Models : *These models are used to describe a situation in written or spoken language in the form of letters, words and sentences.*

   (b) *Mathematical* Models : *These models use mathematical symbols, letters, numbers and mathematical operators to represent relationship among various variables of the system to explain its behaviour and properties.*

1.4.2 *Classification Based on Purpose and Nature*

1. **Descriptive** Models : **The reports of surveys, questionnaire results, inferences of the observations etc., are used in such models to describe the situation. These also include the models such as plant layout diagram, block diagram of an algorithm etc.**

2. **Predictive** Models : **These models are results of the quiery such as "what will follow if this occurs or does not occur?" e.g. Preventive maintenance schedules.**

3. **Normative Model** (or Optimisation Models) : **These models are designed to provide 'optimal' solution to the problems subject to certain limitations on the use of resources or meeting the requirements or at the conditions that normally exist.** *e.g.* **Linear Programming Problem.**

1.4.3 *Based on Certainty*

1. **Deterministic** Model : **If all the parameters of decision variables, constants and their functional relationship are known (or assumed to be known) with certainty, then the model is said to be deterministic. e.g. Certain inventory models, games with saddle points.**

2. **Probabilistic** (or Stochastic) Model : **This is the model in which at least one parameter or decision variables is a random variable. e.g. Probabilistic inventory models, games without saddle points, queuing models.**

1.4.4 *Based on Time Reference*

1. **Static** Model : **These models present a system at a specified time, which do not account for changes over certain period of time. e.g. Replacement of machines when money value is not changing with time.**

2. **Dynamic** Model : **Time is considered as one of the variables and the impact of changes generated by time is accounted while selecting the optimal course of action. e.g. Machine replacement model when money value is changing with time.**

1.4.5 *Based on Method of Solution*

1. **Analytical** Model : **These have a specific mathematical structure and hence can be solved by analytical or mathematical techniques. e.g. Any optimisation model, such as inventory models, waiting lines etc.**

2. **Iterative** Models : **In these models, the solution is obtained from the conclusion of previous step. e.g. Simplex method for LPP, dynamic programming.**

3. **Simulation** Models : **Though these models have a mathematical structure, they can not be solved by applying mathematical techniques. Instead, a simulation model is essentially a computer assistant experimentation on a mathematical structure of a real life problem under certain assumptions over a period of time. e.g. Monte-Carlo simulation, use of random numbers, forecasting models etc.**

1.5 *Application Models of Operations Research*

Basic Operation Research Application Models are discussed here below

1. **Allocation** Models : **Allocation models are used to allocate resources to activities in such a way that sum measure of effectiveness (objective function) is optimized. e.g. Linear Programming Problem (Simplex, Transportation, Assignment etc.,) and Non–linear programming.**

2. **Inventory** Models : **Inventory models are used to determine how much to order and when to place an order so as to optimize the inventory costs such as order cost, carrying cost, shortage cost etc.,**

3. **Competitive** (Game Theory) Models : **These models are used to characterize the behaviour of two or more opponents (called players) who compete for the achievement of conflicting goals. These models are classified according to number of competitors, sum of loss and gain and type of strategy.**

4. **Network** Models : **These models are applied to management (planning and scheduling) of large scale projects, PERT/CPM help in identifying trouble spots in a project by critical path and to determine time-cost trade off, resource allocation and updating of activity times.**

5. **Sequencing** Models : **This arises whenever there is a problem in finding a sequence or order in a number of tasks performed by a number of service facilities. Here, the total time to process all the jobs on all the machines is optimized.**

6. **Waiting Line** (Queuing) Models : **These are used to establish a trade off between cost of providing service and the waiting time of a customer in queuing system by using the probabilities and averages.**

7. **Replacement** Models : **To decide the optimal time to replace equipment, for instance when equipment deteriorates or fails, these models are applied.**

8. **Dynamic** Programming Model : **These may be considered as an out growth of mathematical programming, which involves optimization of multi stage, inter–related decision process.**

9. **Markov** Chain Models : **Used for analysing a system which changes over a period of time among various possible out comes or states.**

10. **Monte-Calro** Simulation : **Simulation is a quantitative procedure, which describes process by developing a model of that process and then conducting a series of organised experiments to predict the behaviour of the process over time. To find out how the real process would react to certain changes, we can produce these changes in our model and simulate the reaction of the real process to them.**

1.6 *Significance and Applications of OR in Industrial Problems*

The Operations Research approach becomes very significant in industrial problems particularly, when the resouces are scarce or/and when a balance is to be brought between conflicting goals where there are many alternative courses of actions available to the decision maker. Let us consider one such example given below.

Suppose, a decision has to be taken regarding inventory management. The objective of production manager is based on quality of material and availability on time. While a finance manager thinks on the line of minimising the costs. Therefore he would prefer to order a party who can supply at lowest costs or can offer some discount. The materials manager wishes to safeguard production from stockouts and against demand fluctuations etc. Therefore he would like to order in such way. But the stores manager will be particular about accommodating them in his stores and other related problems in protecting the materials while storage and so on. Thus all these managers are trying minimise the cost or maximise profit to the organisation. Though every manager is thinking in the interest of organisation growth, their objectives vary according to their specification.

In view of the situations like above a manager has to derive the decisions which should consist.

1. A total system *orientation*

2. Scientific *approach*

3. Realistic *nature*

4. *Models should be able to be expressed quantitatively.*

5. Systematic *method.*

OR is the science that is embedded with suitable blend all the above features.

Thus, OR can play a significant role in bringing a balance among different interdisciplinary people to managerial problems.

Some Areas of Applications :

1. Design of aircraft and aerospace structures for minimum weight.

2. Finding the optimal trajectories of space vehicles.

3. Design of civil engineering structures like frames, foundations, bridges, towers, chimneys and dams for minimum cost.

4. Optimum design of linkages, cams, gears, machine tools and other mechanical components.

5. Selection of machining conditions in metal cutting processes for minimum production cost.

6. Design of material handling equipment like conveyors trucks and cranes for minimum cost.

7. Design of pumps, turbines and heat transfer equipment for maximum efficiency.

8. Optimum design of electrical machinery like motors, generators and transformers.

9. Optimum design of electrical networks.

10. Shortest route taken by salesman visiting different cities.

11. Optimum production planning, controlling and scheduling.

12. Analysis of statistical data and building empirical models from experimental results to obtain the accurate representation of the physical phenomenon.

13. Optimum design of chemical processing equipment and plants.

14. Design of optimum pipe line networks for process industries.

15. Selection of site for an industry.

16. Inventory control to minimise inventory costs such as ordering cost, carrying costs, shortage costs etc.

17. Planning of maintenance and replacement of equipment to reduce the operating costs.

18. Allocation resources of services among several activities to maximise the benefit.

19.

Controlling the waiting and idle items and queuing in production lines to reduce the costs.

20. Planning the best strategy to obtain maximum profit in the presence of competitor.

1.7 Characteristic features of Operations Research Model (or Merits of Operations Research Models)

Some significant features of Operations Research Model are given below.

1.7.1 Distinct Direction in Decision Making

Operations Research model provides a clear and distinct direction to the managers in decision making and problem solving. A major premise of Operations Research is that decision making irrespective of the situation involved, can be considered as a general systematic process.

1.7.2 Scientific Approach

Operations Research employs scientific reasoning to its problems. Therefore the managers can confidently implement their decisions. Even if they fail after taking a decision provided by Operations Research, they will have scientific and evidential basis to plead in their support.

1.7.3 Objective Orientation

Operations Research is oriented to locate the best possible or optimal solution to the problem. As the approach itself is embedded with setting of the goal or objectives, it becomes easy to use this as a measure to compare the alternative courses of action.

1.7.4 Inter - Disciplinary Team Approach

Operations Research is inter disciplinary in nature and therefore needs a team approach. It is a blend of the aspects of various disciplines such as economics, physics, physiology, sociology, anatomy, engineering, technology, mathematics, statistics and management. This feature keeps Operations Research on the common berth to all sectors of people and builds an espirit-de-corps. It can also provide a solution acceptable to all the people.

1.7.5 Compatible to Digital Computer

Perhaps the use of digital computer has become an integral part of the Operations Research approach to decision making. There are several software packages developed with the help of Operations Research approach to problems with high volume and complexive in nature.

1.7.6 Consideration of All Factors

Operations Research takes into account of the goals (objective function) of the organisation with all bottlenecks or hurdles (constraint set) and the feasibility (conditions of variables). This feature provides a manager to take a decision that can keep himself or his organisation on a competitive edge.

1.8 *Limitations (Demerits) of Operations Research*

Operations Research, though widely used has got certain limitations. These are given below :

1.8.1 *Impossibility and Infeasibility*

Operations Research takes care of all the factors in choosing the best alternative. In modern society and in real life these factors are numerous and establishing relation among these is either impossible or infeasible in many cases. Thus many problems will be left unsolved.

1.8.2 *For Intangible Situations*

Operations Research can provide solution to those problems in which all the factors can be quantifiable. But it can not provide any solution to intangible variables and attributes which are qualitative in nature such as a human factors etc.

1.8.3 *Distance Between Managers (Decision Maker) and Operation Researcher*

Operations Research is specialist's job and requires the knowledge of mathematics, statistics etc. The researcher thus may not be aware of the business problem and the aspect of flexibility cannot be clarified by him.

At the same time the managers may fail to understand the complex working of the OR. Some times they may not even find time to notice any misconceptions. Thus there is a large gap built up between one who provides solutions and one who uses them. This leads to confusion and poses lot of problems during implementation or practice. Therefore it may result in utter failure even with a successful formula.

Review Questions

1. Discuss the phases of Operations Research.

2. Discuss the characteristics and limitations of Operations Research.

[Mech. 96, 95/C, 97/S, 98/P, 99/C]

3. Discuss the types and characteristics of models used in Operations Research.

[CSE - 98/S]

4. What are the different types of models used in Operations Research? Mention general methods of solving Operations Research models.

[Mech. 95/S, 96/S, CSE 96, 98]

5. Explain engineering applications to optimization techniques. **[EEE 95, 97, 98/S]**

6. "Operations Research provides bad solutions, otherwise which would have been the worst." Comment.

7. Discuss the merits and demerits of Operations Research models.

8. Discuss the strengths and weakness of Operations Research approach.

9. Discuss the significance of OR in solving managerial problems. **[CSE 2000]**

10. Explain the characteristic features of OR **[CSE 2000/S]**

11. Write short note on 'General methods of solving OR models'

[JNTU - Mech./Prod./Chem. - 2001/S]

12. Write critical essay on the definition and scope of OR **[JNTU - CSE/E.Com. E 2001]**

13. Define Operations Research **[OU - MBA - 90, Dec. 95, Sep 2001,]**

14. State three distinguishing characteristics of OR

[MBA - OU - May 91 Dec. 2000, Sep. 2000]

15. Explain the role of OR in management decision making **[MBA - OU - March 99]**

16. What is "team approach" in OR **[MBA - OU - July 2000]**

17. Write any three definitions of OR **[MBA - OU - Nov. 94]**

Objective Type Questions

1. Which of the following models does not use probabilities.

(a) inventory models   (b) game theory models
(c) queuing models   (d) linear programming

2. Which of the following belongs to Operations Research model classified on the basis of time reference.

    (a) predictive model   (b) normative model

    (c) dynamic model   (d) simulation

3. In which of the following models, we do not use alphabet

    (a) mathematical model   (b) descriptive

    (c) verbal   (d) none of the above

4. Deterministic models are based on

    (a) time frame   (b) degree of certainty

    (c) method of solution   (d) nature or purpose of use

5. Iconic model uses

    (a) scaled version   (b) expectation

    (c) synergy   (d) time reference

6. The Operations Research model represents a system, but does not physically resemble the components of the system is

    (a) iconic   (b) analogue   (c) symbolic   (d) normative

7. Game theory without saddle points belongs to _____ model of OR

    (a) static   (b) dynamic   (c) deterministic   (d) probabilistic

8. Replacement models when money value not changing with time can be considered as

    (a) static   (b) dynamic   (c) probabilistic   (d) simulative

9. Which of the following is not a strength of Operations Research.

    (a) objective orientation   (b) distinct decision making

    (c) team approach   (d) gap between the user and OR model designer

10. Which of the following is not a limitation of Operations Research.

    (a) qualitative aspects

    (b) gap between decision maker and model maker

    (c) involvement multiple constraints

    (d) scientific approach

11. The Operations Research widely employed in marketing problems is

    (a) goal programming   (b) game theory

    (c) dynamic programming   (d) linear programming

12. "If this occurs, what will follow?" This question is connected with _____ model.

(a) normative   (b) iconic   (c) predictive   (d) descriptive

13. The scaled version of a real object is _____

    (a) iconic model   (b) descriptive

    (c) normative model   (d) static model

14. Find odd man out.

    (a) iconic model   (b) verbal model

    (c) mathematical model   (d) probabilistic model

15. The failures of decisions based on Operations Research modeling to managerial problems is attributed to

    (a) model makers ignorance   (b) managers inefficiency

    (c) decision makers fault   (d) understanding between decision maker and model maker

16. In which of the following, deterministic models are not found.

    (a) games and competitive strategies   (b) inventories

    (c) waiting line models   (d) replacement models

17. Plant layout diagram is an example as best fit in

    (a) probabilistic model   (b) dynamic model

    (c) descriptive model   (d) simulative model

18. Demand forecasting can be fit as an example of

    (a) static model   (b) simulative model

    (c) deterministic model   (d) symbolic model

19. Consider these following two groups :

1. Degree of certainty	( )	A.	Iterative
2. Time reference	( )	B.	Normatives
3. Application	( )	C.	Stochastic
4. A method of solution	( )	D.	Verbal
5. Symbolism	( )	E.	Static

    The correct matching is

    (a) 1-C, 2-E, 3-B, 4-D, 5-A   (b) 1-C, 2-E, 3-B, 4-A, 5-D

    (c) A-4, B-3, C-4, D-1, E-2   (d) A-3, B-4, C-2, D-5, E-1

20. Operations Research models can not work in the case of

    (a) tangible variables   (b) quantitative factors

   (c) human factors   (d) decision maker knows OR.

21. Match the following:

    1. Iconic model   ( )   A. Explanation
    2. Dynamic model   ( )   B. Words × letters
    3. Predictive model   ( )   C. Changes with time
 4. Stochastic model   ( )   D. Expectation through clues
 5. Verbal model   ( )   E. Scaled version
 6. Descriptive   ( )   F. Comparative
                                         G. Involvement of risk

Fill in the Blanks

1. The scaled version of a real object can be termed as _____ model of Operations Research.

2. Operations Research problems take into consideration of the objective function, constraints and _____.

3. According to Saaty " Operations Research is choosing _____ decision otherwise which would have been _____.

4. The goodness or optimality of an Operations Research problem is checked by _____.

5. The static and dynamic models of Operations Research are based on _____ reference.

Answers

Objective Type Questions :				
1. (d)	2. (c)	3. (d)	4. (b)	5. (a)
6. (b)	7. (d)	8. (a)	9. (d)	10. (d)
11. (b)	12. (c)	13. (a)	14. (d)	15. (d)
16. (c)	17. (c)	18. (b)	19. (d)	20. (c)

Match the Following :		
1. (e)	2. (c)	3. (d)
4. (g)	5. (b)	6. (a)

Fill in the Blanks :		
1. iconic	2. conditions of variables	3. the bad, the worst
4. objective function	5. time	

2 Chapter

Formulation of Linear Programming Problems

CHAPTER AT A GLANCE

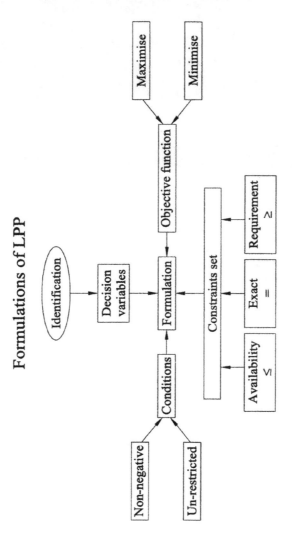

2.0 *Introduction*

In the previous chapter we have learnt that after collection of relevant data, it is to be translated into appropriate mathematical or Operations Research model. One of such models applied to production and allocations is linear programming in which the variables are linearly related. This process of translation is called formulation. In this chapter we learn how to formulate the Linear Programming Problem (LPP).

2.1 *Formulation*

The formulation of relevant data in Linear Programming Problem (LPP) is carried out in the following four steps :

Step 1 : Selection of decision variables.

Step 2 : Setting the objective function.

Step 3 : Identification of constraint set.

Step 4 : Writing the conditions of variables.

These are explained as follows :

Step 1 : **Selection of** Variables :

In the given data, firstly the variables are to be identified. These are also called decision variables or design variables.

A variable in the context of Operations Research problems is often seen in the following ways.

    (a) Number of different types of products to be manufactured per day (or per week) in a manufacturing problem (product mix).

    (b) Number of products to be sold in a sales problem.

    (c) Number of resources (such as men / machine / material) required to do certain job in an allocation problem.

    (d) Number of units of certain model or type to be bought at lowest cost.

    (e) Number of ingredients to be mixed or used in production process problem.

    (f) Number of units to be ordered, in the purchasing problem

    (g) Number of units to be advertised, in a marketing problem

    (h) Number of units of fertilizers or ingradients in a fertilizer to be used in an agricuturing problem.

    (i) Number of tablets / capsules to purchase in a medical / patients problem.

Step 2 : **Setting** Objective Function :

This is to set the goal in the problem. There are only two types of objective functions in Operations Research.

These are : *Maximisation (of profits) and Minimisation (of costs).*

A maximisation type of objective function is set if the data reveals the profits in the problem. The production, sales, revenue, output etc., are to be maximised.

A minimisation type of objective function is set if the data problem is given in terms of costs. The resources (and their utilisation) are to be minimised. The time utilised at given cost, loss, wastage, shortage, spoilage, costs etc., are some other terms where we think of minimisation.

Step 3 : **Identifying the** Constraint Set:

A constraint set is a set of the hurdles or limitations in achieving the objective.

The different types of constraints we come across in linear programming problem are given below :

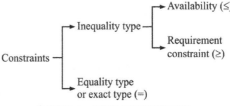

FIGURE 2.1 : TYPES OF CONSTRAINTS

1. *Availability* Constraint : *This is related by an inequality, less than or equal to (≤) and is used when the availability of resource is limited or when maximum limit is imposed.*

Examples :
   (a) *Suppose you are cooking some curry dish and you have to put some salt in it, say 10 gm. You will be constrained to put less than or equal to 10 gm only, but excess is not allowed.*
   (b) *Suppose you have only Rs 100/- in you wallet and went to a restaurant. You will be constrained to order worth less than or equal to Rs. 100/- only. Here x ≤ 100 where x is rupees spent.*
   (c) *You have a machine on which you can utilise 8 hours a day. Thus you can have a maximum of 48 hours a week (of 6 days).*
   Now number of hours you can utilise on the machine (say x) ≤ 48.

2. *Requirement* Constraint : *This constraint is used to show a relation 'greater than or equal to' type of inequality. This constraint is used when certain minimum limit is imposed. Here, more than certain limit is permitted but lesser quantity is not allowed.*

Examples :
   (a) *You are to buy a ready made trouser or skirt. You will be constrained to choose greater than or equal to the required length.*
   (b) *You have a production unit at which you have an order for 50 units. Now if the number of units to be produced is x, then x ≥ 50.*
   (c) *Suppose you are pricing a product then you will say it should be greater than or equal to its (manufacturing and transport) costs.*

3. *Exact* Constraint : *This is the constraint which can deviate to neither side. This is represented in the equation (=) form.*

Examples :
   (a) *You have to order the lens power for your spectacles. Here you will not permit lesser or greater than your specified power.*
   (b) *When you purchase the shoe, you neither allow 'less than' nor 'greater than' your feet size.*
   (c) *The doctor prescribes certain quantity of antibiotic that must be the exact dosage.*
   (d) *Certain design specification in manufacturing will be exact kind.*

Step 4 : **Writing** Conditions of Variables :

Here the conditions of the decision variables will be predetermined. The conditions often found in LPP are in two kinds. These are non-negative and unrestricted. A non-negative conditions is represented by 'greater than equal to zero' (≥ 0).

The above steps are illustrated with the following example.

ILLUSTRATION 1 ──

Sainath and Co manufactures two brands of products namely Shivnath and Harinath. Both these models have to under go the operations on three machines lathe, milling and grinding. Each unit of Shivnath gives a profit of Rs. 45 and requires 2 hours on lathe, 3 hours on milling and 1 hour on grinding. Each unit of Harinath can give a profit of Rs. 70 and requires 3, 5, and 4 hours on lathe, milling and grinding respectively. Due to prior commitment, the use of lathe hours are restricted to a maximum of 70 hours in a week. The operators to operate milling machines are hired for 110 hours / week. Due to scarce availability of skilled man power for grinding machine, the grinding hours are limited to 100 hours/week. Formulate the data into on LPP.

Solution :

Step 1 : **Selection of Variables :**

In the above problem, we can observe that the decision is to be taken on how many products of each brand is to be manufactured. Hence the quantities of products to be produced per week are the decision variables.

Therefore we assume that the number of units of product Shivnath brand produced per week = x_1.

The number of units of product of Harinath brand produced per week = x_2.

Step 2 : **Setting Objective :**

In the given problem the profits on the brands are given.

Therefore objective function is to maximise the profits.

Now, the profit on each unit of Shivnath brand	=Rs. 45.
Number of units of Shivnath to be manufactured	= x_1
$\therefore$ The profit on x_1 units of Shivnath brand	= $45\,x_1$
Similarly, the profit on each unit of Harinath brand	= Rs. 70
Number of units of Harinath brand to be manufactured	= x_2
$\therefore$ The profit on x_2 units of Harinath brand	= $70\,x_2$
The total profit on both brands	= $45x_1 + 70x_2$

This total profit (say z) is to be maximised.

Hence, the objective function is to

$$\boxed{\text{Maximise } z = 45x_1 + 70x_2}$$

Step 3 : **Identification of Constraint Set :**

In the above problem, the constraints are the availability of machine hours.

1. Constraint on Lathe Machine : Each unit of *Shivnath* brand requires 2 hours/ week

So x_1 units of *Shivnath* brand requires $2x_1$ hours / week.

Each unit of Harinath brand requires 3 hours / week
and So x_2 units of *Harinath* brand require $3x_2$ hours / week.

Total lathe hours utilised for both the brands is $2x_1 + 3x_2$
and this cannot exceed 70 hours/ week.

∴ $\boxed{2x_1 + 3x_2 \le 70}$

(Constraint on availability of lathe hours due to prior commitment)

2. Constraint on Milling Machine : ***Milling hours required for each unit of Shivnath brand*** = **3 hours/week.**

∴ For x_1 units = $3x_1$ hours/week

Milling hours required for each unit of *Harinath* brand = 5 hours/week.

∴ For x_2 units = $5x_2$

Total milling hours = $3x_1 + 5x_2$

This can not be more than 110

∴ $\boxed{3x_1 + 5x_2 \le 110}$

(Constraint on availability of milling machine hours due to hiring)

3. Constraint on Grinding Machine :

One unit of *Shivnath* needs one hour/week and x_1 units need x_1 hours/week

One unit of *Harinath* needs 4 hours/week and x_2 units need $4x_2$ hours/week.

Total grinding hours = $x_1 + 4x_2$ and this cannot be greater than 100 hours.

∴ $\boxed{x_1 + 4x_2 \le 100}$

(Constraint on availability of grinding hours due to scarcity of skilled labour)

Step 4 : **Writing** Conditions of Variables :

Both x_1 and x_2 are the number of products to be produced. There can not exist any negative production. Therefore x_1 and x_2 can not assume any negative values (i.e., non negative)

Mathematically

$\boxed{x_1 \ge 0 \text{ and } x_2 \ge 0}$

Step 5 : **Summary :**

Maximise $Z = 45x_1 + 70x_2$

Subject to $2x_1 + 3x_2 \leq 70$

$3x_1 + 5x_2 \leq 110$

$x_1 + 4x_2 \leq 100$

$x_1 \geq 0$ and $x_2 \geq 0$

Some More Examples :

ILLUSTRATION 2 ────────────────────────────

Formulate the following problem as an LP Problem:
A firm engaged in producing 2 models X_1, X_2 performs 3 operations Painting.
Assembly and Testing. The relevant data are as follows:

Model	*Unit Sales Price*	*Hours Required for each unit*		
		Assembly	*Painting*	*Testing*
Model X_1	*Rs. 50*	*1.0*	*0.2*	*0.0*
Model X_2	*Rs. 80*	*1.5*	*0.2*	*0.1*

Total number of hours available are:
For Assembly   *600*
For Painting   *100*
For Testing   *30*
Determine weekly production schedule to maximise revenue. [JNTU Mech. 98/8]

Solution :

Note : *Only Formulation part of the solution is given here*

Variables :

Let the weekly produced units of model $X_i = x_i$

i.e., Let the no. of units produced per week in Model $X_1 = x_1$ and let the no. of units produced per week in model $X_2 = x_2$

Objective Functio :

Unit sales price for model $X_1 =$ Rs. 50/- .

For x_1 units, weekly sales revenue $= 50\, x_1$.

Similarly, unit sales price for model $X_2 =$ Rs. 80/-.

For x_2 units, weekly sales revenue $= 80\, x_2$.

Total sales revenue per week $= 50\, x_1 + 80\, x_2$.

Hence the objective function is to –

$$\boxed{\text{Maximise } Z = 50\, x_1 + 80\, x_2}$$

Constraint Set :

For assembly,

one unit of model X_1 requires 1.0 hrs.

$\therefore$ x_1 units require $1.0\,x_1$ hrs.

one unt of model X_2 requires 1.5 hrs.

x_2 units of model X_2 requires $1.5\,x_2$ hrs.

Total assembly hours per week $= 1.0\,x_1 + 1.5\,x_2$.

This cannot exceed 600 hrs since only 600 hrs are available for assembly.

$\therefore$ $\boxed{1.0\,x_1 + 1.5\,x_2 \le 600}$

(Constraint on availability of assembly hours)

Similarly,

For painting

x_1 units require $0.2\,x_1$ hrs

and x_2 units require $0.2\,x_2$ hrs

$\therefore$ $\boxed{0.2\,x_1 + 0.2\,x_2 \le 100}$

(Constraint on availability of painting hours)

For Testing

$$0.0\,x_1 + 0.1\,x_2 \le 30$$

or

$\boxed{0.1\,x_2 \le 30}$

(Availability constraint on testing hours)

Conditions of Variables :

As the problem is to determine weekly production schedule and we cannot have a negative production.

Therefore,

$\boxed{x_1 \ge 0 \ \text{ and } \ x_2 \ge 0}$

Summary of Foramulation :

maximize $Z = 50\,x_1 + 80\,x_2$

subject to $1.0\,x_1 + 1.5\,x_2 \le 600$

$0.2\,x_1 + 0.2\,x_2 \le 100$

$0.1\,x_2 \le 30$

$x_1,\, x_2 \ge 0$

ILLUSTRATION 3 ───

> *A firm manufactures two products in three departments. Product A contributes*
> *Rs. 5/- unit and requires 5 hrs. in dept. M, 5 hrs. in dept. N and one hour in dept. P.*
> *Product B contributes Rs. 10/- unit and requires 8 hrs. in dept. M, 3 hrs in dept N*
> *and 8 hrs in dept P. Capacities for departments M, N, P are 48 hours per week.*
> *Find out optimal product mix using Simplex model.* **[JNTU CSE 98]**

Soluiton :

Note : *Formulation part of the solution is given here, for solution by simplex ref. illustration 18, chapter 4*

The problem is summarised below.

	A	B	Capacity
M	5	8	48
N	5	3	48
P	1	8	48
Unit contribution	5	10	

Variables :

Let quantity produced of product $A = X_1$

Let quantity produced of product $B = X_2$

Objective Function :

Profit contribution of each unit of $A = Rs.\ 5$

Profit contribution of X_1 units of $A = 5\,X_1$

Profit contribution of each unit of $B = Rs.\ 10$

Profit contribution of X_2 units of $B = 10\,X_2$

Total profit $= 5\,X_1 + 10\,X_2$

Therefore objective function is to $\boxed{\textbf{\textit{maximise } Z = 5 X_1 + 10 X_2}}$

Constraint Set :

In Department M

Time required by each unit of product $A = 5$ hrs

Time required by X_1 units of product $A = 5\,X_1$

Time required by each unit of product $B = 8$

Time required by X_2 units of product $B = 8\,X_2$

Total hours available in dept. M = 48 hrs per week

∴ $\boxed{5\,X_1 + 8\,X_2 \le 48}$

   (Constraint on capacity in dept. M)

Similarly in department N,

$\boxed{5\,X_1 + 3\,X_2 \le 48}$

   (Constraint on capacity in dept. N.)

and in department P,

$$\boxed{X_1 + 8 X_2 \le 48}$$

(Constraint on capacity in department P)

Conditions :

X_1 and $X_2 \ge 0$

Summary :

Max $Z = 5 X_1 + 10 X_2$

Subject to $5 X_1 + 8 X_2 \le 48$

$5 X_1 + 3 X_2 \le 48$

$X_1 + 8 X_2 \le 48$

$X_1 , X_2 \ge 0$

ILLUSTRATION 4 ————————————————————————————

> *Food X contains 6 units of Vitamins A per gram and 7 units of Vitamin B per gram and costs 12 paise per gram. Food Y contains 8 units of Vitamin A per gram and 12 units of Vitamin B and costs 20 paise per gram. The daily minimum requirements of Vitamin A and Vitamin B are 100 units and 120 units respectively. Find the minimum cost of product mix. Use simplex method.*   **[JNTU B.Tech. (ECE) 97]**

Solution :

Note : *Formulation part of the solution is given here, refer illustration - 7, chapter 4 for its solution by simplex method*

Selection of Variables :

Let the No. of grams produced in type $X = x_1$

Let the No. of grams produced in type $Y = x_2$

Objective Function :

Cost of each gram of type $X =$ 12 paise

Cost of x_1 gms of type $X = 12x_1$ ps.

Cost of each gm of type $Y = 20$ ps.

Cost of x_2 gms of type $Y = 20 x_2$ ps.

Total cost $Z = 12x_1 + 20x_2$

∴   Objective function is to $\boxed{\text{minimize } Z = 12 x_1 + 20 x_2}$

Constraint Set :

For Vitamin A :

Availability of vitamin A in each gm of food X = 6 units

Availability of vit-A in x_1 gms of food $X = 6 x_1$

Availability of vit-A in each gms of food $Y = 8$ units

Availability of vit-A in x_2 gms of food $Y = 8\,x_2$

Total vitamin $A = 6x_1 + 8x_2$

Daily minimum vitamin A required $= 100$ units

$$\boxed{6x_1 + 8x_2 \geq 100}$$

(Requirement constraint on vitamin A)

For Vitamin B :

Availability of vitamin B in each gm of food $X = 7$ units

Availability of vit-B in x_1 gms of food $X = 7\,x_1$

Availability of vit-B in each gms of food $Y = 12$ units

Availability of vit-B in x_2 gms of food $Y = 12\,x_2$

Total vitamin $B = 7\,x_1 + 12\,x_2$

Daily minimum requirement of vitamin $B = 120$ units

Hence $\boxed{7x_1 + 12\,x_2 \geq 120}$

(Constraint on requirement of vitamin – B).

Conditions of Variables :

x_1 and x_2 cannot be negative

$\therefore$ $\boxed{x_1 \geq 0}$ & $\boxed{x_2 \geq 0}$

Summary :

$$\min z = 12\,x_1 + 20\,x_2$$

subject to $6\,x_1 + 8\,x_2 \geq 100$

$7\,x_1 + 12\,x_2 \geq 120$

$x_1,\, x_2 \geq 0$

ILLUSTRATION 5

M/s. ABCL company manufactures two types of cassettes, a video and audio. Each video cassette takes twice as long to produce one audio cassette, and the company would have time to make a maximum of 2000 per day if it is produced only audio cassettes. The supply of plastic is sufficient to produce 1500 per day to both audio and video cassettes combined. The video cassette requires a special testing and processing of which there are only 6000 hrs. per day available. If the company makes a profit of Rs. 3/- and Rs. 5/- per audio and video cassette respectively, how many of each should be produced per day in order to maximize the profit? [JNTU 2003 (Set-1)]

Solution :

Let   No of audio cassettes to be produced per day $= x_1$

No. of video casettes to be produced per day $= x_2$

Objective Function :

Profit per one audio casette $=$ Rs. 3

Profit on x_1 audio casettes $= 3x$

Profit on each video casette $= 5$

Profit on x_1 video casettes $= 5x_2$

Total profit $= 3x_1 + 5x_2$

It is to be maximized

∴ Objective function is to Maximize

$$Z = 3x_1 + 5x_2$$

Constraint Set :

(i) Time Constraint :

As time of production of video is twice to that of audio cassettes, if x_1 audio are produced $\dfrac{x_2}{2}$ videos can be produced. Thus in terms of audio casette times, we get

$$x_1 + \frac{x_2}{2} \leq 2000 \qquad \text{(constraint on production time)}$$

$$\Rightarrow \quad 2x_1 + x_2 \leq 4000$$

(ii) Plastic Constraint :

$$x_1 + x_2 \leq 1500$$

(iii) Testing & Processing Time Constraint :

$$x_2 \leq 6000$$

Conditions   Both $x_1 \geq 0$ and $x_2 \geq 0$

Summary   Maximize $Z = 3x_1 + 5x_2$

subject to $2x_1 + x_2 \leq 4000$

$$x_1 + x_2 \leq 1500$$

$$x_2 \leq 6000$$

$$x_1, x_2 \geq 0$$

ILLUSTRATION 6 ―――

> *A manufacturing firm producing refrigerators has given a contract to supply 50 units at the end of the first month, 50 at the end of the second month, and 50 at the end of the third. The cost of producing 'X' no. of refrigerators in any month is given by 'X²'. The firm can produce more number of refrigerators in any month and carry them to the next month. However, a holding cost of Rs. 20 per unit is charged for any refrigerator carried over from month to the next. Assuming there is no initial inventory, determine the no. of refrigerators to be produced in each month so as to minimize the total cost, using dynamic programming approach.* **[JNTU B.Tech (ECE) 97/S]**

Solution :

Note : *Formulation part is discussed here. For complete solution refer chapter 12, illustration 6*

Selection of Variables :

Let, x_1 = No. of refrigerators to be manufactured in first month

x_2 = No. of refrigerators to be manufactured in second month

x_3 = No. of refrigerators to be manufactured in third month.

Objective Function :

Now, Given the production cost of X units manufactured $= X^2$,

we have,   Production cost in first month $= x_1^2$

Production cost in second month $= x_2^2$

and   Production cost in third month $= x_3^2$

and there is no initial inventory, hence no holding in first month.

But, the holding cost in second month if x_1 exceeds 50,

$$= 20 \, (x_1 - 50)$$

also, the holding cost in third month if $x_1 + x_2$ exceeds 100 is,

$$= 20 \, (x_1 + x_2 - 100)$$

Hence, the total cost

$$= x_1^2 + x_2^2 + x_3^2 + 20 \, (x_1 - 50) + 20 \, (x_1 + x_2 - 100)$$

∴   Objective function is to

Minimize $Z = x_1^2 + x_2^2 + x_3^2 + 40 \, x_1 + 20 \, x_2 - 3000$

Constraint Set :

According to contracts of supply,

$x_1 \geq 50$ *(Constraint on supply for first month)*

$x_1 + x_2 \geq 100$ *(Constraint on supply for first two months)*

Assumed that no stock is to be left out after third month.

$x_1 + x_2 + x_3 = 150$ *(Constraint on supply for the three months)*

Condition of Variables :

All non negative i.e., $\boxed{x_1, x_2, x_3 \geq 0.}$

Summary :

Minimise $z = x_1^2 + x_2^2 + x_3^2 + 40\,x_1 + 20\,x_2 - 3000$

subject to $x_1 \geq \ 50$

$x_1 + x_2 \geq 100$

$x_1 + x_2 + x_3 = 150$

$x_1, x_2, x_3 \geq 0.$

ILLUSTRATION 7 ———

A company wants to purchase at most 180 units of a product. There are two types of the product, M_1 and M_2 available. M_1 occupies $2\,ft^3$, cost \$ 12 and the company makes a profit of \$3. M_2 occupies $3\,ft^3$, cost \$15 and the company makes a profit of \$4. If the budget is \$15,000 and the warehouse has $3000\,ft^3$ for product.

Solve the problem using Simplex method.

[JNTU B.Tech. (EEE) 98]

Solution :

Note : *Solution for (i) is discussed here*

Selection of Variables :

Let the no. of products purchased in type $M_1 = x_1$

and the no. of products pruchased in type $M_2 = x_2$

Objective Function :

Profit on each unit of product of type M_1 = $ 3

Profit on x_1 units of product of type M_1 = $ 3 x_1

Profit on each unit of product of type M_2 = $ 4

Profit on x_2 units of product of type M_2 = $ 4 x_2

Total profit $Z = 3x_1 + 4x_1$

∴ Objective function is **maximize $Z = 3 x_1 + 4 x_2$**

Constraint Set :

Space Constraint :

Each unit of product M_1 occupies $2 ft^3$ space.

∴ x_1 units of product M_1 occupy $2 x_1 ft^3$

Similarly, each unit of product M_2 occupies $3 ft^3$

x_2 units of product M_2 occupy $3x_2 ft^3$

Total space for both products $2 x_1 + 3x_2$

Available space $= 3000 ft^3$

∴ $2 x_1 + 3x_2 \leq 3000$ *(Constraint on availability of space)*

Budget Constraint :

Each unit of product M_1 costs $12

∴ x_1 units of product M_1 cost $ 12 x_1

and each unit of product M_2 cost $ 15

∴ x_2 units of product M_2 cost $ 15 x_2

Total cost $= 12 x_1 + 15 x_2$

Available budget $= $ 15, 000

∴ $12 x_1 + 15 x_2 \leq 15000$ *(Constraint on availability of budget)*

Condition of Variables :

both x_1 and x_2 can not be negative

$\therefore$ $\boxed{x_1 \geq 0 \text{ and } x_2 \geq 0}$

Summary :

Maximise $Z = 3x_1 + 4x_2$

subject to

$$2x_1 + 3x_2 \leq 3000$$

$$12x_1 + 15x_2 \leq 15000$$

$$x_1, x_2 \geq 0$$

ILLUSTRATION 8 ────────────────────────────

An oil refinery can blend three grades of crude oil to produce quality P and Q petrol. Two possible blending processes are available. For each production run, the older process uses 5 units of crude A, 7 units of crude B and 2 units of crude C to produce 9 units of P and 7 units of Q. The newer process uses 3 units of crude A, 9 units of B and 4 units of crude C to produce 5 units of P and 9 units of Q petrol. Because of prior contract commitments the refinery must produce at least 300 units of P and at least 300 units of Q for the next month. It has available 1500 units of crude A, 1900 units of crude B and 1000 units of crude C. For each unit of P the refinery receives Rs. 60 while for each unit of Q it receives Rs. 90. Find out the linear programming formulation so as to maximise the revenue.

Solution :

Selection of Decision Variables :

x_1 = No. of production runs of the older process

x_2 = No. of production runs of the newer process.

Objective Function :

No. of units of petrol P in older process for each production run = 9

No. of units of petrol P in older process for x_1 production runs = $9x_1$

No. of units of petrol P in newer process for each production runs = 5

No. of units of petrol P in newer process for x_2 production runs = $5x_2$

Total petrol $P = 9x_1 + 5x_2$

Profit on petrol $P = 60(9x_1 + 5x_2)$.

Similarly, for petrol Q,

No. of units per production run in older process = 7

No. of units for x_1 production runs = $7x_1$.

No. of units for each production run in newer process = 9

No. of units for x_2 production runs in newer process = $9\,x_2$

No. of units of petrol $Q = 7\,x_1 + 9x_2$

Profit on petrol $Q = 90\,(7x_1 + 9x_2)$

Total profit for both P and $Q = 60\,(9x_1 + 5x_2) + 90\,(7x_1 + 9x_2)$

∴ Objective function is

Maximize $Z = 1170\,x_1 + 1110\,x_2$.

Constraint Set :

For Petrol P :

No. of units of petrol produced in both processes is $9x_1 + 5x_2$. This should be at least 500 units.

$$\therefore\quad 9x_1 + 5x_2 \geq 500$$

(Requirement constraint on contract commitment of P)

For Petrol Q :

No. of units of petrol produced in both processes is $7x_1 + 9x_2$. This should be at least 300 units.

$$\therefore\quad 7x_1 + 9x_2 \geq 300$$

(Requirement constraint on constract commitment of Q).

For Crude A :

For older process, each production run requires 5 units.
For older process, x_1 production runs require $5x_1$ units.

For newer process, each production run requires 3 units.

For newer process, x_2 production runs require $3x_2$ units.

Total crude A for both processes $5x_1 + 3x_2$.

This can not be more than 1500.

$$\therefore\quad 5x_1 + 3x_2 \leq 1500$$

(Constriant on availability of crude A)

For Crude B :

Similarly, $7x_1 + 9x_2 \leq 1900$

(Availability constraint on curde B).

For Crude C :

Similarly, $2x_1 + 4x_2 \leq 1000$

(Availability constraint on crude C)

Conditions :

No. of production runs can not be negative.

Hence $x_1 \geq 0$ and $x_2 \geq 0$

Summary :

Maximise $Z = 1170 \, x_1 + 1110 \, x_2$

Subject to $9x_1 + 5x_2 \geq 500$

$7x_1 + 9x_2 \geq 300$

$5x_1 + 3x_2 \leq 1500$

$7x_1 + 9x_2 \leq 1900$

$2x_1 + 4x_2 \leq 1000$

$x_1 , x_2 \geq 0$

ILLUSTRATION 9 ─────────────────────────────────────

An advertising company has to plan their advertising strategy through the different media viz, TV, radio and newspaper. The purpose of advertising is to reach maximum number of potential customers. The cost of an advertisement in TV, Radio and newspaper are Rs. 3,000/-; Rs. 2000/- and Rs. 2500 respectively. The average expected potential customers reached per unit by 20000 of which 15000 are female customers. These figures with radio are 60000 and 40000 and with newspaper 25000 and 12000 respectively. The company has a maximum budget for advertising as Rs. 50000/- only. It is proposed to advertise through TV or radio between 6 and 10 units and at least 5 advertisements should appear in news paper. Further, it decides that at least 1,00,000 exposures should take place among female customers. Budget of advertising newspapers is limited to Rs. 25000 only. Formulate into LPP.

Solution :

Decision Variables :

Number of advertisements by   TV $= x_1$

Number of advertisements by   Radio $= x_2$

Number of advertisements by   Newspaper $= x_3$

Objective Function :

To maximise potential customers.

∴   **Max. $Z = 20000x_1 + 60000x_2 + 25000x_3$**

Constraint Set :

(i) *Budget Constraint :*

$3000x_1 + 2000x_2 + 2500x_3 \leq 50000$

or $30x_1 + 20x_2 + 25x_3 \leq 500$

or $6x_1 + 4x_2 + 5x_3 \leq 100$

(ii) *Female Customer Constraint :*

$15000x_1 + 40000x_2 + 12000x_3 \leq 100000$

or $15x_1 + 40x_2 + 12x_3 \leq 100$

(iii) Cost Constraint on Advertisements in News Paper :

$$2500x_3 \le 25000$$

or $\quad x_3 \le \mathbf{10}$

(iv) Constraint on Number of Advertisement in News Paper :

$$x_3 \ge \mathbf{5}$$

(v) Constraint on Number of Advertisements in TV :

$$6 \le x_1 \le 10$$

or $\quad x_1 \ge \mathbf{6}, \ x_1 \le \mathbf{10}$

(vi) Constraint on number of advertisement in radio :

$$6 \le x_2 \le 10$$

or $\quad x_2 \ge \mathbf{6}$ and $x_2 \le \mathbf{10}$

Conditions :

All the variables are non negative, since there will be no negative number of exposures

$$\therefore \quad x_1, x_2, x_3 \ge \mathbf{0}$$

Summary :

$$\text{Max} \quad Z = 20000x_1 + 60000x_2 + 25000x_3$$

$$\text{Subject to} \quad 6x_1 + 4x_2 + 5x_3 \le 100$$

$$15x_1 + 40x_2 + 12x_3 \le 100$$

$$x_3 \le 10; \qquad x_3 \ge 5$$

$$x_1 \ge 6; \qquad x_1 \le 10$$

$$x_2 \ge 6; \qquad x_2 \le 10$$

$$\text{and} \quad x_1, x_2, x_3 \ge 0$$

ILLUSTRATION 10

Mr. Raju has received his retirement benefits such as provident fund, gratuity etc. He wants to invest this money in Government securities, term deposits, in banks, investment on company deposits, equity shares and house construction. The return on investment (ROI), the time limits and risk on a five point scale is given in the following tabular form.

S.No.	Type of Investment	Return	No. of Years	Risk
1.	*Govt. Securities*	*8%*	*7*	*1*
2.	*Company deposits*	*15%*	*4*	*3*
3.	*Term deposits*	*14%*	*3*	*2*
4.	*Equity shares*	*18%*	*6*	*5*
5.	*House construction*	*25%*	*10*	*1*

He decides that the average risk should not exceed 3 and his funds should not be locked up for more than 10 years. He also has necessarily to invest 30% on house construction. Help Mr. Raju to plan his investment so as to maximise his ROI.

Solution :

Decision Variables :

Let the percentages of investments on these five factors be $x_1, x_2 \ldots x_5$

Objective Function :

Maximise $Z = 8x_1 + 15x_2 + 14x_3 + 18x_4 + 25x_5$

Subject to constraints

$$7x_1 + 4x_2 + 3x_3 + 6x_4 + 10x_5 \leq 10$$

(Constraint on term of investment)

$$x_1 + 3x_2 + 2x_3 + 5x_4 + x_5 \leq 3$$

(Constraint on risk)

$$x_5 \geq 0.3$$

(Constraint on investment on house construction)

$$x_1 + x_2 + x_3 + x_4 + x_5 = 1$$

(Total percentage is 100)

Conditions of Variables :

$$x_1, x_2, x_3, x_4, x_5 \geq 0$$

ILLUSTRATION 11 ────────────────────────────

A pregnant woman is advised to take Iron, Zinc and Folic acid, at least in the quantities 100 mg, 150 mg and 100 mg per day respectively. There are two types of medicines available in the form of a 500 mg tablet and a 250 mg capsule. Each tablet is composed of 30 mg of Iron, 45 mg of Zinc and 40 mg of Folic acid while a capsule contains the three ingradients as 20, 15, 25 mg respectively. The cost of each tablet is Rs. 3/- while that of each capsule is Rs. 2/-. Determine the number of tablets and capsules to be purchased so that the total cost of medicines is minimized.

Solution :

Selection of Variables :

Let , Number of tablets of 500 $mg = x_1$

Number of capsules of 250 $mg = x_2$

Objective Function :

Each tablet costs Rs. 3/-, x_1 tablets will cost $3x_1$

Each capsule costs Rs. 2/-, x_2 capsules will cost $2x_2$

Total cost $= 3x + 2x_2$

∴ Objective function is to

Minimize $Z = 3x_1 + 2x_2$

Constraints :

Iron Constraint :

Iron in each tablet = 30 mg

Iron content in x_1 tablets = $30x_1$

Iron content in each capsule = 20 mg

Iron in x_2 capsules = 20 x_2

Total Iron = $30x_1 + 20x_2$

∴ $30x_1 + 20x_2 \geq 100$

or $\mathbf{3x_1 + 2x_2 \geq 10}$

(Constraint on daily min. requirement of iron)

Similarly, the constraints on zinc and folic acid requirements can be written as

$45x_1 + 15x_2 \geq 150$

or $\mathbf{3x_1 + x_2 \geq 10}$

and $40x_1 + 25x_2 \geq 100$

or $\mathbf{8x_1 + 5x_2 \geq 20}$

Conditions :

x_1, x_2 being number of tablets capsules can not be negative (i.e., $\mathbf{x_1, x_2 \geq 0}$)

Summary :

Minimise $Z = 3x_1 + 2x_2$

Subject to $3x_1 + 2x_2 \geq 10$

$3x_1 + x_2 \geq 10$

$8x_1 + 5x_2 \geq 20$

$x_1, x_2 \geq 0.$

ILLUSTRATION 12 ————————————————————————————————

Medison Nursing Home works 24 hours a day. In a ward, it requires the attendance of nurses as given below:

From	To	Number of Nurses required
00.00	*04.00*	*2*
04.00	*08.00*	*3*
08.00	*12.00*	*5*
12.00	*16.00*	*4*
16.00	*20.00*	*7*
20.00	*0.00*	*3*

Consider that any nurse can work only for 8 hours in a day. Formulate the information in an LPP so as to schedule with minimum man power.

(JNTU B.Tech Mech. & Met. 95/S)

Solution :

Decision Variables :

x_1 = Number of nurses added at   00.00

x_2 = Number of nurses added at   04.00

x_3 = Number of nurses added at   08.00

x_4 = Number of nurses added at   12.00

x_5 = Number of nurses added at   16.00

x_6 = Number of nurses added at   20.00

Objective Function :

The objective is to minimise the man power requirement. Therefore we must minimise the total number of nurses required in all the periods.

$\therefore$ **Min.** $Z = x_1 + x_2 + x_3 + x_4 + x_5 + x_6$

Constraints :

At any four hour period, the number of nurses added and continuing from the previous shift should be at least equal to their requirement in the corresponding period.

Also, the nurses have to work for eight hours and the periods are split in 4 hour periods, any two consecutive periods can be clubbed to arrange the number of required nurses. Thus we get,

Constraint on period 00.00 to 04.00

Number of nurses continuing from previous period (i.e., added at 20.00) = x_6

Number of nurses added at 00.00 = x_1

Total number of nurses = $x_6 + x_1$ and this should be at least 2 in 00.00 to 04.00.

Therefore $x_6 + x_1 \geq 2$ *(Constraint on requirement from 0.00 to 04.00)*

Similarly $x_1 + x_2 \geq 3$ *(Constraint on requirement from 04.00 to 08.00)*

$x_2 + x_3 \geq 5$ *(Constraint on requirement from 08.00 to 12.00)*

$x_3 + x_4 \geq 4$ *(Constraint on requirement from 12.00 to 16.00)*

$x_4 + x_5 \geq 7$ *(Constraint on requirement from 16.00 to 20.00)*

$x_5 + x_6 \geq 3$ *(Constraint on requirement from 20.00 to 0.00)*

Conditions :

Since the no. of nurses can not be negative and cannot be a fraction value, we write the constraints as

$x_i \geq 0$ and integral value or x_i is positive integer where i = 1, 2, ... 6

ILLUSTRATION 13 ───

A farmer has 100 acre farm and can sell all tomatoes, lettuce or radishes that he can raise in the farm. The prices of these are Rs. 1.00 per kg, 0.75 per head and Rs. 2/- per kg respectively. The average yield per acre is 2,000 kg of tomatoes, 3000 heads of lettuce and 1000 of radishes. Fertilizer is available at Rs. 0.50 per kg and the amount required per acre is 100 kg each for tomatoes and lettuce and 50 kg for radishes. Labour required to sow, cultivate and harvest per acre is 5 man days for tomatoes and radishes and 6 man days for lettuce. A total of 400 man days of labour at Rs. 20/- per man day is available.

Optimise the total profit of the farmer.

$\mathcal{S}olution$:   **(Only formulation is shown here).**

Decision Variables :

Let number of acres allocated to tomatoes $= x_1$

Number of acres allocated to lettuce $= x_2$

Number of acres allocated to radish $= x_3$

Objective Function :

Here, we have to maximize the farmer's profits which is sales minus costs.

Sales :

Land for tomatoes $= x_1$ acres.

Yield of tomatoes $= 2000 x_1$

Tomatoes total sale $= 2000 x_1 \times 1.00$

$$= 2000 x_1$$

Similarly, sales of lettuce $= 0.75 \times 3000\, x_2$

and sales of radish $= 2.00 \times 1000 x_3$

Total sales $= 2000 x_1 + 2250 x_2 + 2000 x_3$

Labour Cost :

Man days needed for tomatoes and radishes	= 5 per acre
Land for tomatoes and radishes	$= x_1 + x_3$ acres.
Total man days	$= 5\,(x_1 + x_3)$
Man days needed for lettuce	= 6 per acre
Land for lettuce	$= x_2$ acres
∴ Man-days for lettuce	$= 6x_2$
Total man days	$= 5x_1 + 6x_2 + 5x_3$
Unit labour cost	= Rs. 20 per man-day
∴ Total labour cost	$= 20\,(5x_1 + 6x_2 + 5x_3)$
	$= 100x_1 + 120x_2 + 100x_3$

Fertilizer Cost :

Requirement for tomatoes and lettuce $= 100\, kg\,/$ acre

Land of tomatoes and lettuce $= x_1 + x_2$

∴   Fertilizer required for tomatoes $= 100\,(x_1 + x_2)$

Requirement for radish $= 50\, kg\,/$ acre

Land of radish $= x_3$ acres

Fertilizer needed for radish $= 50x_3$

Total fertilizer $= 100x_1 + 100x_2 + 50x_3$ kg

Cost of fertilizer $=$ Rs. 0.50 per kg

Total cost on fertilizer $= 0.50\,(100x_1 + 100x_2 + 50x_3)$

                                      $= 50x_1 + 50x_2 + 25x_2$

Now profit $=$ sales - costs

                   $=$ sales revenue - [labour cost + fertilizer cost]

                   $= (2000x_1 + 2250x_2 + 2000x_3)$

$$- \left\{ (100x_1 + 120x_2 + 100x_3) + (50x_1 + 50x_2 + 25x_3) \right\}$$

                   $= 1850x_1 + 2080x_2 + 1875x_3$

∴   The objective function is to

Maximise $Z = 1850x_1 + 2080x_2 + 1875x_3$

Constraints :

Land Constraint :

       Total land used $= x_1 + x_2 + x_3$

       Land available $= 100$ acres

       ∴ $x_1 + x_2 + x_3 \leq \mathbf{100}$

       (Constraint on availability of land)

Labour Constraint :

       Total labour days required $= 5x_1 + 6x_2 + 5x_3$ man-days

       Available labour days $= 400$ man-days

       ∴ $\mathbf{5x_1 + 6x_2 + 5x_3 \leq 400}$

       (Constraint on availability of man-days)

Conditions :

       Since land cannot be negative

               $x_1, x_2, x_3 \geq \mathbf{0}$

Summary :

       Max $Z = 1850\,x_1 + 2080x_2 + 1875x_3$

       Subject to $x_1 + x_2 + x_3 \leq 100$

                     $5x_1 + 6x_2 + 5x_3 \leq 400$

                     $x_1 \geq 0, \, x_2 \leq 0, \, x_3 \geq 0$

ILLUSTRATION 14 ————————————————————————————————

> *Old hens can be bought at Rs. 20 each and young ones at Rs. 50 each. The old hens lay 3 eggs per week and the young ones lay 5 eggs per week, each egg being worth of Rs. 1.50 ps. A hen (young or old) costs Rs. 1.50 per week to feed, I have only Rs. 800 to spend for hens, how many of each kind should I buy to give a profit of at least Rs. 60/- per week, assuming that I cannot house more than 20 hens.*
>
> **[JNTU B.Tech. Mech.]**

Solution :

Decision Variables :

Number of old hens $= x_1$

Number of young hens $= x_2$

Objective Function :

Number of eggs each old hen lays $= 3$

Number of eggs laid by x_1 old hens $= 3x_1$

Number of eggs each young hen lays $= 5$

Number of eggs that x_2 young hens lay $= 5x_2$

Total eggs $= 3x_1 + 5x_2$

Price of each egg $= 1.50$

∴ Sales revenue on eggs $= 1.5 \,(3x_1 + 5x_2)$

                                        $= 4.5x_1 + 7.5x_2$

Total expenditure for feeding $(x_1 + x_2)$

hens @ Rs. 1.50 per week $= 1.5 \,(x_1 + x_2)$

∴ Profit = sales - expenditure

$= (4.5x_1 + 7.5x_2) - (1.5x_1 + 1.5x_2)$

$= 3x_1 + 6x_2$

Hence, the objective function is to **Maximise $Z = 3x_1 + 6x_2$**

Constraints :

Profit Constraint :

The profit should be at least Rs. 60

∴ $3x_1 + 6x_2 \geq 60$

Constraint of Money :

Cost of old hen $=$ Rs. 20

Cost of x_1 old hens $= 20x_1$

Cost of each young hen $=$ Rs. 50

Cost of x_2 young hens $= 50x_2$

Total cost on hens $= 20x_1 + 50x_2$

Since I have Rs. 800/- only,

    $20x_1 + 50x_2 \leq 800$ or $\mathbf{2x_1 + 5x_2 \leq 80}$

(Constraint on availability of money)

Constraint of Space :

Total number of hens $= x_1 + x_2$

Available space is sufficient for 20 hens

$\therefore \quad x_1 + x_2 \leq 20$

(Constraint on availability of space)

Conditions :

It is not possible to purchase negative quantity of hens

Therefore, $x_1 \geq 0, \; x_2 \geq 0$

Summary :

Maximise $Z = 3x_1 + 6x_2$

Subject to $3x_1 + 6x_2 \geq 60$

$\qquad\qquad 2x_1 + 5x_2 \leq 80$

$\qquad\qquad x_1 + x_2 \leq 20$

and $x_1 \geq 0, x_2 \geq 0.$

2.2 *Major Assumption of LPP*

The following are the assumptions (Charateristics) of LPP.

1. **Proportionality : A primary requirement of LPP is that the objective function and all constraints, are linear and hence these are directly proportional. Linearity implies that the product of variables (like** $x_1 x_2$**), powers of variables (like** x_1^2, x_1^3 **etc.), combinations (e.g.** $ax_1 + b \log x_2$**) and exponentials (e.g.** e^x **,** a^x **etc.) are not allowed.**

2. **Additivity : The LPP provide concept of additivity in the quantities having same units. For example a machine needs** t_1 **&** t_2 **hrs. for product** A **and** B **respectively then total time can be said to** $t_1 + t_2$ **for both products on the machine. However, this additivity does not hold good for the quantities of different chemical composition.**

3. **Multiplicativity : LPP provides proportionality and thus if one item costs Rs. 10, then 10 items will cost Rs. 100/-.**

4. **Divisibility : LPP permits fractional values in addition to the integral values.**

5. **Deterministic : All parameters in LPP are assumed to be known exactly. However actual production may depend on chance also.**

6. **Sensitivity : Further, the problem can be extended by sensitivity analysis to check the post optimal situations.**

7. **Decision Making** by Conditions : **The conditions on the answers stated at the beginning of the problem will aid the decision making. The assumptions such as non negativity or unrestricted variables can influence the answers.**

Practice Problems

1. A mining company is taking a certain kind of ore from two mines X and Y. The ore is divided into 3 quality groups A, B and C. Every week the company has to supply 240 tonnes of A, 160 tonnes of B and 440 tonnes of C. The cost per day for running the mine X is Rs. 3000, while it is Rs. 2000 for the mine Y. Each day, X will produce 60 tonnes of A, 20 tonnes of B and 40 tonnes of C. The corresponding figures for Y are 20,20 and 80.

   Develop the most economical production plan by finding the number of days for which the mines X and Y should work per week.   **[June 2003 (Set-2)]**

   Answer :   cost = Rs. 18000/-

2. A person requires 10, 12 and 12 units of chemicals A, B and C respectively for his gardens. A liquid product contains 5, 2 and 1 units of A, B and C respectively per jar. A dry product contains 1, 2 and 4 units of A, B and C per carton. If the liquid product sells for Rs. 3 per jar and the dry product sells for Rs. 2 per carton, how many of each should be purchased to minimize the cost and meet the requirements? Formulate the above problem as a LPP and solve it by graphical method.   **[June 2003 (Set-3)]**

   Answer :   cost = Z_{min} = Rs. 13

3. A company has two bottling plants, one located at Bangalore and the other at Mysore. Each plant produces three brands of drinks A, B and C. Bangalore plant can produce (in one day) 1500, 3000 and 2000 bottles of A, B and C respectively. The capacity of Mysore plant remains, 1500, 1000 and 5000 bottles per day of A, B and C respectively. A market survey indicates that during the month of April there will be demand of 20,000 bottles of A, 40,000 bottles of B and 44,000 bottles of C. the operating cost per day for the plant at Bangalore is Rs. 600/- while the operating cost per day for the plant at Mysore is Rs. 400/- For how many days each plant be run in April so as to minimize the production cost while still meeting the demand?   **[June 2003 (Set-4)]**

   Answer :   cost of operation is Rs. 8800/-

4. Hema Pens Ltd. produces two types of pens namely; Supermo(S) and Economy pens (E). The net profits on these types are Rs. 10/- and Rs. 6/- respectively. Raw material required for Supermo(S) is twice as that of Economy (E). The supply of raw materials is sufficient only for 1000 pens of E per day. Pen S requires a special nib and Pen E requires ordinary nib. There are 400 special nibs, 700 ordinary nibs available in stock. Formulate for the daily product mix so as to maximise the total profit of company.   **[Dr. BRAOU MBA - 2004 Assign.]**

   Answer :   Max $Z = 10\,x_1 + 6\,x_2$;

   $$\text{S.t.} \quad \frac{x_1}{2} + x_2 \le 1000$$

   $$x_1 \le 400 \;;$$

   $$x_2 \le 700$$

   $$x_1, x_2 \ge 0 \quad \text{and integers}$$

5. Sreeja & Co wishes to plan its advertising strategy. There are two media under consideration, Siti cable and Popular channel. Siti cable has a reach of 2000 potential customers and Popular channel has a reach of 3000 potential customers. The cost per appearance of one minute is Rs. 6000 and Rs. 9000 in Siti and Popular respectively. The budget of Sreeja is Rs. 80,000 per month. There is an important requirement that the total reach for the income group under Rs. 60,000 per annum should not exceed 3000 potential customers. The reach in Siti cable and Popular channel for this income group is 300 and 150 potential customers. How many appearances of one minute advertisements should Sreeja plan so as to maximise the total reach?

Answer :   Max $Z = 2000\,x_1 + 3000\,x_2$

   S.t. $6000\,x_1 + 9000\,x_2 \le 80000$

   $300\,x_1 + 150\,x_2 \le 3000$

   $x_1,\ x_2 \ge 0$

6. Himaja Electrical Equipments (P) Ltd is engaged in the production of high power and low power transformers using three operations : mould making, coil assembly and joining & final assembly. A high power transformer is sold at Rs. 80,000 while low power transformer is sold at Rs. 3,000/-. The costs of HPT and LPT are Rs. 3000 and Rs. 2000 respectively. The time required in hours in each operation, and their weekly availability are given below.

Operation	Transformer		Weekly Availability
	High - Power	Low - Power	
Mould making	3 hours	1 hours	40
Coil assembly	16	3	150
Joining & final assembly	6	2	100

Formulate the problem into an LPP so as to maximise the profits.

Answer : Profit = sales price - cost

   Max $Z = 5000\,x_1 + 1000\,x_2$

   S.t. $3x_1 + x_2 \le 40$; $16\,x_1 + 3\,x_2 \le 150$

   $6\,x_1 + 2\,x_2 \le 100$; $x_1, x_2 \ge 0$

Note : *Third constraint is redundant since it is already covered by first constraint*

7. The manager of an oil refinery has to decide on the optimal mix of two possible blending processes of which the inputs and outputs per production run are given as follows :

Process	Inputs		Outputs	
	Crude A	Crude B	Gasoline X	Gasoline Y
I	5	3	5	8
II	4	5	4	4

The maximum amount available of crude *A* and *B* are 200 units and 150 units respectively. Market requirements show that at least 100 units of petrol *X* and 80 units of petrol *Y* must be produced. The profit per production run from process I and process II are Rs. 300 and Rs. 400 respectively. Formulate the problem as a linear programming model.

[OU - 87, OU - MBA 90, Sep. 2001]   [JNTU Mech - 95]

Answer : $x_1, x_2 =$ No. of units from Process I & II respectively.

Max $Z = 300 x_1 + 400 x_2$

Subj. to $5x_1 + 4x_2 \leq 200$; $3x_1 + 5x_2 \leq 150$;

$5x_1 + 4x_2 \geq 100$; $8x_1 + 4x_2 \geq 80$; $x_1, x_2 \geq 0$

8. MIDHANI produces an alloy of specific gravity 0.98, chromium content not more than 8% and melting point below 450° C. Three raw materials M_1, M_2 and M_3 are used to manufacture this alloy. The properties of these materials is as follows :

Property	Property of R/M		
	M_1	M_2	M_3
Specific gravity	0.92	0.97	1.04
Chromium content	7%	13%	16%
Melting point	440° C	490° C	480° C

Costs of the raw materials are Rs. 90, Rs. 100 and Rs. 80. Formulate to determine the proportions that will minimise the total costs.

Answer : Percentage of RM's are x_1, x_2, x_3

Max $90 x_1 + 100 x_2 + 80 x_3$; S.t. $0.92 x_1 + 0.97 x_2 + 1.04 x_3 = 0.98$

$440 x_1 + 490 x_2 + 480 x_2 = 450$

$\dfrac{7}{100} x_1 + \dfrac{13}{100} x_2 + \dfrac{16}{100} x_3 \leq \dfrac{8}{100}$; $x_1 + x_2 + x_3 = 1$ (i.e., 100%)

$x_1, x_2, x_3 \geq 0$

9. Mona Food Company produces four types of foods whose contents are given below :

Food type	Content per unit			Cost per unit in Rs.
	Protiens	Fats	Carbohydrates	
Wheat - apple	3	2	6	45
Wheat - banana	4	2	4	40
Wheat - honey	8	7	7	75
Wheat - orange	6	5	4	65
Minimum Requirement	80	40	70	

Formulate into LPP.

Answer : Min $Z = 45 x_1 + 40 x_2 + 75 x_3 + 65 x_4$

S.t $3 x_1 + 4 x_2 + 8 x_3 + 6 x_4 \geq 80$; $2x_1 + 2 x_2 + 7 x_3 + 5x_4 \geq 40$

$6 x_1 + 4 x_2 + 7 x_3 + 4 x_4 \geq 70$; $x_1, x_2, x_3, x_4 \geq 0$

10. Sanja & co. manufactures three products A, B and C by using three machines Milling, Lathe and Grinder. The information related to this manufacture is given in the following table. Formulate for LP.

Machine	Product A	Product B	Product C	Available hours per week
Milling machine	8	2	3	250
Lathe	4	3	1	150
Grinder	2	1	1	100
Profit contribution	20	6	8	

Answer : Max $\quad Z = 20\, x_1 + 6\, x_2 + 8\, x_3$

S.t. $\quad 8\, x_1 + 2\, x_2 + 3\, x_3 \leq 250$; $\qquad 4x_1 + 3x_2 + x_3 \leq 150$

$\qquad\qquad 2\, x_1 + x_2 + x_3 \leq 100$; $\qquad\qquad x_1, x_2, x_3 \geq 0$

11. Mother Love Engineering College has to decide its monthly investment plan for Rs. 10,000 on sports and games. It has three disciplines Mechanical Engineering (ME), Computer Science Engineering (CSE) and Electronics and Computer Engineering (ECE). Each discipline has got about 60% boys and 40% girls. The probability of winning in the inter college sports meet in various events is as follows.

Discipline	ME(boys)	ME(girls)	CSE(boys)	CSE(girls)	ECE(boys)	ECE(girls)
Probability	0.3	0.25	0.35	0.4	0.5	0.45

It is the policy of the college that at least 40% of funds are to be invested on ME, and not more than 35% on CSE or ECE branches. Also at least 30% is to be invested on girls. Formulate the problem into an LPP.

Answer : Let x_{1B} investment on ME boys,

$\qquad\qquad x_{1G}$ investment on ME girls,

$\qquad\qquad x_{2B}$ investment CSE boys,

$\qquad\qquad x_{2G}$ investment CSE girls and so . . . on.

and maximise the total probability of winning

Max $\quad Z = 0.3\, x_{1B} + 0.25\, x_{1G} + 0.35\, x_{2B} + 0.4\, x_{2G} + 0.5\, x_{3B}$

S.t. $\quad x_{1B} + x_{1G} + x_{2B} + x_{2G} + x_{3B} + x_{3G} \leq 10000$

$\qquad\quad x_{1B} + x_{1G} \geq 4000$; $\;\; x_{2B} + x_{2G} \leq 3500$

$\qquad\quad x_{3B} + x_{3G} \leq 3500$; $\;\; x_{1G} + x_{2G} + x_{3G} \geq 3000$

$\qquad\quad x_{1B} + x_{2B} + x_{3B} \leq 7000$

$\qquad\quad x_{1B}, x_{1G}, x_{2B}, x_{2G}, x_{3B}, x_{3G} \geq 0$

12. A second hand automobile dealer wishes tó maximises his profits on his sales and purchase of cars. He can purchase three types of cars Maruthi 800CC, Maruthi-Zen and Maruthi-1000CC at Rs. 50,000, Rs. 70,000 and Rs. 80,000 respectively. These can be sold at Rs. 60,000; Rs. 80,000 and Rs. 1,05,000 respectively. The sales probability of the cars in this month are 0.7, 0.8 and 0.6 respectively. For every two cars of Zen, he should buy one car of 800CC or 1000CC and he has 10 lakhs to invest. How should be invest? Give formulation only.

Answer : x_1 = investment on 800 CC

   x_2 = investment on zen and

   x_3 = investment on 1000 CC,

   Profit = $\sum$ (sales - profit) sales prob.

   Max $Z = 10000 \, x_1 \times 0.7 + 10000 \, x_2 + 0.8 + 250000 \, x_3 \times 0.6$

   or Max $z = 7000 \, x_1 + 8000 \, x_2 + 15000 \, x_3$

   S.t. $70000 \, (2x_2) + 80000 \, x_3 \leq 1000000$ or $14 \, x_2 + 8 \, x_3 \leq 100$

   $x_1, x_2, x_3 \geq 0$

13. An Engineering College needs 22 to 30 Professors/Lecturers daily depending on the teaching loads and lab work distributions. The requirements are as follows:

Time Period	Number of Professors/Lecturers needed
9 am to 11 am	22
11 am to 1 pm	30
1 pm to 3 pm	25
3 pm to 5 pm	23

The College has employed 24 staff members on permanent rolls and needs some part time professors. The part time professors must put in exactly 4 hours per day but can start at any time between 9 am to 1 pm. Permanent employees (full time lecturers) work for 9 am to 5 pm but are allowed one hour for lunch. Assume half of full time lecturer take lunch at 12 noon while other half at 1 pm and thus full timers provide 35 hours per week of teaching time. The university limits the college to engage part time hours to a maximum of 50% of the day's total requirement. Part timers earn Rs. 280 per day while full timers earn Rs. 900 in salary and benefits on an average. The management has to set a schedule so as to minimise the total daily manpower costs.

Answer :   Let x_1 = full time lecturers/professors

   x_2 = part times starting at 9 am

   x_3 = PTs adding at 11 am and

   x_4 = PTs adding at 1 pm

   Min $Z = 900 \, x_1 + 280 \, (x_1 + x_2 + x_3)$

S.t $x_1 + x_2 \geq 22$ for 9 am to 11 am

$\frac{1}{2} x_1 + x_2 + x_3 \geq 30$ for 11 am to 1 pm

$\frac{1}{2} x_1 + x_3 + x_4 \geq 25$ for 1 pm to 3 pm

$x_1 + x_4 \geq 23$ for 3 pm to 5 pm

$4 (x_1 + x_2 + x_3) \leq 0.50 (22 + 30 + 25 + 23)$

(part - timers should not exceed 50% of total hrs).

$x_1, x_2, x_3, x_4 \geq 0$

14. Mr. Raju, a retired Govt. officer, has recently received his retirement benefits viz. PF, gravity etc. He is contemplating as two how much he should invest in various alternatives open to him so as to maximise ROI. The investment alternatives and and subjective estimate of risk on five point scale is given. The relevant data about the investment, return and no. of years blocked is as follows.

Investment	Return	No. of Years	Risk
Govt. Securities	6%	12	1
Company deposits	13%	3	3
Time deposits	10%	5	2
Equity shares	20%	6	5
House construction	25%	10	1

Mr. Raju wants invest to maximise. ROI so that the risk is not more than 4 and funds should not be blocked for more than 15 years. He should necessarily invest at least 28% in house construction. Formulate the data as an LPP.

[OU - MBA Dec. 2000]

Review Questions

1. Explain the structure of an LPP with examples.

2. Discuss different types of constraints that occur in LPP

3. Discuss the significance of condition of variables in LPP

4. How do you set objective function of an LPP

5. Define linear programming problem. Give example

[OU - MBA Feb 93, M 95, A99]

6. Discuss applications of LPP   **[JNTU CSE 95, Mech. 97/C, 98/P, 99/S,**
OU - MBA - M 95, F 93, A 94, M 92, S 2001, J 2000]

Objective Type Questions

1. Which of the following is not a part of an LPP problem

   (a) objective function
   (b) constraint set
   (c) condition set
   (d) end corrections

2. The availability constraint is represented with

   (a) $<$
   (b) $\leq$
   (c) $>$
   (d) $\geq$

3. A constraint represented by '=' sign is

   (a) availability, constraint
   (b) requirement constraint
   (c) exact constraint
   (d) no constraint

4. If a maximum limit is imposed on a resource usage then it is known as

   (a) availability constraint
   (b) requirement constraint
   (c) exact constraint
   (c) inside constraint

5. A product P contributes a profit of Rs. 50 and product Q contributes Rs. 60. This is represented in objective functions Z as _____ for x_P and x_Q products manufactured.

   (a) min. $Z = 50\, x_P + 60\, x_Q$
   (b) max. $Z = 60\, x_P + 50\, x_Q$
   (c) min. $Z = 60\, x_P + 50\, x_Q$
   (d) max. $Z = 50\, x_P + 60\, x_Q$

6. The product A requires 2 man hours, B requires 3 hrs. and C requires 4 man hrs. If the company works for 6 day week with only one shift of 8 hrs, then this is represented as _____ (Assume x_1, x_2, x_3 units are manufactured in A, B and C types respectively)

   (a) $2x_1 + 3x_2 + 4x_3 \leq 56$
   (b) $2x_1 + 3x_2 + 4x_3 \geq 8/6$
   (c) $2x_1 + 3x_2 + 4x_3 \geq 48$
   (d) $2x_1 + 3x_2 + 4x_3 \leq 48$

7. A company has to decide how many units of each product to be produced. It estimates Rs. 60 per unit of product X and Rs. 80 per unit of product Y. It incurs cost of Rs. 40 on X and Rs. 30 on Y to manufacture. The objective function can be written as

   (a) max. $Z = 60\, x + 80\, y$
   (b) max. $Z = 20\, x + 50\, y$
   (c) max. $Z = 30\, x + 40\, y$
   (d) min. $Z = 40\, x + 30\, y$

8. If a product X_1 contributes a profit of Rs. 75 per unit and product X_2 costs Rs. 50 per unit, the objective function $f(x_i)$ is written as_____

   (a) min. $f(x) = -75\, x_1 - 50\, x_2$
   (b) max. $f(x) = 75x_1 + 50\, x_2$
   (c) max. $f(x) = 75\, x_1 - 50\, x_2$
   (d) max. $f(x) = 25\, x_i$

9. The non-negative conditions for variables x_1 and x_2 can be represented as

    (a) $x_1 > 0, \; x_2 > 0$ (b) $x_1 = x_2 \geq 0$

    (c) $x_1 \geq 0, \; x_2 \geq 0$ (d) Any of the above

10. A furniture manufacturer produces x_1 chairs and x_2 tables every week. He has an order for 50 chairs and 60 tables in a week. This is represented as

    (a) $x_1 + x_2 \geq 110$ (b) max. $50\,x_1 + 60\,x_2$

    (c) $x_1 \leq 50, \; x_2 \leq 60$ (d) $x_1 \geq 50, \; x_2 \geq 60$

11. If your pocket money is Rs. 100/- per day, assuming Rs. x you spend on a day, formulate it.

    (a) max. $100\,x$ (b) min. $100\,x$

    (c) $x \geq 100$ (d) $x \leq 100$

12. You are purchasing glass for spectacles. The variable is the power of the glass to suit eye sight and size of the glass. The constraint used in this regard is _____

    (a) less than or equal to   (b) greater than or equal to

    (c) exact type   (d) can not be represented

13. Production of x pens cost Rs. 100 and that of y pencils cost Rs. 120. If the company wishes to optimise unit costs, the objective function is

    (a) min. $100\,x + 120\,y$ (b) min. $\dfrac{x}{100} + \dfrac{y}{120}$

    (c) max. $-\dfrac{100}{x} - \dfrac{120}{y}$ (d) $x \leq 100, \, y \leq 120$

14. 10 units of product x_1 costs Rs. 40 while 5 units of product x_2 costs Rs. 25. The objective function of this information will be

    (a) min. $-10\,x_1 - 5\,x_2$ (b) min. $40\,x_1 + 25\,x_2$

    (c) min. $-4x_1 - 5x_2$ (d) min. $4x_1 + 5x_2$

15. A belts of type - I requires leather twice as that of type - II. Total leather available is sufficient for 300 belts of type II. If x_1 and x_2 are the number of belts of type I & II to be produced respectively, then the constraint is

    (a) $100\,x_1 + 200\,x_2 \leq 300$ (b) $200\,x_1 + 100\,x_2 \leq 300$

    (c) $x_1 = 2x_2 \leq 300$ (d) none of the above

16. Which of the following is usually minimised

    (a) sales   (b) profits

    (c) operating time   (d) revenue

17. The requirement constraint is shown with a mathematical relationship as _____ in LPP formulation

    (a) $ax_1 + bx_2 \leq c$ (b) $ax_1 + bx_2 \geq c$

    (c) $ax_1 + bx_2 = c$ (d) $c \geq ax_1 + bx_2$

18. Cost of 5 pens is equal to the cost of 8 pencils. I have to maximise the number of pens and pencils (say x_1 and x_2 respectively) when I have Rs. 100/-. The problem is formulated as _____

    (a) max. $x_1 + x_2$ subject to $5x_1 + 8x_2 \leq 100$

    (b) max. $x_1 + x_2$ subject to $5x_1 + 8x_2 \geq 100$

    (c) max. $5x_1 + 8x_2$ subject to $x_1 + x_2 = 100$

    (d) max. $x_1 + x_2$ subject to $\dfrac{x_1}{5} + \dfrac{x_2}{8} \leq 100$

19. A machine is to be stopped for at least 30 min, if it continuously runs for 2 hours. If x is the number of machine hours to be planned on this machine in a 8 hour day, we have constraint as

    (a) $x \leq 8$ (b) $x \leq 6$

    (c) $x \leq 6.5$ (d) $x \geq 6.5$

20. A variable unrestricted in sign can assume the values as

    (a) only positive   (b) only negative

    (c) zero   (d) any value.

Fill in the Blanks

1. The objective function in an LPP will be in the form of _____ or _____

2. The objective function chosen for an LPP when costs are given is _____ type.

3. A constraint, mathematically represented as $ax + by = c$ is called _____ type constraint

4. If a minimum limit is imposed on production of a certain commodity, the constraint is said to be _____ type

5. The cost a product X is Rs. 5 per unit while that of product Y is Rs. 8 per unit. If x and y are the units to be manufactured, the objective function is _____

6. Krishna has Rs. 100 in his purse and went to a restaurant. He orders x_1 number of units of worth Rs. 10 each and x_2 number of units of worth Rs. 20 each. Then constraint is written as _____

7. The number of units of salt to put in a curry dish, to maximise its taste is an example for _____ type of constraint

8. A company has to spend Rs. 200 per unit of advertisement in New paper while Rs. 400 per unit in T.V. If the budget is Rs. 10,000 then the constraint is written as _____

9. Product M costs Rs. 10 and product N cost Rs. 40 per unit manufactured. The profit contribution are Rs. 50 and 30 respectively. The objective function is _____

10. Products P and Q cost Rs. 10 and 20 each to produce. The objective function is given as max. $Z = -10 x_1 - 20 x_2$ where $x_1 =$ _____ $x_1 =$ _____

Answers

Objective Type Questions :				
1. (d)	2. (b)	3. (c)	4. (a)	5. (d)
6. (d)	7. (b)	8. (c)	9. (c)	10. (d)
11. (d)	12. (c)	13. (c)	14. (d)	15. (d)
16. (c)	17. (b)	18. (d)	19. (c)	20. (d)

Fill in the Blanks :	
1. Maximisation, Minimisation	2. Minimisation
3. Exact or equality	4. Requirement
5. min. $Z = 5x_1 + 8x_2$	6. $10 x_1 + 20 x_2 \leq 100$
7. Availability or $\leq$	8. $200 x_1 + 400 x_2 \leq 10000$
9. max. $z = 40 x_1 - 10 x_2$	10. No. of products of P to be produced, No. of products of Q to be produced

3

Chapter

Graphical Solution of LPP

CHAPTER OUTLINE

CHAPTER AT A GLANCE

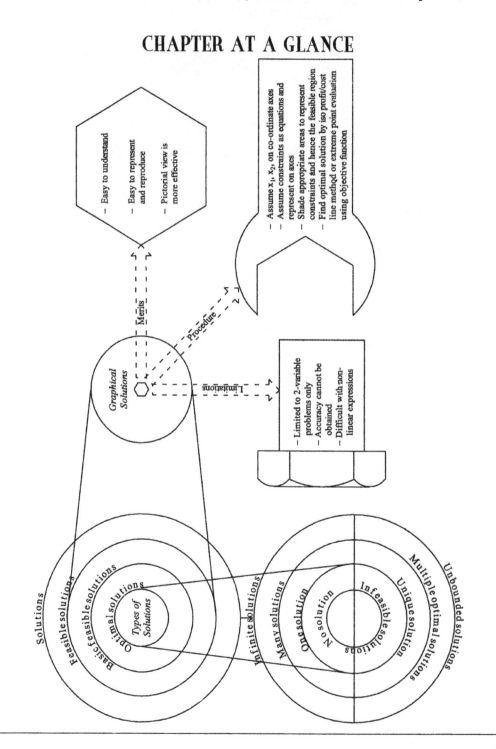

- Easy to understand
- Easy to represent and reproduce
- Pictorial view is more effective

- Assume x_1, x_2, on co-ordinate axes
- Assume constraints as equations and represent on axes
- Shade appropriate areas to represent constraints and hence the feasible region
- Find optimal solution by iso profit/cost line method or extreme point evaluation using objective function

Merits

Procedure

Graphical Solutions

Limitations

- Limited to 2-variable problems only
- Accuracy cannot be obtained
- Difficult with non-linear expressions

Solutions

Feasible solutions

Basic feasible solutions

Optimal solutions

Types of Solutions

Infinite solutions

Many solutions

One solution

No solution

Infeasible solutions

Unique solution

Multiple optimal solutions

Unbounded solutions

3.0 *Introduction*

In the previous chapter you have learnt how to formulate a given information into a linear programming problem. In this chapter you will learn how to solve these problems using graphs.

3.1 *Merits and Demerits of Graphical Solutions*

Graphical solutions are easier to understand and reproduce. Also a pictorial view is always a better representation. Thus graphical solutions have gained prominence in Operations Research.

However, graphical solutions have certain limitations such as :

1. Limited to the problems of two decision variables only.

2. Accuracy can not be obtained.

3. Some times it is difficult to represent certain expressions, particularly in the case of non linear expressions.

3.2 *Graphical Solution Procedure*

1. Assuming x_1 and x_2 (the decision variables) to be represented on X and Y axes respectively check the conditions of variables and accordingly prepare the graph sheet.

   If $x_1 \geq 0$, $x_2 \geq 0$, your graph will be in first quadrant, if x_1 is unrestricted and $x_2 \geq 0$, the solution will be I and II quadrants, if $x_1 \geq 0$ and x_2 is unrestricted the solution will be in I and IV quadrants. If both are unrestricted, the solution may be any where (in any quadrant).

   [It is better to leave three units at the bottom and also at the left side so that your graphical solution will be clearly represented].

   Divide the scale approximately on X and Y axes such that you can represent all its values. It is always better to have same scale on both axes.

2. Assume the constraints as equations and find any two points for each equation so that the equation can be represented as a straight line on graph.

   [It is always easy to assume $x_2 = 0$ to find x_1 (say a) and $x_1 = 0$ to find x_2 (say b) and thus you draw line connecting (a,0) and (0,b)]

3. Similarly, draw all the constraint lines.

4. Shade the appropriate areas as given by the constraints. If the constraint is $\leq$ type shade the area towards origin. If the constraint is $\geq$ type shade the area away from origin. If the constraint is '=' type, do not shade any area, and the line itself is the region.

Note : *The above rules are applicable for the lines which have positive constant at right hand side, otherwise the above rules are to be reversed. Refer section 3.5 for details.*

5. Identify the feasible region by locating the area satisfying all constraints that is common with all the constraint areas. This region will have the optimal solutions. If there is no common area possible then the solution is infeasible.

6. Now, to find optimal solution we have two methods :

   (a) Iso-profit (or Iso-cost) line method.

   (b) Extreme point evaluation method.

   (a) ***Iso profit line*** method for maximisation objective [Iso cost ***line in the case of minimisation objective]:*** In this method, first assuming zero profits or zero costs, the objective function line is drawn. (Find any one point giving any value to x_1 and find x_2. Connect this point to origin and extend). Any line parallel to this line belongs to the family of objective functions. Among all these family of lines locate the one that will be touching the feasible region at the farthest point (in the case of maximisation) or the one which touches at nearest points (in the case of minimisation).Identify these points and read their X and Y co-ordinates.

   Find the Z_{max} or Z_{min} (as the case may be) by substituting these values in the objective function.

   *(b) **Extreme Point** Evaluation Method : **In this method, after locating the feasible region, read all the extreme (corner) points. Substitute these values in objective function. Identify the point(s) that yields maximum value or minimum value (as the case may be). Hence the solution.***

   [This method is easier (comparatively) but not reliable in some cases. Therefore students are advised to use the Iso-profit/cost line method and may check the answer by using extreme point evaluation]

   These are illustrated through the solved examples to follow.

   Illustrations of Graphical Method :

ILLUSTRATION 1 ————————————————————————————————

Solve graphically

   Maximise $Z = 8x_1 + 6x_2$

   Subject to $2x_1 + x_2 \leq 72$

                      $x_1 + 2x_2 \leq 48$

                      $x_1 \geq 0, x_2 \geq 0$ [JNTU B.Tech Mech 93]

Solution :

Step 1 : The graph sheet is prepared selecting x_1 on X - axis, and x_2 on Y - axis. It is clear from the conditions ($x_1 \geq 0$, $x_2 \geq 0$) that we get the figure in first quadrant.

Step 2 & 3 : Write inequations as equations and locate any two points on each equation.

$$2x_1 + x_2 \leq 72 \text{ written as } 2x_1 + x_2 = 72$$

If $x_1 = 0$, $x_2 = 72 \Rightarrow$ point is A (0, 72)

and $x_1 = 36$ if $x_2 = 0 \Rightarrow$ point is B (36, 0)

Again $x_1 + 2x_2 \leq 48$ is written as $x_1 + 2x_2 = 48$

If $x_1 = 0$, $x_2 = 24 \Rightarrow$ point is C (0, 24)

and $x_1 = 48$ if $x_2 = 0 \Rightarrow$ point is D (48, 0)

Draw these lines on the graph sheet as AB with the points A (0, 72), B (36, 0) and CD with points C (0, 24) and D (48, 0)

Step 4 : Shade the corresponding areas that represents the inequations, i.e. shade the area below AB towards origin since the inequation is $\leq$ type, similarly below the line CD

Step 5 : Locate the common area that can satisfy all the constraints and name it as "*feasible region*". In the following graph it represented as *OCEB*

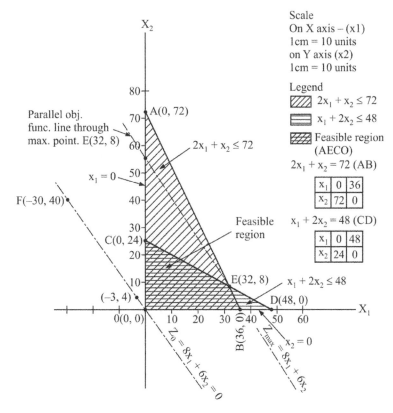

FIGURE 3.1 :

Step 6 : **(a) Iso-Profit Line** Method : **The objective function line is first drawn through origin treating profit $Z = 0$.**

$\therefore$ $Z_0 \equiv 8x_1 + 6x_2 = 0$ passes through $O(0, 0)$

Since if $x_1 = 0$, $x_2 = 0$

Also, if $x_2 = 40 \Rightarrow x_1 = -30$

$\therefore$ $F(-30, 40)$ lies on objective function line.

$\therefore$ OF represents $8x_1 + 6x_2 = 0$.

Represent this line with 'dot and dash' to distinguish from the constraint lines.

Now draw a parallel line by using set - squares so as to touch at the most maximum point on feasible region. This is $E(32, 8)$.

Thus $Z_{max} = 8(32) + 6(8) = 256 + 48$

$$= \textbf{304 units.}$$

(b) Extreme Point Evaluation :

The evaluation is shown in a tabular form given below :

Extreme Point	X_1	X_2	$8x_1$	$6x_2$	$Z = 8x_1 + 6x_2$
O	0	0	0	0	0
C	0	24	0	144	144
E	32	8	256	48	**304** $\leftarrow Z_{max}$
B	36	0	288	0	288

Solution :

$$X_1 = 32; \quad X_2 = 8$$
$$Z_{max} = 304$$

ILLUSTRATION 2 ──────────────────────────────

Find the minimum value of $Z = 4x_1 + 2x_2$ subject to the constraints:

$x_1 + 2x_2 \geq 2,$ $3x_1 + x_2 \geq 3,$ $4x_1 + 3x_2 \geq 6$ and

$x_1, x_2 \geq 0$ by graphical method.   **[JNTU Mech./Prod.Chem. 2000]**

Solution :

First convert the inequality into equality and find the minimum two co-ordinates to draw a straight line in the graph as follows:

Constraint 1 : $x_1 + 2x_2 = 0$ Constraint 2 : $3x_1 + x_2 = 3$

x_1	0	2
x_2	1	0

x_1	0	1
x_2	3	0

Constraint 3 : $4x_1 + 3x_2 = 6$ Objective function line $4x_1 + 2x_2 = 0$

x_1	0	-3
x_2	2	6

x_1	0	-1
x_2	0	2

Assume $Z = 0$ i.e., $4x_1 + 2x_2 = 0$ to find two co-ordinates for this equation.

Now the geometrical interpretation of the problem is given in graph. (Fig. 3.2)

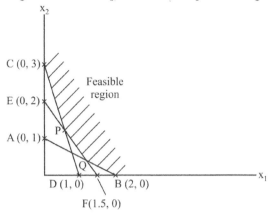

FIGURE 3.2 :

The minimum value of Z is found by a parallel lines to $Z = 0$. The minimum point which it touches the feasible region is solution i.e., at P (0.55, 1.25).

$$\text{Min } Z = 4X_1 + 2X_2 = 4 \times 0.55 + 2 \times 1.25 = 4.7$$

Note : *Observe the shadings given in illustration 1 and 2. In illustration 1 the shading is done below the constraint lines since they are of ≤ type, while in illustration 2 the constraint are ≥ type, therefore the areas above these lines are shaded. Similarly in the illustration 1 objective function is maximisation type therefore we take highest point on the feasible region that objective function can touch, whereas in illustration 2 we take the lowest point touched by objective function on the feasible region since it is to minimise.*

ILLUSTRATION 3

An oil refinery can blend two grades of crude oil to produce quality P and Q petrol. Two possible blending processes are available. For each production run, the older process uses 5 units of crude A, 7 units of crude B and 2 units of crude C to produce 9 units of P and 7 units of Q. The newer process uses 3 units of crude A, 9 units of B and 4 units of crude C to produce 5 units of P and 9 units of Q petrol. Because of prior contract commitments the refinery must produce at least 500 units of P and at least 300 units of Q for the next month. It has available 1500 units of crude A, 1900 units of crude B and 1000 units of crude C. For each unit of P the refinery receives Rs. 60 while for each unit of Q it receives Rs. 90. Find out the linear programming formulation so as to maximize the revenue and solve graphically or other wise.

Solution :

Refer Illustration 7 of chapter - 2 for formulation of LPP,

Summary : The above information is formulated as given below :

Maximise $Z = 1170\, x_1 + 1110\, x_2$

Subject to

$$9x_1 + 5x_2 \geq 500$$
$$7x_1 + 9x_2 \geq 300$$
$$5x_1 + 3x_2 \leq 1500$$
$$7x_1 + 9x_2 \leq 1900$$
$$2x_1 + 4x_2 \leq 1000$$
$$x_1, x_2 \geq 0$$

The feasible region is formed by two constraints only. These are $9x_1 + 5x_2 \geq 500$ and $7x_1 + 9x_2 \leq 1900$.

The other three constraints are redundant constraints.

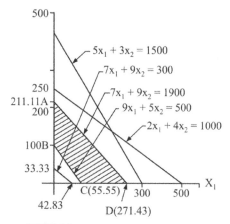

FIGURE 3.3 :

Solution by Extreme Points :

Point	X_1	X_2	$1170\ X_1$	$1110\ X_2$	$Z = 1170\ X_1 + 1110\ X_2$
A	0	211.11	0	234332.1	234332.1
B	0	100	0	111000	111000
C	55.55	0	64993.5	0	64993.5
D	271.43	0	317573.1	0	$317573.1 \leftarrow Z_{\max}$

Solution :

$$X_1 = 271.43 \qquad X_2 = 0 \qquad Z_{\max} = 317573.1$$

3.3 *Redundant Constraint*

It is the constraint without which there will not be any effect on the feasible region.

Ex : 1. Refer solution of above problem. Three constraints given by $7x_1 + 9x_2 \geq 300$, $2x_1 + 4x_2 \leq 1000$ and $5x_1 + 3x_2 \leq 1500$ are redundant.

2. Suppose you have a constraint $x_1 + x_2 \geq 2$ and another constraint $x_1 + x_2 \geq 10$, then the first one is redundant since second constraint if gets satisfied, automatically satisfies the first.

3. If $2x_1 + 3x_2 \leq 50$ and $4x_1 + 6x_2 \leq 37$, the former is redundant.

ILLUSTRATION 4 ────────────────────────────

Use graphical method to solve the following LP problem:
Maximize $Z = 3x_1 + 4x_2$
Subject to the constraints:
$$2x_1 + x_2 \leq 40$$
$$2x_1 + 5x_2 \leq 180$$
$$X_1, x_2 \geq 0$$ **[JNTU B. Tech (Mech) 1998/S, OU - MBA July 2000]**

Solution :

Consider a set of rectangular cartesian axes OX_1 and OX_2 in $OX_1 X_2$ plane. Plot the lines by considering the inequations as equations.

$2x_1 + x_2 = 40$ is ploted by taking two points A (20, 0) and D (0, 40) and $2x_1 + 5x_1 = 180$ is ploted by taking the points B (0, 36) and E (90 , 0).

Now, shade the region to represent the inequations (below the line for ≤) and to find the solutions space or feasible region. The feasible region is $OACB$ as shown in graph.

The four vertices (corner pionts) form a convex region. These are $O(0, 0)$, $A(20,0)$, B (0, 36), C (2.5, 35).

Now, on substituting these points in the objective function $Z = 3x_1 + 4x_2$,

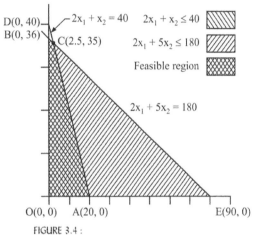

$$z_o = 0 + 0 = 0$$

$$z_A = 3 \times 20 + 0 = 60$$

$$z_B = 0 + 4 \times 36 = 60$$

$$z_c = 3 \times 2.5 + 4 \times 35 = 147.5$$

z_c is maximum,

Hence, the optimum solution is

$$x_1 = 2.5$$

$$x_2 = 35$$

$$z_{max} = 147.5$$

FIGURE 3.4 :

ILLUSTRATION 5

Find the minimum value of $Z = -x_1 + 2x_2$
Subject to the constraints
$$-x_1 + 3x_2 \leq 10$$
$$x_1 + x_2 \leq 6$$
$$x_1 - x_2 \leq 2$$
$$x_1, x_2 \geq 0$$
Use graphical method   **[JNTU B.Tech (ECE) 1997]**

Solution :

Rewriting each constraint as follows : (in the equation form)

$$-x_1 + 3x_2 = 10$$

$$x_1 + x_2 = 6$$

$$x_1 - x_2 = 2$$

$$x_1, x_2 \geq 0$$

Draw line for each constraint by first treating it as a linear equation. Then use the inequality condition of each constraint to make the feasible regions as shown in the figure.

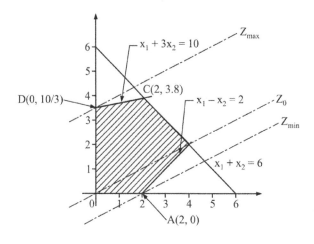

FIGURE 3.5 :

The coordinates of extreme points of the feasible region are

$$O = (0, 0)$$
$$A = (2, 0)$$
$$B = (4, 2)$$
$$C = (2, 3.8)$$
$$D = \left(0, \frac{10}{3}\right)$$

The value of the objective function at each of these extreme points is as follows:

Extreme Point	Coordinates (x_1, x_2)	Objective Function Value $Z = -x_1 + 2x_2$
O	(0, 0)	$-1(0) + 2(0) = 0$
A	(2, 0)	$-1(2) + 2(0) = -2$
B	(4, 2)	$-1(4) + 2(2) = 0$
C	(2, 3.8)	$-1(2) + 2(3.8) = 5.6$
D	$\left(0, \frac{10}{3}\right)$	$-1(0) + 2\left(\frac{10}{3}\right) = \frac{20}{3}$

The minimum value of the objective function $Z = -2$ occurs at the extreme point (2,0). Hence the optimal solution to the given LP problem is $x_1 = 2$, $x_2 = 0$ and Min $Z = -2$.

3.4 *Different Types of Solutions*

As we have seen the structure of an LPP is composed of three main parts viz objective function, set of constraint and the conditions of variables. With reference to these parts, we report the following types of solutions.

1. Solutions.

2. Feasible solutions.

3. Basic feasible solutions.

4. Optimal solutions.

1. Solutions : *All those values of variable which satisfy the conditions given are solutions:* Thus if the conditions are $x_1 \geq 0$, $x_2 > 0$ (non - negative), then all the values in first quadrant (covered by positive X– axis and positive Y - axis) are the solutions.

Similarly, $x_1 \geq 0$ (X-axis) and x_2 unrestricted yields the solutions in first and fourth quadrants of graph, and so on.

2. Feasible Solution : *All the solutions which satisfy all the conditions of variables as well as all the constraints are feasible solutions.*

We can notice here that if all the constraints are exact type, we may get '*point solution*'. One exact and another inequality constraint will yield '*line of solutions*'. All the inequality constraints will generate an area of feasible solutions often referred to as '*feasible region*'.

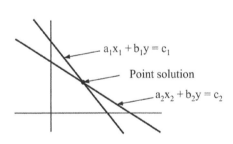

(a) Point solution with two exact constraints (point size feasible region)

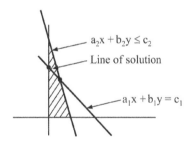

(b) A line of solution with one exact constraint (line feasible region)

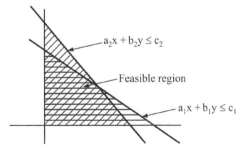

(c) A feasible region with inequality constraints

FIGURE 3:6 :

3. Basic Feasible Solutions : *The values of variables represented by the points along the border lines of feasible region are basic feasible solutions.*

If m non identical equation with 'n' variable ($m < n$) exist in a problem, then keeping ($n - m$) variables constant, usually at zero, the values of the variables yield a solution called '*basic solution*'. As it satisfies all the constraints it can be called a region, it can also be called a 'basic feasible solution'. Therefore each selection of ($n - m$) variables from n variables gives raise to '$\left({}^{n}C_{n-m} \right)$' basic feasible solutions' (BFS)

Of all these BFS, the one with which we start working out the problem is 'initial basic feasible solution'. (IBFS).

Most commonly, the IBFS will be choosen to start at the worst case of the solution set so as not to miss to examine any solution. Thus a solution with zero profit or nil production or the values of decision variables as zeros (i.e., origin) in graphical solution will be IBFS.

4. Optimal Solution : The solutions which satisfy all the conditions of variables, all the constraints and the objective function are 'optimal basic feasible solutions (OBFS)' or simply 'optimal solutions'.

Schematically :

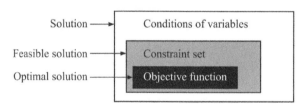

FIGURE 3.7 : TYPES OF SOLUTIONS

3.5 *Locating Feasible Region*

Most of the students of Operations Research are complaining that they are getting confused in locating a feasible region. Some tips are given below to remove their confusion.

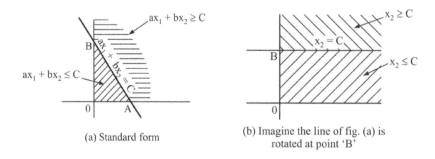

(a) Standard form   (b) Imagine the line of fig. (a) is rotated at point 'B'

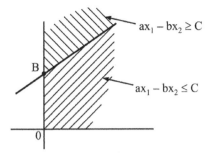

(c) Imagine line in fig. (a) is further rotated at 'B'

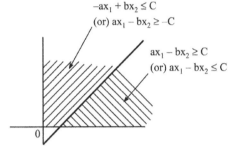

(e) Imagine line in fig. (a) is further rotated at point 'A'

FIGURE 3.8

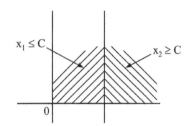

(d) Imagine line of fig. (a) is rotated at 'A'

Note:
In a simple way youcan rember as
1. *If area is comming towards origin, it is $\leq$ type constraint, and*
2. *If area is going away from origin, it is $\geq$ type constraint provided the right hand side of the inequation is with positive sign.*

3.6 *Types of Feasible Regions*

The area consisting of set of feasible solutions exist in several ways. These are described below :

1. **With Reference to** Feasibility :

    (a) *Infeasible* Region : *If there exists no feasible region (solution) at all, it is infeasible region.*

    If there is no common area or line or point that satisfies all the constraints, the solution is infeasible as no feasible region exists.

    (b) *Feasible* Region : *It is a point or line or region that satisfies the set of constraints.*

2. **With Reference to Number of** Feasible Points :

    (a) *Infeasible* Region : *As discussed above.*

    (b) *Point Feasible* Region : *If only one point exits as feasible region e.g., Two equality constraints intersecting at a point and satisfying the conditions of variables (or three or more equality constraints concurrent at a point).*

    (c) *Line Feasible* Region : *It is a feasible region in the form of a line. It is found in case at least one equality constraint exist in the set.*

(d) *Area* Feasible Region : *It is the area surrounded by many inequality constraints.*

(e) *Volume* Feasible Region : *It is the feasible region found when there are three variables by which three planes are formed. It is three dimensional.*

3. **With Reference to Shape :**

(a) *Convex* Feasible Region : *It is a region formed by a collection of points such that every point on a line segment dram selecting any two points of the collection, falls in the collection. This means that if P and Q are any two points in a collection, the line segment PQ is also in the collection.*

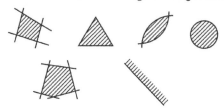

FIGURE 3.9 :   CONVEX FEASIBLE REGION

(b) *Concave* Feasible Region : *It is a region formed by a collection of points such that at least one point on a line segment drawn selecting any two points of the collection, does not fall in the collection. This means that if P and Q are any two points is a collection, at least one point on line segment PQ does not fall in the collection/region.*

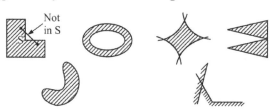

FIGURE 3.10 :   CONCAVE FEASIBLE REGION

4. **With Reference to its Boundaries :**

(a) *Bounded Feasible* Region : *This feasible region will have a distinct boundary line. Most times the boundaries are connected by the co-ordinate axes and some availability (≤ type) constraints.*

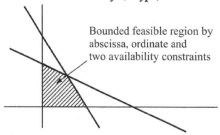

Bounded feasible region by abscissa, ordinate and two availability constraints

FIGURE 3.11 :   BOUNDED FEASIBLE REGION

(b) Island Feasible Region : *This is also a bounded feasible region but not connected by the co-ordinate axes.*

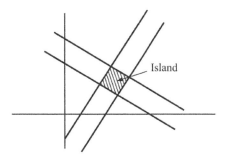

FIGURE 3.12 : ISLAND FEASIBLE REGION

(c) Unbounded Feasible Region : *This feasible region contains no boundary on atleast one side. Usually it is observed with all requirement. (i.e., ≥ type) constraints.*

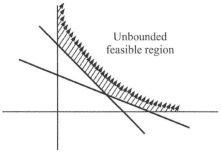

FIGURE 3.13 : UNBOUNDED FEASIBLE REGION

3.7 *Types of Optimal Solutions*

The optimal solutions can be categorised into the following types :

1. Infeasible solutions.

2. Unique solutions.

3. Multiple optimal solutions.

4. Unbounded solutions.

1. Infeasible Solution : When feasible region does not exist, the solution we get is infeasible.

In graphical solution it is found when one constraint is availability (≤) type and the other is requirement (≥) type and these two can not produce any common area (non intersecting) in the specific quadrant (such as $x_1 \geq 0$, $x_2, \geq 0$ indicates first quadrant).

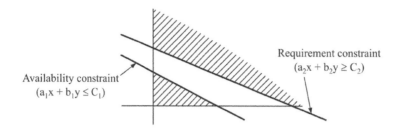

FIGURE 3.14 : INFEASIBLE SOLUTION

   2. Unique Solution : **If the optimal solution is one and only one, then the solution is unique.**

   Thus the feasible region will yield one maximum or minimum solution values in the objective function at only one point.

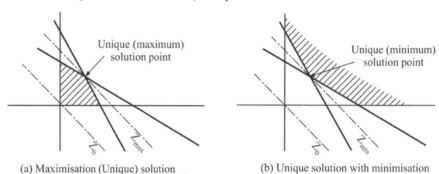

(a) Maximisation (Unique) solution   (b) Unique solution with minimisation

FIGURE 3.15 : UNIQUE SOLUTIONS

   3. Multiple Optimal Solution : **When several optimal solutions exist, the solutions are said to be multiple optimal solutions or alternate optimal solutions.**

   In graphical solutions they exists when the objective function is parallel (having same slope) to one of the constrains, provided there exists a feasible region.

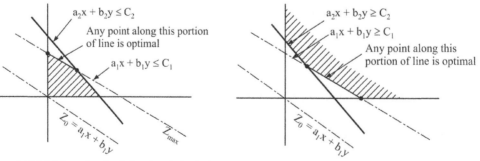

(a) Multiple optional solution (maximisation)   (b) Multiple optional solution (minimisation)

FIGURE 3.16 : MULTIPLE OPTIMAL SOLUTIONS OR ALTERNATE OPTIMAL SOLUTIONS

4. Unbounded Solution : **If a distinct and finite solution can not be found or the solution exists at infinity, the solution is said to be unbounded.**

In graphical solution, unbounded solutions are obtained if the feasible region is unbounded (formed by requirement constraint i.e., ≥ type) while the objective function is maximisation.

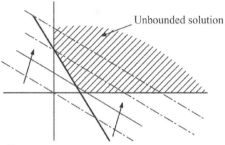

FIGURE 3.17 : UNBOUNDED SOLUTION

(Since it has to be taken to infinity to locate maximum value we have no finite or unbounded solution)

Solved Problems

Infeasible Solutions

ILLUSTRATION 6

Solve Graphically:

Maximise	$Z = 50x_1 + 60x_2$
Subject to	$x_1 + x_2 \leq 12$
	$2x_1 + 3x_2 \geq 60$
	$x_1, x \geq 0$

[JNTU B.Tech (Mech) 96/CCC]

Solution :

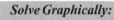

$x_1 + x_2 = 12$

x_1	0	12
x_2	12	0

$2x_1 + 3x_2 = 60$

x_1	0	30
x_2	20	0

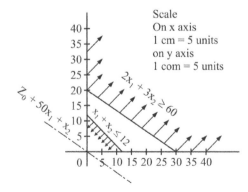

FIGURE 3.18 :

Unique Solutions

In addition to the above illustrations (1 to 5), one more is given below.

ILLUSTRATION 7

> Minimize $Z = 2x_1 + 3x_2$
> Subject to $x_1 + x_2 \leq 4,$ $3x_1 + x_2 \geq 4$
> $x_1 + 5x_2 \geq 4,$ $0 < x_1 \leq 3,$ $0 \leq x_2 \leq 3$
>
> (*Remark : In this problem the conditions of variables are given in constraints*).

Solution :

Points for $x_1 + x_2 = 4$ are $(0, 4)$, $(4, 0)$

Points for $3x_1 + x_2 = 4$ are $(0, 4)$, $\left(\dfrac{4}{3}, 0\right)$

Points for $x_1 + 5x_2 = 4$ are $\left(0, \dfrac{4}{5}\right)$, $(4, 0)$

and $x_1 = 3$ is a line parallel to $X_2 - axis$ i.e., Y axis passing through $(3, 0)$

Similarly $x_2 = 3$ is a line parallel to $X_1 - $ axis i.e., $X - $ axis passing through $(0, 3)$

Also, $0 \leq x_1$ or $x_1 \geq 0$ and $0 \leq x_2$ or $x_2 \geq 0$ represent the condition that the solution exists in first quadrant.

Now these can be plotted on the graph as shown below :

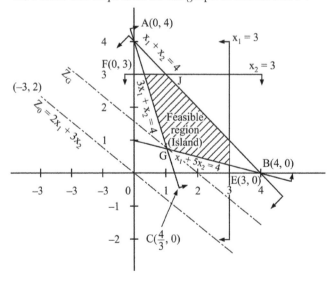

FIGURE 3.19 :

Feasible region is area covered by *GHIJK*

Iso - Cost Line Method : **Bringing** $Z = 2x_1 + 3x_2$ **to the minimum point of feasible region it touches at** G.

Therefore solution is $G\left(\dfrac{8}{7}, \dfrac{4}{7}\right),$ $Z_{min} = 4$

Extreme Point Evaluation :

At $\quad G\left(\dfrac{8}{7}, \dfrac{4}{7}\right),$ $\qquad Z_G\ 2 \times 3 + 3 \times \dfrac{1}{5} = 6.6$

At $\quad H\left(3, \dfrac{1}{5}\right),$ $\qquad Z_H = 2 \times 3 + 3 \times 1 = 9$

At $\quad I\,(3, 1),$ $\qquad Z_I = 2 \times 1 + 3 \times 3 = 11$

At $\quad J\,(1, 3),$ $\qquad Z_J = 2 \times \dfrac{1}{3} + 3 \times = 9.67$

At $\quad K\left(\dfrac{1}{3}, 3\right)$ $\qquad Z_K = 2 \times \dfrac{1}{3} + 3 \times 3 = 9.67$

$\qquad Z_G$ is $Z_{\min}$, i.e., 4

At $\quad x_1 = \dfrac{8}{7}$ and $x_2 = \dfrac{4}{7}$

Multiple Optimal Solutions

ILLUSTRATION 8

Maximise $\quad 15x_1 + 9x_2$

Subject to $\quad 10x_1 + 6x_2 \leq 60, \quad x_1 + 2x_2 \leq 10, \qquad x_1, x_2 \geq 0$

Solution :

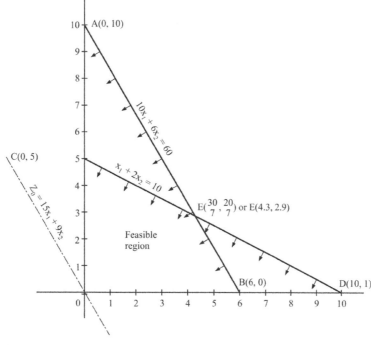

FIGURE 3.20 :

As $Z_0 \equiv 15x_1 + 9x_2$ is parallel to $10x_1 + 6x_2 = 60$, it fits exactly on this line when its iso-profit line is taken to represent maximum. Therefore any point along the line segment BE is a solution.

Hence there are Multiple optimal solutions to this problems

Extreme Point Evaluation :

At $O(0, 0)$

$$Z_O = 15(0) + 9(0) = 0$$

At $C(0, 5)$

$$Z_C = 15(0) + 9(5) = 45$$

At $B(6, 0)$

$$Z_B = 15 \times 6 + 9 \times 0 = 90 \text{ (Maximum)}$$

At $E\left(\dfrac{30}{7}, \dfrac{20}{7}\right)$

$$Z_E = 15 \times \frac{30}{7} + 9 \times \frac{20}{7}$$

$$= \frac{630}{7} = 90 \text{ (Maximum)}$$

As the values at B and E are showing maximum, we can infer that any point along the line BE is optimal and hence the solution is multiple optimal.

Remarks :

   (i) *You may be misled if you misread the values of E in the graph.*

   Suppose you read as

$$x_1 = 4.3 \text{ and } x_2 = 2.9$$

$$Z_E = 15(4.3) + 9(2.0) = 64.5 + 26.1$$

$$= 90.6 \approx 91$$

Thus it may give a wrong anser as 'unique solution' at E with $Z_{max} = 90.6$ or 91. Therefore it is very important to read exact values (which is very difficult in graphical solutions). However, the exact values can be obtained by solving the equations to find the point of intersection.

   (ii) *In view of the above, Students are advised to choose the Iso-profit/Iso-cost line method to eliminate such error.*

   (iii) *Observe the slope of objective function* $\left(-\dfrac{15}{9} = \dfrac{-5}{3}\right)$ *and that of first constraint* $\left(i.e., \dfrac{-10}{6} = -\dfrac{5}{3}\right)$. *As these are same by looking at the problem, we can predict the multiple optimal solutions.*

(iv) *The above remark fails if there is no chance of getting feasible region. (Infeasible solution)*

(v) *The remark (iii) also fails if the objective function is to minimise in this case,*

Since $Z_{\min} = 0$ *at origin is most minimum.*

Unbounded Solution

ILLUSTRATION 9

Solve:
    Maximum $Z = 3x_1 + 4x_2$
    Subject to $x_1 - x_2 \le 1$
                  $-x_1 + x_2 \le 2$
                  $x_1 + x_2 \ge 0$ [JNTU (CSE) 3/4 - 2001]

Solution :

$x_1 - x_2 = 1$

x_1	0	1
x_2	-1	0

$-x_1 + x_2 = 2$

x_1	0	-2
x_2	2	0

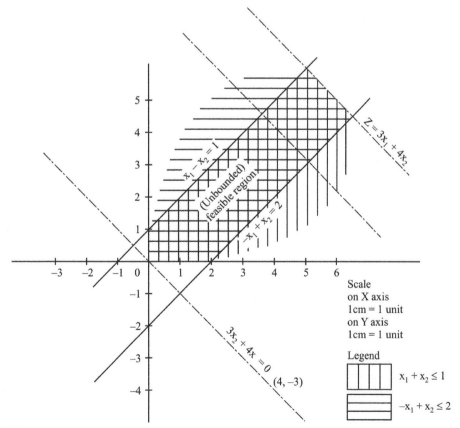

FIGURE 3.21 :

Since there is no boundary on one side and the maximum value of $Z = 3x_1 + 4x_2$ is found at this side only, the solution is found at infinity. Thus there is no finite solution to this problem. In other words, the solution is unbounded.

Remarks :

(i) *In the above problem, if the both equations are greater than or equal to ($\geq$) type i.e., $x_1 - x_2 \geq 1$ and $-x_1 + x_2 \geq 2$ then the solution will be infeasible.*

(ii) *Pupils may get confused between unbounded (or no finite) solution and the infeasible solution (i.e., no solution). In unbounded solution, the solution exists at infinity, this indicates infinite scope of improvement (without any finite value) through its objective function, while in infeasible solution there exits 'no' solution on in other words, it is impossible to have any solution to the problem.*

Practice Problems

Infeasible Solutions ▬▬▬▬▬▬▬

Solve Graphically :

1. Maximise $Z = 8x_1 + 10x_2$

    Subject to $2x_1 + 3x_2 \geq 12$

    $x_1 + x_2 + 4$

    $x_1, x_2 \geq 0$

2. Maximise $Z = 10x_1 + 15x_2$

    Subject to $4x_1 + 6x_2 \geq 240$

    $x_1 + x_2 \leq 40$

    $4x_1 + 3x_2 \geq 90$

    $x_1, x_2 \geq 0$

3. Minimise $Z = 3x_1 + 4x_2$

    Subject to $x_1 - x_2 \geq 1$

    $-x_1 + x_2 \geq 2$

    $x_1, x_2 \geq 0$

4. Minimise $Z = x_1 + 2x_2$

    Subject to $x_1 - x_2 \geq 3$

    $-x_1 + x_2 \geq 4$

    $x_1 \geq 0, x_2 \geq 0$

5. Solve graphically, the following LPP if possible.

$x_1 - 3x_2 \leq 6;$ $\qquad$ $x_1 + 2x_2 \geq 4$

$x_1 - 3x_2 \geq -6$ $\quad$ and $\qquad$ $x_1, x_2 \geq 0$

With an objective function as

(a) Minimise $\qquad 2x_1 + x_2$

(b) Maximise $\qquad 2x_1 + 3x_2$

Answer : $\qquad$ (a) 4 $\quad$ (b) Unbounded.

Unique Solution ━━━━━━━━━━━━━━━━━━━━━━━━━━━━━━━

Solve Graphically :

1. Maximise $\quad Z = 9x_1 + 16x_2$

$\qquad$ Subject to $\quad x_1 + 4x_2 \leq 80$

$\qquad\qquad\qquad\qquad 2x_1 + 3x_2 \leq 90$

$\qquad\qquad\qquad\qquad x_1 \geq 0, x_2 \geq 0$

Answer : $Z_{\max} = 440$ at $x_1 = 24, x_2 = 14$

2. Minimise $\quad Z = 3x_1 + 2x_2$

$\qquad$ Subject to $5x_1 + x_2 \geq 10$

$\qquad\qquad\qquad\quad x_1 + x_2 \geq 6$

$\qquad\qquad\qquad\quad x_1 + 4x_2 \geq 12$

$\qquad\qquad\qquad\quad x_1, x_2 \geq 0$

Answer : $\qquad Z_{\min} = 13$ at $x_1 = 1, x_2 = 5.$

3. Solve graphically the following LPP.

Maximise : $Z = 3x_1 + 2x_2$

subject to the constraints

$\qquad 2x_1 + x_2 \leq 1$

$\qquad x_1 + x_2 \leq 3$

$\qquad x_1 \leq 2$ and

$\qquad x_1, x_2 \geq 0$ $\qquad\qquad\qquad$ [JNTU - Mech./Prod./Chem. - 2001/S]

4. Aditi wishes to invest Rs. 12000 in Bank Saving Certificates (BSC) and National Saving Bonds (NSB). According to rules at least Rs 1000 and at least 2000 should be invested on B.S.C and N.S Bonds respectively. If the rate of interest on B.S.C is 8% p.a. and that on NSB is 10% p.a., how should Aditi invest her money to earn highest income. Solve graphically or otherwise.

Answer : BSC Rs. 1000 on NSB Rs 11000

$\qquad$ Max. interest = 1180p.a. total income = Rs. 3180

5. Beetel manufacturers produce two types of telephone sets, viz with cord and cordless. The cord type requires 2 hours and cordless requires four hours to assemble. The company has at the most 800 hours per day to assemble and its packaging department at the most can pack 300 sets per day. If the company sells cord type at Rs. 600 and the cordless at Rs. 1000, how may sets of each type should be produced to attain highest sales. Solve graphically.

Answer : cord type $= 200$, cordless $= 100$; Max sales $=$ Rs. 2,20,000

6. With the constraints

$$x_1 - x_2 \leq 1, \quad x_1 + x_2 \geq 3, \quad x_1 \geq 0, \quad x_2 \geq 0 \text{ and the objective as}$$

(a) Minimise $Z = 3x_1 + 2x_2$

(b) Maximise $Z = 3x_1 + 3x_2$

(c) Minimise $Z = 3x_1 + 3x_2,$

Solve the LPP graphically

Answer : (a) $x_1 = 2, x = 1$; *Min Z* $= 8$

(b) Unbounded solution.

(c) Multiple optimal solution (along the line $x_1 + x_2 = 3$)

$$x_1 = 2, \quad x_2 = 1 \; Z_{min} = 9$$

or $x_1 = 1, x_2 = 2 \; Z_{min} = 9$ or $x_1 = 1.5, x_2 = 1.5$ and so on

Multiple Optimal Solution

1. Solve graphically. Is the feasible region convex or concave?

Maximise $Z = 4x_1 + 3x_2$

Subject to $8x_1 + 6x_2 \leq 48$

$2x_1 \leq 9$

$x_2 \leq 6$

$x_1, x_2 \geq 6$

Answer : Multiple optimal solutions

$x_1 = 4.5, x_2 = 2, Z_{max} = 24$; or $x_1 = 1.5, x_2 = 6, Z_{max} = 24$

The feasible region is convex.

2.   (a) Use graphical method to solve the following LPP.

(b) Examine whether the LPP has any alternative solution, other than you found.

(c) Also check whether feasible regions is convex/concave and bounded/unbounded.

$$\text{Minimise} \quad Z = 20x_1 + 30x_2$$
$$\text{Subject to} \quad x_1 \geq 5$$
$$2x_2 \geq 7$$
$$4x_1 + 6x_2 \leq 24$$
$$x_1 \geq 0, \; x_2 \geq 0$$

(d) What will be the solution if the objective function in the above problem is to maximise instead of minimisation.

Answer : (a) $x_1 = 5$, $x_2 = 2/3$, $Z_{min} = 120$

(b) The solution is multiple optimal solutions an alternative solution can be $x_1 = 0.75$, $x_2 = 3.5$, $Z_{min} = 120$

(c) The feasible region is concave type.

(d) If the objective function is $Z_{max} = 20\,x_1 + 30\,x_2$ then the solution is unbounded i.e., no finite solution.

3. Use the following constraints to solve given objective functions of an LPP.

$x_1 + x_2 \leq 1$, $x_1 + x_2 \geq 3$; $x_1 \geq 0$, $x_2 \geq 0$ and objective functions as

(a) Min $z = 3\,x_1 + 2\,x_2$

(b) Min $3x_1 + 3x_2$

(c) Max $Z = 3x_1 + 3x_2$

(d) Min $Z = x_1 + 3x_2$

Answer : (a) Unique solution $x_1 = 0$, $x_2 = 3$, $Z_{min} = 6$

(b) Multiple optimal solutions $x_1 = 0$, $z_2 = 3$, $Z_{min} = 9$;

$x_1 = 2$, $x_2 = 1$ $Z_{min} = 9$;

(c) Unbounded solution

(d) $x_1 = 2$, $x_2 = 3$; $Z_{min} = 5$

4. Solve $Z_{max} = 0.4\,x_1 + 0.2\,x_2$

S.t $2x_1 + x_2 \leq 72$

$x_1 + 2x_2 \leq 48$

$x_1, x_2 \geq 0$

Answer : Multiple optimal solutions $x_1 = 32$, $x_2 = 8$ or $x_1 = 36$, $x_2 = 0$

$Z_{max} = 14.4$

5. Use graphical methods to solve the following LPP. What type of feasible region do you get ?

$$\text{Max } Z = 5x + 4y$$

Subject to

$$x + y \leq 40$$
$$x \leq 20$$
$$15x + 12y = 300$$
$$x \geq 0 \ , y \geq 0$$

Answer : Multiple optimal solution any point along the line $15\,x + 12y = 300$
$Z_{\max} = 100 \ at \ x = 20, \ y = 0 \ \text{or} \ x = 0 \ , y = 20 \ , x = 10, \ y = 12.5$, and so on.
The feasible region is 'a line' only

Unbounded Solution ━━━━━━━━━━━━━━━━━━━

Use graphical method to solve the following LPP

1. Maximise $Z = 6x_1 + 4x_2$

Subject to $x_1 - x_2 \leq 1$
$$3x_1 - 2x_2 \leq 6;$$
$$x_1, x_2 \geq 0$$

2. Maximise $Z = 30x_1 + 50x_2$

Subject to $2x_1 + x_2 \geq 7$
$$x_1 + x_2 \geq 6$$
$$2x_1 + 6x_2 \geq 18$$
$$x_1 \geq 0, x_2 \geq 0$$

3. Maximise $Z = 3x_1 + 2x_2$

Subject to $-2x_1 + 3x_2 \leq 9$
$$3x_1 - 2x_2 \leq 20$$
$$x_1 \geq 0; x_2 \geq 0$$

4. Minimise $Z = 3x_1 - 2x_2$

Subject to $2x_1 + x_2 \leq 1$
$$x_1 + x_2 \geq 5$$
$$x_1 \geq 0, x_2 \geq 0$$

5. Maximise $Z = 2x_1 + 3x_2$

Subject to $4x_1 + 6x_2 \geq 240$
$$x_1 + x_2 \leq 40$$
$$4x_1 + 3x_2 \geq 180$$
$$x_1, x_2 \geq 0$$

Remarks :

The solution the above problem (5) seems to be multiple optimal, but it is unbounded.

Additional Problems

Solve graphically

1. Minimise $Z = 30x_1 + 20x_2$;

    Subject to $x_1 + x_2 \leq 8$

                      $6x_1 + 4x_2 \geq 12$

                      $5x_1 + 8x_2 \geq 20$

                      $x_1 \leq 6$

    and $x_1, \ x_2 \geq 0$

2. Minimise $Z = 2000x_1 + 3000x_2$

    Subject to $6x_1 + 9x_2 \leq 100$

                  $2x_1 + x_2 \leq 20$

                  $x_1, \ x_2 \geq 0$ **[OU - MBA - Sep 98]**

3. Maximise $Z = 6x_1 + 7x_2$

    Subject to $3x_1 + 9x_1 \geq 36$

                  $6x_1 + 2x_2 \geq 24$

                  $2x_1 + 2x_2 \geq 16$

                    $x_1, \ _2 \geq 0$ **[OU - MBA - Nov. 94]**

4. Maximise $Z = x_1 + x_2$

    Subject to $2x_1 + x_2 = 5$

                  $3x_1 - x_2 = 6$

                  $x_1, \ x_2$ unrestricted   **[OU - MBA Apr. 98, Sep. 98]**

5. Provide graphical solution to

    Minimise $Z = 1000x + 800y$

    Subject to $6x + 2y \geq 12$

                  $2x + 2y \geq 8$

                  $4x + 12y \geq 24$

                  $x, y \geq 0$ **[OU - MBA - Dec. 20000]**

6. Maximise $Z = x + y$

    Subject to $2x + y = 5$

                  $3x - y = 6$

                  $x, \ y \geq 0$ **[OU - MBA - Dec. 2000]**

7. Maximise $Z = 3x_1 + 5x_2$

    Subject to $x_1 + 4x_2 \leq 9$

                  $2x_1 + 3x_2 \leq 11$

                    $x_1, x_2 \geq 0$ **[OU - MBA - Sep. 98, Dec. 2000]**

8. Minimise $Z = 20x_1 + 14x_2$

Subject to $x_1 + x_2 \leq 18$

$5x_1 + 3x_2 \geq 10$

$4x_1 + 6x_2 = 20$

$x_1 \geq 0, \ x_2$ unrestricted   **[OU - MBA - May 94]**

9. Maximise $Z = 20 x_1 + 40 x_2$

Subject to $36x_1 + 6x_2 \geq 108$

$3x_1 + 12x_2 \geq 36$

$20x_1 + 10x_2 \geq 100$

$x_1, x_2 \geq 0$ **[OU - MBA Feb. 93]**

10. Maximise $Z = x_1 + 3x_2$

Subject to $3x_1 + 6x_2 \leq 8$

$5x_1 + 2x_2 \leq 10$

$x_1, \ x_2 \geq 0$ **[OU - MBA - Mar. 98]**

11. Minimise $Z = 3x_1 + 5x_2$

Subject to $x_1 + x_2 = 200$

$x_1 \leq 80; \ x_2 \geq 80$

$x_1, \ x_2 \geq 0$ **[OU - MBA - May 95]**

12. Maximise $Z = 2x_1 + 5x_2$

Subject to $x_1 + 4x_2 \leq 24$

$3x_1 + x_2 \leq 21$

$x_1 + x_2 \leq 9$

$x_1, x_2 \geq 0$ **[OU - MBA - May 95]**

13. Minimise $Z = 2500 x_1 + 3500 x_2$

Subject to $50x_1 + 60x_2 \geq 2500$

$100x_1 + 60x_2 \geq 3000$

$100x_1 + 200 x_2 \geq 7000$

$x_1, \ x_2 \geq 0$ **[OU - MBA -May 91, Dec. 94, Dec. 95]**

14. Maximise $Z = 3x_1 - x_2$

Subject to $2x_1 + x_2 \geq 2$

$x_1 + 3x_2 \leq 3$

$x_2 \leq 4$

$x_1, x_2 \geq 0$ **[OU - MBA - May 92]**

15. Maximise $\quad Z = 4x_1 + x_2$

    Subject to $\quad 3x_1 + 2x_2 \geq 6$

    $\qquad\qquad -3x_1 - x_2 \leq -4$

    $\qquad\qquad 4x_1 + 5x_2 = 10$

    x_1 is unrestricted and $x_2 \geq 0$ $\qquad\qquad$ **[OU - MBA - May 92]**

16. A firm can produces two types of furniture viz. chairs and tables. The contribution for chairs and tables is Rs. 200 and Rs. 300 respectively. Both are processed on three machines M_1, M_2 and M_3. The time required by each product and total time available per week are given below.

Machine	Chair	Table	Available Hours
M_1	3	3	36
M_2	5	2	50
M_3	2	6	60

Formulate the LP and solve it, using graphical method determine optimal production. $\qquad\qquad$ **[OU - MBA - Sep. 99]**

17. What do you understand by graphic method? give its limitations use the graphical method to find the amounts of factor A and factor B to form a product which must weight 50 kgs at a minimum total cost. At least 20 kgs of A and no more than 40 kgs of B can be used. The cost of A is Rs. 100 per kg of B Rs. 25 per kg. $\qquad\qquad$ **[OU -MBA - March 99]**

18. An animal feed company must produce 200 kgs of a mixture consisting of ingredient X_1 and X_2. The ingredient X_1 cost Rs. 8 per kg and X_2 Rs. 5 per kg. Not more than 80 kg of X_1 can be used and at lest 60 kg of X_2 must be used. Find the minimum cost mixture. Use graphical method.

$\qquad\qquad$ **[OU - MBA - March 98]**

19. Three magazines P, Q, R are available for advertisement. Their exposure is number of buyers multiplied by the subsequent number of advertisements. Budget availability is Rs. One lakh manager feeds that P should have at least 2 adds each. Formulate LP model and solve it using graph.

$\qquad\qquad$ **[OU - MBA - July 2000]**

20. Solve graphically

    Maximise $Z = 4x_1 + 5x_2$

    Subject to $x_1 + x_2 \geq 1$

                      $-2x_1 + x_2 \leq 1$

                      $4x_1 + 2x_2 \leq 1$

    and $x, x_2 \geq 0$ **[OU - MBA - Sept. 98]**

21. Use penalty method to minimize $Z = 3X + 2.5\ Y$
    Subject to $2X + 4Y \geq 40$
                  $3X + 2Y \geq 50$
                  $X, Y \geq 0$ **[JNTU (Mech./Prod.) Nov. 2002 (Set - 3)]**

Review Questions

1. Discuss the merits and demerits of graphical solutions in operations research

2. What is meant by redundant constraint? Explain with examples.

3. Discuss different types of solutions with reference to graphical methods in linear programming problems.

4. Discuss various types of optimal solutions to LPP through graphical methods. How do you identify them

5. Discuss different types of feasible regions in graphical solutions to LPP

6. Discuss different types of constraints and explain how you represent them in graphical solutions of LPP

7. How do you identify the following solution in graphical methods ?

   (a) Multiple optimal solutions

   (b) Infeasible solutions

   (c) unbounded solution   **[OU - MBA - May 95, May 92]**

8. What is an unbounded solution ? How is it different from infeasible solution
                                                   [OU - MBA - March, 98, March 99]

9. What is convex set? Explain with suitable examples.   **[OU - MBA - Nov. 94]**

10. How are unrestricted variables treated in graphical solutions to LPP.
                                                   [OU - MBA - Dec. 95]

Objective Type Questions

1. The solution for an LPP with two exact constraints and no inequality constraints will be

   (a) infeasible   (b) unbounded

   (c) multiple optimal   (d) unique point size solution

2. The feasible region will be _____ type if one of the constraints is exact type in a two variable LPP

   (a) point   (b) line

   (c) area   (d) volume

3. In graphical solutions, if x is unrestricted and $y \geq 0$, we get the solution in _____

   (a) 1st quadrant   (b) 1st & 2nd quadrant

   (c) 1st & 4th quadrant   (d) any quadrant

4. In which of the following cases we do not maximise the objective function

   (a) sales   (b) profits

   (c) contribution   (d) costs

5. If every point of a line drawn with any two points in feasible region fall in the feasible region the region is said to be

   (a) convex   (b) concave

   (c) island   (d) infeasible

6. In an LPP, if the values given to the variable satisfy conditions and constraints but not the objective function, then the solution is

   (a) feasible solution   (b) optimal solution

   (c) infeasible solution   (d) we can not say

7. The feasible region in the form of a ring is _____

   (a) convex   (b) concave

   (c) concavo-convex   (d) convexo-concave

8. The number of basic feasible solutions in a feasible region will be

   (a) finite   (b) infinite

   (c) zero   (d) we can not say

9. The solution for max. $Z = 10 x_1 + 6 x_2$ subject to $5x_1 + 3x_2 \leq 30$; $x_1 + 2x_2 \leq 18$; $x_1, x_2 \geq 0$ is _____

   (a) unique　　　　　　　　　　　　　　(b) infeasible

   (c) multiple optimal　　　　　　　　　(d) unbounded

10. Which of the following constraints has no effect on solution space.

    (a) exact constraint　　　　　　　　　(b) availability constraint

    (c) requirement constraint　　　　　(d) redundant constraint

11. There will be no solution in

    (a) unbounded solutions　　　　　　　(b) infeasible solutions

    (c) multiple solutions　　　　　　　　(d) unique solutions

12. Which of the following solutions is independent of objective function

    (a) unique　　　　　　　　　　　　　　(b) unbounded

    (c) infeasible　　　　　　　　　　　　(d) multiple

    Consider the graphical solution figure given below.

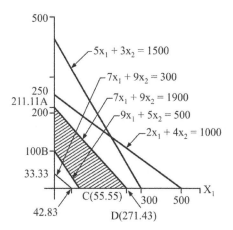

13. In the above graphical solution, which of the following is not redundant.

    (a) $7x_1 + 9x_2 \geq 300$　　　　　　　　(b) $5x_1 + 3x_2 \leq 1500$

    (c) $2x_1 + 4x_2 \leq 1000$　　　　　　　(d) $9x_1 + 5x_2 \geq 500$

14. In the above figure, if objective function is to maximise $500 x_1 + 300 x_2$, then we get

    (a) infeasible solution　　　　　　　　(b) unbounded solution

    (c) unique solution　　　　　　　　　(d) multiple solution.

15. For the above figure which of the following objective function can give multiple optimal solution

(a) $1170 x_1 + 1110 x_2$ (b) $1000 x_1 + 600 x_2$

(c) $700 x_1 + 900 x_2$ (d) $1000 x_1 + 2000 x_2$

16. The get multiple optimal solutions in LPP which of the following is necessary condition.

(a) objective function is perpendicular to one of the constraint
(b) any two constraints are parallel
(c) no two constraints are parallel
(d) objective function is parallel to one of the constraints.

17. The number of optimal solutions of LPP for a maximizing objective function on unbounded feasible region is

(a) infeasible   (b) unique
(c) multiple   (d) infinite

18. The solution for the LPP : max. $Z = 20 x_1 + 30 x_2$ subject to $5x_1 + 6x_2 \leq 15$; $2x_1 + 3x_2 \geq 200$; $x_1, x_2 \geq 0$ may be _____ solution

(a) multiple optimal   (b) unbounded
(c) infeasible   (d) unique

19. The solution for max. $Z = 3x_1 + 4x_2$ subject to $x_1 - x_2 \leq 1, x_1 + x_2 \geq 2$ will be

(a) infeasible   (b) unique
(c) multiple   (d) unbounded

20. Which of the following is not a limitation in graphical solutions

(a) limited to two variables only
(b) accuracy is not possible
(c) infeasible solutions can not be predicted
(d) multiple solutions can not be predicted easily.

Fill in the Blanks

1. If a convex set of feasible solutions of an LPP is a bounded solution space then its extreme points are said to be _____ solution.

2. If a convex set of feasible solution of an LPP is a bounded feasible region then it will have at least one _____ solution.

3. _____ solution is nothing to do with objective function of LPP.

4. _____ constraint has no effect on feasible region

5. A solution is said to be _____ if it satisfies the conditions, constraints and objective function

6. Any set of values, which satisfy the conditions of LPP is a _____

7. Graphical solutions are limited to the problems with _____ variables only

8. If multiple optimal solution exists to an LPP, at least one of the constraints will have same slope as that of _____

9. If an LPP has infinite number of solutions, it is said to have _____ solution

Answers

Objective Type Questions :				
1. (d)	2. (b)	3. (b)	4. (d)	5. (a)
6. (a)	7. (b)	8. (a)	9. (c)	10. (d)
11. (b)	12. (c)	13. (d)	14. (c)	15. (c)
16. (d)	17. (d)	18. (c)	19. (d)	20. (c)

Fill in the Blanks :	
1. basic feasible	2. optimal
3. infeasible	4. redundant
5. optimal	6. solution
7. two	8. objective function
9. unbounded	

4 Chapter

Simplex Method

CHAPTER AT A GLANCE

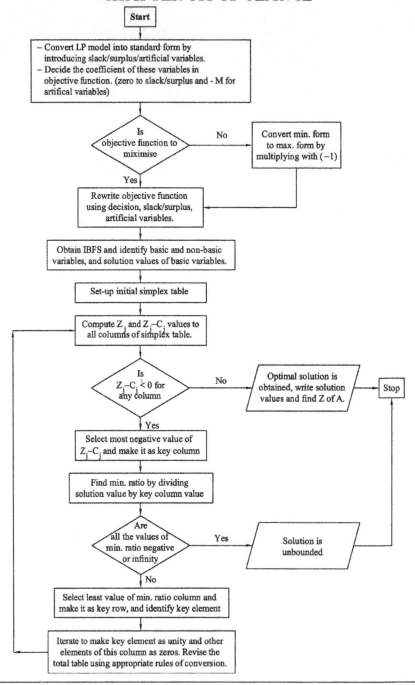

4.0 *Introduction*

Simplex algorithm (a step by step or iterative approach) was originally proposed by G.B Dantzig in 1948.

It starts at a basic (the worst case) level of the problem.

At each step it projects the improvement in the objective function over its previous step. Thus, the solution becomes optimum when no further improvement is possible on the objective function.

4.1 *Simplex Algorithm*

The algorithm goes as follows :

Step 1 : Formulation of LPP :

   – Selection of decision variables

   – Setting of objective function.

   – Identification of constraint set

   – Writing the conditions of variables.

Step 2 : Convert constraints into equality form.

   – Add slack variable if constraint is ≤ type.

   – Subtract surplus and add an artificial variable if the constraint is ≥ type.

   – Add an artificial variable if constraint is exact (=) type.

Step 3 : Find, Initial Basic Feasible Solution (IBFS)

   – If m non identical equations have n variables ($m < n$) including all decision, slack / surplus and artificial variables, we get m number of variables basic and ($n - m$) variables non basic (i.e, equated to zero).

   – First make all decision variables (and surplus) as non basic i.e., equate to zero to identity the IBFS.

   – Find solution values for basic variables.

C_B	BV	C_j SV	C_j $x_i, S \& A$		Minimum ratio	Remarks
Contribution of basic variable in objective function	Basic variables	Solution values	y_i B O D Y Key Element KE		Most minimum ratio of SV/key co. value	← Key row
		Z_j	Sum of products of C_B and y_i			
		$Z_j - C_j$	Most negative value			

Key column

Step 4 : Construct, Initial Simplex Tableau as given above with the following notations.

C_B : Coefficient of basic variable in the objective functions
(or contribution of basic variable)

BV : Basic variable (from IBFS)

SV : Solution value (from IBFS)

C_j : Contribution of jth variable or coefficient of each variable (j^{th}) in objective function.

$Z_j - C_j$: Net contribution.

Step 5 : Find 'out going' and 'incoming' variables.

– Find Z_j by summation of products of C_B and y_i for each column.

– Computed $Z_j - C_j$ value for each column

– To find key coloumn use most negative value of $Z_j - C_j$.

– Variable in key coloumn is 'in coming variable' or 'entering' variable.

– The variable of key row is 'out going' or 'exiting' variable.

– Find the minimum ratio of solution value to corresponding key coloumn value to identify key row.

– The cross section of key column and key row is key element with which the next iteration is carried out.

Step 6 : Re-write next tableau as per given set of rules.

– Replace the exiting variable from the *basis* with the entering variable along with its coefficient (or contribution).

– You have to make key element as unity (i.e. 1) and other elements in the key column as zeros.

– To make key element as unity, divide the whole key row by the key element. This is supposed as the new row in the place of key row in the next iteration table.

– To find other rows of next iteration table, use this new row. By appropriate adding or subtracting entire new row in the old rows, make other elements of the key column as zeros.

– Accordingly the solution value gets improved.

Step 7 : Check whether all the values of $Z_j - C_j$ are positive. If all are positive, the optimal solution is reached. Write the solution values and find Z_{opt}.
(i.e., Z_{max} or Z_{min} as the case may be).

If $Z_j - C_j$ values are still negative, again choose most negative among these and go to step 5 and repeat the iteration till all the values of $Z_j - C_j$ become positive.

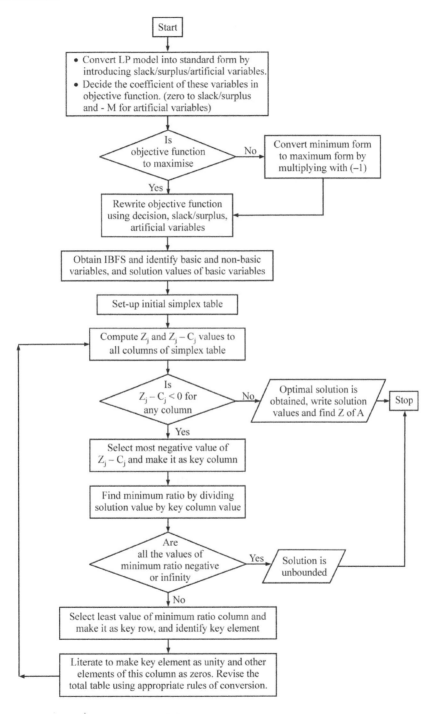

FIGURE 4.1 : FLOW CHART OF SIMPLEX METHOD

The above algorithm is illustrated through the following numerical example.

ILLUSTRATION 1

> *Maximise* $Z = 6x_1 + 9x_2$
> *Subject to* $2x_1 + 2x_2 \leq 24$
> $x_1 + 5x_2 \leq 44$
> $6x_1 + 2x_2 \leq 60$
> *and* $x_1, x_2 \geq 0$ *by using simplex method.*   [JNTU MECH, ECE -1999/C]

Step 1 : **Foromulation :**

The given problem is already formulated in canonical form.

Step 2 : Writing standard form by introducing slack variables;

(Given problem is in canonical form i.e., maximising objective function, constraints ≤ type and all non negative conditions)

Express given problem in the standard form i.e., objective function is to be maximised (existing already), equality constraints (to be converted) and non-negative conditions of variables (aready existing).

∴   Standard form of the given problem is

Maximise $Z = \quad 6x_1 + 9x_2 + 0.s_1 + 0.s_2 + 0.s_3$

$$2x_1 + 2x_2 + s_1 = 24 \qquad \ldots (1)$$

$$x_1 + 5x_2 + s_2 = 44 \qquad \ldots (2)$$

$$6x_1 + 2x_2 + s_3 = 60 \qquad \ldots (3)$$

and $x_1 \geq 0, x_2 \geq 0, s_1 \geq 0, s_2 \geq 0, s_3 \geq 0$

Where s_1, s_2 and s_3 are slack variables added left side of equation (1), (2) and (3) respectively so as to make the inequations (≤) as equations. Also the condition of these variables must be non-negative.

Step 3 : Now, we have three equations with five variables viz., x_1, x_2, s_1, s_2 and s_3 to get the Initial Basic Feasible Solution (IBFS), we have to fix the values of any two of these five variables. We can do this in 5_{c_2} ways, i.e., 10 types of solutions.

We choose x_1 and x_2 at zero since the decision is to be made on them and we start at most minimum possible value of x_1 and x_2 ($x_1 \geq 0$, $x_2 \geq 0$) and this is the most worst case from where we iterate to improve the objective function to its maximum value.

Thus IBFS is

   $x_1 = 0, x_2 = 0$

[We now call x_1, x_2 as non basic variables]

⇒ $s_1 = 24$, $s_2 = 44$, $s_3 = 60$

[We now call s_1, s_2 and s_3 as basic variables].

[Now, if the non-basic variables are brought into basis by the iteration, then the optimal solution can be attained].

Step 4 : Construct Initial Simplex Tableau - 1

INITIAL ITERATION TABLEAU

$$\text{Max} \quad Z = \quad 6\,x_1 \; + \; 9\,x_2 \; + \; 0S_1 + 0S_2 \; + \; 0S_3$$

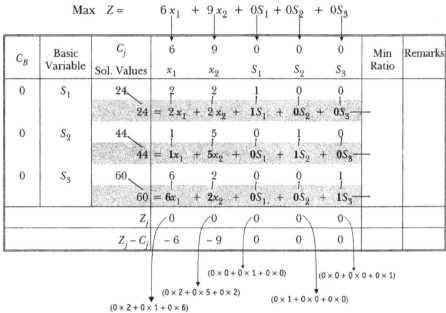

Step 5 : Identify key column (most negative $Z_j - C_j$) and key row (min. ratio) and hence the incoming and outgoing variable respectively.

ITERATION TABLEAU - I

		C_j	6	9	0	0	0	Min	Remarks
C_B	BV	SV	x_1	x_2	S_1	S_2	S_3	Ratio	
0	S_1	24	2	2	1	0	0	24/2	
0	S_2	44	1	5	0	1	0	44/5	Min Key row
0	S_3	60	6	2	0	0	1	60/2	
		Z_j	0	0	0	0	0		
Out going variable		$Z_j - C_j$	– 6	– 9	0	0	0		

In coming Variable

Key column

Step 6 : Re-write the next iteration tableau - II with the given set of rules :

[We can notice in iteration tableau that the values of body i.e. y_i, the coefficient matrix for basic variables is in the form of a unit matrix. When s_1, s_2 and s_3 are Basic Variables, we find unit matrix under their respective columns. When s_2 is replaced with x_2 now, we should make a unit matrix with s_1, x_2 and s_3, the basic variables in the next iteration tableau. To make this improvement, we follow the following rules.]

Rule 1 : Re-writting new row in the place of key row.

- The out going variable (s_2) is replaced by incoming variable (x_2) in the basis. Obviously the corresponding contribution i.e., C_B ('0' in old row) changes (as 9 in new row).

- Now, key element (5 in old row) is to transformed as unity. Therefore the total row is divided by 5.

Thus $R_2^N \to \dfrac{1}{5} R_2^O$ [Read as New R_2 is changed as one fifth of old R_2]

Summarily,

We have to make this as '1', so divide entire row by '5'

Replaced

Old R_2 is 0 s_2 44 1 **5** 0 1 0

New R_2 is 9 x_2 $\dfrac{44}{5}$ $\dfrac{1}{5}$ 1 0 $\dfrac{1}{5}$ 0

Rule 2 : Re-writing other new rows.

- Other basic variables and their contribution (C_B) remain same i.e., figures in first two columns are unaltered except for the replacement as given in rule 1 discussed above..

- The other values in the key column must possess zeros therefore, subtract or add R_2^N appropriate number of times so as to get zeros.

- Thus entire row is re-written by adding or subtracting the new row found from rule 1, 'm' number of times if m is the vaue in old row.

- Thus here R_1 is transformed as $R_1^N \to R_1^O - 2R_2^N$
  (Read as new R_1 is changed as old R_1 minus two times new R_2)

Entire row is to be transformed with same rule

We subtract this '1' two times to make zero

We have to make this zero. So subtract key row 2 times

It is illustrated here below :

Old R_1 is 0 S_1 24 2 **2** 1 0 0

New R_2 is 9 x_2 $\dfrac{44}{5}$ $\dfrac{1}{5}$ 1 0 $\dfrac{1}{5}$ 0
(obtained from rule 1)

New R_1 is 0 S_1 $24 - 2 \times \dfrac{44}{5}$ $2 - 2 \times \dfrac{1}{5}$ $2 - 2 \times 1$ $1 - 2 \times 0$ $0 - 2 \times \dfrac{1}{5}$ $0 - 2 \times 0$

   $\dfrac{32}{5}$ $\dfrac{8}{5}$ **0** 1 $-\dfrac{2}{5}$ 0

Similarly, $R_3^N \to R_3^O - 2\,R_2^N$

New R_3 is 0 S_3 $60 - \dfrac{2 \times 44}{5}$ $6 - \dfrac{2 \times 1}{5}$ $2 - 2 \times 1$ $0 - 2 \times 0$ $0 - 2 \times \dfrac{1}{5}$ $1 - 2 \times 0$

i.e. 0 S_3 $\dfrac{212}{5}$ $\dfrac{28}{5}$ **0** 0 $-\dfrac{2}{5}$ 1

These are written in tableau - II given below and the above steps are repeated.

ITERATION TABLEAU - II

In coming Variable

C_B	B.V	C_j / SV	6 x_1	9 x_2	0 S_1	0 S_2	0 S_3	Min Ratio	Remarks
0	S_1	$\dfrac{32}{5}$	$\dfrac{8}{5}$	0	1	$-\dfrac{2}{5}$	0	$\dfrac{32/5}{8/5}$	$R_1^N \to R_1^0 - 2\,R_2^N$ Key row
9	x_2	$\dfrac{44}{5}$	$\dfrac{1}{5}$	1	0	$\dfrac{1}{5}$	0	$\dfrac{44/5}{1/5}$	$R_2^N \to R_2^0/5$
0	S_3	$\dfrac{212}{5}$	$\dfrac{28}{5}$	0	0	$-\dfrac{2}{5}$	1	$\dfrac{212/5}{28/5}$	$R_3^N \to R_3^0 - 2R_2^N$
		Z_j	$\dfrac{9}{5}$	9	0	$\dfrac{9}{5}$	0		
		$Z_j - C_j$	$-\dfrac{21}{5}$	0	0	$\dfrac{9}{5}$	0		

Key element

Out going variable

Key column

$0 \times 0 + 9 \times 1 + 0 \times 0$

$0 \times \dfrac{8}{5} + 9 \times \dfrac{1}{5} + 0 \times \dfrac{28}{5}$

$0 \times 1 + 9 \times 0 + 0 \times 0$

$0 \times \left(\dfrac{-2}{5}\right) + 9 \times \dfrac{1}{5} + 0 \left(\dfrac{-2}{5}\right)$

$0 \times 0 + 9 \times 0 + 0 \times 1$

ITERATION TABLEAU - III

C_B	B.V	C_j / SV	6 x_1	9 x_2	0 S_1	0 S_2	0 S_3	Min Ratio	Remarks
6	x_1	$\dfrac{32}{5} \times \dfrac{5}{8}$ $= 4$	$\dfrac{8}{5} \times \dfrac{5}{8}$ $= 1$	$0 \times \dfrac{5}{8}$ $= 0$	$1 \times \dfrac{5}{8}$ $= 5/8$	$-\dfrac{2}{5} \times \dfrac{5}{8}$ $= -1/4$	$0 \times 5/8$ $= 0$		$R_1^N \to R_1^0 \big/ \dfrac{8}{5}$ or $R_1^N \to \dfrac{5}{8} R_1^0$
9	x_2	$\dfrac{44}{5} - \dfrac{1}{5} \times 4$ $= 8$	$\dfrac{1}{5} - \dfrac{1}{5}$ $\times 1 = 0$	$1 - \dfrac{1}{5} \times 0$ $= 1$	$0 - \dfrac{1}{5} \times \dfrac{5}{8}$ $= -\dfrac{1}{8}$	$\dfrac{1}{5} - \dfrac{1}{5}\left(-\dfrac{1}{4}\right)$ $= \dfrac{1}{4}$	$0 - \dfrac{1}{5} \times 0$ $= 0$		$R_2^N \to R_2^0 - \dfrac{1}{5} R_1^N$
0	S_3	$\dfrac{212}{5} - \dfrac{28}{5} \times 4$ $= 20$	$\dfrac{28}{5} - \dfrac{28}{5} \times 1$ $= 0$	$0 - \dfrac{28}{5} \times 0$ $= 0$	$0 - \dfrac{28}{5} \dfrac{5}{8}$ $= -7/2$	$-\dfrac{2}{5} - \dfrac{28}{5}\left(\dfrac{-1}{4}\right)$ $= 1$	$1 - \dfrac{28}{5} \times 0$ $= 1$		$R_3^N \to R_3^0 - \dfrac{28}{5} R_1^N$
		Z_j	6×1 $+ 9 \times 0$ $+ 0 \times 0$	6×0 $+ 9 \times 0$ $+ 0 \times 0$	$6 \times 5/8$ $+ 9(-1/8)$ $+ 0(-7/2)$	$6(-\tfrac{1}{4})$ $+ 9(\tfrac{1}{4})$ $+ 0 \times 1$	6×0 $+ 9 \times 0$ $+ 0 \times 1$		
		$Z_j - C_j$	$6 - 6 = 0$	$9 - 9 = 0$	$\dfrac{21}{8} - 0$ $= \dfrac{21}{8}$	$\dfrac{3}{4} - 0$ $= \dfrac{3}{4}$	$0 - 0$ $= 0$		

Iteration tableau - III is re-written with simplified values.

C_B	B.V	C_j / SV	6 / x_1	9 / x_2	0 / S_1	0 / S_2	0 / S_3	Min Ratio	Remarks
6	x_1	4	1	0	$\dfrac{5}{8}$	$-\dfrac{1}{4}$	0		
9	x_2	8	0	1	$-\dfrac{1}{8}$	$\dfrac{1}{4}$	0		
0	S_3	20	0	0	$-\dfrac{7}{2}$	1	1		
		Z_j	6	9	$\dfrac{21}{8}$	$\dfrac{3}{4}$	0		
		$Z_j - C_j$	0	0	$\dfrac{21}{8}$	$\dfrac{3}{4}$	0		All positive, so $Z_{\max} = 96$

As the values of $Z_j - C_j$ are positive for all the variables, the optimisation is reached.

Therefore, the solution is

$$\left.\begin{array}{l} x_1 = 4 \\ x_2 = 8 \\ S_3 = 20 \end{array}\right\} \text{Basic}$$

$$\left.\begin{array}{l} S_1 = 0 \\ S_2 = 0 \end{array}\right\} \text{Non-basic}$$

$$\begin{aligned} Z_{\max} &= 6\,x_1 + 9\,x_2 + 0 \cdot S_1 + 0 \cdot S_2 + 0 \cdot S_3 \\ &= 6 \times 4 + 9 \times 8 + 0 \times 0 + 0 \times 0 + 0 \times 20 \\ &= 24 + 72 = 96 \end{aligned}$$

Remarks :

(i) *From the above solution values, we can understand that 20 units of third resource are left unutilised while first and second resources are fully utilized.*

(ii) *Students are advised not to express the fractions in the decimal form, since these will be normally simpler calculations. If they are expressed in decimal form, it may affect the accuracy.*

Note : (i) *While calculating minimum ratio, if you get negative value or infinity ignore them.*

(ii) *A minimisation objective function can be changed to maximisation by multiplying with (– 1). You can observe these in the illustrative example to follow.*

Let us consider an example with minimisation objective function.

ILLUSTRATION 2 ───

Minimise	$Z = 2x_1 \; 6x_2 + 4x_3$
Subject to	$3x_1 - x_2 + 3x_3 \le 7$
	$-2x_1 + 4x_2 \le 12$
	$-4x_1 + 3x_2 + 5x_3 \le 10$
	$x_1, \; x_2, \; x_3 \ge 0$

$Solution$:

Step 1 : **Formulation :** It is already formulated but it is not in the canonical form with reference to its objective function.

Therefore we convert into canonical form.
Since, Minimising $Z = (-1)$ Maximising Z
(Mini - Max theorem),
we rewrite the problem as,
Maximise $\quad Z = -2x_1 + 6x_2 - 4x_3$

$$\text{Subject to} \quad 3x_1 - x_2 + 3x_3 \le 7$$
$$-2x_1 + 4x_2 \le 12$$
$$-4x_1 + 3x_2 + 5x_3 \le 10$$
$$x_1, x_2, x_3 \ge 0$$

Step 2 : To write in standard form by introducing slack variables,
$$3x_1 - x_2 + 3x_3 + S_1 = 7$$
$$-2x_1 + 4x_2 + S_2 = 12$$
$$-4x_1 + 3x_2 + 5x_3 + S_3 = 10$$
and maximise $Z = -2x_1 + 6x_2 - 4x_3 + 0.S_1 + 0.S_2 + 0.S_3$
$$x_1, \; x_2, \; x_3, S_1, S_2, S_3 \ge 0$$

Step 3 : In the above problem, we have three (3) equations and six (6) variables. Therefore we fix three of them in basis by assigning zero value to the other three (non- basic) variables.

This can be done in 6_{C_3} or $\left(\dfrac{6.5.4}{1.2.3} = 20 \right)$ ways, leading to 20 different types of solutions. Of them, we choose the initial basic feasible solution (IBFS) as the decision variables at non - basis i.e., assuming zero values. [This indicates the unutilized resources when the production is nil or stand still].

Thus when $x_1 = 0$, $x_2 = 0$ & $x_3 = 0$ the solution values (SV) for $S_1 = 7$, $S_2 = 12$ and $S_3 = 10$

Step 4 & 5 : Setting initial simplex tableau and identifying key column, key row and hence key element. Also locate in coming and out going variables.

ITERATION TABLEAU - I

In coming Variable

C_B	B.V	C_j	-2	6	-4	0	0	0	Min Ratio	Remarks
		SV	x_1	x_2	x_3	S_1	S_2	S_3		
0	S_1	7	3	-1	3	1	0	0	$\dfrac{7}{-1} = -ve$ (Ignore)*	
0	S_2	12	-2	4	0	0	1	0	$\dfrac{12}{4} = 3$	Key row
				Key element						
0	S_3	10	-4	3	5	0	0	1	$\dfrac{10}{3} > 3$	
		Z_j	0	0	0	0	0	0		
		$Z_j - C_j$	2	-6	4	0	0	0		

Out going variable

Key column

* Ignore if minimum ratio is negative or infinity.

Step 6 : The tableau - I is iterated to give tableau - II, according to given remarks in tableau - II

(Find R_2^N by $\dfrac{1}{4} R_2^0$ and then R_1^N and R_3^N using R_2^N).

ITERATION - TABLEAU - II

In coming Variable

C_B	B.V	C_j	-2	6	-4	0	0	0	Min Ratio	Remarks
		SV	x_1	x_2	x_3	S_1	S_2	S_3		
0	S_1	10	$\dfrac{5}{2}$	0	3	1	$\dfrac{1}{4}$	0	4 (key row)	$R_1^N \to R_1^0 + R_2^N$
			Key element							
6	x_2	3	$-\dfrac{1}{2}$	1	0	0	$\dfrac{1}{4}$	0	$-ve$ (ignore)	$R_2^N \to \dfrac{1}{4} R_2^0$
0	S_3	1	$-\dfrac{5}{2}$	0	5	0	$-\dfrac{3}{4}$	1	$-ve$ (ignore)	$R_3^N \to R_3^0 - 3 R_2^N$
Out going variable		Z_j	-3	6	0	0	$3/2$	0		
		$Z_j - C_j$	-1	0	4	0	$3/2$	0		

Key column

ITERATION TABLEAU - III (first find R_1^N by $\frac{2}{5} R_1^0$ and then R_2^N and R_3^N)

C_B	B.V	C_j SV	-2 x_1	6 x_2	-4 x_3	0 S_1	0 S_2	0 S_3	Min Ratio	Remarks
-2	x_1	4	1	0	6/5	2/5	1/10	0		$R_1^N \to \frac{2}{5} R_1^0$
6	x_2	5	0	1	3/5	1/5	3/10	0		$R_2^N \to R_2^0 + \frac{1}{2} R_1^N$
0	S_3	11	0	0	8	1	$-\frac{1}{2}$	1		$R_3^N \to R_3^0 + \frac{5}{2} R_1^N$
		Z_j	-2	6	6/5	2/5	8/5	0	$Z_{\max} = +22$	
		$Z_j - C_j$	0	0	26/5	2/5	8/5	0	$Z_{\min} = -22$	

All the values of $Z_j - C_j$ are positive, therefore optimality is reached.

Solution is

$$x_1 = 4, x_2 = 5, x_3 = 0$$

(Also $S_1 = 0$, $S_2 = 0$, $S_3 = 11$ i.e., unutilized resource)

$$Z_{\max} = -2 \times 4 + 6 \times 5 - 4 \times 0 + 0 \times 0 + 0 \times 0 + 0 \times 11$$
$$= -8 + 30 = 22$$
$$\therefore \quad Z_{\min} = -22$$

(at $x_1 = 4$, $x_2 = 5, x_3 = 0$)

Now, let us consider a case, with minimisation objective and mixed type of constraints (one of the cosntraints is $\geq$ type)

ILLUSTRATION 3

> *Min* $Z = x_1 - x_2 - 2x_3$
> *S.t* $\quad 3x_1 + x_2 + 3x_3 \leq 7$
> $\quad\quad -x_1 + 2x_2 \geq -6$
> $\quad\quad 4x_1 + 3x_2 + 5x_3 \leq 10$
> $\quad\quad x_1, x_2, x_3 \geq 0$

Solution:

Step 1: **Formulation :** Already formulated but not in canonical form.

Converting into canonical form.

$$\text{Max} \quad Z = -x_1 + x_2 + 2x_3$$
$$\text{S.t.} \quad 3x_1 + x_2 + 3x_3 \leq 7$$
$$x_1 - 2x_2 \leq 6$$
$$4x_1 + 3x_2 + 5x_3 \leq 10$$
$$x_1, x_2, x_3 \geq 0$$

(Here, obj. function and second constraint are multiplied by - 1)

Step 2 : Writing in standard form by introducing slack variables.

$$3x_1 + x_2 + 3x_3 + S_1 = 7$$

$$x_1 - 2x_2 + S_2 = 6$$

$$4x_1 + 3x_2 + 5x_3 + S_3 = 10$$

And obj. fun : is max $Z = -x_1 + x_2 + 2x_3 + 0S_1 + 0.S_2 + 0.S_3$

$$x_1, x_2, x_3 \geq 0, \ S_1, S_2, S_3 \geq 0$$

Step 3: Finding IBFS

There are three equations and six variables

∴ Three of them are kept in basis and three in non basis (we get 6_{C_3} types of solutions)

IBFS : non basic $x_1 = 0, \ x_2 = 0, \ x_3 = 0$

basic $S_1 = 7, \ S_2 = 6, \ S_3 = 10$

Step 4 & 5: Write initial tableau, identify key column, key row & key element, and in coming & out going variables

ITERATION TABLEAU - I :

C_B	B.V	C_j / S.V.	-1 / x_1	$+1$ / x_2	$+2$ / x_3	0 / S_1	0 / S_2	0 / S_3	Min Ratio	Remarks
0	S_1	7	3	$+1$	3	1	0	0	$\dfrac{7}{3}$	If Min ratio is -ve or ∞, it is ignored
0	S_2	6	-1	-2	0	0	1	0	$\dfrac{6}{0} = \infty$ (ignore)	
0	S_3	10	4	3	5	0	0	1	$\dfrac{10}{5} = 2$ ← Key row	
		Z_j	0	0	0	0	0	0		
		$Z_j - C_j$	$+1$	-1	-2	0	0	0		

Key element

↑ Key column

Step 6 : Iterate Tableau - I, according to given remarks in tableau - II

◢ **ITERATION - TABLEAU - II :**

C_B	B.V	C_j S.V.	-1 x_1	1 x_2	2 x_3	0 S_1	0 S_2	0 S_3	Min Ratio	Remarks
0	S_1	1	$\frac{3}{5}$	$-\frac{4}{5}$	0	1	0	$-\frac{3}{5}$		$R_1^N \to R_1^0 - 3\,R_3^N$
0	S_2	6	1	-2	0	0	1	0		$R_2^N \to R_2^0$ (no change)
2	x_3	2	$\frac{4}{5}$	$\frac{3}{5}$	1	0	0	$\frac{1}{5}$		$R_3^N \to \frac{1}{5} R_3^0$
		Z_j	$\frac{8}{5}$	$\frac{6}{5}$	2	0	0	$\frac{2}{5}$	Since $Z_j - C_j \geq 0\ \forall\ x_j$ and S_j we have $Z_{max} = 4$	
		$Z_j - C_j$	$\frac{13}{5}$	$\frac{1}{5}$	0	0	0	$\frac{2}{5}$	and $Z_{min} = -4$	

As $Z_j - C_j$ values are positive for all the variables, optimality is reached. Optimal solution is

$$x_1 = 0, \quad x_2 = 0, \quad x_3 = 2$$
$$\left(S_1 = 1, \quad S_2 = 6, \quad S_3 = 0\right)$$
and $\quad Z_{max} = -0 + 0 + 2 \times 2 + 0 \times 1 + 0 \times 6 + 0 \times 0$
$$= 4$$
$$Z_{min} = -4 \text{ at } x_1 = 0, x_2 = 0 \text{ and } x_3 = 2$$

Remarks :

Converting a 'greater than or equal to' ($\geq$) type constraint into 'less than or equal to' ($\leq$) type constraint is not applicable to all the cases. Particularly,

(i) When all the constraints are 'greater than or equal to' type, this method is not suitable, and

(ii) This is applicable when right hand side of inequation is negative.

In all such cases we may have to choose either BIG M or 2 phase method to be discussed in the sections to follow.

4.2 *Some Special Cases*

We have learnt how we get multiple optimal solutions, infeasible solutions and unbounded solutions by graphical solution method. Let us now observe how we identify such solutions in simplex method.

4.2.1 *Unbounded Solution*

In some cases if the value of a variable is increased indefinitely, the constraints are not violated. This indicates that the feasible region is unbounded at least in one direction. Therefore, the objective function value can be increased indefinitely. This means that problem has been poorly formulated or concieved.

In simplex method, this can be noticed if $Z_j - C_j$ value is negative to a variable (entering) which is notified as key column and the ratio of solution value to key column value is either negative or infinity (both are to be ignored) to all the variables. This indicates that no variable is ready to leave the basis, though a variable is ready to enter. We cannot proceed further and the solution is unbounded or not finite.

This is illustrated through the following example.

ILLUSTRATION 4

Maximise	$Z = 4x_1 + 3x_2$
Subject to	$x_1 \le 5$
	$x_1 - x_2 \le 8$
	$x_1, x_2 \ge 0$

[IGNOU - MBA - 1997
BRAOU - MBA - 1998]

Solution :

Let us introduce slack variables to express inequalities;

Maximise $Z = 4x_1 + 3x_2 + 0.s_1 + 0.s_2$

Subject to $x_1 + s_1 = 5$

$x_1 - x_2 + s_2 = 8$

$x_1, x_2 \ge 0, s_1, s_2 \ge 0$

IBFS $x_1 = 0$, $x_2 = 0$ (Non basic)

$s_1 = 5$, $s_2 = 8$ (basic)

With usual steps it is solved through the following iterations.

ITERATIONS TABLEAU - i

Entering Variable

C_B	BV	C_j / SV	4 / x_1	3 / x_2	0 / S_1	0 / S_2	Min ratio	Remarks
0	S_1	5	1	0	1	0	$\frac{5}{1} = 5$	Key row
0	S_2	8	1	−1	0	1	$\frac{8}{1} = 8$	
Leaving variable		Z_j	0	0	0	0		
		$Z_j - C_j$	−4	−3	0	0		

Key row

ITERATION TABLEAU - II

$$R_1^N \to R_1^0; \quad R_2^N \to R_2^N - R_1^0$$

Entering Variable

C_B	BV	C_j SV	4 x_1	3 x_2	0 S_1	0 S_2	Min ratio	Remarks
4	x_1	5	1	0	1	0	$\dfrac{5}{0} = \infty$	to be ignored
0	S_2	3	0	–1	–1	1	$\dfrac{3}{-1} = -ve$	to be ignored
		Z_j	0	0	0	0	No variable is ready to leave the basis	
		$Z_j - C_j$	0	–3	4	0	Solution is UNBOUNDED	

Key column

From the above tableau - II, it is clear that x_2 is entering variable into the basis (key column variable whose $Z_j - C_j$ is negative) but no variable is ready to leave since the ratio of solution vaue to key column value is infinity for R_1 and negative for R_2 both of which are to be ignored. Thus we cannot proceed further because key row (leaving variable) cannot be found. Thus, this problem yields **no finite solution** or in other words, an **unbounded** solution.

4.2.2 *Multiple Optimal Solutions*

When the objective function is parallel to one of the constraints, the multiple optimal solutions may exist. As we have seen from graphical solutions, that the optimal solution exists at the extreme point on the feasible region, the multiple optimal solutions will be noticed on at least two points of the binding constraint parallel to that of objective function. Thus in simplex method also at least two solution can be found.

This alternate optima is identified in simplex method by using the following principle.

After reaching the optimality, if at least one of the non-basic (decision) variables possess a zero value in $Z_j - C_j$, the multiple optimal solutions exist.

This is illustrated through the following numerical example.

ILLUSTRATION 5

Maximise	$Z = 3x_1 + 6x_2$
Subject to	$x_1 + x_2 \leq 5$
	$x_1 + 2x_2 \leq 6$
	$x_1, x_2 \geq 0$

Solution :

Converting inequalities into equations, we get

$$\text{Max } Z = 3x_1 + 6x_2 + 0.S_1 + 0.S_2$$

Subject to

$$x_1 + x_2 + S_1 = 5$$
$$x_1 + 2x_2 + S_2 = 6$$
$$x_1, x_2, S_1, S_2 \geq 0$$

IBFS : $S_1 = 5, \quad S_2 = 6$ (Basic)

$$x_1 = 0, \quad x_2 = 0 \text{ (Non basic)}$$

ITERATION TABLEAU - I:

Entering Variable

C_B	BV	C_j / SV	3 / x_1	6 / x_2	0 / S_1	0 / S_2	Min ratio	Remarks
0	S_1	5	1	1	1	0	5	
0	S_2	6	1	2	0	1	3	Key row
		Z_j	0	0	0	0		
		$Z_j - C_j$	-3	-6	0	0		

Key element (on the 2 in x_2 column, S_2 row)

Leaving variable — S_2. Key column — x_2.

ITERATION TABLEAU - II:

Entering Variable

C_B	BV	C_j / SV	3 / x_1	6 / x_2	0 / S_1	0 / S_2	Min ratio	Remarks
0	S_1	2	$\frac{1}{2}$	0	1	$-\frac{1}{2}$	4 Key row	$R_1^N \rightarrow R_1^0 - R_2^N$
6	x_2	3	$\frac{1}{2}$	1	0	$\frac{1}{2}$	6	$R_2^N \rightarrow \frac{1}{2} R_2^0$
		Z_j	3	6	0	3	Since $Z_j - C_j \geq 0 \; \forall$ all variables,	
		$Z_j - C_j$	0	0	0	3	Optimal solution is $Z_{\max} = 18$	

Key element (on the $\frac{1}{2}$ in x_1 column, S_1 row). Leaving variable. Key Column — x_1.

From the above Iteration Tableau - II the optimality is already reached since, $Z_j - C_j \geq 0$ i.e., positive for all the variables. Therefore the solution is

$$x_1 = 0, \quad x_2 = 3,$$
$$\text{and } Z_{\max} = 3 \times 0 + 6 \times 3 = 18$$

However, the value of $Z_j - C_j = 0$ for the non basic decision variable x_1 in tableau - II. This indicates alternative optimal solution. Therefore choosing first column as key column, if we further iterate, we get the following tableau - III

ITERATION TABLEAU - III :

C_B	BV	C_j / SV	3 / x_1	6 / x_2	0 / S_1	0 / S_2	Min ratio	Remarks
3	x_1	4	1	0	2	-1	5	$R_1^N \to 2\,R_1^0$
6	x_2	1	0	1	-1	1	3	$R_2^N \to R_2^0 - \dfrac{1}{2}\,R_1^N$
		Z_j	3	6	0	3	$Z_j - C_j \geq 0$ for all variable and hence	
		$Z_j - C_j$	0	0	0	3	Optimal solution is $Z_{max} = 18$	

This tableau - III yields another solution as,

$$x_1 = 4, \quad x_2 = 1$$

$$Z_{max} = 3 \times 4 + 6 \times 1 = 18$$

We can observe that $Z_j - C_j$ value will be always zero for basic variables in any simplex tableau [observe for S_1 & S_2 in Tableau - I , for S_1 & x_2 in Tableau - II and for x_1 & x_2 in tableau - III they have $Z_j - C_j$ values zero .] However, in Tableau - II, apart from basic variables S_1 and x_2 , the other variable x_1 is also showing $Z_j - C_j = 0$. This indicates that x_1 is also worth coming into basis, as it is behaving similar to any basic variables. Thus it provides an alternate optimal solution.

However, if all the basic variables are not replaced by decision variables it does not mean any multiple solution unless this condition (i.e., $Z_j - C_j = 0$) is obeyed by any non - basic decision variable.

4.2.3 *Unique Solution*

The problem solved in illustration 1 to 3   are the examples of this type of solution. Further, one more example is shown below.

ILLUSTRATION 6 ────────────────────────────────────

Mayukha & company produces two types of toys. Labour time required for first type is twice that of the seconds. If the first type alone is produced, the capacity is sufficient to produce 1000 numbers per day. The demand potential is 300 and 500 toys. The profits per toy are Rs. 10 and Rs.8 respectively. Formulate the L.P problem and solve it.   [JNTU (Mech/Prod.) 2001,2003]

Solution:

Step 1 : **Formulation :**

Variables: Let the no. of toys produced in first type $= x_1$

and the no. of toys produced in 2nd type $= x_2$

Objective function: Profit on each first type toy = Rs.10

Profit on x_1 no. of 1st type toy = $10\,x_1$

Profit on each second type toy = Rs.8

Profit on x_2 no. of 2nd type toy = $8\,x_2$

Total profit = $10\,x_1 + 8\,x_2$

$\therefore$ Objective function: maximize $Z = 10\,x_1 + 8\,x_2$

Constraints:

(i) *Labour time constraint:* As labour time of first toy is twice that of second, if x_1 of first type is produced, $2\,x_2$ of second can be produced.

$\therefore$ Total $x_1 + 2x_2 \leq 1000$ (constraints on labour time)

(ii) *Demand potential constraints:*

$$x_1 \leq 300, \qquad x_2 \leq 500 \text{ (constraint on demand)}$$

Since negative production is not possible

$$x_1, x_2 \geq 0$$

Summary   Max $Z = 10\,x_1 + 8\,x_2$

subject to $x_1 + 2\,x_2 \leq 1000$

$$x_1 \leq 300$$

$$x_2 \leq 500$$

$$x_1, x_2 \geq 0$$

Step 2 : Conversion into equations by introducing slack variables (standard form)

$$\text{Max } Z = 10\,x_1 + 8\,x_2 + 0 \cdot S_1 + 0 \cdot S_2 + 0 \cdot S_3$$

$$x_1 + 2\,x_2 + S_1 = 1000,$$

$$x_1 + S_2 = 300,$$

$$x_2 + S_3 = 500$$

$$x_1, x_2, S_1, S_2, S_3 \geq 0$$

Step 3 : IBFS :   Total variables = 5

Total equations = 3

$\therefore$ Basic variables = 3

i.e., $S_1 = 1000, S_2 = 300$ and $S_3 = 500$

Non basic variables = $5 - 3 = 2$

i.e., $x_1 = 0$ and $x_2 = 0$

No. of basic solutions = $5C_2$ or $5C_3$ i.e. 10

ITERATION TABLE - I :

C_B	BV	C_j / SV	10 / x_1	8 / x_2	0 / S_1	0 / S_2	0 / S_3	Min Ratio	Remarks
0	S_1	1000	1	2	1	0	0	1000	
0	⟨S_2⟩	300	[1]	0	0	1	0	300 ⟵	
0	S_3	500	0	1	0	0	1	∞ (Ignore)	
		Z_j	0	0	0	0	0		
		Z_j-C_j	−10	−8	0	0	0		

ITERATION TABLE - II :

C_B	BV	C_j / SV	10 / x_1	8 / x_2	0 / S_1	0 / S_2	0 / S_3	Min Ratio	Remarks
0	S_1	700	0	[2]	1	−1	0	350 ⟵	$R_{1_N} \rightarrow R_1^o - R_2^N$
10	x_1	300	1	0	0	1	0	∞ (Ignore)	$R_2^N \rightarrow R_2^o$
0	S_3	500	0	1	0	0	1	500	$R_3^N \rightarrow R_3^o$
		Z_j	10	0	0	10	0		
		Z_j-C_j	0	−8	0	10	0		

ITERATION TABLE - III :

C_B	BV	C_j / SV	10 / x_1	8 / x_2	0 / S_1	0 / S_2	0 / S_3	Min Ratio	Remarks
8	x_2	350	0	1	1/2	−1/2	0		$R_1^N \rightarrow \frac{1}{2} R_1^o$
10	x_1	300	1	0	0	1	0		$R_2^N \rightarrow R_2^o$
0	S_3	150	0	0	−1/2	+1/2	1		$R_3^N \rightarrow R_3^o - R_1^N$
		Z_j	10	8	4	6	0	$x_1 = 300$	$Z_{max} = 5800$
			0	0	4	6	0	$x_2 = 350$	

Since $Z_j - C_j \geq 0$ ∀ variable, optimal solution is arrived.

∴ $\qquad\qquad\qquad x_1 = 300,$

$$x_2 = 350$$
$$Z_{max} = 10 \times 300 + 8 \times 350$$
$$= 3000 + 2800$$
$$= \text{Rs. } 5800/-$$

(unutilized resources $S_3 = 150$ units).

4.2.4 *Infeasible Solution*

There may not exist any solution to certain LPP. This in LPP jargon is said to be infeasible solution. In this type of solution, there exists no feasible region (refer graphical solutions). We do not get any infeasible solution with all constraints as ' less than or equal to type'.

This will be explained later. However, to get an awareness, consider the following example.

Maximise $Z = 2x_1 + 3x_2$

Subject to $x_1 \leq 5$

$$x_1 - x_2 \geq 10$$

$$x_1 \geq 0, \quad x_2 \geq 0$$

Solution :   Converting the problem into canonical form,

Max , $Z = 2x_1 + 3x_2$

Subject to $x_1 \leq 5$

$$-x_1 + x_2 \leq -10$$

$$x_1 \geq 0, \quad x_2 \geq 0$$

Now introduce slack variables, to get equations

$$x_1 + S_1 = 5$$

$$-x_1 + x_2 + S_2 = -10$$

$$x_1, x_2, S_1 \text{ and } S_2 \geq 0$$

IBFS : $\left.\begin{array}{l} x_1 = 0 \\ x_2 = 0 \end{array}\right\}$ Non basic

$\left.\begin{array}{l} S_1 = 5 \\ S_2 = -10 \end{array}\right\}$ Basic

From the condition $S_2 \geq 0$, S_2 can not take value -10, therefore no IBFS can exist with such less than or equal to type constraints.

So also, this method of converting '$\geq$' type consraint to '$\leq$' type with a negative sign can not yield any result and not suitable in such cases, as said earlier.

Practice Problems

1. Max. $Z = 2x_1 + 5x_2$

   S.t. $2x_1 + x_2 \leq 430$

   $2x_2 \leq 460$

   $x_1, x_2 \geq 0$ **[JNTU (CSE) 97/S]**

Answer : Unique solution : $x_1 = 100$, $x_2 = 230$ and Max $Z = 1350$

2. A firm manufacturers two products in three departments. Product *A* contributes Rs. 5/- per unit and requires 5 hrs. in dept. *M*, 5 hrs. in dept. *N* and one hour in dept. *P*. Product *B* contributes Rs. 10/– per unit and requires 8 hrs. in dept. *M*, 3 hrs in dept N and 8 hrs in dept *P*. Capacities for departments *M*,*N*,*P* are 48 hours per week. Find out optimal product mix using Simplex model.

   [JNTU (CSE) 98]

Answer : Unique solution : $x_1 = 0$ $x_2 = 6$, $Z_{max} = 60$

   (Refer illustration 3 of chapter 2 for formulation)

3. Formulate the following problem as an LP Problem:

   A firm engaged in producing 2 models X_1, X_2 performs 3 operations – Painting, Assembly and Testing. The relevant data are as follows :

Model	Unit Sale Price	Hours Required for each unit		
		Assembly	Painting	Testing
Model X_1	Rs. 50	1.0	0.2	0.0
Model X_2	Rs. 80	1.5	0.2	0.1

   Total number of hours available are :

   For Assembly   600
   For Painting   100
   For Testing   30   **[JNTU (Mech.) 98/S]**

   Determine weekly production schedule to maximise revenue.

Answer : Unique solution : $x_1 = 150$, $x_2 = 300$

   Max $Z = 31,500$

   (Refer illustration 2 of chapter 2 for formulation)

4. Use simplex method to solve the following LP problem and verify by graph.
   Maximize $Z = 3x_1 + 4x_2$

   Subject to the constraints :

   $2x_1 + x_2 \leq 40$; $2x_1 + 5x_2 \leq 180$; $x_1, x_2 \geq 0$ **[JNTU (Mech) 98, OU - MBA Jully 2000]**

Answer : Unique solution $x_1 = 2.5$ $x_2 = 35$

   $Z_{max} = 147.5$

   (Refer illustration 4 of chapter 3 for graphical solution)

5. Find the minimum value of $Z = -x_1 + 2x_2$

   Subject to the constraints

   $$-x_1 + 3x_2 \leq 10$$

   $$x_1 + x_2 \leq 6$$

   $$x_1 - x_2 \leq 2$$

   $$x_1, x_2 \geq 0$$

   Solve graphically.

   Use simplex method to verify its graphical solution.   **[JNTU (ECE) 97]**

   Answer : Unique solution $x_1 = 2$, $x_2 = 0$;   Min $= -2$

   (Refer illustration 4 of chapter 3 for graphical solution)

6. Using simplex method,

   Maximize $Z = 5x_1 + 4x_2$

   Subject to constraints $4x_1 + 5x_2 \leq 10$

   $$3x_1 + 2x_2 \leq 9$$

   $$8x_1 + 3x_2 \leq 12$$

   $$x_1, x_2 \geq 0$$ **[JNTU (ECE) 97]**

   Answer : Unqiue solution $x_1 = \dfrac{15}{4}$, $x_2 = \dfrac{8}{7}$; Max $Z = \dfrac{653}{28}$

7. Put the following problem into a standard form of L.P.P :
   Maximise $f(x) = 2x_1 + 6x_2$
   subject to $-x_1 + x_2 \leq 1$
   $$2x_1 + x_2 \leq 2$$
   $$x_1 \geq 0, x_2 \geq 0$$

   Hence solve the problem by graphical method or otherwise. **[JNTU (EEE) 97/S]**

   Answer :   Unique solution $x_1 = \dfrac{1}{3}$ $x_2 = \dfrac{4}{3}$; $Z_{max} = \dfrac{26}{3}$

8. A company wants to purchase at most 180 units of a product. There are two types of the product, M_1 and M_2 available. M_1 occupies $2\,ft^3$, cost \$12 and the company makes a profit of \$3. M_3 occupies $3\,ft^3$, cost \$15 and the company makes a profit of \$4. If the budget is \$15,000 and the warehouse has $3000\,ft^3$ for product.

   (i) Set up the problem as a linear programming problem.

   (ii) Solve the problem using simplex method.   **[JNTU (EEE) 98]**

   Answer : (ii) Unique solution $x_1 = 0$, $x_2 = 100$; $Z_{max} = \$\,4000$

   (for (i), Refer illustration 7 of chapter 2)

9. Max. $Z = 45 x_1 + 70 x_2$

$$\text{S.t.} \quad 2 x_1 + 3 x_2 \leq 70$$
$$3 x_1 + 5 x_2 \leq 110$$
$$x_1 + 4 x_2 \leq 100$$
$$x_1 \geq 0, \quad x_2 \geq 0$$

Answer : Unique solution $x_1 = 20, \quad x_2 = 10 \quad Z_{max} = 1600$

10. Max $Z = 8 x_1 + 6 x_2$

$$\text{S.t.} \quad 2 x_1 + x_2 \leq 72$$
$$x_1 + 2 x_2 \leq 48$$
$$x_1 \geq 0, \quad x_2 \geq 0 \qquad \text{[JNTU - Mech. 93]}$$

Answer : Unique solution $x_1 = 32, \quad x_2 = 8 \quad Z_{max} = 304$

(Refer illustration 1 of chapter 3 for graphical solution)

11. Max $Z = 5x_1 + 6x_2 + 8 x_3$

$$\text{S.t.} \quad 2 x_1 + x_2 + 2x_3 \leq 50$$
$$x_1 + 6 x_2 + 2 x_3 \leq 50$$
$$x_1 + 2 x_2 + x_3 \leq 26$$
$$x_1, \ x_2, \ x_3 \geq 0$$

Answer : Unique Solution $x_1 = 0, \ x_2 = 0, \ x_3 = 25; \ \text{Max } Z = 200$

12. Max $Z = 2 x_1 + 4 x_2 + 2 x_3$

$$\text{S.t.} \quad 2 x_1 + x_2 - x_3 \leq 2$$
$$- 2 x_1 + x_2 - 5 x_3 \geq - 6$$
$$4 x_1 + x_2 + x_3 \leq 6$$
$$x_1, \ x_2, \ x_3 \geq 0$$

Answer : Unique solution, $x_1 = 0, \ x_2 = 4, \ x_3 = 2 ; \ \text{Max } Z = 20$

13. Maximise $Z : 3x_1 + 2x_2 ;$

$$\text{S.t constraints} \quad 2x_1 + x_2 \leq 2$$
$$3x_1 + 4x_2 \geq 12 \text{ and}$$
$$x_1, \ x_2 \geq 0 \qquad \text{[JNTU - Mech/Prod/Chem. 2001/S]}$$

14. A small paint factory produces three types of paints as follows :

Paint	Production in kg/day	Profit, units /kg
1	X	10
2	Y	4
3	Z	1

Due to limitations on the processing machinary, the following constraints are to be met with :

$$X + Y \leq 5$$

$$2X + Y + Z \leq 20$$

Using the simplex tableau, find out the product values of all the three paints, which will maximise the profit; and the value of the profit. **[JNTU - CSE/E Comp. E 2001]**

Unbounded Solutions

1. Solve by simplex method

Max $Z = 3 x_1 + 4 x_2$

S.t. $x_1 - x_2 \leq 1$

         $- x_1 + x_2 \leq 2$

          $x_1, x_2 \geq 0$ **[JNTU (CSE) 2001]**

Answer : Unbounded solution

2. Max $Z = 6 x_1 + 4 x_2$

S.t. $x_1 - x_2 \leq 1$

         $3 x_1 - 2 x_2 \leq 6$

        $x_1 \geq 0, \ x_2 \geq 0$

Answer : Unbounded or no finite solution

3. Max $Z = 3 x_1 + 2 x_2$

S.t. $2 x_1 - 3 x_2 \geq -9$

         $3 x_1 - 2 x_2 \leq 20$

            $x_1, x_2 \geq 0$

Answer : Unbounded or no finite solution

4. Maximise $Z = - 2 x_1 + 3 x_2$

                $x_1 \leq 5$

         $\dfrac{x_1}{3} - \dfrac{x_2}{2} \leq 1$

         $x_1, \ x_2 \geq 0$

Hint : In final table at $x_1 = 5$, $x_2 = 9$ and Max $Z = 15$ but $Z_j - C_j = -\dfrac{3}{2}$.

All the elements in key column (second column) are negative.
Hence solution is unbounded

5. (a) Min $Z = -6x_1 + 2x_2$

S.t. $x_1 - \dfrac{1}{2} x_2 \le 1$; $x_1 \le 4$

and $x_1, x_2 \ge 0$

(b) What is your comment if obj. function is Min $Z = 6x_1 - 2x_2$

Answer : (a) Unbounded or no finite solution.

(b) At $x_1 = 4$, $x_2 = 6$, Max $Z = 12$, But all the elements in second column are negative. The feasible region is unbounded but the optimal solution is bounded.

6. Show that the LPP

Max $Z = 4x_1 + x_2 + 3x_3 + 5x_4$
Subject to $4x_1 - 6x_2 - 5x_3 - 4x_4 \ge -20$,
$\quad\quad\quad 3x_1 - 2x_2 + 4x_3 + x_4 \le 10$,
$\quad\quad\quad 8x_1 - 3x_2 + 3x_3 + 2x_4 \le 20$
$\quad\quad\quad x_1, x_2, x_3, x_4 \ge 0$
Has an unbounded solution

Multiple Optimal Solutions

1. Max $Z = 4x_1 + 3x_2$

S.t. $8x_1 + 6x_2 \le 48$

$\quad\quad 2x_1 \le 9$

$\quad\quad\quad x_2 \le 6$

$x_1 \ge 0$, $x_2 \ge 0$

Answer : Multiple optimal solution $\quad x_1 = 4.5$, $x_2 = 2$, $Z_{max} = 24$

$\quad\quad\quad\quad\quad\quad\quad\quad\quad\quad\quad\quad\quad\quad x_1 = 1.5$, $x_2 = 6$, $Z_{max} = 24$

2. Max $Z = 0.4x_1 + 0.2x_2$

S.t $\quad 2x_1 + x_2 \le 72$

$\quad\quad x_1 + 2x_2 \le 48$

$\quad\quad\quad x_1, x_2 \ge 0$

Answer : Multiple optimal solution $\quad x_1 = 32$, $x_2 = 8$ or

$\quad\quad\quad\quad\quad\quad\quad\quad\quad\quad\quad\quad\quad x_1 = 36$, $x_2 = 0$ for Max $Z = 14.4$

3. Max $Z = 5x + 4y$

S.t. $x + y \leq 40$

$x \leq 20$

$15x + 12y \leq 300$

$x \geq 0, \ y \geq 0$

Answer : Multiple optimal solution $x = 20, \ y = 0$ or

$x = 0, \ y = 20, \ Z_{max} = 100$

4. Max $Z = 6x_1 + 2x_2 + 4x_3$

S.t. $2x_1 + 3x_2 + x_3 \ \leq 28$

$3x_1 + x_2 + 2x_3 \leq 24$

$x_1 + 2x_2 + 3x_3 \ \leq 35$

$x_1 \geq 0, \ x_2 \geq 0, x_3 \geq 0$

Answer : Multiple optimal solution $x_1 = 8, \ x_2 = 0, \ x_3 = 0$; Max $Z = 48$

5. Max $Z = 6x_1 + 3x_2$

S.t $2x_1 + 3x_2 \leq 8$

$x_1 + x_2 \leq 6$

$x_2 \leq 3$

$x_1, \ x_2 \geq 0$

Answer : Multiple optimal solution $x_1 = 4, x_2 = 0$

or $x_1 = \dfrac{5}{2}, \ x_2 = 3$ and Max $Z = 24$

6. Show that the LPP Max $Z = 0.4 x_1 + x_2$

S.t. $2x_1 + x_2 \leq 50 \ ; \ 2x_1 + 5x_2 \leq 100$

$2x_1 + 3x_2 \leq 90; \ x_1 \geq 0, x_2 \geq 0$

Has multiple optimal solutions.

Additional Problems

1. Max $Z = x_1 - x_2 + 3x_3$

S.t. $x_1 + x_2 + x_3 \ \leq 10$

$2x_1 - x_3 \ \leq 2$

$2x_1 - 2x_2 + 3x_3 \leq 0$

$x_1, x_2, x_3 \ \geq 0$

2. Use simplex to maximise $Z = 5x_1 + 2x_2 + 3x_3$

   S.t. $x_1 + 2x_2 + 2x_3 + x_4 = 8$

   $3x_1 + 4x_2 + x_3 + x_5 = 7$

   $x_j \geq 0$ where $j = 1, 2, 3, 4, 5$

3. Max $Z = 4x_1 + 3x_2 + 4x_3 + 6x_4$

   S.t $x_1 + 2x_2 + 2x_3 + 4x_4 \leq 80$

   $2x_1 + 2x_3 + x_4 \leq 60$

   $3x_1 + 3x_2 + x_3 + x_4 \leq 80$

   $x_j \leq 0$ where $j = 1$ to 4

4. Max $Z = 4x_1 + 5x_2 + 9x_3 + 11x_4$

   S.t. $x_1 + x_2 + x_3 + x_4 \leq 15$

   $7x_1 + 5x_2 + 3x_3 + 2x_4 \leq 120$

   $3x_1 + 5x_2 + 10x_3 + 15x_4 \leq 100$

   $x_j \geq 0, j = 1$ to 4

5. Max $Z = x_1 - 3x_2 + 2x_3$

   S.t. $3x_1 - x_2 + 2x_3 \leq 7$

   $x_1 - 2x_2 \geq -6$

   $-4x_1 + 3x_2 + 8x_3 \leq 10$

   $x_1, x_2, x_3 \geq 0$

6. Max $Z = 5x_1 + 4x_2$

   S.t. $4x_1 + 5x_2 \leq 10$

   $3x_1 + 2x_2 \leq 9$

   $8x_1 + 3x_2 \leq 12$

   $x_1 \geq 0$

   $x_2 \geq 0$

7. Max $Z = x_1 + 2x_2 + 3x_3$

   S.t. $x_1 + 2x_2 + 3x_3 \leq 10$

   $x_1 + x_2 \leq 5$

   $x_1, x_2, x_3 \geq 0$

$$\text{Max } Z = 2\,x_1 + 4\,x_2 + x_3 + x_4$$

$$\text{S.t} \qquad x_1 + 3\,x_2 + x_4 \le 4$$

$$2\,x_1 + x_2 \le 3$$

$$x_2 + 4\,x_3 + x_4 \le 3$$

$$x_1, x_2, x_3, x_4 \ge 0$$

9. $\text{Max } 15\,x_1 + 6\,x_2 + 9\,x_3 + 2\,x_4$

$$\text{S.t.} \qquad 2\,x_1 + x_2 + 5\,x_3 + 0.6\,x_4 \le 10$$

$$3\,x_1 + x_2 + 3\,x_3 + 0.25\,x_4 \le 12$$

$$7\,x_1 + x_4 \le 35$$

$$x_j \ge 0 \quad \text{where } j = 1, 2, 3, 4$$

10. $\text{Max } Z = 2\,x_1 + 5\,x_2 + 6\,x_3$

$$\text{S.t.} \qquad 5x_1 + 6\,x_2 - x_3 \le 3$$

$$-2\,x_1 + x_2 + 4\,x_3 \le 4$$

$$x_1 - 5\,x_2 + 3\,x_3 \le 1$$

$$-3\,x_1 - 3\,x_2 + 7\,x_3 \le 6$$

$$\text{and} \quad x_1, x_2, x_3 \ge 0$$

11. $\text{Max } Z = 3\,x_1 + 4\,x_2 + x_3 + 7\,x_4$

$$\text{S.t.} \qquad 8\,x_1 + 3\,x_2 + 4\,x_3 + x_4 \le 7$$

$$2\,x_1 + 6\,x_2 + x_3 + 5\,x_4 \le 3$$

$$x_1 + 4\,x_2 + 5\,x_3 + 2\,x_4 \le 8$$

$$\text{and} \qquad x_1, x_2, x_3, x_4 \ge 0$$

12. $\text{Max } Z = 3\,x_1 + 4\,x_2 + x_3$

$$\text{S.t.} \qquad x_1 + 2\,x_2 + 3\,x_3 \le 90$$

$$2\,x_1 + x_2 + x_3 \le 60$$

$$3\,x_1 + x_2 + 2\,x_3 \le 80$$

$$\text{and} \qquad x_1, x_2, x_3 \ge 0$$

13. $\text{Max } Z = .2\,x_1 - 4\,x_2 + 4\,x_3$

$$\text{S.t.} \qquad x_1 + 5\,x_2 - 5\,x_3 \le 8$$

$$-6\,x_1 + 7\,x_2 - 7\,x_3 \le 3$$

$$\text{and} \qquad x_1 \ge 0, \quad x_3 \ge 0$$

14. Max $Z = 5 x_1 + 3 x_2$

    S.t.
    $$x_1 + x_2 \leq 2$$
    $$5 x_1 + 2 x_2 \leq 10$$
    $$3 x_1 + 8 x_2 \leq 12$$
    and
    $$x_1, \ x_2, \ \geq 0$$

15. Min $Z = x_1 - 3 x_2 + 2 x_3$

    S.t.
    $$3 x_1 - x_2 + 2 x_3 \leq 7$$
    $$- 2 x_1 + 4 x_2 \leq 12$$
    $$- 4 x_1 + 3 x_2 + 8 x_3 \leq 10$$
    and
    $$x_1, x_2, x_3 \ \geq 0$$ [IAS - Main 93]

16. Max $Z = 2x_1 - x_2 + x_3$

    S.t.
    $$3x_1 - 2x_2 + 2x_3 \leq 15$$
    $$- x_1 + x_2 + x_3 \leq 3$$
    $$x_1 - x_2 + x_3 \leq 4$$
    $$x_1, x_2, x_3 \ \geq 0$$ [OU - MBA Apr. 99]

17. Max $Z = 23x_1 + 32x_2$

    S.t.
    $$10x_1 + 6x_2 \leq 2500$$
    $$5x_1 + 10x_2 \leq 200$$
    $$x_1 + x_2 \leq 500$$
    $$x_1, x_2 \ \geq 0$$ [OU - MBA - Apr 94, Sep. 99]

18. Solve : Max $Z = 400x_1 + 300x_2 + 200x_3$

    S.t.
    $$40x_1 + 20x_2 + 30x_3 \leq 600$$
    $$10x_1 + 20x_2 + 20x_3 \leq 400$$
    $$30x_1 + 20x_2 + 10x_3 \leq 80$$
    $$x_1, x_2, x_3 \ \geq 0$$ [OU - MBA 99/S]

19. Use simplex and verify by graph
    Max $Z = x_1 + 3x_2$

    S.t.
    $$3x_1 + 6x_2 \leq 8$$
    $$5x_1 + 2x_2 \leq 10$$
    $$x_1, x_2 \geq 0$$ [OU - MBA - Apr 98]

4.3 *Simplex Method : Using Surplus & Artificial Variables*

So far we have solved the linear programming problems with less than or equal to (≤) or availability constraints using simplex method. In all these problems we introduced 'slack variables' to convert the constraint inequalities into equalities. Even if the constraint is 'greater than or equal to type', we converted to less than or equal to type by multiplying with − 1 through out (changing the sign). But this method can be applied if the right hand side of the inequation is already negative.

For example : $x_1 - x_2 \geq -1$ can be written as $- x_1 + x_2 \leq 1$, but it is not suitable for the cases like $x_1 - x_2 \geq 1$ which on conversion becomes $- x_1 + x_2 \leq - 1$ and troubles us while calculating minimum ratio (since − *ve* min. ratio is ignored). Therefore this method is not suitable in all cases. In such cases we go for one of the following methods.

1. Big - M Method or Penalty Method

2. Two - Phase Method

These are explained here below.

4.3.1 *Big -M Method*

This is also called *Charnes Penalty Method.*

When all or some of the constraints are 'greater than or equal to' type, then we choose this method to solve the LPP. The methodology is shown as a flow chart given below :

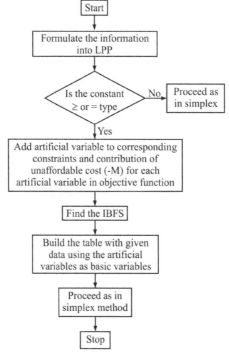

FIGURE 4.2 :   FLOW CHART OF BIG - M METHOD

Algorithm : *For Big-M or Penalty Method :*

Step 1: Formulate the given information into an LPP

Step 2 : Convert the inequations to equations by introducing 'SURPLUS' variables (subtract) followed by adding an 'ARTIFICIAL' variable (for ≥) or add slack (for ≤) and rewrite the objective function in maximisation form with zero contribution of each surplus or slack variables but '- M' for each artificial variable. The ' - M' indicates high penalty.

Step 3 : Find IBFS by keeping artificial variables and slack (if any) in the basis and others equated to zero (non- basis)

Step 4 : Develop initial simplex tableau - I and proceed as usually. i.e., find Z_j which is sum of product of C_B value ($-M$ for artificial variables and zero for slack variables) and value against corresponding column in the coefficient body matrix (y_i). Then find $Z_j - C_j$ to identify key column (most negative value of $Z_j - C_j$) and thus the corresponding 'entering variable'. Find minimum ratio dividing solution value with key column value to identify key row and thus the corresponding 'leaving variable'.

Step 5 : Iterate to develop next simplex tableau.

    – Replace leaving variable with entering variable along with its C_B.

    – Make key element (intersection element of key column and key row) as unity. i.e., Divide entire key row of iteration tableau - I by key element (except it's C_B value i.e., first column), to find the corresponding new row of iteration tableau - II.

    – Other elements of key column of iteration tableau - I are to be made as zero. To make this add or subtract required number of times the entire key row of Iteration tableau - II in the elements of corresponding rows of iteration tableau - I

    – Find Z_j, $Z_j - C_j$ and minimum ratio etc., to continue further and repeat (step 4 & 5) till all the values of $Z_j - C_j$ become positive.

Note : *In this method when an artificial variables is replaced, it's column also is to be deleted from the tableau for further iteration.*

Step 6 : Obtain final iteration tableau and write solution value and hence find the optimum solution in objective function.

The above algorithm is illustrated through the following numerical examples.

ILLUSTRATION 7 ——————————————————————

Food X contains 6 units of Vitamin A per gram and 7 units of Vitamin B per gram and costs 12 paise per gram. Food Y contains 8 units of Vitamin A per gram and 12 units of Vitamin B and costs 20 paise per gram. The daily minimum requirements of Vitamin A and Vitamin B are 100 units and 120 units respectively. Find the minimum cost of product mix. Use simplex method.

(JNTU - ECE - 7)

Solution :

Refer illustration 4 of chapter 2 for its formulation

Summary for Formulation :

$$\text{Min} \quad z = 12\, x_1 + 20\, x_2$$

$$\text{S.t.} \quad 6\, x_1 + 8\, x_2 \geq 100$$

$$7\, x_1 + 12\, x_2 \geq 120$$

$$x_1, x_2 \geq 0$$

Conversion of inequations to equations by introducing surplus variables & artificial variables.

$$6x_1 + 8x_2 - S_1 + A_1 = 100$$

$$7x_1 + 12\, x_2 - S_2 + A_2 = 120 \qquad S_1, S_2 \geq 0$$

Revised Objective Function :

$$\text{Min } Z = 12\, x_1 + 20\, x_2 - OS_1 - OS_2 + MA_1 + MA_2$$

$$\therefore \quad \text{Max. } Z = -12\, x_1 - 20\, x_2 + OS_1 + OS_2 - MA_1 - MA_2$$

Initial Basic Feasible Solution :

Basic variables are A_1, A_2 and others are non basic.

$$A_1 = 100 \qquad A_2 = 120 \qquad S_1 = 0 \qquad S_2 = 0 \qquad X_1 = 0 \qquad X_2 = 0$$

ITERATION TABLEAU · I :

C_B	B.V	C_j / S.V	-12 / x_1	-20 / x_2	0 / S_1	0 / S_2	$-M$ / A_1	$-M$ / A_2	Min Ratio	Remarks
$-M$	A_1	100	6	8	-1	0	1	0	$\dfrac{100}{8}$	
$-M$	A_2	120	7	12	0	-1	0	1	$\dfrac{120}{12}$ ← Key row	
		Z_j	$-13\,M$	$-20\,M$	M	M	$-M$	$-M$		
		$Z_j - C_j$	$-13M + 12$	$-20\,M + 20$	M	M	0	0		

Incoming Variable → (x_2) Key column

Out going variable ← (A_2)

Note : *12 is the key element, R^N indicates new or current row value and R^0 is to be read from previous table. A_2 becomes non basic and x_2 becomes basic variable. Also A_2 will be removed in the next iteration.*

ITERATION TABLEAU - II :

Incoming Variable

C_B	B.V	C_j S.V	-12 x_1	-20 x_2	0 S_1	0 S_2	$-M$ A_1	Min. Rat.	Remarks
$-M$	A_1	20	$\dfrac{4}{3}$	0	-1	$\dfrac{2}{3}$	1	$\dfrac{15}{}$ $\leftarrow$ K R	$R_1^N \rightarrow R_1^0 - 8R_2^N$
-20	X_2	10	$\dfrac{7}{12}$	1	0	$-\dfrac{1}{12}$	0	$\dfrac{120}{7}$	$R_2^N \rightarrow \dfrac{R_2^0}{12}$
		Z_j	$-\dfrac{4M}{3} - \dfrac{140}{12}$	-20	M	$-\dfrac{2M}{3} + \dfrac{20}{12}$	$-M$		
Out going variable		$Z_j - C_j$	$-\dfrac{4}{3}M + \dfrac{1}{3}$	0	M	$-\dfrac{2M}{3} + \dfrac{5}{3}$	0		

Key column

Note : *A_1 becomes non basic and x_1 becomes basic in next iteration. And A_1 will be removed.*

ITERATION TABLEAU - III :

C_B	B.V	C_j S.V	-12 x_1	-20 x_2	0 S_1	0 S_2	Remarks
-12	x_1	15	1	0	$-\dfrac{3}{4}$	$\dfrac{1}{2}$	$R_1^N \rightarrow \dfrac{3}{4} R_1^0$
-20	x_2	$\dfrac{5}{4}$	0	1	$\dfrac{7}{16}$	$-\dfrac{3}{8}$	$R_2^N \rightarrow R_2 - \dfrac{7}{12} R_1^N$
		Z_j	-12	-20	$\dfrac{1}{4}$	$\dfrac{3}{2}$	$Z_{max} = +205$
		$Z_j - C_j$	0	0	$\dfrac{1}{4}$	$\dfrac{3}{2}$	$Z_{max} = -205$

Since $Z_j - C_j \geq 0$ for all variables, the optimisation is reached.

The solution is

$$x_1 = 15 \qquad x_2 = \frac{5}{4}$$

$$\text{Max} \quad Z = -12 \times 15 - 20 \times \frac{5}{4} = -205$$

or $\quad$ Min $\quad Z = 12 \times 15 + 20 \times \dfrac{5}{4} = 180 + 25$

$$= 205$$

Optimal solution : Min Z = 205 at $x_1 = 15$, $x_2 = \dfrac{5}{4}$, $S_1 = 0$, $S_2 = 0$

ILLUSTRATION 8 (With mixed constraints - with maximising case) ———————————

> *Use Big-M method to solve the following LPP:*
>
> $\qquad$ *Maximise* $\qquad z = 3x_1 - x_2$
> $\qquad$ *Subject to* $\qquad 2x_1 + x_2 \geq 2$
> $\qquad\qquad\qquad\qquad x_1 + 3x_2 \leq 3$
> $\qquad\qquad\qquad\qquad x_2 \leq 4$
> $\qquad$ *and* $\qquad\qquad x_1, x_2 \geq 0$ $\qquad\qquad$ [JNTU B.Tech. (Mech) 99]

Solution :

$\qquad$ Conversion of inequalities into equalities

$$2\,x_1 + x_2 - S_1 + A_1 = 2$$
$$x_1 + 3\,x_2 + S_2 = 3$$
$$x_2 + S_3 = 4$$

Object Function :

$\qquad$ Max. $Z = 3\,x_1 - x_2 - OS_1 + OS_2 + OS_3 - MA_1$

$\qquad$ IBFS : $x_1 = 0, x_2 = 0, S_1 = 0; A_1 = 2, S_2 = 3, S_3 = 4$

ITERATION TABLEAU - I :

In coming Variable

C_B	B.V	C_j S.V	3 x_1	-1 x_2	0 S_1	0 S_2	0 S_3	$-M$ A_1	Min ratio
$-M$	A_1	2	2	1	-1	0	0	1	1 ←
0	S_2	3	1	3	0	1	0	0	3
0	S_3	4	0	1	0	0	1	0	∞
		Z_j	$-2M$	$-M$	M	0	0	$-M$	
		$Z_j - C_j$	$-2M - 3$	$-M + 1$	M	0	0	0	

Out going variable

Key column

In coming Variable

ITERATION TABLE II :

C_B	B.V	C_j S.V	3 x_1	-1 x_2	0 S_1	0 S_2	0 S_3	Min ratio	Remarks
3	x_1	1	1	1/2	$-1/2$	0	0	-2 (ignore)	$R_1^N \to \frac{1}{2} R_1^0$
0	S_2	2	0	5/2	1/2	1	0	4 ←	$R_2^N \to R_2^0 - R_1^N$
0	S_3	4	0	1	0	0	1	∞ (ignore)	$R_3^N \to R_3^0$
		Z_j	3	3/2	$-3/2$	0	0	A_1 is removed from tableau since it is replaced by x_1	
Out going variable		$Z_j - C_j$	0	5/2	$-3/2$	0	0		

Key column

ITERATION TABLE - III :

C_B	B.V	C_j S.V	3 x_1	-1 x_2	0 S_1	0 S_2	0 S_3	Min ratio	Remarks
3	x_1	3	1	3	0	1	0		$R_1^N \to R_1^0 + \frac{1}{2} R_2^N$
0	S_1	4	0	5	1	2	0		$R_2^N \to 2 R_2^0$
0	S_3	4	0	1	0	0	1		$R_3^N \to R_3^0$
		Z_j	3	9	0	3	0		$Z_{max} = 9$
		$Z_j - C_j$	0	10	0	3	0		

Since $Z_j - C_j \geq 0$ for all variables, the above solution is optimal.

The solution is $x_1 = 3$, $x_2 = 0$, $S_1 = 4$, $S_2 = 0$, $S_3 = 4$

$$Z_{max} = 3 \times 3 - 1 \times 0 + 0 \times 4 + 0 \times 0 + 0 \times 4 = 9$$

Optimal solution : $Z_{max} = 9$ at $x_1 = 3$; $x_2 = 0$

ILLUSTRATION 9 (With mixed constraints including equality constraint)

Solve the following using Linear Programming Method.

Maximise $Z = -4x_1 - 3x_2$

Subject to $3x_1 + x_2 = 3$

$3x_1 + 4x_2 \geq 6$

$x_1 + x_2 \leq 4$

$x_1, x_2 \geq 0$ [JNTU B.Tech. (CSE) 97]

Solution :

Converting the constraint inequations into equations.

$3 x_1 + x_2 + A_1 = 3$

$3 x_1 + 4x_2 - S_2 + A_2 = 6$

$x_1 + x_2 + S_1 = 4$

$x_1 , x_2 \geq 0, S_1 \ S_2 , A_1 , A_2 \geq 0$

IBFS :
B.V : $A_1 = 3, A_2 = 6$ and $S_1 = 4$
NBV : $x_1 = 0, x_2 = 0$ and $S_2 = 0$

Assigning the largest negative price $- M$ to the artificial variables A_1 , A_2 the objective function becomes Max. $Z = - 4x_1 - 3x_2 + 0S_1 + 0S_2 - MA_1 - MA_2$.

ITERATION TABLEAU - I :

Entering Variable

C_b	B.V	C_j S.V	-4 x_1	-3 x_2	0 S_1	0 S_2	$-M$ A_1	$-M$ A_2	Min. Ratio	Remarks
$-M$	A_1	3	3	1	0	0	1	0	1	Key row
$-M$	A_2	6	3	4	0	-1	0	1	4/3	
0	S_1	4	1	1	1	0	0	0	6	
		$Z_j - C_j$	$-6M+4$	$-5M+3$	0	$+M$	0	0		

Leaving variable   Key column

ITTERATION TABLEAU - II :

Entering Variable

C_b	B.V	C_j S.V	-4 x_1	-3 x_2	0 S_1	0 S_2	$-M$ A_2	Min. Ratio	Remarks
-4	x_1	1	1	1/3	0	0	0	3	$R_1^N \to R_1^0/3$
$-M$	A_2	3	0	3	0	-1	1	1/3 ←	$R_2^N \to R_2^0 - 3R_1^N$
0	S_1	3	0	2/3	1	0	0	15/2	$R_3^N \to R_3^0 - R_1^N$
		$Z_j - C_j$	0	$\dfrac{(-3M+5)}{3}$	0	$+M$	0		A_1 col. is deleted

Leaving variable   Key column

ITERATION TABLEAU - III :

C_B	B.V	C_j S.V	-4 x_1	-3 x_2	0 S_1	0 S_2	Min. Ratio	Remarks
-4	x_1	2/3	1	0	0	1/9	$-ve$	$R_1^N \to R_1^0 - \dfrac{1}{3} R_2^N$
-3	x_2	1	0	1	0	$-1/3$	1	$R_2^N \to \dfrac{1}{3} R_2^0$
0	S_1	7/3	0	0	1	$+2/9$	43/2	$R_3^N \to R_3^0 - \dfrac{2}{3} R_2^N$
		$Z_j - C_j$	0	0	0	5/9	A_2 col. is deleted	$Z_{max} = -7/3$

As $\quad Z_j - C_j \geq 0$ for all variables, the optimal solution is

$$x_1 = \frac{2}{3}, \ x_2 = 1$$

$$S_1 = \frac{7}{3}, \ S_2 = 0$$

$$\text{Max. } Z = -4\left(\frac{2}{3}\right) - 3(1) = -\frac{17}{3}.$$

Significance of Artificial Variable :

In case of 'Less than or equal to type' inequality, we convert into equality by adding a 'slack' variable.

Ex : $3x_1 + 2x_2 \le 15$ is converted as

$3x_1 + 2x_2 + S_1 = 15$ so that $S_1 \ge 0$ provided $x_1 \ge 0$ and $x_2 \ge 0$.

But in 'greater than or equal to type' inequality for instance, as, $5x_1 + 6 x_2 \ge 9$ and $x_1 \ge 0$ $x_2 \ge 0$ we have to subtract a variable called 'surplus' variable and thus $5x_1 + 6x_2 - S_2 = 9$ and $x_1, x_2 \ge 0$. Here S_2 can not be negative since the negative nature of S_2 violates the inequation. Therefore is $S_2 \ge 0$. Now as $x_1 \ge 0$ and 0 $x_2 \ge 0$, at no production or 'stand still conditions' i.e., $x_1 = 0$ and $x_2 = 0$, makes the equation inconsistent with reference to the condition that $S_2 \ge 0$ (If $x_1 = 0$ & $x_2 = 0$ $\Rightarrow$ $S_2 = -9$ which is violated by $S_2 \ge 0$). To solve this riddle we have to add another variable in the equation called 'Artificial variable' (say 'A') so that $A \ge 0$. And A is supposed as very large (the can never be zero) that costs unaffordably. In Big - M we consider its cost as 'M' a very high cost and concentrate on its removal while in two-phase method, though we assume unit negative cost, fundamentally we proceed to remove these artificial variables in phase - I with out which we will not enter phase - II. If 'Artificial variable' could not be removed in phase - I, we declare there itself that the solution is infeasible.

4.3.2 *Two - Phase Method*

As the name suggests, we solve LPP in two phases here. The objective function is split into two sub-objectives, one with artificial variables (and slack if any) only, while the second with the decision variables. If the phase - I yields a solution we proceed to phase - II, otherwise we conclude at phase - I as infeasible solution.

Advantages of Two Phase Method Over Big-M Method :

The advantages of this method over Big - M method are.

1. It is easier to calculate as it does not involve 'M' and all are numericals.

2. It can give the solution at the first phase itself if the LPP is infeasible. We need not go through second phase.

3. In the case of a digital computer, it is not possible to get the solution by Big-M where as 2 - phase method can be applied.

The Flow Chart for Two Phase Method :

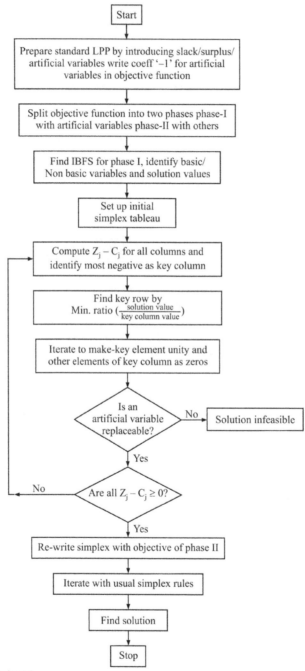

FIGURE 4.3 : FLOW CHART

The Algorithm of Two-Phase Method :

The following steps are involved in 2 - phase method.

Phase - I :

Step 1: Formulation of Problem : After formulating the information to an LPP with usual method.

   (a) If the objective function is to maximise proceed to (d)

   (b) Minimisation objective function is to be converted into maximisation form. This is done by multiplying with '–1' through out.

   (c) Convert inequality form of constraints to equality form by using artificial, surplus or slack variables.

   (d) Rewrite the objective function (say Z_1) by assigning '–1' value to the coefficients of Artificial variables (instead of $-M$ as in Big M method) and 'Zero to all the other variables.

Step 2 : Initial Tableau :

   (a) Find IBFS, by assigning zero value to decision and surplus variables. Find the solution values for artificial variables and slack (if any) variables.

   (b) Develop initial tableau with usual notations.

Step 3 : Iteration :

   (a) Examine the optimality condition by checking if there is any negative value for $Z_j - C_j$. [Z_j is summation of products C_B value with corresponding column coefficient].

   (b) Find 'key column' whose $Z_j - C_j$ is most negative. Note its corresponding variable as 'the entering variable'. This replaces the leaving variable in the next iteration along with its C_j value.

   (c) Find 'key row' with minimum ratio by dividing the solution value by key column value. Note the variable of this row as 'leaving variable'. This is replaced by entering variable along with its contribution in the next iteration.

   (d) Iterate with usual method, i.e., to obtain unity at the place of key element and other elements as zeros in its column, thus find the next tableau.

Step 4 : Basic Feasible Solution (BFS) :

   (a) Repeat step - 3 until the values of $Z_j - C_j$ for all the columns are positive.

   (b) Obtain the solution called ' Basic feasible solution (BFS)'. This table may be noted as Auxiliary LPP Solution.

(c) If $Z_j - C_j$ is still negative and Artificial variables are not replaced, stop and conclude that the LPP has '*infeasible solution*'.

(d) Note that when an artificial variable is replaced, it should be deleted from the table (even its corresponding column also)

Phase - II :

Step 1 : Resetting the Problem :

(a) Reset the objective function now, without using of artificial variable say (Z_2). Use the original contribution of the variables in 'Maximisation objective function' for all the other variables.

(b) Use constraint Coefficients as obtained from Auxiliary LPP in phase - I , i.e., the final (BFS) table of LPP is written as it is, with re-writing the contributions of all the variables and removing artificial variables.

Step 2 : Initial Tableau :

(a) Obtain initial tableau with the help of the final tableau of phase - I (Aux. LPP). Replace the coefficient of variables in objective function with original values i.e., all C_j and C_B values are re-written according to the reset objective function.

(b) Check optimality condition and all the process is as same as in step - 3 of phase - I

Step 3 : Iteration :

Iterate the tableau of step - 2 of phase II obtained above with usual method as mentioned in step - 3 of phase - I

Step 4 : Optimal Basic Feasible Solution :

Obtain final tableau with all values of $Z_j - C_j$ positive (for all columns) and write the solution value set. Find optimal value by substituting in the objective function of phase - II.

This is illustrated here below.

ILLUSTRATION 10 ──

Let us consider the example in illustration - I of Big - M method.

After formulation of the information, we get

   Min $Z = 12x_1 + 20x_2$

   S.t. $6x_1 + 8x_2 \geq 100$

                   $7x_1 + 12x_2 \geq 120$

                   $x_1 \geq 0, x_2 \geq 0$ [JNTU B.Tech (ECE) 97]

Now conversion of inequality constraints into equality form by introducing surplus and artificial variables.

$$6x_1 + 8x_2 - S_1 + A_1 = 100$$

$$7x_1 + 12x_2 - S_2 + A_2 = 120$$

$$x_1 \geq 0, \ x_2 \geq 0, \ S_1 \geq 0, \ S_2 \geq 0, \ A_1 \geq 0, \ A_2 \geq 0$$

Revised objective function

Max. $Z = -12x_1 - 20x_2 - 0S_1 - 0S_2 - 1A_1 - 1A_2$

Split the objective function into two sub-objectives

Max. $Z_1 = 0.x_1 - 0.x_2 - 0.S_1 - 0.S_2 - A_1 - A_2$

$\qquad = -A_1 - A_2 \qquad$ (for phase - I)

and Max. $Z_2 = -12x_1 - 20x_2 - 0S_1 - 0S_2$

$\qquad\qquad\qquad\qquad$ (for phase - II)

Now for phase - I , we have to solve

Max. $Z_1 = -A_1 - A_2$

subject to $\quad 6x_1 - 8x_2 - S_1 + A_1 = 100$

$$7x_1 + 12x_2 - S_2 + A_2 = 120$$

$$x_1, x_2, S_1, S_2, A_1, A_2 \geq 0$$

IBFS is $x_1 = 0, \ x_2 = 0, \ S_1 = 0, \ S_2 = 0$ (non - basis)

$A_1 = 100$ and $A_2 = 120$ (basis)

ITERATION (INITIAL) TABLEAU - I :

(Max. $Z_1 = 0x_1 + 0x_2 - 0S_1 + 0S_2 - A_1 - A_2$)

Entering variable

C_B	B.V	C_j	0	0	0	0	− 1	− 1	Min Ratio	Remarks
		S.V	x_1	x_2	S_1	S_2	A_1	A_2		
− 1	A_1	100	6	8	− 1	0	1	0	$\dfrac{100}{8}$	
− 1	A_2	120	7	12	0	− 1	0	1	$\dfrac{120}{12}$ ← Key row	
		Z_j	− 13	− 20	1	1	− 1	− 1		
		$Z_j - C_j$	− 13	− 20	1	1	0	0		

Key element

Leaving variable

Key column

C_B	B.V	C_j / S.V	x_1	x_2	S_1	S_2	A_1	A_2 column is deleted as it is replaced in basis	Min Ratio	Remarks
			0 (Entering variable)	0	0	0	-1			
-1	A_1	20	$\boxed{\dfrac{4}{3}}$ (Key element)	0	-1	$\dfrac{2}{3}$	1		$\dfrac{20 \times 3}{4}$	$R_1^N \to R_1^0 - 8\,R_2^N$
0	x_2	10	$\dfrac{7}{12}$	1	0	$-\dfrac{1}{12}$	0		$\dfrac{10 \times 12}{7}$	$R_2^N \to R_2^0/12$
		Z_j	$-\dfrac{4}{3}$	0	1	$-\dfrac{2}{3}$	-1			
	(Leaving variable)	$Z_j - C_j$	$-\dfrac{4}{3}$	0	1	$-\dfrac{2}{3}$	0			

Key column

C_B	B.V	C_j / S.V	x_1	x_2	S_1	S_2	A_1 column is removed from tableau as it is replaced in basis	Min Ratio	Remarks
			0	0	0	0			
0	x_1	15	1	0	$-\dfrac{3}{4}$	$\dfrac{1}{2}$			$R_1^N \to \dfrac{3}{4} R_1^0$
0	x_2	$\dfrac{5}{4}$	0	1	$\dfrac{7}{16}$	$-\dfrac{3}{8}$			$R_2^N \to R_2^0 - \dfrac{7}{12} R_1^N$
		Z_j	0	0	0	0			
		$Z_j - C_j$	0	0	0	0			

As $Z_j - C_j \geq 0$ for all the variables and all artificial variables are replaced, phase - I yields solution. Therefore we proceed to phase - II

Phase - II :

Objective function Max. $Z_2 = -12\,x_1 - 20\,x_2 - 0S_1 - 0S_2$

C_B	B.V	C_j / S.V	-12 / x_1	-20 / x_2	0 / S_1	0 / S_2	Min Ratio	Remarks
-12	x_1	15	1	0	$-\dfrac{3}{4}$	$\dfrac{1}{2}$		
-20	x_2	$\dfrac{5}{4}$	0	1	$\dfrac{7}{16}$	$-\dfrac{3}{8}$		
		Z_j	-12	-20	$\dfrac{1}{4}$	$\dfrac{3}{2}$		
		$Z_j - C_j$	0	0	$\dfrac{1}{4}$	$\dfrac{3}{2}$		

As $Z_j - C_j \geq 0$ for all variables, the optimization is reached. And the solution is

$$x_1 = 15 \quad x_2 = \frac{5}{4}$$

$$Z_{min} = 12 \times 15 + 20 \times \frac{5}{4}$$

$$= 180 + 25$$

$$= 205$$

ILLUSTRATION 11

Solve the following LLP

Maximise $Z = 10X + 15Y$

Subject to $Y \geq 3: X - Y \geq 0; Y \leq 12:$

$X + Y \leq 30; X \leq 20$ and $X, Y \geq 0$ [JNTU (CSE) 2003 (Set - 1)]

Solution :

Rewriting the objective function and constraint set in standard form by introducing slack, surplus and artificial variables, we get,

Maximise $Z = 10X + 15Y$

Subject to

$$Y - S_1 + A_1 = 3$$
$$X - Y - S_2 + A_2 = 0$$
$$Y + S_3 = 12$$
$$X + Y + S_4 = 30$$
$$X + S_5 = 20$$

$X, Y \geq 0, S_1, S_2, S_3, S_4, S_5 \quad \geq 0$ and $A_1, A_2 \geq 0$

(where X, Y are decision variables, S_1 & S_2 are surplus variables, S_3, S_4 & S_5 are variables and A_1 & A_2 are artificial variables.)

Phase I :

For phase I, we consider the objective function as Max. $Z_1 = -A_1 - A_2$

$Y - S_1 + A_1 = 3; \quad X - Y - S_2 + A_2 = 0; \quad Y + S_3 = 12,$

$X + Y + S_4 = 30; \quad X + S_5 = 20$ and

$X, Y \geq 0, S_j \geq 0, A_1, A_2 \geq 0 \quad (j = 1, 2, 3, 4, 5)$

IBFS : Basic variables : $A_1 = 3, A_2 = 0, S_3 = 12, S_4 = 30, S_5 = 20$

Non basic variables : $X = Y = S_1 = S_2 = 0$

ITERATION TABLEAU - I :

C_B	B.V	C_j $S.V$	0 X	0 Y	0 S_1	0 S_2	0 S_3	0 S_4	0 S_5	-1 A_1	-1 A_2	Min. Ratio
-1	A_1	3	0	1	-1	0	0	0	0	1	0	$3/1 \leftarrow$ KR
-1	A_2	0	1	-1	0	-1	0	0	0	0	1	$- ve$ (Ignore)
0	S_3	12	0	1	0	0	1	0	0	0	0	12/1
0	S_4	30	1	1	0	0	0	1	0	0	0	30/1
0	S_5	20	1	0	0	0	0	0	1	0	0	∞ (ignore)
Leaving variable		Z_j	-1	1	1	1	0	0	0	-1	-1	
		$Z_j - C_j$	-1	1	0	0	0	0	0	0	0	

Key column

Note : *We can choose X or Y as key column here, but better to choose Y and if X is taken, it may lead to confusions.*

ITERATION TABLEAU - II :

C_B	B.V	C_j $S.V$	0 X	0 Y	0 S_1	0 S_2	0 S_3	-1 S_4	-1 S_5	-1 A_2	Min. Ratio.	Remarks
0	Y	3	0	1	-1	0	0	0	0	0	∞	$R_1^N \to R_1^0$
-1	A_2	3	1	0	-1	-1	0	0	0	1	$3/1 \leftarrow$	$R_2^N \to R_2^0 + R_1^N$
0	S_3	9	0	0	-1	0	1	0	0	0	∞	$R_3^N \to R_3^0 - R_1^N$
0	S_4	27	1	0	-1	0	0	1	0	0	27/1	$R_4^N \to R_4^0 - R_1^N$
0	S_5	20	1	0	0	0	0	0	1	0	20/1	$R_5^N \to R_5^0$
Leaving variable		Z_j	-1	0	1	1	0	0	0	-1		
		$Z_j - C_j$	-1	0	1	1	0	0	0	0		

Key column

Note : *In the above iteration, we can observe that there is no need of changing R_1 and R_5 since the desired 1 and 0 are already available*

ITERATION TABLEAU - III :

C_B	B.V	C_j $S.V$	0 X	0 Y	0 S_1	0 S_2	0 S_3	0 S_4	0 S_5	Min. Ratio.	Remarks
0	Y	3	0	1	-1	0	0	0	0		$R_1^N \to R_1^0$
0	X	3	1	0	-1	-1	0	0	0		$R_2^N \to R_2^0$
0	S_3	9	0	0	-1	0	1	0	0		$R_3^N \to R_3^0$
0	S_4	24	0	0	0	1	0	1	0		$R_4^N \to R_4^0 - R_2^N$
0	S_5	17	0	0	1	1	0	0	1		$R_5^N \to R_5^0 - R_2^N$
		Z_j	0	0	0	0	0	0	0		
		$Z_J - C_j$	0	0	0	0	0	0	0	This is now called auxilary simplex tableau	

Since $Z_j - C_j = 0$ for all the variable phase - I computation is complete at this stage. Both artificial variables have been replaced from the basis. Therefore we proceed to phase - II now. Now for phase - II, we take above auxilary simplex tableau with all numericals 'as is' except for the values of C_j. These are taken from the new objective function of second phase.

Phase II :

— Entering variable

Objective function : Max. $Z_2 = 10X + 15Y$.

C_B	B.V	C_j $S.V$	10 X	15 Y	0 S_1	0 S_2	0 S_3	0 S_4	0 S_5	Min. Ratio.	Remarks
15	Y	3	0	1	-1	0	0	0	0	$-ve$	
10	X	3	1	0	-1	-1	0	0	0	$-ve$	
0	S_3	9	0	0	-1	0	1	0	0	$-ve$	
0	S_4	27	0	0	0	1	0	1	0	∞	
0	S_5	17	0	0	1	1	0	0	1	$20/1 \leftarrow$	
		Z_j	10	15	-25	-10	0	0	0		
		$Z_J - C_j$	0	0	-25	-10	0	0	0		

Key column

Leaving variable

C_B	B.V	C_j / S.V	10 X	15 Y	0 S_1	0 S_2	0 S_3	0 S_4	0 S_5	Min. Ratio.	Remarks
15	Y	20	0	1	0	1	0	0	1		$R_1^N \rightarrow R_1^0 + R_5^N$
10	X	20	1	0	0	0	0	0	1		$R_2^N \rightarrow R_2^0 + R_5^N$
0	S_3	26	0	0	0	1	1	0	0		$R_3^N \rightarrow R_3^0 + R_5^N$
0	S_4	24	0	0	0	1	0	1	0		
0	S_1	17	0	0	1	1	0	0	1		$R_5^N \rightarrow R_5^0$
		Z_j	10	15	0	15	0	0	25		$Z_{max} = 10 \times 20$
		$Z_j - C_j$	0	0	0	15	0	0	25		$+ 15 \times 20$ $= 500$

Since all the values of $Z_j - C_j$ are positive (≥ 0), we arrived, at optimal solution.

The optimal solution is $X = 20$; $Y = 20$

$$Z_{max} = 10 \times 20 + 15 \times 20 = 200 + 300 = 500$$

Special Cases in Simplex Methods :

1. Infeasible Solutions :

In Big-M method, we find infeasible solution if all the artificial variable do not leave, the basis.

Similarly, in two - phase method also, if phase - I does not yield any solution since all artificial variables can not be replaced, the solution is infeasible.

These are illustrated through the following numerical examples.

ILLUSTRATION 12 ————————————————————————

$$\text{Max.} \qquad Z = 4x_1 + 3x_2$$
$$3x_1 + 4x_2 \le 6$$
$$5x_1 + 6x_2 \ge 15$$
$$x_1, x_2 \ge 0$$

Solution :

By Big - M method :

Conversion to equality

Max. $Z = 4x_1 + 3x_2 + 0S_1 + 0S_2 - MA$

S.t. $\quad 3x_1 + 4x_2 + S_1 = 6$

$\quad 5x_1 + 6x_2 - S_2 + A = 15$

$\quad x_1, x_2, S_1, S_2, A \ge 0$

IBFS $\quad x_1 = 0, \ x_2 = 0, \ S_2 = 0 \qquad$ (Non-basis)

$\quad S_1 = 6 \quad A = 15 \qquad\qquad$ (Basis)

ITERATION TABLEAU · I :

Entering variable

C_B	B.V	C_j / S.V	4 / x_1	3 / x_2	0 / S_1	0 / S_2	$-M$ / A	Min Ratio	Remarks
0	S_1	6	3	4	1	0	0	$\frac{6}{4} = \frac{3}{2}$	Key row
$-M$	A	15	5	6	0	-1	1	$\frac{15}{6} = \frac{5}{2}$	
	Leaving variable	Z_j	$-5M$	$-6M$	0	M	$-M$		
		$Z_j - C_j$	$-5M-4$	$-6M-3$	0	M	0		

Key column — Entering variable

C_B	B.V	C_j / S.V	4 / x_1	3 / x_2	0 / S_1	0 / S_2	$-M$ / A	Min Ratio	Remarks
0	S_1	6	$\frac{3}{4}$	1	$\frac{1}{4}$	0	0	$2 \xleftarrow{\text{Key row}}$	$R_1^N \to R_1^0/4$
$-M$	A	15	$\frac{1}{2}$	0	$-\frac{3}{2}$	-1	1	12	$R_2^N \to R_2^N - 6R_1^N$
	Leaving variable	Z_j	$-\frac{M}{2} + \frac{9}{4}$	3	$\frac{3}{4} + \frac{3M}{2}$	M	$-M$		
		$Z_j - C_j$	$-\frac{M}{2} - \frac{7}{4}$	0	$\frac{3M}{2} + \frac{3}{4}$	M	0		

Key column

◢ **ITERATION TABLEAU - III :**

C_B	B.V	C_j / S.V	4 / x_1	3 / x_2	0 / S_1	0 / S_2	$-M$ / A	Min Ratio	Remarks
4	x_1	2	1	$\dfrac{4}{3}$	$\dfrac{1}{3}$	0	0		$R_1^N \rightarrow \dfrac{4R_1^0}{3}$
$-M$	A	5	0	$-\dfrac{2}{3}$	$-\dfrac{5}{3}$	-1	1		$R_2^N \rightarrow R_2^0 - \dfrac{1}{2} R_1^N$
		Z_j	4	$\dfrac{2M}{3}+\dfrac{16}{3}$	$\dfrac{5M}{3}+\dfrac{4}{3}$	M	$-M$		
		$Z_j - C_j$	0	$\dfrac{2M}{3}+\dfrac{7}{3}$	$\dfrac{5M}{3}+\dfrac{4}{3}$	M	0		

Now all the values of $Z_j - C_j \geq 0$ but artificial variable did not vanish. Hence the Solution is **Infeasible :**

(For the above problem if solution is written it is $x_1 = 2$, $A = 5$, $x_2 = 0$, $S_1 = 0$ and $S_2 = 0$ $Z_{\max} = 8 - 5M$ which is not feasible as M is enormously high value that is unaffordable)

Now, we verify this by 2 - phase method.

By Two - Phase Method :

The above problem is done by two - phase method as follows.

For phase - I Max. $Z_1 = -A$

For phase - I Max. $Z_2 = 4x_1 + 3x_2 + 0S - 0S_2$

(And the constraints are converted to equality, as same as in the above problem)

Phase - I :

ITERATION TABLEAU I :

Entering variable

C_B	B.V	C_J / S.V	4 / x_1	3 / x_2	0 / S_1	0 / S_2	$-M$ / A	Min Ratio	Remarks
0	S_1	6	3	4	1	0	0	$\dfrac{6}{4}$	Key row
-1	A	15	5	6	0	-1	1	$\dfrac{15}{6}$	
		Z_j	-5	-6	0	1	-1		
		$Z_j - C_j$	-5	-6	0	1	0		

Leaving variable

Key column

ITERATION TABLEAU II :

Entering variable

C_B	B.V	C_j S.V	4 x_1	3 x_2	0 S_1	0 S_2	$-M$ A	Min Ratio	Remarks
0	x_2	$\frac{3}{2}$	$\frac{3}{4}$	1	$\frac{1}{4}$	0	0	$2 \leftarrow \frac{\text{Key}}{\text{row}}$	$R_1^N \to R_1^0/4$
-1	A	6	$\frac{1}{2}$	0	$-\frac{3}{2}$	-1	1	12	$R_2^N \to R_2^N - 6\,R_1^N$
		Z_j	$-\frac{1}{2}$	0	$\frac{3}{2}$	1	-1		
		$Z_j - C_j$	$-\frac{1}{2}$	0	$\frac{3}{2}$	1	0		

Key element (under x_1 column, $\frac{3}{4}$)

Leaving variable

Key column

ITERATION TABLEAU - III :

C_B	B.V	C_j S.V	4 x_1	3 x_2	0 S_1	0 S_2	$-M$ A	Remarks
0	x_1	2	1	$\frac{4}{3}$	$\frac{1}{3}$	0	0	$R_1^N \to R_1^0 \times \frac{4}{3}$
-1	A	5	0	$-\frac{2}{3}$	$-\frac{5}{2}$	-1	1	$R_2^N \to R_2^0 - \frac{1}{2}\,R_1^N$
		Z_j	0	$\frac{2}{3}$	$\frac{5}{3}$	1	-1	
		$Z_j - C_j$	0	$\frac{2}{3}$	$\frac{5}{3}$	1	0	

As all the values of $Z_j - C_j \geq 0$, we stop here. But the artificial variable 'A' has not been replaced. Therefore phase - I does not yield any solution.

Solution is Infeasible

Unbounded Solution :

When an LPP does not give finite solution values of variables, such solution is unbounded. In such cases x_j can take very high values without violating the constraints and conditions. This is interpreted in final tableau of simplex as all ratios will be either negative or infinity so that key row can not be generated and no variable does leave the basis.

This is already shown under ordinary simplex model and now it is illustrated with Big - M and 2 - phase methods.

ILLUSTRATION 13 ─────────────────────────────────

Maximise	$Z = 3x_1 + 2x_2$
Subject to	$x_1 - x_2 \leq 1$
	$x_1 + x_2 \geq 3$
	$x_1, x_2 \geq 0$

Solution :

        Solution by Big - M

            Conversion :

                Max. $Z = 3x_1 + 2x_2 + 0.S_1 - 0.S_2 - MA$

           S.t. $x_1 - x_2 + S_1 = 1$

                  $x_1 + x_2 - S_2 + A = 3$

                  $x_1, x_2 \geq 0, \; S_1 \geq 0, S_2 \geq 0, A \geq 0$

           IBFS

                $x_1 = 0, x_2 = 0, S_2 = 0$ (non basis)

                $S_1 = 1 \; A = 3$ (basis)

ITERATION TABLEAU - I :   *Entering variable*

C_B	B.V	C_j S.V	3 x_1	2 x_2	0 S_1	0 S_2	$-M$ A	Min Ratio	Remarks
0	S_1	1	1	-1	1	0	0	1	
$-M$	A	3	1	1	0	-1	1	12	
		Z_j	$-M$	$-M$	0	M	$-M$		
		$Z_j - C_j$	$-M-3$	$-M-2$	0	M	0		

Leaving variable   Key column

ITERATION TABLEAU - II :   *Entering variable*

C_B	B.V	C_j S.V	3 x_1	2 x_2	0 S_1	0 S_2	$-M$ A	Min Ratio	Remarks
3	x_1	1	1	-1	1	0	0	$-ve$	$R_1^N \to R_1^0$
$-M$	A	2	0	2	-1	-1	1	$1 \xleftarrow{\text{Key column}}$	$R_2^N \to R_2^0 - R_1^N$
		Z_j	3	$-2M$	M	M	$-M$		
		$Z_j - C_j$	0	$-2M$	M	M	0		

Leaving variable   Key column

ITERATION TABLEAU - III :

C_B	BV	C_j / SV	3 / x_1	2 / x_2	0 / S_1	0 / S_2	Min. Ratio	Remarks
3	x_1	2	1	0	$\dfrac{1}{2}$	$-\dfrac{1}{2}$	$-ve$	$R_1^N \to R_1^N + R_2^N$
2	x_2	1	0	1	$-\dfrac{1}{2}$	$-\dfrac{1}{2}$	$-ve$	$R_2^N \to \dfrac{1}{2} R_2^0$
		Z_j	3	2	$\dfrac{1}{2}$	$-\dfrac{5}{2}$		

As $Z_j - C_j$ is still negative for at least one of the variables, but the min. ratio is not possible (– ve for both x_1, x_2 variables), we can not proceed further and has no finite solution.

Solution is Unbounded

Note : *Students may try the same problem with two phase method and check.*

ILLUSTRATION 14

> **Max.** $\quad Z = 4x_1 + 3x_2$
>
> **S.t.** $\;x_1 - 2x_2 \geq -10$
>
> $\qquad x_1 - x_2 \geq -12$
>
> $\qquad x_1, x_2 \geq 0$

conversion :

Objective Function of Phase - I :

Max. $Z = -A_1 - A_2$

S.t. $\quad x_1 - 2x_2 - S_1 + A_1 = -10$

$\qquad x_1 - x_2 - S_2 + A_2 = -12$

$\qquad x_1, x_2, S_1, S_2, A_1, A_2 \geq 0$

Objective Function of Phase - II :

Max. $Z = 4x_1 + 3x_2$

$\qquad - 0.S_1 - 0.S_2$

S.t. Auxiliary LPP Tableau values

Phase - I :

ITERATION TABLEAU - I :

C_B	BV	C_j / SV	4 / x_1	3 / x_2	0 / S_1	0 / S_2	-1 / A_1	-1 / A_2		Remarks
-1	A_1	-10	1	-2	-1	0	1	0	$-ve$	
-1	A_2	-12	1	-1	0	-1	0	1	$-ve$	
		Z_j	-2	3	1	1	1	1		
		$Z_j - C_j$	-2	3	1	1	0	0		

As $Z_j - C_j \leq 0$ for x_1, it is entering variable, but no basic (artificial) variable is ready to leave since min. ratio can not be found (*–ve* for both rows). We can not proceed further

Solution is Unbounded

Note : If the above problem is converted into canonical form as

Max. $Z = 4\,x_1 + 3\,x_2$

S.t. $-x_1 + 2\,x_2 \leq 10$

    $-x_1 + x_2 \leq 12$

    $x_1, x_2 \geq 0$

We can do it with out any surplus and artificial variable but simply by introducing slack variable. But there also we get same problem and the solution is not finite (unbounded). (students are advised to verify).

Multiple Optimal Solutions :

As studied in the earlier topics, the same condition if seen in simplex tableau, the solution is multiple optimal. (Refer Sec. 4.3.2)

We know that in an optimal solution, we have

1. $Z_j - C_j = 0$ for all basic variables.

2. $Z_j - C_j \geq 0$ for non-basic variables

In addition to the above two points, a multiple optimal solution is found if,

3. $Z_j - C_j = 0$ for at least one of the non-basic variables

4. Z_{opt} remains same for all solutions. Usually all the decision variables will not enter the basis but $Z_j - C_j \geq 0$ for all variables here.

ILLUSTRATION 15 ───

Minimise $Z = 20\,x_1 + 40\,x_2$

Subject to $x_1 + 2\,x_2 \geq 3$

      $x_1 + x_2 \leq 4$

      $0 \leq x_1 \leq \dfrac{5}{2}$

      $0 \leq x_2 \leq \dfrac{3}{2}$

Solution :

Conversion into equalities by introducing slack, surplus and artificial variables and rewriting the objective function :

Objective function Max. $Z = -20\,x_1 - 40\,x_2 - 0.S_1 + 0.S_2 + 0.S_3 + 0.S_4 - MA_1$

$$x_1 + 2\,x_2 - S_1 + A_1 = 3$$

$$x_1 + x_2 + S_2 = 4$$

$$x_1 + S_3 = \frac{5}{2}$$

$$x_2 + S_4 = \frac{3}{2}$$

$$x_1, x_2 \geq 0,\ S_1 \geq 0,\ S_2, S_3, S_4 \geq 0, A_1 \geq 0$$

IBFS : $x_1 = 0,\ x_2 = 0,\ S_1 = 0$ (non basic)

$$A_1 = 3,\ S_2 = 4,\ S_3 = \frac{5}{2} \text{ and } S_4 = \frac{3}{2} \text{ (basic)}$$

ITERATION TABLEAU - I :

C_B	BV	C_j / SV	-20 / x_1	-40 / x_2	0 / S_1	0 / S_2	0 / S_3	0 / S_4	$-M$ / A_1	Min. Ratio	Remarks
$-M$	A_1	3	1	2	-1	0	0	0	1	$\frac{3}{2}$	Key row
0	S_2	4	1	1	0	1	0	0	0	$\frac{4}{1}$	Tie
0	S_3	$\frac{5}{2}$	1	0	0	0	1	0	0	∞ neglect	
0	S_4	$\frac{3}{2}$	0	1	0	0	0	1	0	$\frac{3}{2}$	Key row
		Z_j	$-M$	$-2M$	M	0	0	0	$-M$		
		$Z_j - C_j$	$-M+20$	$-2M+40$	M	0	0	0	0		

Entering variable — x_2

Key element (2). Key column (x_2). Leaving variable.

There is a tie for key row in this problem, which is called 'DEGENERACY' in simplex tableau. In such cases, we choose arbitrarily. However, while deciding the leaving variable we give preference to artificial variable when tie exists between artificial, slack and decision variable. [We prefer slack if Tie exists between a slack and decision and we choose arbitrarily if there is tie between two artificial or two stacks or two decision variable]. (This is explained in next section clearly).

ITERATION TABLEAU 2 :

C_B	BV	C_j SV	-20 $\widehat{x_1}$	-40 x_2	0 S_1	0 S_2	0 S_3	0 S_4	Min. Ratio	Remarks
-40	x_2	$\dfrac{3}{2}$	$\dfrac{1}{2}$	1	$-\dfrac{1}{2}$	0	0	0	3	$R_1^N \to \dfrac{1}{2} R_1^0$
0	S_2	$\dfrac{5}{2}$	$\dfrac{1}{2}$	0	$+\dfrac{1}{2}$	1	0	0	5	$R_2^N \to R_2^0 - R_1^N$
0	$\widehat{S_3}$	$\dfrac{5}{2}$	1	0	0	0	1	0	$\dfrac{5}{2}$ Key row	$R_3^N \to R_3^0$
0	S_4	0	$-\dfrac{1}{2}$	0	$\dfrac{1}{2}$	0	0	1	$-ve$	$R_4^N \to R_4^0 - R_1^N$
		Z_j	-20	-40	$+20$	0	0	0		
Leaving variable		$Z_j - C_j$	0	0	20	0	0	0		

Entering variable

Key column (for verifying multiple optimal solution)

From the above table as $Z_j - C_j \geq 0$ for all the variables, the optimality is reached. Thus the solution values are

$$x_1 = 0, \quad x_2 = \frac{3}{2}$$

$$\text{and} \quad Z_{\max} = -20 \times 0 - 40 \times \frac{3}{2}$$

$$= -60$$

$$\text{or} \quad Z_{\min} = 60 \text{ units}$$

Check for Multiple Optimal Solution :

However, the above tableau indicates possibility of getting an alternate optimal solution. Since, the solution is obtained without replacement by all decision variables (i.e., x_1 has not entered basis), and $Z_j - C_j = 0$ for x_1 column, this indicates is a scope for multiple optimal solution. We find this by re-iterating Tableau - II by selecting x_1 column as key column and S_3 as leaving variable.

ITERATION TABLEAU 3 :

C_B	BV	C_j / SV	-20 / x_1	-40 / x_2	0 / S_1	0 / S_2	0 / S_3	0 / S_4	Min. Ratio	Remarks
-40	x_2	$\dfrac{1}{4}$	0	1	$-\dfrac{1}{2}$	0	$-\dfrac{1}{2}$	0		$R_1^N \to R_1^0 - \dfrac{1}{2} R_3^N$
0	S_2	$\dfrac{5}{4}$	0	0	$\dfrac{1}{2}$	1	$-\dfrac{1}{2}$	0		$R_2^N \to R_2^0 - \dfrac{1}{2} R_3^N$
-20	x_1	$\dfrac{5}{2}$	1	0	0	0	1	0		$R_3^N \to R_3^0$
0	S_4	$\dfrac{5}{4}$	0	0	$\dfrac{1}{2}$	0	$\dfrac{1}{2}$	1		$R_4^N \to R_4^0 + \dfrac{1}{2} R_3^N$
		Z_j	-20	-40	$+20$	0	0	0	$Z_{max} = -60$	
		$Z_j - C_j$	0	0	20	0	0	0	$Z_{min} = 60$	

From the above tableau - III, we get $Z_j - C_j \geq 0$ for all the variables, therefore the optimal solution is reached. The alternate optimal solution is

$$x_1 = \frac{5}{2} \qquad x_2 = \frac{1}{4}$$

$$Z_{max} = -20 \times \frac{5}{2} - 40 \times \frac{1}{4} = -50 - 10 = -60$$

$$\therefore \qquad Z_{min} = 60$$

4.4 *Degeneracy in Simplex Method*

While solving an LPP, the situations may arise (i) in which there is a tie between two or more basic variables for leaving the basis·i.e., equal minimum ratios or (ii) one or more basic variables in the "solution values" column become equal to zero. (Such problems are ignored intentionally in the problems dealt so far in this book, with an intention to explain here separately). These two cases are called Degeneracies in simplex problem.

Degeneracy may occur at the initial tableau or at any subsequent iterations.

A degeneracy is detected by the following points.

Case (i) : **Beale's Cycling :**

If one are more basic variables contain solution value as 'zero', then the iteration in simplex tableau repeat (cycle) indefinitely without arriving at an optimal solution.

Suppose a simplex tableau has x_1, x_2, S_1, S_2 and S_3 of which S_1, S_2 and S_3 are in basis. Now suppose x_1 replaces S_1 in first iteration, x_2 replaces x_1 in second iteration, S_1 replaces x_2 in third iteration and so on. This will continue $x_1 \to S_1$; $x_2 \to x_1$ and $S_1 \to x_2$ as cycle and can not yield any optimal solution.

Beale has first detected this type of problem and is explained in illustration 16.

Case (ii) : **Tie for Leaving Variable From the Basis :**

Another interesting case of degeneracy is found with same minimum ratio for two or more basic variables leaving the basis. This automatically raises the confusion in selection of key row. In such cases selection may be arbitrary.

This degeneracy can also be resolved by the same method as given for cycling problem. However, more simplified method to avoid more calculations and minimise the number of iterations to arrive the optimal solutions are given below. [One example can be observed in illustration - 15]

Resolution of Degeneracy : (for Cycling Problem)

The degeneracy, when occurs particularly due to cycling, it is resolved by the following rules.

Rule 1 : Divide the coefficients of slack variables (in the simplex tableau where degeneracy is detected) by corresponding positive numbers of the key column in the row, starting from left to right.

Rule 2 : The row which contains smallest ratio comparing from left to right column thus becomes the key row.

Resolution of Degeneracy : (In Case of Tie)

In general the selection of key row is made arbitrary if minimum ratio is same. (But this selection may increase number of iterations. Therefore use the following rules).

This are shown as tabular form given below.

S.No.	Tie for Leaving Variables	Preference to Select as Key Row with
1.	Between artificial & slack variable	Artificial variable
2.	Between artificial & decision variable	Artificial variable
3.	Between slack & decision variable	Slack variable
4.	Between surplus & decision variable	Surplus variable
5.	Between surplus & slack variable	Surplus variable
6.	Between artificial & artificial variable	Arbitrary variable
7.	Between slack & slack variable	Arbitrary variable
8.	Between decision & decision variables	Arbitrary variable

Thus order of preference is summarised as $A > S > X$ where A is for artificial, S for slack (or surplus in rare cases) and X for decision variables. (To remember note that they are in alphabetical order of preference). And arbitrary selection for like variables such as A & A, S & S, or X & X.

The above degeneracies can be observed in the following problems.

ILLUSTRATION - 16 :

> *Use degeneracy principles to solve the following LLP.*
>
> *Max.* $\quad Z = 3x_1 + 9x_2$
>
> *S.t.* $\qquad x_1 + 4x_2 \leq 8$
>
> $\qquad\qquad x_1 + 2x_2 \leq 4$
>
> $\qquad\qquad x_1, \, x_2 \geq 0$

Solution :

Adding slacks to convert into equalities, we get

Max. $\qquad Z = 3\,x_1 + 9\,x_2 + 0.S_1 + 0.S_2$

S.t. $\qquad\qquad x_1 + 4\,x_2 + S_1 \;= 8$

$\qquad\qquad\qquad x_1 + 2\,x_2 + S_2 \;= 4$

$\qquad\quad x_1, x_2 \geq 0, \; S_1, S_2 \;\geq 0$

IBFS : $\qquad S_1 = 8, \; S_2 = 4 \;$ (basic)

$\qquad\qquad x_1 = 0, \; x_2 = 0 \;$ (non-basic)

ITERATION TABLEAU - I :

C_B	BV	C_j / SV	3 / x_1	9 / x_2	0 / S_1	0 / S_2	Min.. Ratio $\left(\dfrac{SV}{KCV}\right)$	Remarks
0	S_1	8	1	4	1	0	$\dfrac{8}{4} = 2$	Tie for out going variable
0	S_2	4	1	2	0	1	$\dfrac{4}{2} = 2$	
		Z_j	0	0	0	0		
Tie in out going		$Z_j - C_j$	-3	-9	0	0		

Incoming variable — x_2

Key column

As there is a tie for out going variable, a degeneracy is indicated. Therefore we use degeneracy rules by rearranging the tableau - 1.

(S_1 and S_2 are brought to left and x_1 & x_2 to right in body matrix)

REARRANGE TABLEAU · I :

Entering variable

C_B	BV	C_j / SV	0 / S_1	0 / S_2	3 / x_1	9 / x_2	Min.. Ratio $\left(\dfrac{FCV}{KCV}\right)$	Remarks
0	S_1	8	1	0	1	4	$\dfrac{4}{1}=4$	
0	S_2	4	0	1	1	2	$\dfrac{0}{2}=0$ ← Key row	
		Z_j	0	0	0	0		
Leaving variable		Z_j-C_j	0	0	-3	-9		

Key column

Now applying rules of degeneracy, minimum ratio is found by dividing first column value (FCV i.e., S_1) by key column value (KCV i.e., x_2), we get among $\left(\dfrac{4}{1},\dfrac{0}{2}\right)$ least is 0, therefore S_2 is leaving variable. Further iteration is as usual

◢ ITERATION TABLEAU · II :

C_B	BV	C_j / SV	0 / S_1	0 / S_2	3 / x_1	9 / x_2	Min.. Ratio	Remarks
0	S_1	0	1	-2	-1	0		$R_1^N \to R_1^0 - 4\,R_2^N$
9	x_2	2	0	$\dfrac{1}{2}$	$\dfrac{1}{2}$	1		$R_2^N \to \dfrac{1}{2}\,R_2^0$
		Z_j	0	$\dfrac{9}{2}$	$\dfrac{9}{2}$	9		
		Z_j-C_j	0	$\dfrac{9}{2}$	$+\dfrac{3}{2}$	0		

From the above tableau - 2 ,

Since $Z_j - C_j \geq 0$ for all the variables, we have the optimal solution as follows.

$$x_1 = 0, \quad x_2 = 2$$

$$Z_{\max} = 3 \times 0 + 9 \times 2 = 18$$

ILLUSTRATION 17

A firm manufacturers two products in three departments. Product 'A' contributes Rs. 5 per unit and requires 5 hrs. in dept. M, 5 hrs. in dept. N and one hour in dept. P. product 'B' contributes Rs. 10 per unit requires 8 hrs. in dept. M, 3 hrs in dept. N and 8 hrs in dept. P. Capacities for departments M,N,P are 48 hours per week. Find out optimal product mix using Simplex model. Give dual formulation also.

[JNTU-CSE-98]

Solution :

(Refer illustration 3 of chapter 2 for formulation)

Summary of Formulation :

$$\text{Max. } Z = 5\,X_1 + 10\,X_2$$

$$\text{Subject to} \quad 5\,X_1 + 8\,X_2 \leq 48$$

$$5\,X_1 + 3\,X_2 \leq 48$$

$$X_1 + 8\,X_2 \leq 48$$

$$X_1, X_2 \geq 0$$

Conversions to Equations by Adding Slack Variables :

$$5\,X_1 + 8\,X_2 + S_1 = 48$$

$$5\,X_1 + 3\,X_2 + S_2 = 48$$

$$X_1 + 8\,X_2 + S_3 = 48$$

$$X_1, X_2 \geq 0 ; \quad S_1, S_2, S_3 \geq 0$$

Initial Solution :

Non basic variables : ($X_1 = 0$, $X_2 = 0$)

Basic variables : $S_1 = 48$, $S_2 = 48$, $S_3 = 48$

ITERATION TABLEAU - I :

Entering variable

C_B	BV	C_j / S.V	5 / x_1	10 / x_2	0 / S_1	0 / S_2	0 / S_3	Min.. Ratio	Remarks
0	S_1	48	5	8	1	0	0	$\frac{48}{8}$	Key row
0	S_2	48	5	3	0	1	0	$\frac{48}{3}$	Tie
0	S_3	48	1	8	0	0	1	$\frac{48}{8}$	
		Z_j	0	0	0	0	0		
Leaving variable (Selected arbitrarily)		$Z_j - C_j$	– 5	– 10	0	0	0		

Key column

[The above table has same min. ratio for two rows i.e., degeneracy. As both variables are slack, we can select arbitrarily or otherwise use degeneracy principle as min. ratio of $\frac{8}{1}, \frac{3}{0}, \frac{8}{0}$ i.e., $\frac{8}{1}$, hence choose first row].

ITERATION TABLEAU - II :

C_B	BV	C_j / S.V	5 / x_1	10 / x_2	0 / S_1	0 / S_2	0 / S_3	Min.. Ratio	Remarks
10	x_2	6	$\frac{5}{8}$	1	$\frac{1}{8}$	0	0		$R_1^N \to R_1^0/8$
0	S_2	30	$\frac{25}{8}$	0	$-\frac{3}{8}$	1	0		$R_2^N \to R_2^0 - 3\,R_1^N$
0	S_3	0	-1	0	-1	0	1		$R_3^N \to R_3^0 - 8\,R_1^N$
		Z_j	$\frac{50}{8}$	10	$\frac{10}{8}$	0	0		
		$Z_j - C_j$	$\frac{10}{8}$	0	$\frac{10}{8}$	0	0		

As all the values of $Z_j - C_j$ are positive the optimal solution is obtained as $X_1 = 0$, $X_2 = 6$; $S_1 = 0$, $S_2 = 30$, $S_3 = 0$ and

$$Z = 5(0) + 10\,(6) + 0 \times 0 + 0 \times 30 + 0 \times 0 = Rs.\ 60$$

The Dual Formulation for the Above Problem : (This is explained in next section 4.6)

Objective function is

$$\text{Min.}\ Z = 48\,W_1 + 48\,W_2 + 48\,W_3$$

Subject to $5\,W_1 + 5\,W_2 + W_3 \geq 5$

$$8\,W_1 + 3\,W_2 + 8\,W_3 \geq 10$$

$$W_1,\ W_2 \text{ and } W_3 \geq 0.$$

4.4.1 Tie for entering Variable (i.e., Key Column)

In simplex method, some times tie may occur with same most negative value of $Z_j - C_j$ for two or more columns. This tie for key column or entering variable selection can be made arbitrary. However, the number of iterations may be reduced by selecting according to the order of preference as decision variable, slack variable/surplus variable.

..e. • If tie is between decision and surplus / slack variable prefer decision variable.

- If tie is in between two decision variables or two slack/surplus variables choose arbitrarily.

- There will be no chance of getting a tie with artificial variable since they will be selected in basis of initial tableau and when they leave basis, these will vanish from the tableau.

?.g. : Refer to Game theory - Linear Programming method of Solving Rectangular games.

4.4.2 *Unrestricted Variables in LPP*

Usually an LPP is assumed to have non-negative condition of variables. But in practical situations it may contain one or more variable with no restrictions i.e., the variable may be positive or negative or zero. Such variables are known as 'unrestricted variables'.

Since the use of simplex method necessarily requires the conditions of variables to be non-negative at each iteration, we convert the unrestricted variable into a restricted non-negative variable by splitting into two parts one as positive part and the second as negative part. The positive part is any way non-negative while – negativeof negative part may be taken as again positive i.e., non-negative condition. Summarily, the unrestricted variable is written as difference of two non-negative variables.

If x_1 is unrestricted, let $x_1 = x_1' - x_1''$ where x_1' and $x_1'' \geq 0$,

4.4.3 *Equality Constraints in LPP (or Mixed Constraints)*

We have learnt that an LPP, the constraints may be either exact (equality) or availability (less than or equal to) or requirement (greater than or equal to) type.

Since the simplex method demands the constraints to be in equality form for the computation, use the following rules to make the inequality constraints into equality type.

1. When the constraint is 'less than or equal to' ($\leq$) type, add 'SLACK' variable.

2. When the constraint is grater than or equal to type ($\geq$), we subtracted 'SURPLUS' variable. But to satisfy the non-negative condition of surplus variable we introduce an 'ARTIFICIAL' variable

3. When the constraint is exactly 'equal to' type, it becomes necessary to add·an 'ARTIFICIAL' variable since the exact constraint becomes inconsistent if all

the decision variables of the constraint assume 'zero' values or non-basic (If at least one of the variables of exact constraint is chosen in basis with coefficient as unity i.e., unit vector we need not, add an artificial variable. But it is always safe to add artificial variable for equality constraint).

Summarising we get

Constraint Type	Variables Used
$\leq$	Add slack ($+ S$)
$\geq$	Subtract surplus and add artificial ($- S + A$)
$=$	Add artificial ($+ A$)

The points dealt in the above two topics are illustrated with a numerical example given below.

ILLUSTRATION 18

$$Minimise \quad Z = 2x_1 + x_2$$
$$S.t. \quad 3x_1 + x_2 = 3$$
$$4x_1 + 3x_2 \geq 6$$
$$x_1 + 2x_2 \leq 4$$
$$x_1 \geq 0, \quad x_2 \quad unrestricted.$$

Solution :

The above problem is converted into equalities by adding artificial, surplus and slack variables. Also, as x_2 is unrestricted, it is re-written as difference of two variables x_2' and x_2'' such that $x_2 = x_2' - x_2''$ where $x_2' \geq 0$ and $x_2'' \geq 0$

$$3x_1 + x_2' - x_2'' + A_1 = 3$$

$$4x_1 + 3x_2' - 3x_2'' - S_2 + A_2 = 6$$

$$x_1 + 2x_2' - 2x_2'' + S_3 = 4$$

Revised Objective Function :

$$Z = -2x_1 - x_2' + x_2' - 0.S_2 + 0.S_3 - MA_1 - MA_2$$

where $\quad x_1 \geq 0, \ x_2' \geq 0, \ x_2'' \geq 0, \ S_2 \geq 0, \ S_3 \geq 0$

IBFS : $\quad x_1 = 0, \ x_2' = 0, \ x_2'' = 0, \ S_2 = 0 \quad$ (Non basic)

$$A_1 = 3, \ A_2 = 6, \ S_3 = 4 \qquad \text{(Basic)}$$

ITERATION TABLEAU - 1 :

Incoming variable

C_B	BV	C_j / SV	-2 / x_1	-1 / x_2'	$+1$ / x_2''	0 / S_2	0 / S_3	$-M$ / A_1	A_2	Min. Ratio	Remarks
$-M$	A_1	3	3	1	-1	0	0	1	0	$\frac{3}{3}$	
$-M$	A_2	6	4	3	-3	-1	0	0	1	$\frac{6}{4}$	
0	S_3	4	1	2	-2	0	1	0	0	$\frac{4}{1}$	
Out going variable		Z_j	$-7M$	$-4M$	$4M$	M	0	$-M$	$-M$		A_1 is deleted for next iteration
		$Z_j - C_j$	$-7M+2$	$-4M+1$	$4M-1$	M					

Key column

ITERATION TABLEAU - 2 :

Incoming variable

C_B	BV	C_j / SV	-2 / x_1	-1 / x_2'	$+1$ / x_2''	0 / S_2	0 / S_3	$-M$ / A_2	Min. Ratio	Remarks
-2	x_1	1	1	$\frac{1}{3}$	$-\frac{1}{3}$	0	0	0	3	$R_1^N \to \frac{1}{3} R_1^0$
$-M$	A_2	2	0	$\frac{5}{3}$	$-\frac{5}{3}$	-1	0	1	$\frac{6}{5}$	$R_2^N \to R_2^0 - 4 R_1^N$
0	S_3	3	0	$\frac{5}{3}$	$-\frac{5}{3}$	0	1	0	$\frac{9}{5}$	$R_3^N \to R_3^0 - R_1^N$
Out going variable		Z_j	0	$-\frac{5M}{3}-\frac{2}{3}$	$\frac{5M}{3}+\frac{2}{3}$	M	0	$-M$		A_2 column is deleted for next iteration
		$Z_j - C_j$	0	$\frac{-5M+1}{3}$	$\frac{+5M-1}{3}$	M	0	0		

Key column

ITERATION TABLEAU - 3 :

C_B	BV	C_j / SV	-2 / x_1	-1 / x_2'	1 / x_2''	0 / S_2	0 / S_3	Min. Ratio	Remarks
-2	x_1	$\frac{3}{5}$	1	0	0	$\frac{1}{5}$	0		$R_1^N \to R_1^0 - \frac{1}{3} R_2^N$
-1	x_2	$\frac{6}{5}$	0	1	-1	$-\frac{3}{5}$	0		$R_2^N \to \frac{3}{5} R_2^0$
0	S_3	1	0	0	0	1	1		$R_3^N \to R_3^0 - \frac{5}{3} R_2^N$
		Z_j	-2	-1	1	$\frac{1}{5}$	0		$Z_j - C_j \geq 0$ for all variables
		$Z_j - C_j$	0	0	0	$\frac{1}{5}$	0		$\therefore Z_{max} = -12/5$ $Z_{min} = 12/5$

Since $Z_j - C_j \geq$ for all variables in the above (final) simplex tableau, the optimal solution is obtained as follows.

$$x_1 = \frac{3}{5} \ x_2' = \frac{6}{5} \ x_2'' = 0$$

Max. $\quad Z = -2\left(\frac{3}{5}\right) - \frac{6}{5} = -\frac{12}{5} \ $ or $ \ $ Min. $Z = \frac{12}{5}$

Note : *In the above result it can be further understood that x_2 always takes a positive value though it is unrestricted since x_2'' takes zero value.*

4.5 *Concept of Duality in Simplex*

One of the most important discoveries in linear programming revealed the presence of duality that disclosed the fact that an LPP has associated with another LPP. Thus for the same data we can express two LPP problems, one of which is called 'primal' and the other 'dual'. The relationship between primal and dual is very intimate and useful that eased in obtaining the solution for the LPP in other way when one is difficult.

The above concept is explained through an illustration given below.

ILLUSTRATION 19 —————————————————————————————

Prasanna Industries Ltd. manufactures two types of chairs under the trade names mayurasan and Sinhasan. Eeach Mayurasan contributes a profit of Rs. 40 while that of Singasan is 50. For manufacturing each Mayurasan, it requires 2 units of raw material and 8 hrs. of labour while each Sinhasan requires 3 units of raw materials and 4 hrs. of labour. The firm has 30 units of raw material and 45 hrs. of labour available per day. Formulate t he information as an LPP as primal as well as dial.

Solution :

The above information can be formulated in two ways (say Primal and Dual).

Case (i) : To maximise the profits on the chairs subject to the availability constraints of the raw material and labour provided the decision is to be taken on the number of chairs of each type to be produced (decision variables). Obviously the variables can not assume negative values (production units can not be negative) and hence we use non-negative condition to the decision variable. Let us say this problem as '*primal*'

Case (ii) : To minimise the utility units or cost (that automatically raises profits)* without compromising with the minimum profit. Thus the decision variables in this case will be number of units of raw material to be used and number of hrs. of labour to the used so that at least the desired profit is obtained. Here also the variable are non-negative. [Variables need not be non-negative always in all cases. This depends on the constraints of its primal problem and vice-versa]. This problem is called '*dual*' to the above primal [we can take either way i.e., if case (ii) is considered as primal then case (i) becomes its dual].

Minimising cost is the other way of maximising, the profits. For example : A product can be produced in 8 hours that gives a profit of Rs. 100. Now, if you can make two jobs in 8 hrs. we get Rs. 200 profit and on the other hand if a job can be completed in 4 hrs. (which gives profit of Rs. 100), it leaves use productive 4 hrs. worth of producing another Rs. 100.

Let us now formulate the above problems simultaneously in two ways.

Case (i) - Primal	Case - (ii) - Dual
Step - 1 : *Selection of Variables :*	
No. of units of Mayurasan to be produced $= x_1$ No. of units of Sinhasan to be produced $= x_2$	Cost of each unit of raw material to manufacture $= w_1$ Cost of each unit of hrs. of labour to be utilized $= w_2$
Step 2 : *Setting Objective Function*	
Profit on each unit Mayurasan = Rs. 40	Cost on each unit of R/M $= w_1$
No. of units of Mayurasan produced $= x_1$	No. of units of R/M available = 30
Profit on x_1 units of Mayurasan $= 40x_1$	Cost on 30 units of R/M $= 30w_1$
Profit on each unit of Sinhasan = Rs. 50	Cost on each unit of labour $= w_2$
No. of units of Sinhasan produced $= x_2$	The labour hours available = 45
Profit on x_2 units of Sinhasan $= 50x_2$	Cost on labour $= 45w_2$
Total profit on both chairs $= 40x_1 + 50x_2$	Total cost $= 30w_1 + 45w_2$
This profit is to be maximised	This cost is to be minimised.
$\therefore$ Obecjtive function is to $\qquad$ maximise $Z_p = 40x_1 + 50x_2$	$\therefore$ Objective function is to $\qquad$ minimise $Z_d = 30w_1 + 45w_2$
Step 3 : *Identification of Constraint Set*	
(i) *Constraint on availability raw material*	(i) *Constraint on required profits on Mayurasan*
$\qquad$ Each Mayurasan requires 2 units of R/M	$\qquad$ R/M required per unit = 2 units
$\qquad$ x_1 units of Mayurasan require $2x_1$ units of of R/M	$\qquad$ Reqd. profit on R/M per unit Mayurasan $= 2w_1$
$\qquad$ Each Sinhasan requires 3 units of R/M	$\qquad$ Labour required per unit = 8 hrs.
$\qquad$ x_2 units of Sinhasan require $3x_2$ units of R/M	$\qquad$ Reqd. profit on labour per unit Mayurasan $= 8w_2$
$\qquad$ Total R/M required $= 2x_1 + 3x_2$	$\qquad$ Total reqd profit/unit Mayurasan $= 2w_1 + 8w_2$
$\qquad$ This can not exceed 30 available units	$\qquad$ This reqd profit should produce not less than 40
$\qquad$ $\therefore$ $2x_1 + 3x_2 \le 30$	$\qquad$ $\therefore$ $2w_1 + 8w_2 \ge 40$
$\qquad$ (Constraints on availability of R/M)	$\qquad$ (Constraint min. profit requirement on Mayurasan)
(ii) Constraint on Labour	(ii) Constraint on per unit profit of Sinhasan
$\qquad$ In similar way we can obtain the constraint as	$\qquad$ Similarly we get
$\qquad$ $8x_1 + 4x_2 \le 45$	$\qquad$ $3w_1 + 4w_2 \ge 50$
$\qquad$ (constraint on availability of labour)	$\qquad$ (Constraint min. reqd. profit per unit of Sinhasan)
Step - 4 : *Condition of Variables*	
x_1 and x_2 (the no. of units to be produced) can not be negative. $\qquad$ $\therefore$ $x_1 \ge 0, x_2 \ge 0$	w_1, w_2 (costs of production) are not negative in this case $\qquad$ $\therefore$ $w_1 \ge 0, w_2 \ge 0$

Summary :

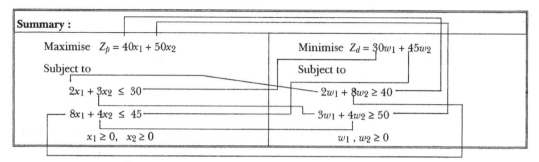

Maximise $Z_p = 40x_1 + 50x_2$ Minimise $Z_d = 30w_1 + 45w_2$

Subject to   Subject to

$2x_1 + 3x_2 \leq 30$ $2w_1 + 8w_2 \geq 40$

$8x_1 + 4x_2 \leq 45$ $3w_1 + 4w_2 \geq 50$

$x_1 \geq 0, \quad x_2 \geq 0$ $w_1, w_2 \geq 0$

Comparison of Solution to the Primal and its Dual :

ILLUSTRATION 20

Consider the following pair of dual problems: Compare their solutions by simplex method.

	Primal		**Dual**
Max.	$Z_x = 40x + 50x_2$	Max.	$Z_w = 30w_1 + 45w_2$
S.t.	$2x_1 + 3x_2 \leq 30$	S.t.	$2w_1 + 8w_2 \geq 40$
	$8x_1 + 4x_2 \leq 45$		$3w_1 + 4w_2 \geq 50$
	$x_1, x_2 \geq 0$		$w_1, w_2 \geq 0$

Solution

Primal	**Dual**
Rewriting primal problem in std. form	Rewriting the dual problem in std. form

Primal

Max. $Z_x = 40x_1 + 50x_2 + 0S_1 + 0S_2$

S.t. $2x_1 + 3x_2 + S_1 = 30$

$8x_1 + 4x_2 + S_2 = 45$

$x_1, x_2, S_1, S_2 \geq 0$

B.V. : S_1, S_2
NBV $= x_1, x_2$

Dual

Max. $Z = -30w_1 - 45w_2 - MA_1 - MA_2 - 0S_1 - 0S_2$

$2w_1 + 8w_2 - S_1 + A_1 = 40$

$3w_1 + 4w_2 - S_2 + A_2 = 50$

$w_1, w_2, S_1, S_2, A_1, A_2 \geq 0$

B.V. $= A_1$ & A_2
NBV $= x_1, x_2$
S_1 and S_2

Primal table

C_B	B.V	$C_j \to$ SV	40 x_1	50 x_2	0 S_1	0 S_2	Min. Ratio
0	(S_1)	30	2	$\boxed{3}$	1	0	$\dfrac{30}{3} \leftarrow$
0	S_2	45	8	4	0	1	$\dfrac{45}{4}$
	$Z_j - C_j$		-40 $\uparrow$	-50	0	0	M/R
50	x_2	10	$\dfrac{2}{3}$	1	$\dfrac{1}{3}$	0	$\dfrac{30}{2}$
0	S_2	5	$\boxed{\dfrac{16}{3}}$	0	$-\dfrac{4}{3}$	1	$\dfrac{15}{16} \leftarrow$
	$Z_j - C_j$		$-\dfrac{20}{3}$ $\uparrow$	0	$\dfrac{50}{3}$	0	M/R

Dual table

C_B	B.V	$C_j \to$ SV	-30 w_1	-45 w_2	0 S_1	0 S_2	$-M$ A_1	$-M$ A_2	Min. Ratio
$-M$	(A_1)	40	2	$\boxed{8}$	-1	0	1	0	$\dfrac{40}{8} \leftarrow$
$-M$	A_2	50	3	4	0	-1	0	1	$\dfrac{50}{4}$
	$Z_j - C_j$		$+30$ $-5M$ $\uparrow$	$+45$ $-12M$	M	M	0	0	M/R
-45	w_2	5	$\dfrac{2}{8}$	1	$-\dfrac{1}{8}$	0	X	0	$\dfrac{40}{2}$
$-M$	A_2	30	$\boxed{2}$	0	$+\dfrac{1}{2}$	-1	X	1	$\dfrac{30}{2} \leftarrow$
	$Z_j - C_j$		$\dfrac{150}{8}$ $-2M$ $\uparrow$	0	$\dfrac{45}{8}$ $-M/2$	M	X	0	M/R

						Inversed							
50	x_2	$\left(\dfrac{75}{8}\right)$	0	1	$\dfrac{1}{2}$	$-\dfrac{1}{8}$		-45	w_2	$\left\langle\dfrac{5}{4}\right\rangle$	0	1	$-\dfrac{3}{16}$	$\dfrac{1}{8}$	X	X
40	x_1	$\left(\dfrac{15}{16}\right)$	1	0	$-\dfrac{1}{4}$	$\dfrac{3}{16}$		-30	w_1	$\langle 15\rangle$	1	0	$\dfrac{1}{4}$	$-\dfrac{1}{2}$	X	X
	$Z_j - C_j$		0	0	$\langle 15\rangle$	$\left\langle\dfrac{5}{4}\right\rangle$	$\begin{array}{c}Z_{max}\\=\dfrac{2025}{4}\end{array}$		$Z_j - C_j$		0	0	$\left(\dfrac{15}{16}\right)$	$\left(\dfrac{75}{8}\right)$	X	X	$\begin{array}{c}Z_{min}\\=-\dfrac{18225}{4}\end{array}$

Since all values of $Z_j - C_j$ are positive, optimal solution of primal is obtained as $x_1 = 15/16$, $x_2 = 75/8$ and shadow prices are 15 and 5/4 $Z_{max} = 40 \times 15/16 + 50 \times 75/8 = \dfrac{2025}{4}$	Since, all $Z_j - C_j$ values are positive, optimal solution for dual is reached and the solution is $w_1 = 15$, $w_2 = 5/4$; and shadow prices are 15/16, 75/8 $\quad Z_{max} = -30 \times 15 - 45 \times 5/4 = -\dfrac{18225}{4}$ $\therefore\ Z_{min} = \dfrac{18225}{4}$

Remarks : From above comparison, we can conclude that the solution to a primal problem of linear programming can always, provide a solution to its dual, check the relationship described by arrows carefully to understand to comparison.

4.5.1 *Conversion of Primal to Dual or Dual to Primal*

The following steps (rules) will enable the conversion of any primal problem into its dual and vice-versa.

Step 1 :   (a) If in the given problem, objective function is in maximisation form proceed to step 2.

(b) Convert the objective function of the given problem to maximisation form if it is minimisation form. This conversion can be done by multiplying with -1

Ex : Min. $Z = x_1 + 9x_2 + x_3$ converts to Max. $Z = -2x_1 - 9x_2 - x_3$

Step 2 :   (a) If all constraints have "less than or equal to" ($\leq$) sign, go to step - 3

(b) If a constraint has "greater than or equal to" ($\geq$) sign, convert it to "less than or equal to "

Ex : (i) $x_1 + 4x_2 + 2x_3 \geq 5$ is rewritten as $- x - 4x_2 - 2x_3 \leq - 5$

(ii) $3x_1 + x_2 - 2 x_3 \geq 4$ is rewritten as $- 3x_1 - x_2 + 2x_3 \leq - 4$

(c) If a constraint has an "equality" sign (=), split this into two constraints in two opposite inequalities.

Ex : $2x_1 + 3 x_2 = 5$ is split into $2 x_1 + 3 x_2 \leq 5$ and $2 x_1 + 3 x_2 \geq 5$

Then these are re-written using step - 2 (b)
$$2 x_1 + 3 x_2 \leq 5$$
$$\text{and} - 2 x_1 - 3 x_2 \leq - 5$$

Step 3 :   (a) If all the variables are non-negative i.e., $x_j \geq 0$, then proceed to step 4

(b) If any variable is unrestricted replace it by difference of two non-negative variables (These variables are to be replaced in the entire problem)

Ex :   If x_j is unrestricted, change it as $x_j' \geq 0$, $x_j'' \geq 0$ where $x_j = x_j' - x_j''$

Step 4 : Now we have standard problem in primal. To write its dual we use the following principles.

(a) Take a different set of decision variables. The number of variables in dual will be equal to the number of equations in primal (and vice versa).

(b) We have to minimise the objective function in dual (primal in maximisation from)

(c) The objective function is written with the constants of inequality constraints as coefficients of decision variables.

(d) Consider a matrix of coefficients of constraint inequations of primal. Transpose the rows and columns of coefficients matrix of primal to write constraint inequations of dual.

(e) Write $\geq$ sign inequality for dual constraints.

(f) Constants of inequality of dual are drawn from coefficients of variables in objective function of primal.

Step 5 : Conditions of variables are unaltered i.e.,non negative

Step 6 : Rearrange the dual formulation if required. For example if a constraint has negative sign on its right side (i.e., for constant), multiply entire inequality with (−1) and convert the constraint into '$\leq$' type. Similarly, re-arrange the conditions of variables if required.

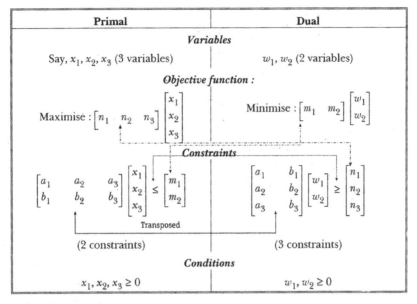

ILLUSTRATION 21 ————————————————————————

Write the dual of the following primal LPP.

Maximise $Z_p = 50x_1 + 60x_2$

Subject to $2x_1 + x_2 \leq 300$

$3x_1 + 4x_2 \leq 509$

$4x_1 + 7x_2 \leq 812$

$x_1, x_2 \geq 0$

Solution :

Step 1 : Objective function is max. form therefore proceed to step - 2

Step 2 : All constraints are $\leq$ type so, proceed to step - 3.

Step 3 : Condition of variables are all non-negative, hence go to step 4

Step 4 :   (a) Number of variables should be 'three' since there are three constraints in primal (say w_1, w_2, w_3)

   (b) Objective function is to be minimised (since it is in max. form in primal)

   (c) Coeff. of variables in dual objective function are taken from constant of primal constraints viz. 300, 509 and 812

   $\therefore$ objective function is Min. $Z_D = 300\, w_1 + 509\, w_2 + 812\, w_3$

   (d) Coefficients of constraints set in primal are transposed in dual constraint set.

   Thus transpose of $\begin{bmatrix} 2 & 1 \\ 3 & 4 \\ 4 & 7 \end{bmatrix}$ is $\begin{bmatrix} 2 & 3 & 4 \\ 1 & 4 & 7 \end{bmatrix}$

(e) Constraints in dual take '$\geq$' form since primal has all '$\leq$' constraints

(f) Dual constraints are taken from coefficients of objective function of primal, viz. 50 and 60

∴ constraint set is

$$2\,w_1 + 3\,w_2 + 4\,w_3 \geq 50$$

$$w_1 + 4\,w_2 + 7\,w_3 \geq 60$$

Step 5 : All the dual variables take non-negative condition.

i.e., $w_1 \geq 0$, $w_2 \geq 0$ and $w_3 \geq 0$

Summary :

Primal	*Dual*
Max $Z_P = 50\,x_1 + 60\,x_2$	Min. $Z_D = 300\,w_1 + 509\,w_2 + 812\,w_3$
S.t. $2\,x_1 + x_2 \leq 300$	S.t. $2\,w_1 + 3\,w_2 + 4\,w_3 \geq 50$
$3\,x_1 + 4\,x_2 \leq 509$	$w_1 + 4\,w_2 + 7\,w_3 \geq 60$
$4\,x_1 + 7\,x_2 \leq 812$	$w_1,\ w_2,\ w_3 \geq 0$
$x_1, x_2 \geq 0$	

ILLUSTRATION 22 ──────────────────────────────────

> *Write dual for*
> *Minimise $Z = 2x_1 + 9x_2 + x_3$*
> *S.t.* $x_1 + 4x_1 + 2x_3 \geq 5$
> $3x_1 + x_2 + 2x_3 \geq 4$
> $x_1, x_2, x_3 \geq 0$ **[JNTU (Mech) 2001]**

Solution :

Step 1 : Convert Minimising objective function to maximasation form

i.e., Max $Z_p = -2x_1 - 9x_2 - x_3$

Step 2 : Convert '$\geq$' inequality to $\leq$ type

$$-x_1 - 4\,x_2 - 2\,x_3 \leq -5 \cdot$$

$$-3x_1 - x_2 - 2\,x_3 \leq -4$$

Step 3 : All variables are non-negative

$$x_1, x_2, x_3 \geq 0$$

Step 4 : For dual,

(a) Variables must be two only (since primal has two constraints) say w_1, w_2

(b) Objective function is to minimise

(c) Coefficients are constants of constraints in primal

i.e., -5 and -4

$\therefore$ objective function is   Min. $Z_D = -5\,w_1 - 4\,w_2$

or   Max. $Z_D = 5\,w_1 + 4\,w_2$

(d) Coefficients of primal constrains are transposed.

i.e., $\begin{bmatrix} -1 & -4 & -2 \\ -3 & -1 & -2 \end{bmatrix}$ is transposed as $\begin{bmatrix} -1 & -3 \\ -4 & -1 \\ -2 & -2 \end{bmatrix}$

(e) All take $\geq$ form

(f) Constraints are taken from coefficient of variables in objective function
i.e., -2, -9 and -1

$\therefore$ $-3\,w_1 - 3\,w_2 \geq -2$ or $w_1 + 3\,w_2 \leq 2$

$-4\,w_1 - w_2 \geq -9$ $4\,w_1 + w_2 \leq 9$

$-2\,w_1 - 2\,w_2 \geq -1$ $2\,w_1 + 2\,w_2 \leq 1$

(g) Both variables take non negative conditions

i.e., $w_1,\ w_2 \geq 0$

Summary :

Primal	Standard Primal	Dual to Standard Primal	Dual (Re-arranged)
Min. $Z_P = 2x_1 + 9x_3 + x_3$	Max. $Z_P = -2x_1 - 9x_2 - x_3$	Min. $Z_D = -5\,w_1 - 4w_2$	Max. $Z_D = 5w_1 + 4w_2$
S.t. $x_1 + 4x_2 + 2x_3 \geq 5$	S.t. $-x_1 - 4x_2 - 2x_3 \leq -5$	S.t. $-w_1 - 3w_2 \geq -2$	S.t. $w_1 + 3w_2 \leq 2$
$3x_1 + x_2 + 2x_3 \geq 4$	$-3x_1 - x_2 - 2x_3 \leq -4$	$-4w_1 - w_2 \geq -9$	$4w_1 + w_2 \leq 9$
$x_1, x_2, x_3 \geq 0$	$x_1, x_2, x_3 \geq 0$	$-2\,w_1 - 2w_2 \geq -1$	$2w_1 + 2w_2 \leq 1$
		$w_1, w_2 \geq 0$	$w_1, w_2 \geq 0$

ILLUSTRATION 23 ——————————————————————

Write the dual of the following primal
   Min. $Z = 4x_1 + 3x_2$

   S.t. $3x_1 + x_2 = 3$

   $3x_1 + 4x_2 \geq 6$

   $x_1 + x_2 \leq 4$

   $x_1 + x_2 \geq 0$ *[JNTU - CSE - 1997]*

Solution :

   Step 1 : Converting objective function min. to max. form

   Max. $Z_P = -4x_1 - 3x_2$

Step 2 : Written all constraints in '≤' form – $3x_1 - x_2 \leq -3$; $3x_1 + x_2 \leq 3$

$$-3x_1 - 4x_2 \leq -6; \quad x_1 + x_2 \leq 4$$

Step 3 : Condition of all variables are in non-negative form, hence proceed

Step 4 : Now, for dual we have,

   (a) 4 variables say w_1', w_1'', w_2 and w_3 (since there are 4 constraint in primal ref. step 2)

   (b) Objective function is to minimise, since primal is to maximse (ref. step -1)

   (c) Coefficient of obj. function of dual are $-3, +3, -6$ and 4 (the constants in constraints of primal)

∴ **Objective function** is

   Min. $Z_D = -3w_1' + 3w_1'' - 6w_2 + 4w_3$

   Min. $Z_D = -3(w_1' - w_1'') - 6w_2 + 4w_3$

or   Max. $Z_D = +3(w_1' - w_1'') + 6w_2 - 4w_3$

   (d) Constraint set transposing

$$\begin{bmatrix} -3 & -1 \\ +3 & +1 \\ -3 & -4 \\ 1 & 1 \end{bmatrix} \quad \text{to} \quad \begin{bmatrix} -3 & +3 & -3 & 1 \\ -1 & +1 & -4 & 1 \end{bmatrix}$$

   (e) Constants are -4 and -3 (Since these are coefficients in primal objective function)

   ∴ **Constraint set** is

$$-3w_1' + 3w_1'' - 3w_2 + w_3 \geq -4$$

   i.e., $-3(w_1' - w_1'') - 3w_2 + w_3 \geq -4$

   and $-w_1' + w_1'' - 4w_2 + w_3 \geq -3$

   i.e., $-(w_1' - w_1'') - 4w_2 + w_3 \geq -3$

   or $3(w_1' - w_1'') + 3w_2 - w_3 \leq 4$

           $(w_1' - w_1'') + 4w_2 - w_3 \leq 3$

   (f) All variables are non-negative

   ∴ $w_1', w_1'', w_2, w_3 \geq 0$

   Assuming $w_1' - w_1'' = w_1$ and hence w_1 unrestricted

   We get the dual as

   Max.. $Z_D = 3w_1 + 6w_2 - 4w_3$

   S.t. $3w_1 + 3w_2 - w_3 \leq 4$

          $w_1 + 4w_2 - w_3 \leq 3$

          w_1 unrestricted, $w_2, w_3 \geq 0$

Summary :

Primal (x_1, x_2)	Standard form of Primal (x_1, x_2)	Dual (w_1', w_1'', w_2, w_3)	Dual re-arranged (w_1, w_2, w_3)
Min. $Z_P = 4x_1 + 3x_2$	Max. $Z_P = -4x_1 - 3x_2$	Max. $Z_D = -3w_1' + 3w_1''$ $-6w_2 + 4w_3$	Max. $Z_D = 3w_1 + 6w_2 - 4w_3$ or Min. $Z_D = -3w_1 - 6w_2 + 4w_3$
S.t. $3x_1 + x_2 = 3$	S.t. $-3x_1 - x_2 \leq -3$ $3x_1 + x_2 \leq 3$	S.t. $-3w_1' + 3w_1'' - 3w_2$ $+ w_3 \geq -4$	S.t $3w_1 + 3w_2 - w_3 \leq 4$
$3x_1 + 4x_2 \geq 6$ $x_1 + x_2 \leq 4$	$-3x_1 - 4x_2 \leq -6$ $x_1 + x_2 \leq 4$	$-w_1' + w_1'' - 4w_2$ $+ w_3 \geq -3$	$w_1 + 4w_2 - w_3 \leq 3$
$x_1, x_2 \geq 0$	$x_1, x_2 \geq 0$	$w_1', w_1'', w_2, w_3 \geq 0$	w_1 unrestricted $w_2, w_3 \geq 0$

Remarks : *1. In the above final dual the objective function can also be Min.
$Z_D = 3w_1 - 6w_2 + 4w_3$ or Max. $Z_D = -3w_1 + 6w_2 - 4w_3$ since w_1 is unrestricted.*

2. For a constraint of equality form in primal, we get corresponding variable unrestricted in dual.

3. For an unrestricted variable in primal, we get corresponding constraint as equality in its dual.

ILLUSTRATION 24 ───────────────────────────

Find the dual of

Max. $Z_p = 3x_1 + x_2 + 2x_3 - x_4$

S.t. $2x_1 - x_2 + 3x_3 + x_4 = 1$

$x_1 + x_2 - x_3 + x_4 = 3$

$x_1, x_2, x_3 \geq 0, x_4$ **unrestricted** [Dr. BRAOU (MBA)-Assignment 98]

Solution :

With usual steps, writing the standard form of primal we get

Max. $\quad Z_P = 3x_1 + x_2 + 2x_3 - x_4' + x_4''$

S.t $\quad 2x_1 - x_2 + 3x_3 + x_4' - x_4'' \leq 1$ $\Big\}$ equality constraint is re-arranged

$-2x_1 + x_2 - 3x_3 - x_4' + x_4'' \leq -1$

$x_1 + x_2 - x_3 + x_4' - x_4'' \leq 3$ $\Big\}$ equality constraint is re-arranged

$-x_1 - x_2 + x_3 - x_4' + x_4'' \leq -3$

$x_1, x_2, x_3, x_4', x_4'' \geq 0$ where $x_4' - x_4'' = x_4$ (since x_4 is unrestricted)

Dual form is

Min. $Z_D = w_1' - w_1'' + 3w_2' - 3w_2''$

S.t. $\quad 2w_1' - 2w_1'' + w_2' - w_2'' \geq 3$

$-w_1' + w_1'' + w_2' - w_2'' \geq 1$

$$3w_1' - 3\,w_1'' - w_2' + w_2'' \geq 2$$
$$w_1' - w_1'' + w_2' - w_2'' \geq -1$$
$$-w_1' + w_1'' - w_2' - w_2'' \geq 1$$
$$w_1', w_1'', w_2' \text{ and } w_2'' \geq 0$$

Re-arranging with assumptions as $w_1 = w_1' - w_1''$ and $w_2 = w_2' - w_2''$ by which w_1 and w_2 will become unrestricted we get the dual revised as,

Min. $Z_D = w_1 + 3\,w_2$ S.t. $2\,w_1 + w_2 \geq 3$ $-w_1 + w_2 \geq 1$ $3\,w_1 - w_2 \geq 2$ $w_1 + w_2 \geq -1$ $-w_1 - w_2 \geq 1$ w_1 and w_2 unrestricted	This can be further reduced to Min. $Z_D = w_1 + 3w_2$ St. $2w_1 + w_2 \geq 3$ $-w_1 + w_2 \geq 1$ $3w_1 - w_2 \geq 2$ $w_1 + w_2 = -1$ (inlieu of two inequations) w_1 and w_2 unrestricted.

4.5.2 *Comparison Between Primal & Dual*

1. *With Reference to Formulation :*

	Primal	Dual
1.	If objective function is in **Maximisation** form	Objective function is in **Minimisation** form
2.	If objective function is in **Minimisation** form	Objective function is in **Maximisation** form
3.	Coefficients of decision variables in objective function	Constants of constraints set
4.	Constants in constraint set	Coefficients of decision variables in objective function
5.	Number of decision variables	Number of constraints
6.	Number of constraints	Number of decision variables
7.	Coefficients of variables in constraints row wise	Coefficients of variables in constraints column wise
8.	Coefficients of variables in constraints column wise	Coefficients of variables row wise
9.	≤ type constraints	≥ type constraints
10.	≥ type constraints	≤ type constraint
11.	Equality type (=) constraint	Corresponding variable unrestricted
12.	Unrestricted variable	Corresponding constraint is equality (=) type

2. With Reference to Solution :

S.No.	Primal	Dual
1.	Solution is taken from solution values column of final primal tableau (simplex)	Solution is taken $Z_j - C_j$ row of final simplex tableau of primal
2.	Unbounded solution	Infeasible solution
3.	Infeasible solution	Unbounded solutions
4.	Solution values	Shadow prices
5.	Shadow prices	Solution values

4.5.3 *Advantages & Applications of Duality*

Duality will be more advantageous in the following cases.

1. Sometimes dual problem solution may be easier than primal solution particularly, when number of decision variables is considerably less than slack/surplus variables.

2. In the areas like economics, it is highly helpful in obtaining future decision in the activities being programmed.

3. In physics, it is used in parallel circuit & series circuit theory.

4. In game theory, dual is employed by Column player who wishes to minimise his maximum loss while his opponent while Row player applies primal to maximise his minimum gains. However, if one problem is solved, the solution for other also can be obtained from the simplex tableau (Refer game theory : LPP method)

5. When a problem does not yield any solution in primal, it can be verified with dual.

6. Economic interpretations can be made and shadow prices can be determined enabling the managers to take further decisions.

4.5.4 *Shadow Price*

The shadow price of a resource is the unit price that is equal to increase in profit to be realised by one additional unit of the resource. (or)

It is the change in the optimum value of the objective function per unit increase of the resource.

The shadow price can be determined from final simplex tableau using the values of $Z_j - C_j$, in primal problem.

In the illustration - 21 of this chapter, we conclude that

1. **For Primal :** A unit increase of resource covered by first constraint raises the profit of x_1 by 15 units and a unit increase in resource covered by second constraint raises profit of x_2 by 5/4 units. Thus shadow prices of x_1 and x_2 are 15 and 5/4 units separately.

2. **For Dual :** A unit slash in requirement of first constraint reduces the cost of w_1 by 5/16 units while a unit slash on second constraint requirements reduces cost of w_2 by 75/8 units. Thus shadow prices of w_1 and w_2 are 15/16 and 75/8 respectively.

Practice Problems

Solve the following LP problems by two-phase method.

1. Max. $Z = 3 x_1 - x_2$

   Subject to the constraints

   $$2 x_1 + x_2 \geq 2 ; \qquad x_1 + 3 x_2 \leq 2$$
   $$x_2 \leq 4 ; \qquad x_1, x_2 \geq 0$$

   [OU - MBA May 92]

   Answer : $x_1 = 2 ; x_2 = 0;$ Max. $Z = 6.$

2. Max.. $Z = 2 x_1 + 3 x_2 + 5 x_3,$

   Subject to : $3 x_1 + 10 x_2 + 5 x_3 \leq 15;$ $\qquad x_1 + 2 x_2 + x_3 \geq 4;$
   $$33 x_1 - 10 x_2 + 9 x_3 \leq 33 ; \qquad x_1, x_2, x_3 \geq 0$$

   Answer : Infeasible solution since artificial variables can not be removed.

3. Max. $Z = 5 x_1 + 8 x_2$ Subject to $3x_1 + 2 x_2 \geq 3 ;$

   $$x_1 + 4 x_2 \geq 4 ; \qquad x_1 + x_2 \leq 5 ; \quad x_1, x_2 \geq 0$$

   Answer : $x_1 = 0; x_2 = 5$ and Max. $Z = 40$

4. Max. $Z = 5 x_1 - 2 x_2 + 3 x_3$

   Subject to : $2 x_1 + 2 x_2 - x_3 \geq 2 ;$ $\qquad 3 x_1 + 4 x_2 \leq 3$
   $$x_2 + 3 x_3 \leq 5 ; \qquad x_1, x_2, x_3 \geq 0$$

   Answer : $x_1 = \dfrac{23}{3} ; x_2 = 5 ; x_3 = 0 ;$ Max.. $Z = 85/3$

5. Max. $Z = x_1 + 1.5 x_2 + 2x_3 + 5 x_4$ with the constraints

   $$3 x_1 + 2 x_2 + 4 x_3 + x_4 \leq 6 ; \qquad 2 x_1 + x_2 + x_3 + 5x_4 \leq 4 ;$$
   $$2 x_1 + 6 x_2 - 8 x_3 + 4 x_4 = 0 ; \qquad x_1 + 3 x_2 - 4 x_3 + 3 x_4 = 0$$
   $$x_j (j = 1, 2, 3, 4) \geq 0$$

   Answer : $x_1 = 1.2 ; x_2 = 0; x_3 = 0.9, x_4 = 0,$ Max. $Z = 19.8$

6. Max. $Z = 3x_1 + 2x_2 + x_3 + 4x_4$

Subject to $\quad 4x_1 + 5x_2 + x_3 - 3x_4 = 5 ; \qquad 2x_1 - 3x_2 - 4x_3 + 5x_4 = 7$

$\qquad x_1 + 4x_2 + 2.5x_3 - 4x_4 = 6 ; \qquad\qquad x_1, x_2, x_3 \geq 0$

Answer : Infeasible solution

7. Minimise $Z = x_1 - 2x_2 - 3x_3$

Subject to $\quad -2x_1 + x_2 + 3x_3 = 2 ; \qquad 2x_1 + 3x_2 + 4x_3 = 1$

$\qquad x_j \geq 0, \quad j = 1, 2, 3,$

Answer : Here all $Z_j - C_j \geq 0$, but artificial variable A_1 does not vanish from the basis. Hence the given LPP has infeasible solution.

8. Solve the following problems by simplex method adding artificial variables.

Max. $Z = 2x_1 + 5x_2 + 7x_3$

Subject to $\quad 3x_1 + 2x_2 + 4x_3 \leq 100 ; \qquad x_1 + 4x_2 + 2x_3 \leq 100$

$\qquad x_1 + x_2 + 3x_3 \leq 100 ; \qquad\qquad x_1, x_2, x_3 \geq 0$

Answer : $x_1 = 0 ; \ x_2 = \dfrac{50}{3} ; \ x_3 = \dfrac{50}{3} ;$ Max $Z = 200$

9. Max. $Z = 4x_2 + x_2 + 4x_3 + 5x_4$

Subject to $\quad 4x_1 + 6x_2 - 5x_3 + 4x_4 \geq -20 ; \quad 3x_1 - 2x_2 + 4x_3 + x_4 \leq 10 ;$

$\qquad 8x_1 - 3x_2 - 3x_3 + 2x_4 \leq 20 ; \qquad\qquad x_1, x_2, x_3, x_4 \geq 0$

Answer : Unbounded solution

10. Hema & Co. produces three types of pens of types namely sharp, fighter and pride. These products require three qualities of inks In the refills say P_1, P_2 and P_3. The quantities of these refills available are 22, 14 & 14 hundreds respectively. Each of these pens, the refill requirements are

	Sharp (S)	Fighter (F)	Pride (P)
P_1	3	–	3
P_2	1	2	3
P_3	3	2	0
Profit hundreds (Rs. in thousands)	1	4	5

The company makes a profit of one, four & five hundreds on each unit of the products (pens) respectively. How many pens of each type should the company produce to maximize the profits?

Hint :   Formulation is Max.. $Z = x_1 + 4 x_2 + 5 x_3$;

Subject to $3 x_1 + x_3 \leq 22$

$$x_1 + 2 x_2 + 3 x_3 \leq 14$$

$$3x_1 + 2x_2 \leq 14$$

$$x_1, x_2, x_3 \geq 0$$

11. A manufacturer of steel furniture makes three produces chairs, filing cabinets and tables. Three machines (Say P, Q and R) are available on which these products are processed. The manufacturer has 100 hours per week available on each of three machines. The time required by each of three products on three machines is summarised in the following table.

Product	Time Required (in hrs)		
	Machine P	Machine Q	Machine R
Chair	2	2	1
Filing cabinet	2	1	2
Table	–	1	2

The profit analysis shows that the net profit on each chair, filing cabinet and table is Rs. 22, Rs. 30, and Rs. 25 respectively. What should be the weekly production of these products so that the manufacturer's total profit per week is maximised.

Answer : $\dfrac{100}{3}$ chairs, $\dfrac{50}{3}$ tables, $\dfrac{50}{3}$ file cabinets. Max. profit = Rs. 1650.

12. A factory is engaged in manufacturing three products $X, Y,$ and Z which involve lathe work, grinding and assembling. The cutting, grinding and assembling time required for one unit of X are 2, 1 and 1 hours respectively. Similarly they are 3, 2 and 3 hours for one unit of Y and 1 , 3 and 1 hours for one unit of Z and the profits on X, Y and Z are Rs. 2 , Rs. 2 and Rs. 4 per unit respectively. Assuming that there are available 300 hrs of lathe time, 300 hours of grinding time & 240 hours of assembly time, how many units of each products should be produced to maximise profits? Work the problem using simplex methods.

13. A transistor radio company manufacturers four models A, B, C and D which have profit contributions of Rs. 8, Rs. 15, and Rs. 25 on A, B, and C respectively and a loss of Re. 1 on model D per dozen. A dozen units of model A require one hour of manufacturing, two hours for assembly and one hour for packing. The corresponding figures for a dozen units of model B are 2, 1 and 2 and for a dozen units of C are 3, 5 and 1, while a dozen units of D require 1 hour of packing only. During the fourth coming week, the company will be able to make available 15 hours of manufacturing, 20 hours of assembling and 10 hours of packing time. Obtain the optimal production schedule for the company.

Hint : Formulation is

Max. $Z = 8 x_1 + 15 x_2 + 25 x_3 - x_4$

Subject to $\quad x_1 + 2 x_2 + 3 x_3 \leq 15 ; \qquad 2 x_1 + x_2 + 5 x_3 \leq 20$

$\quad x_1 + 2 x_2 + x_3 + x_4 \leq 10 ; \qquad x_1 , x_2, x_3, x_4 \geq 0$

Answer : $x_1 = x_2 = x_3 = \dfrac{5}{2} ; \quad x_4 = 0 ; \quad$ Max $Z =$ Rs. 120

14. A manufacturer has two products P & Q, both of them are produced in two steps by machines M_1 & M_2. The process times per hundred for the products on the machines are :

	M_1	M_2	Contribution (per hundred)
P	4	5	10
Q	5	2	5
Available (hrs)	100	80	

The manufacturer is in a market upswing and can sell as much as he can produce of both products. Formulate and determine optimum product mix using simplex method.

Hint : Formulation of the problem is

Max. $Z = 10 x_1 + 5 x_2 ,$

Subject to $\quad 4 x_1 + 5 x_2 \leq 100$

$\quad 5 x_1 + 2 x_2 \leq 80$

$\quad x_1, \ x_2 \geq 0$

Answer : $x_1 = \dfrac{2000}{17} \approx 1177; \ x_2 = \dfrac{18000}{17} \approx 1059.$

15. A manufacturer of leather belts makes three types of belts A, B and C which are processed on three machines M_1, M_2 & M_3. Belt A requires 2 hours on M_1 3 hours on M_3. Belt B requires 3 hours on M_1 , 2 hours on machine M_2 and 2 hours on machine M_3. Belt C requires 5 hours on M_2 and 4 hours on M_3. There are 8 hours of time per day available on machine M_1 , 10 hours per day available on machine M_2 and 15 hours of time per day on M_3. The profit gained from Belt A is Rs. 3 per unit, from belt B is Rs. 5 per unit, from belt C is Rs. 4 per unit, what should be the daily production of each type of belts so that the profit is maximum.

16. An animal feed company must produce 200 kgs of a mixture consisting of ingredient X_1 & X_2 daily. X_1 cost Rs. 3 per kg and X_2 costs Rs. 8 per kg. No more than 80 kgs of X_1 can be used and at least 60 kgs of X_2 must be used. Find how much of the each ingredient should be used if the company wants to minimize cost?   **[OU - MBA - Apr. 98]**

   Hint : Formulation is Min. $3\,x_1 + 8\,x_2$

   Subject to $x_1 + x_2 = 200;$ $x_1 \le 80$

                          $x_2 \ge 60 ;$ $x_1, x_2 \ge 0$

   Answer : $x_1 = 80$; $x_2 = 120$; Min. cost = Rs. 1200

17. A television company has three major departments for manufacturing of its models supreme and delux. Monthly capacities are given as follows

Department	Per Unit Time Requirements (Hrs)		Hours Available Per Month
	Model Supreme	Model Delux	
I	4.0	2.0	1,600
II	2.5	1.0	1,200
III	4.5	1.5	1,600

The marginal profit of model supreme is Rs. 400 each and that of model delux is Rs. 100 each. Assuming that the company can sell any quantity of either products due to fovourable market conditions; Determine the optimum output for both the models, the highest possible profit for this month and slack time in the three departments.

   Hint : Formulation is : Max. $Z = 400\,x_1 + 100\,x_2$,

   Subject to $4\,x_1 + 2\,x_2 \le 1600 ;$ $\dfrac{5}{2}x_1 + x_2 \le 1200$

                     $\dfrac{9}{2}x_1 + \dfrac{3}{2}x_2 \le 1600 ; x_1, x_2 \ge 0$

   Answer : $x_1 = \dfrac{3200}{9}$; $x_2 = 0$; $x_3 = 0$; Max. $Z = \dfrac{1280000}{9}$

18. A teacher gives his students three long lists of problems with the instructions to submit not more than 100 of them correctly solve, for credit. The problem in the first list are of 5 points each, in second 4 points each and in third 6 points each. On an average 3 minutes are required to solve a problems from first list, 2 min. for a problem from second & 4 min. for a problem from third. The students devote more than $2\frac{1}{2}$ hours of numerical hours. How many problems from each list, a student should solve so as to get the maximum credit.

Answer : Max $Z = 5x_1 + 4x_2 + 6x_3$

Subj to $x + x_2 + x_3 \leq 100; \quad 3x_1 + 2x_2 + 4x_3 \leq 150$

$$x_1, x_2, x_3 \geq 0$$

19. A furniture company manufactures four models of desks. Each desk is first constructed in the carpentry shop and is next sent to the finishing shop where it is varnished, waxed & polished. The number of man hours of labour required in each shop is as follows :

Shop	Desk			
	I	II	III	IV
Carpentry	4	9	7	10
Finishing	1	1	3	40
Profit per item (Rs.)	12	20	18	40

Because of limitation in capacity of the plant, not more than 6,000 man hours can be expected in the carpentry shop and 4,000 in the finishing shop in a month. Assuming that raw materials are available in adequate supply and all desks produced can be sold, determine the quantities of each type of desk to be made for maximum profit of the company.

Hint : Formulation of the problem is :

Max. $Z = 12 x_1 + 20 x_2 + 18 x_3 + 40 x_4$,

Subject to $\quad 4 x_1 + 9 x_2 + 7 x_3 + 10 x_4 \leq 6000$;

$$x_1 + x_2 + 3 x_3 + 40 x_4 \leq 4000 ;$$

$$x_1, x_2, x_3, x_4 \geq 0$$

Answer : $x_1 = \dfrac{4000}{3}$, $x_2 = 0$, $x_3 = 0$, $x_4 = \dfrac{200}{3}$ and Max. $Z =$ Rs $\dfrac{56000}{3}$

20. A manufacture produces three products A, B and C. Each product can be produced on either one of two machines, I & II. The time required to produce 1 unit of each product on a machine is given in the table below.

Product	Time to Produce Limit (Hrs)	
	Machine - I	Machine - II
A	0.5	0.6
B	0.7	0.8
C	0.9	1.05

There are 85 hours available on each machine ; the operating cost is Rs. 5 per hour for M/C I and Rs. 4 per hour for M/C - II and the product requirements are at least 90 units of A, at least 80 units of B and at least 60 units of C. The manufacturer wishes to meet the requirements at minimum cost.

Solve the given linear programming problem by simplex method.

Answer : 150 hrs on m/c II, no time on m/c I and min. cost = Rs. 600.

Additional Problems

1. The post master of a local post office wishes to hire extra helpers during the Deepawali seasons, because of a large increase in the volume of mail handling & delivery. Because of the limited office space and budgetary condition, the number of temporary helpers must not exceed 10. According to the past experience, men can handle 300 letters and 80 packages per day, on the average and women can handle 400 letters and 50 packages per day. The post master believes that the daily volume of extra mail and packages will be no less than 3,400 and 680 respectively. A man receives Rs. 25 a day and a woman receives Rs. 22 a day. How many men and women helpers should be hired to keep the pay-roll at a minimum?

Answer : Min $Z = 25x_1 + 22x_2$

S.t. $x_1 + x_2 \le 10$; $300x_1 + 400x_2 \ge 3400$;

$80x_1 + 50x_2 \ge 680$ and $x_1, x_2 \ge 0$

2. Two products X and Y are processed on three machines M_1, M_2 and M_3. The processing times per units, machine availability and profit per unit are as under.

Machine	Processing Time (Hrs)		Availability (Hrs)
	X	Y	
M_1	2	3	1500
M_2	3	2	1500
M_3	1	1	1000
Profit/unit	10	12	

Formulate the mathematical model, solve it by simplex method & also find the number of hours machine M_3 remains unutilized.

3. A manufacturing firm has discontinued production of a certain unprofitable product line. This created considerable excess production capacity and management is considering to devote this excess capacity to one or more of three products : X, Y and Z. The available capacity on the machines which might limit output is summarized in the table.

The no. of machine hours required for the unit of the respective product is given below.

Machine Type	Productivity (in machine hrs per unit)			Available time in M/C hrs/week
	Product X	Product Y	Product Z	
Milling machine	8	2	3	250
Lathe	4	3	–	150
Grinding	2	–	1	50

The unit profit would be Rs. 20, Rs. 6 and Rs. 8 respectively for products X, Y and Z. Find how much of the each product the firm should produce in order to maximise profit?

Hint : Formulation is : Max. $Z = 20 x_1 + 6 x_2 + 8 x_3$

Subject to $8 x_1 + 2 x_2 + 3 x_3 \leq 250$

$4 x_1 + 3 x_2 \leq 150 ;$

$2x_1 + x_3 \leq 50;$

$x_1, x_2, x_3 \geq 0$

Answer : $x_1 = 0;$ $x_2 = 50, x_3 = 50$; max. $Z = 700$

4. A plant is engaged on the production of two products which are processed through three departments, the number of hours required to finish each is indicated in the table below :

Machine Type	Product		Max.. Hours Available Per week
	A	B	
I	7	8	1600
II	8	12	1600
III	15	16	1600

(a) If the profit for the products is Rs. 6 for a unit of product A but only Rs. 4 for a unit of product B, what quantities per week should be planned to maximize profit.

(b) Capacity can be increased on one department only, in which department should it be done & why? To what extent should be capacity be increased?

(c) If the cost per hour in department I is Rs. 25, in dept II, Rs. 40, & in dept III, Rs. 50; what quantities should be planned to minimize to cost of production?

Answer : (a) $\dfrac{320}{3}$ units of A & no units of B ; max. profit = Rs. 640

(c) $\dfrac{10}{3}$ units of production A & No. unit of B, min. cost = Rs. 20

5. Max. $4x + 10y$

  S.t. $2x + y \leq 50$,

      $2x + 5y \leq 100$,

      $2x + 3y \leq 90$,

      $x, y \geq 0$

Solve graphically

If solved by simplex method, is the solution unique ? Why/why not? if not, give two basic optimal solutions. Also find all the non-basic optimal solutions to the problem.

6. Minimise $200x_1 + 350x_2$

  S.t. $50x_1 + 60x_2 \geq 2500$

       $100x_1 + 60x_2 \geq 3000$

       $100\, x_1 + 200\, x_2 \geq 7000$

        $x_1, x_2 \geq 0$ **[OU - MBA - May 91, Nov. 94, Oct. 95]**

7. Min. $Z = 2x_1 + x_2 - x_3 - x_4$

  S.t. $x_1 - x_2 + 2x_3 - x_4 = 2$

     $2x_1 + x_2 - 3x_3 + x_4 = 6$

     $x_1 + x_2 + x_3 + x_4 = 7$

    and $x_j \geq 0$ **[OU - MBA - May. 92]**

8. Max. $Z = 9x_1 + 15x_2 + 20\, x_3$

  S.t. $3x_1 + 6x_2 + 7x_3 \leq 210$

      $3x_1 + 5x_2 + 4x_3 \leq 240$

      $8x_1 + 10x_2 + 12x_3 \leq 260$

       $x_1, x_2, x_3 \geq 0$ **[OU - MBA- May 91, Nov. 94]**

9. Max. $Z = 5x_1 + 2x_2 + 10\, x_3$

  S.t. $x_1 - x_3 \leq 10$

       $x_2 - x_3 \geq 10$

     $x_1 + x_2 + x_3 \leq 10$

      $x_1, x_2, x_3 \geq 0$ **[OU - MBA May 95]**

10. Min. $Z = 20x_1 + 40x_2$

    S.t. $36x_1 + 6x_2 \geq 108$

    $3x_1 + 12x_2 \geq 36$

    $20x_1 + 10x_2 \geq 100$

    $x_1, x_2 \geq 0$ [OU - MBA Feb. 93]

11. Max. $Z = 3x_1 + 4x_2 + x_3$

    S.t. $x_1 + 2x_2 + 3x_3 \leq 90$

    $2x_1 + x_2 + x_3 \leq 60$

    $3x_1 + x_2 + 2x_3 \leq 80$

    and $x_j \geq 0$ [OU - MBA - Feb 93]

12. Max. $Z = 10x_1 + 12x_2 + 8x_3$

    S.t. $3x_1 + x_2 + 2x_3 \leq 100$

    $x_1 + 4x_2 + x_3 \leq 120$

    $2x_1 + 3x_2 \leq 80$

    $x_1, x_2, x_3 \geq 0$ [OU - MBA - May 94]

Practice Problems in Duality

1. Min. $3x_1 + 4x_2$

   Subject to $2x_1 + 6x_2 \leq 16$; $5x_1 + 2x_2 \geq 20$, $x_1, x_2 \geq 0$

   Answer : Max. $Z = 16w_1 - 20w_2$,

   S.t. $2w_1 - 5w_2 \geq 3$; $6w_1 - 2w_2 \geq 4$; $w_1, w_2 \geq 0$

2. Min. $Z = x + y + z$

   S.t. $x - 3y + 4z = 5$; $x - 2y \leq 3$; $2y - z \geq 4$; $x , z \geq 0$ & y is unrestricted

   Answer : Max. $Z_w = 5w_1 + 3w_2 + 4w_3$

   S.t $w_1 - w_2 \leq 1$; $-3w_1 + 2w_2 + 2w_3 \leq 1$

   $4w_1 - w_3 = 1$; $w_3, w_2 \geq 0$, w_1 is unrestricted

3. Max. $Z = 3a + b + 2c - d$

   S.t $2a - b + 3c + d = 1$; $a + b - c + d = 3$; $a , b , c \geq 0$, d is unrestricted

   Answer : Min. $Z_w = w_1 + 3w_2$

   S.t. $2w_1 + w_2 \geq 3$; $-w_1 + w_2 \geq 1$, $3w_1 - w_2 = 2$

   $w_1 + w_2 = -1$; w_1 & w_2 are unrestricted.

4. Min. $Z = 10 x_1 + 6 x_2 + 2 x_3$

   S.t $\quad x_1 + 5 x_2 + x_3 \geq 1$; $3 x_1 + x_2 - x_3 \geq 2$; $\quad x_1 , x_2 , x_3 \geq 0$

Answer : Max $Z_D = w_1 + 2w_2$

   S.t. $\quad w_1 + 3w_2 \leq 10$; $5w_1 + w_2 \leq 6$; $\quad w_1 - w_2 \leq 2$; $w_1, w_2 \geq 0$

5. Max. $Z = x - y + 3z$

   Subject to $\quad x + y + z \quad \leq 10$; $\qquad\qquad 2x - z \leq 2$

   $\qquad\qquad 2x - 2y + 3z \leq 6$; $\qquad\qquad x, y, z \geq 0$

Answer : Min. $Z_w = 10w_1 + 2 w_2 + 6 w_3$

   S.t. $\qquad w_1 + 2 w_2 + 2 w_3 \geq 1$; $\qquad w_1 - 2w_3 \geq -1$;

   $\qquad w_1 - w_2 + 3 w_3 \geq 3$; $\quad w_1 , w_2 , w_3 \geq 0$

6. Max. $Z = 3 x_1 + x_2 + 4 x_3 + 9 x_5$,

   S.t. $\qquad 4 x_1 - 5x_2 - 9 x_3 + x_4 - 2x_5 \leq 6$;

   $\qquad 2 x_1 + 3 x_2 + 4 x_3 - 5 x_4 + x_5 \leq 9$;

   $\qquad x_1 + x_2 - 5x_3 - 7 x_4 + 11 x_5 \leq 10$;

   $\qquad\qquad x_1 , x_2 , x_3 , x_4, x_5 \geq 0$

Answer : Min. $Z_w = 6 w_1 + 9 w_2 + 10 w_3$

   S.t. $\qquad 4 w_1 + 2 w_2 + w_3 \geq 3$; $\quad - 5 w_1 + 5 w_2 + w_3 \geq 0$;

   $\qquad - 9 w_1 + 4 w_2 - 5 w_3 \geq 4$; $\quad w_1 - 5 w_2 - 7 w_2 \geq 1$;

   $\qquad - 2 w_1 + w_2 + 11 w_3 \geq 9$; $\qquad w_1 , w_2 , w_3 \geq 0$

7. Max. $Z = 6a + 4b + 6c + d$

   S.t. $\qquad 4a + 4b + 4c + 8d = 21$; $\quad 3a + 17b + 80c + 2d \leq 48$;

   $\qquad a, b \geq 0$ & c, d are unrestricted

Answer : Min. $Z_w = 21 w_1 + 48 w_2$

   S.t. $\qquad 4 w_1 + 3 w_2 \geq 6$; $\qquad 4 w_1 + 17 w_2 \geq 4$;

   $\qquad 4 w_1 + 80 w_2 = 6$; $\qquad 8w_1 + 2w_2 = 1$;

   $\qquad w_2 \geq 0$ $\quad w_1$ is unrestricted.

8. Min. $Z_p = x_3 + x_4 + x_5$

   S.t. $\quad x_1 - x_3 + x_4 + x_5 = -2$; $\qquad x_2 - x_3 - x_4 + x_5 = 1$;

   $\qquad x_j \geq 0 \ (j = 1, 2, \ldots 5)$

Answer : Max. $\quad Z_D = +2\,w_1 + w_2$

   S.t. $\qquad w_1 - w_2 \leq 1$; $\qquad w_1 + w_2 \leq -1$

   $\qquad -w_1 + w_2 \geq 1$; $\qquad w_1, w_2 \quad$ are unrestricted

9. Max. $Z = 3x + 2y$

   S.t. $\quad 2x + y \leq 5$; $\qquad x + y \leq 3$;

   $\qquad x, y \geq 0$

Answer : $x = 2$; $\ y = 1$, Max. $Z = 8$; $\ w_1 = 1$, $w_2 = 1$, Min. $Z_w = 8$

   Dual is Min $Z_w = 5w_1 + 3w_2$

   S.t. $\quad 2w_1 + w_2 \geq 3$; $\qquad w_1 + w_2 \geq 2$; $w_1, w_2 \geq 0$

10. (i) Use simplex method to maximize

   $Z = 5x - 2y + 3z$; subject to the restrictions

   $2x + 2y - z \geq 2$; $3x - 4y \leq 3$, $y + 3z \leq 5$

   where x, y, z are non negative variables

   (ii) Verify your solution using duality

Answer : $x = \dfrac{23}{3}$; $y = 5$, $z = 0$, Max. $Z = \dfrac{85}{3}$

11. Max. $Z = 2x + 3y + 5z$

   S.t. $\qquad x + y + z \leq 7$; $\qquad x + 2y + 2z \leq 13$;

   $\qquad 3x - y + z \leq 5$; $\qquad x, y, z \geq 0$

Answer : Min. $Z_D = 7w_1 + 13w_2 + 5w_3$

   S.t. $\ w_1 + w_2 + 3w_3 \geq 2$; $\qquad w_1 + 2w_2 - w_3 \geq 3$;

   $\qquad w_1 + 2w_2 + w_3 \geq 5$; $\qquad w_1, w_2, w_3 \geq 0$

12. Write the dual of the primal problem given below :

   Minimize : $Z = x_1 + x_2$

   S.t. $\qquad 2x_1 + 4x_2 \geq 4$

   $\qquad 2x_1 + x_2 \geq 3$

   $\qquad -x_1 + 3x_2 = 5$

   $\qquad x_1, x_2 \geq 0$

Answer : Max $Z_d = 4w_1 + 3w_2$

S.t. $2w_1 + 2w_2 - w_3 \quad \leq 1$

$4w_1 + w_2 + 3w_3 \quad \leq 1$

$w_1, w_2 \geq 0; w_3$ un restricted   [JNTU Mech/Prod/Chem 2001/S]

13. Apply simplex to Max. $Z : 30z + 23b + 29c$

S.t. $6a + 5b + 3c \leq 26 ; \; 4a + 2b + 5c \leq 7 ; \; a, b, c \geq 0$

Also use duality of the problem from the final table

Answer : $a = 0, b = \dfrac{7}{2} ; \qquad c = 0; \text{max. } Z = \dfrac{16}{2}$

$w_1 = 0, w_2 = \dfrac{23}{2} ; \quad \text{min. } Z_w = \dfrac{16}{2}$

14. Give dual formulation of

Max $Z = 5x_1 + 10 x_2$

S.t. $5x_1 + 8x_2 \leq 48; \qquad 5x_1 + 3x_2 \leq 48$

$x_1 + 8x_2 \quad \leq 48 \qquad\qquad x_1, x_2 \geq 0$ [JNTU (CSE) 98]

Answer : Min. $Z = 48w_1 + 48w_2 + 48w_3$

S.t. $5w_1 + 5w_2 + w_3 \geq 5; \qquad 8w_1 + 3w_2 + 8w_3 \geq 10$

$w_1, w_2, w_3 \geq 0$

15. Apply principle of dual to Max. $Z = 3x_1 + 2x_2$

S.t. $x_1 + x_2 \geq 1 ; x_1 + x_2 \leq 7 ; x_1 + 2x_2 \leq 10 ; x_2 \leq 3; x_1, x_2$ are non - negative

Hint : Min. $Z = w_1 + 7w_2 + 10 w_3 + 3w_4$

S.t. $w_1 + w_2 + w_3 \leq 3; \; w_1 + w_2 + 2w_3 + w_4 \geq 2$ and $w_1, w_2, w_3, w_4 \geq 0$

Answer : Dual $w_1 = 0 ; w_2 = 3, \; w_3 = 0, \; w_4 = 0$ & Min. $Z = 21$

16. Max. $Z = 2 x_1 + x_2$

S.t. $x_1 + 2 x_2 \leq 10 ; x_1 + x_2 \leq 6 ; x_1 - x_2 \leq 2 ; \; x_1 - 2x_2 \leq 1 ; x_1, x_2 \geq 0$

Answer : $x_1 = 4, x_2 = 3, \; \text{Max. } Z = 10$

17. A diet conscious house wife wishes to ensure certain minimum intake of vitamins A, B & C for the family. The minimum daily (quantity) needs of the vitamins A, B and C for the family are respectively 30, 20 & 16 units. For the supply of these minimum vitamin requirements, the house-wife relies on two fresh foods. The first one provides 7, 5, 2 units of the three vitamins per gram respectively & the second provides 2, 4, 8 units of the same three vitamins per gram of the food stuff respectively. The first food stuff costs Rs. 3 per gram and the second Rs. 2 per gram. The problem is how many grams of each food stuff should the house wife buy every day to keep her food bill as low as possible.

   (i) Formulate the LPP

   (ii) Write the 'Dual' problem

   (iii) Solve the primal problem graphically

   (iv) Solve the dual problem using simplex method

   (v) Interpret the dual problem and its solution.

Answer : (i) Min. $Z_x = 3x_1 + 2x_2$,

     Subject to $7x_1 + 2x_2 \geq 30$; $5x_1 + 4x_2 \geq 20$; $2x_1 + 8x_2 \geq 16$ and x , $x_2 \geq 0$

   (ii) Max. $Z_w = 30\,w_1 + 20\,w_2 + 16\,w_3$

     Subject to $7w_1 + 5w_2 + 2w_3 \leq 3$; $\ 2w_1 + 4w_2 + 8w_3 \leq 2$; w_1 , w_2 , $w_3 \geq 0$

   (iii) $w_1 = \dfrac{5}{3}$, $w_2 = 0$, $w_3 = \dfrac{2}{3}$, $Z_w = 14$

   (iv) $x = 4\ \ y = 1$, $Z_{max} = 14$

18. Min. $Z = -2\,x_1 + 3x_2 + 4x_3$

   S.t. $-2x_1 + x_2\ \geq 3$; $-x_2 + x_2\ \geq 3$;

     $-x_2 + 3\,x_2 + x_3\ \geq -1$; $x_1, x_2, x_3\ \geq 0$

Answer : $x_1 = 0$, $\ x_2 = 3$; $x_3 = 0$ Min. $Z = 9$

19. Find the dual of Min. $Z_p = 4x_1 + 3\,x_2 + 3x_3$

   S.t. $x_1 + 2\,x_2 \geq 2$, $3\,x_1 + x_2 + x_3 \geq 4$, $4\,x_3 \geq 1$, $x_1 + x_3 \geq 1$, $x_1, x_2, x_3 \geq 0$

Answer : Max $Z_D = 2w_1 + 4w_2 + w_3 + w_4$

     S.t. $w_1 + 3w_2 + w_4 \leq 4$; $2w_1 + w_2 \leq 3$;

        $w_3 + w_4 \leq 3$; $w_1, w_2, w_3\ w_4 \geq 0$

20. Prove that the dual of the dual is the primal. Write down the dual of the following & solve them.

     Max. $Z = 4\,x_1 + 2\,x_2$

     S.t. $x_1 + x_2 \geq 3$; $x_1 - x_2 \geq 2$, x_1 , $x_2 \geq 0$

Find dual and hence or otherwise solve the above problems. **[OU - MBA Dec. 95]**

Answer : $x_1 = 4$, $x_2 = 2$, $Z_{max} = 18$

21. Find the dual of the problem Max. $Z = 2x - y$

    Subject to $x + y \leq 10$; $-2x + y = 2$; $4x + 3y \geq 12$; $x, y \geq 0$

    Solve the primal by simplex method & deduce from it, the solution to the dual problem.

22. Max. $Z = 8x_1 + 6x_2$

    Subject to $x_1 - x_2 \leq \dfrac{3}{5}$; $x_1 - x_2 \geq 2$, $x_1, x_2 \geq 0$

    Show that both the primal & dual have no feasible solution

23. A dairy has two bottling plants located at A, & B. Each plant bottles up three different kinds of milk i.e., cow, buffallo and goat. The capacities of the two plants in number of bottles per shift in a day are as follows.

Milk	Plant	
	A	B
Cow	2,000	1,000
Buffallo	2,000	3,000
Goat	1,000	1,000

    Market survey shows that the cow, buffallo and goat milk are required at least 14000, 22000 and 1000 bottles per day. The operating costs per shift of running plants A and B are respectively Rs. 9000 and Rs. 6000 only. How many shifts should the firm run each plant per day so that the production cost is minimum meet to the market demand.

    Write the dual of this & give an economic interpretation of the dual variables.

24. Solve by simplex method :

    Min. $Z = \dfrac{15}{2} x_1 + 3x_2$

    Subject to $3x_1 - x_2 - x_3 \geq 3$, $x_1 - x_2 + x_3 \geq 2$, $x_1, x_2, x_3 \geq 0$

    Write the dual of the above problem. What should be the maximum value of the objective function of the dual.

25. (i)   Min. $Z = x_1 + 10x_2$

           S.t. $x_1 - 5x_2 \geq 0$, $x_1 - 5x_2 \geq -5$ & $x_1, x_2 \geq 0$

    (ii)   Max. $Z = -5y$

           S.t. $x + y \leq 1$, $-5x - 5y \leq -10$; $x, y \geq 0$

           Explain how (i) & (ii) are related?

    Answer : One is dual of the other

26. A company makes three products A, B & C out of three materials P_1, P_2, P_3 the three products use units of the three materials accordingly.

	P_1	P_2	P_3	Unit profit (Rs.)
A	1	2	3	3
B	2	1	1	4
C	3	2	1	5
Available (units)	10	12	15	

Determine the product mix to maximize total profit. Solve the primal problem & write the dual & give geometrical interpretation.

27. Use duality, if any to solve the following LPP.

Min. $Z = 15x + 10y$

S.t. $3x + 5y \geq 5$; $5x + 2y \geq 3$; $x, y \geq 0$

Answer : $x = \dfrac{5}{19}, y = \dfrac{16}{19}$, Min. $Z = \dfrac{235}{19}$

28. Use duality to Max. $Z = 6x_1 + 5x_2 - 3x_3 - 4x_4$

S.t. $2x_1 + 3x_2 + 2x_3 - 4x_4 = 24$; $x_1 + 2x_2 \leq 10$

$x_1 + x_2 + 2x_3 + 3x_4 \leq 15$; $x_2 + x_3 + x_4 \leq 8$

$x_1, x_2, x_3, x_4 \geq 0$

Answer : Infeasible solution

29. Solve the dual of the following graphically.

Min. $Z = 10a + 6b + 2c$

S.t. $-a + b + c \geq 1$; $3a + b - c \geq 2$, $a\, b, c \geq 0$

Answer : $a = \dfrac{1}{4}, b = \dfrac{5}{4}, c = 0$ Min. $Z = 10$, check answer with graph of dual

Max. $Z = x_1 + 2x_2$ S.t. $-x_1 + 3x_2 \leq 10$;

$x_1 + x_2 \leq 6$; $x_1 - x_2 \leq 2$;

$x_1, x_2 \geq 0$

30. A firm makes three products A, B & C each product requires production time in each of three departments as shown.

Products	Time Taken (Hrs/Unit)		
	Dept. I	Dept. II	Dept. III
A	3	2	1
B	4	1	3
C	2	2	3

Total time available is 60 hours, 40 hours, and 30 hours, in depts. I, II & III respectively. If product A contributes Rs. 2 per unit & product B and C Rs. 4 and Rs. 250 respectively, determine the optimum product mix

Write the dual of this problem & give its economic interpretation.

31. Find the dual of the following if it exists or otherwise solve it.

Min. $Z = 6x_1 + 5x_2 - 2x_3$

S.t. $x_1 + 3x_2 + 2x_3 \geq 5$; $2x_1 + 2x_2 + x_3 \geq 2$;

       $4x_1 - 2x_2 + 3x_3 \geq -1$;   and $x_1, x_2, x_3 \geq 0$

32. Use dual simplex method to solve the LPP.

Min. $Z = -4x_1 - 6x_2 - 18x_3$

S.t. $x_1 + 3x_3 \quad \geq 3$

        $x_1 - x_2 + x_3 \quad \geq 2$ and

        $x_1, x_2, x_3 \quad \geq 0$ **[JNTU Mech/Prod/Chem 2001/S]**

33. Find the dual and solve the linear programming problem :

Min. $Z = 20x_1 + 14x_2$

S.t. $x_1 + x_2 \leq 18$

            $5x_1 + 3x_2 \geq 10$

            $4x_1 + 6x_2 = 10$

and $x_1 \geq 0$, x_2 unrestricted   **[OU - MBA Apr/May 94]**

34. Find the dual and solve the following LPP.

Min. $Z = 4x_1 + x_2$

S.t. $3x_1 + 2x_2 \geq 6$

           $-3x_1 - x_2 \leq -4$

           $4x_1 + 5x_2 = 10$

x_1 unrestricted and $x_2 \geq 0$ **[OU - MBA, May 92]**

35. Find the dual and solve the following LPP.

Min. $Z = 10x_1 + 8_2$

S.t. $x_1 + 2x_2 \geq 5$

           $2x_1 - x_2 \geq 12$

           $x_1 + 3x_2 \geq 4$

$x_1 \geq 0$ and x_2 unrestricted   **[OU - MBA Aug/Sep. 99]**

36. Find the dual and solve the following

Min. $Z = x_1 - 3x_2 + 2x_3$

S.t. $x_1 + 3x_2 - x_3 \geq 7$

         $-2x_2 + 4x_3 \geq 12$

      $-4x_1 + 3x_2 + 8x_3 \leq 10$ **[OU - MBA Dec 2000]**

37. Solve the dual of the following using simplex method

    Max. $Z = 3x_1 + 5x_2$

    S.t. $\qquad x_1 + 4x_2 \leq 9$

    $\qquad\qquad 2x_1 + 3x_2 \leq 11$

    $\qquad\qquad x_1, x_2 \geq 0$ $\qquad\qquad$ [OU - MBA Apr. 98, Sep. 98, Dec. 2000]

38. Find the dual of following LPP.

    Min. $Z = 15x_1 + 10x_2 + 5x_3$

    S.t. $\qquad x_1 + 2x_2 + 4x_3 \leq 10$

    $\qquad\qquad 3x_1 + 5x_2 + x_3 \geq 12$

    $\qquad\qquad 4x_1 + 7x_2 + 6x_3 = 60$

    $\qquad x_1 \geq 0, \; x_2$ and x_3 unrestricted.

39. Find the optimal solution to the following LPP by solving its dual.

    Min. $Z = 6x_1 + 7x_2$

    S.t. $\qquad\qquad 3x_1 + 9x_2 \geq 36$

    $\qquad\qquad 6x_1 + 2x_2 \geq 24$

    $\qquad\qquad 2x_1 + 2x_2 \geq 16$

    and $\qquad\qquad x_1, x_2 \geq 0$ $\qquad\qquad$ [OU - MBA Nov. 94]

40. Write the dual of the following LPP and interpret it

    Max. $Z = x_1 + x_2$

    S.t. $\qquad 2x_1 + x_2 = 5$

    $\qquad\qquad 3x_1 - x_2 = 6$

    $\qquad x_1, x_2$ unrestricted $\qquad\qquad$ [OU - MBA Apr. 98, Sep. 98, Dec. 2000]

41. Find the dual and solve

    Max. $Z = 3x_1 + 5x_2$

    S.t. $\qquad\qquad 4x_1 + 5x_2 \leq 10$

    $\qquad -3x_1 + 2x_2 \geq -6$

    $\qquad 5x_1 + 3x_2 \quad = 15$

    and $x_1 \geq 0, \; x_2$ unrestricted

42. Find the optimal solution to the following LPP by solving its dual.

    Max. $Z = 80x_1 + 120x_2$

    S.t. $\qquad\qquad x_1 + x_2 \leq 9$

    $\qquad\qquad 20x_1 + 50x_2 \leq 360$

    $\qquad\qquad\qquad x_1 \geq 2$

    $\qquad\qquad\qquad x_2 \geq 3$

    and $\qquad\qquad x_j \geq 0$ for all j $\qquad\qquad$ [OU - MBA , May 91]

43. Formulate dual of the following LPP and solve it using graph

    Max. $Z = 2000x + 3000y$

    S.t. $6x + 9y \leq 100$

                    $2x + y \leq 20$

                    $x, y \geq 0$ **[OU - MBA Sep. 98, Apr. 99, Sep. 2001]**

44. Solve the dual of the following LPP using simplex method verify by graph.

    Max. $Z = 8T + 4C$

    S.t. $4T + 2C \leq 30$

                      $2T + 4C \leq 24$

                        $T, C \geq 0$ **[OU - MBA July 2000]**

45. Find dual of

    Max. $Z = 4x_1 + 5x_2 + 4x_3 - 2x_4$

    S.t. $3x_1 - 2x_2 + 3x_3 + x_4 = 5$

                    $2x_1 + 5x_2 - x_3 + 2x_4 = 6$

                                $x_1, x_2 \geq 0$

                    x_3 & x_4 unrestricted   **[OU - MBA Nov. 94]**

46. Find dual and solve

    Max. $Z = 4x_1 + 5x_2 + 6x_3$

    S.t. $2x_1 + 3x_2 + 4x_3 \leq 12$

                      $4x_1 + 6x_3 + 7x_3 \leq 14$

            x_1, x_2 unrestricted, $x_3 \geq 0$ **[OU - MBA May 95]**

47. Find solution for the dual of

    Max. $Z = 2x_1 + 5x_2$

    S.t. $x_1 + 4x_2 \leq 24$

                              $3x_1 + x_2 \leq 21$

                              $x_1 + x_2 \leq 9$

                              $x_1, x_2 \geq 0$ **[OU - MBA May 95]**

48. Find dual and solve

    Max. $Z = 3x_1 + 4x_2$

    S.t. $x_1 + 2x_2 \leq 4$

                              $-3x_1 - x_2 \leq -6$

                              $4x_1 + 5x_2 = 10$

            x_1, x_2 unrestricted   **[OU - MBA Feb. 93]**

49. Solve the dual of

    Max. $Z = 5x_1 + 3x_2$

    S.t.
    $$3x_1 + x_2 \le 12$$
    $$x_1 + 2x_2 \le 14$$
    $$x_1 + x_2 \le 8$$
    $$x_1, x_2 \ge 0$$

    [OU - MBA Feb. 93]

50. Find solution by solving dual

    Min $Z = 2500\,x_1 + 2000\,x_2 + 500\,x_3$

    S.t.
    $$10x_1 + 5x_2 + x_3 \ge 23$$
    $$6x_1 + 10x_2 + 2x_3 \ge 32$$
    $$x_1, x_2, x_3 \ge 0$$

    [OU - MBA Apr. 94]

Review Questions

1. Give the algorithm of simplex method to solve an LPP. **[ECE - 96/OT]**

2. Discuss different types of solutions you get in LPP. How do you identify the occurrence of such solutions in simplex and graphical methods **[Mech. 95/C]**

3. With a flow chart explain the method of solving LPP by simplex. **[Mech. 96/P]**

4. With a flow chart, explain penalty method of solving LPP

5. Give step by step procedure to solve LPP by BIG -M method. **[EEE - 97/OT/S]**

6. Through a flow chart discuss two-phase method of solving LPP

7. Give 2-phase algorithm for solving linear programming problems

8. Discuss the merits and demerits of 2-phase and Big-M methods.

9. Do you think two-phase method is superior to Big-M method to solve LPP. If so why? If not why not?

10. With reference to simplex, Big-M and two phase methods of solving LPP discuss the following terms.

    (a) Decision variable   (b) Slack variable

    (c) Surplus variable   (d) Artificial variable

    (e) Basic variable   (f) Non-basic variable

    (g) Entering variable   (h) Leaving variable

11. With reference to LP problems discuss the following terms.

    (a) Solutions   (b) Basic solutions

    (c) Basic feasible solutions   (d) Optimal solutions

                                                    [OU - MBA - D'2000, M 92]

12. Discuss the significance of artificial variables in simplex method of solving LPP.

                                                    [OU - MBA - S 99]

13. How do you identify the following solutions in simplex, penalty and 2-phase methods.

    (a) Infeasible solutions   **[OU - MBA - D 90, S, 98, M 92, M 95, J 2000]**

    (b) Unique solutions   **[M 95]**

    (c) Multiple optimal solutions   **[A 94, S 98, N 94, J 60, S 2001, F 93, M 95]**

    (d) Unbounded solutions   **[OU - MBA A 98, M 91]**

14. What is meant by degeneracy in simplex? When does it occur? How do you resolve it?   **[OU - MBA 90 D, M 92, A 94, F 93, S 2001]**

15. With an example discuss the degeneracy in simplex.

16. In the case of tie in entering variable in simplex, how do you resolve? Justify your answer with an example.

17. How do you convert the non-standard simplex into standard one in the following cases? Give suitable examples to each case.

    (a) Objective functions

    (b) Constraint set

    (c) Conditions of variables.

18. How do you deal with the following cases in simplex method of solving an LPP.

    (a) Unrestricted variable   (b) Exact constraints

                                                    [OU - MBA - 95 D , M 91, S '01]

19. What is cycling in simplex? When do you get such problem? How do you proceed in such case?

20. How do you set standard LPP in Big M and 2-phase in the following cases.

    (a) Maximization case of three variables with one ≤ , one = and ≥ constraints and all variables are non negative.

    (b) Minimisation case of three variables in three constraints as one '≤', one '=' and one '≥' constraints and all variables are non negative.

(c) Max. case of two variables in two constraints one '=' and one '≥' type constraints with one of the variables unrestricted and the other non-negative.

(d) Min. case of two variables in two constraints one '=' , one '≥' type constraints and with one of the variables unrestricted and the other non-negative.

21. What is meant by mixed constraints? [OU - MBA - S 99]

Review Questions in Duality

1. What do you understand by the concept of duality in simplex.

2. Discuss the advantages of duality in respect of simplex method of solutions for LP problems. [OU - MBA - S 98]

3. Dual of a dual is primal. Verify this statement with an example. [OU - MBA - N 94]

4. Discuss the principles/rules to follow when a primal problem is to be converted into its dual.

5. Distinguish between the terms of Primal and dual of simplex of LPP.

6. Compare primal and dual characteristics with reference the formulation and solution method. [OU - MBA - M 91, M 92]

7. What do you understand by shadow price? What is its significance in simplex method of solving LPP. [OU MBA - S 98]

8. Write step by step method to convert a non - standard primal problem to its dual.

9. Give the applications of dual problem in LPP.

10. Write a short note on the concept of duality in LPP. [OU - MBA - 94 D, 95 D, M 91, M 92, S 2001]

11. How do you get the solution of a primal problem from final solution tableau of dual?

12. Discuss the significance of primal-dual of LPP with reference to game theory.

13. If a primal has m constraints and n variables, how many constraints and variable will its dual have ? [OU - MBA - A 98]

Objective Type Questions

1. In simplex table, the value of $Z_j - C_j$ for basic variables will be

   (a) negative   (b) positive

   (c) zero   (d) unity

2. The solution of LPP is not an infeasible if

   (a) all minimum ratios are either infinity or negative

   (b) artificial variables are not replaced

   (c) a replacement of variables is cycle

   (d) phase - I does not yield any solution in 2-phase method

3. Which of the following is a degeneracy in the case of simplex

   (a) total supply is equal to total demand

   (b) the number of allocated cells exceed the rows+columns - 1

   (c) the number of occupied cells is less than rows+columns -1

   (d) a tie exists for leaving basic variable.

4. If a primal has unrestricted variable, its dual will have

   (a) corresponding variable unrestricted

   (b) corresponding constant is requirement type

   (c) corresponding constraint is exact type

   (d) the constraint is written by splitting the corresponding variable into two parts.

5. Which of the following will have multiple optimal solutions

   (a) $Z_j - C_j$ is zero for non basic (decision) variables

   (b) cycling of variables in the basis

   (c) all slack variables are not replaced and $C_j - Z_j$ is still negative

   (d) all the artificial variables are not replaced.

6. For a set of 'm' equations in 'n' variables (n>m), the number of solutions will be _____

   (a) m_{c_n} (b) n_{c_m}

   (c) $n_{c_{n-m}}$ (d) infinite

7. If primal problem has unrestricted variables then its dual will have

   (a) $\geq$ type constraints   (b) $\leq$ type constraints

   (c)= type constraints   (d) unrestricted variables

8. If the solution for a primal problem is infeasible, its dual will have _____ solution

    (a) multiple optimal   (b) unbounded

    (c) infeasible   (d) unique

9. For 'm' dissimilar equations with 'n' variables ($n>m$), the number of basic feasible solutions will be

    (a) m_{c_n} (b) n_{c_m}

    (c) $n_{c_{n-m}}$ (d) infinite

10. In a simplex table Key Column is determined by

    (a) most negative value of $C_j - Z_j$
    (b) most negative value of $Z_j - C_j$
    (c) minimum value of (solution value)/(coefficient of variable in constraint)
    (d) none of the above

11. If there is a tie for entering variable, we give preference to

    (a) slack variable   (b) surplus variable
    (c) decision variable   (d) artificial variable

12. If a constraint is less than or equal to type, we add

    (a) slack variable only
    (b) surplus variable only
    (c) surplus variable and artificial variable
    (d) slack variable and artificial variable

13. If there is a tie for Key column selection, then preference is given to _____ variable

    (a) slack   (b) surplus
    (c) artificial   (d) arbitrary

14. If minimum ratio is same for a two rows, preference is given to the row having _____ variable

    (a) slack   (b) surplus
    (c) artificial   (d) arbitrary

15. $Z_j - C_j$ value for basic variable in a simplex tableau will be

    (a) any positive value   (b) zero

    (c) negative value   (d) ≥ 0

16. Simplex tableau yields an optimal solution if

    (a) basic variable has $C_j - Z_j \geq 0$

    (b) all non basic variables have $Z_j - C_j \leq 0$

    (c) all non basic variables have $Z_j - C_j \geq 0$

    (d) all variables have $Z_j - C_j \leq 0$

17. To improve the simplex tableau, we try to make

    (a) basic variable equal to unity   (b) key element equal to unity

    (c) solution value=0   (d) non basic variable =0

18. In simplex tableau we neglect _____ value for minimum ratio

    (a) unity   (b) zero

    (c) infinity   (d) positive

19. In an LPP: Max $Z = 5x_1 + 6x_2$, subject to $2x_1 + 3x_2 \geq 50$, $4x_1 + 3x_2 \geq 100$, the objective function of first phase is _____

    (a) $5x_1 + 6x_2$ (b) $+ 5x_1 + 6x_2 - MA_1 - MA_2$

    (c) $+A_1 + A_2$ (d) $0.x_1 + 0.x_2 - A_1 - A_2$

20. The minimum ratio is neglected if

    (a) zero   (b) negative

    (c) infinity   (d) infinity or negative

21. In simplex tableau, the basic variables will constitute _____ matrix in the body

    (a) null   (b) unit

    (c) inverse   (d) singular

22. If the coefficient of variables shown in the form of matrix in primal and dual, then one is _____ matrix to the other

    (a) unit   (b) inverse

    (c) transpose   (d) null

23. Coefficients of variables in objective function of a primal problem are _____ of dual problem

    (a) coefficient of variables in objective function

    (b) solution values of constraints

    (c) conditions

    (d) zero

24. If primal has three constraints, the dual will have

    (a) three constraints   (b) three variables

    (c) three conditions   (d) any of the above

25. If a dual has three variables in two constraints, its primal will have

    (a) three variables and two constraints
    (b) three conditions and two variables
    (c) two variables and three constraints
    (d) two conditions and three variables

26. The solution values of a dual-solution are _____ of primal solution

    (a) solution values   (b) zeros

    (c) shadow prices   (d) negative values

27. If a set of values satisfy all conditions, constraints, it is called

    (a) feasible solution   (b) optimal solution

    (c) multiple solutions   (d) infeasible solution

28. For an LPP: Max $Z_p = 5 x_1 + 3 x_2$ subject to $2 x_1 + x_2 \geq 50, x_1 + x_2 \leq 30$, $x, \geq 0$, x_2 is unrestricted, dual is to be written. Which of the following is false

    (a) min. $Z_d = 50 w_1 + 30 w_2$ (b) $2w_1 + w_2 \leq 5$

    (c) $w_1 + w_2 \geq 3$ (d) $w_1 \geq 0, \; w_2 \geq 0$

29. For a primal Max $Z = 50 x_1 + 40 x_2$, S.t. $3 x_1 + 4 x_2 \leq 60$; $2 x_1 + 3 x_2 \leq 70$; $x_1 \geq 0, x_2 \geq 0$, the dual is written. Then objective function of dual will be _____

    (a) max. $Z = -50 x_1 - 40 x_2$ (b) min. $Z = 50 x_1 + 40 x_2$

    (c) min. $Z = -60 w_1 - 70 w_2$ (d) max. $Z = -60 w_1 - 70 w_2$

30. Penalties are used in

    (a) Two phase method   (b) Big-M method

    (c) dual simplex   (d) all the above

Fill in the Blanks

1. Dual of a dual is _____

2. Coefficients of variables in objective function of dual problem are _____ of primal problem.

3. If primal is to maximise, dual is to _____

4. A surplus variable is subtracted and Artificial variable is added for _____ type of constraint

5. The basic variable will have $C_j - Z_j =$ _____

6. Solution values of primal solution are _____ of dual solution

7. If there is no solution in I-phase for a simplex problem in 2 - phase method, then final solution is _____

8. If $Z_j - C_j = 0$ for a non basic variable in simplex tableau, we may get _____ solution

9. The number of basic feasible solutions for system of three equations with four variables is _____

10. The set of values that satisfy the conditions and constraint is said be _____

Answers

Objective Type Questions :				
1. (c)	2. (c)	3. (d)	4. (c)	5. (a)
6. (d)	7. (c)	8. (b)	9. (c)	10. (b)
11. (c)	12. (a)	13. (a)	14. (c)	15. (b)
16. (c)	17. (b)	18. (c)	19. (d)	20. (d)
21. (b)	22. (c)	23. (b)	24. (b)	25. (c)
26. (c)	27. (a)	28. (c)	29. (d)	30. (b)

Fill in the Blanks :	
1. primal	2. solution values or constraints of constraints
3. minimise	4. requirement or $\geq$
5. zero	6. shadow prices
7. infeasible	8. multiple or alternate optimal
9. four	10. feasible solution

5 Transportation Problem

Chapter

CHAPTER AT A GLANCE

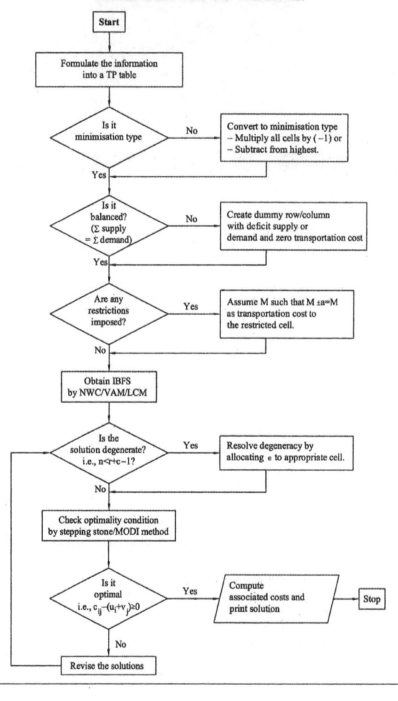

5.0 *Introduction*

Transportation problem is another case of application to linear programming problems, where some physical distribution (transportation) of resources is to be made from one place to another to meet certain set of requirements with in the given availability. The places from where the resources are to be transferred are referred to as sources or origins. These sources or origins will have the availability or capacity or supply of resources. The other side of this transportation i.e. to where the resources are transported are called sinks or destinations such as market centres, godowns etc. These will have certain requirements or demand.

5.1 *Formulation*

Transportation problem is applied to the situations in which a single product is transported from several origins (say O_1, O_2, ... O_m) to several destinations (say $D_1, D_2, \ldots D_m$). Let us assume the cost of transporting a unit product from O_i to D_j is C_{ij} and the no. of units transported be x_{ij}. Let the capacity of O_i be a_i and requirement of D_j be b_j. Then, the transportation problem can be mathematically written as

$$\text{Minimise (Total transportation cost) } Z = \sum_{i=1}^{m} \sum_{j=1}^{n} C_{ij} x_{ij}$$

Subject to the constraints

$$\sum_{j=i}^{M} x_{ij} = a_{ij}, \ i = 1, 2, \ldots \ldots m \text{ (supply or availability constraints)}$$

$$\sum_{i=1}^{M} x_{ij} = b_{ij}, j = 1, 2, \ldots \ldots n \text{ (demand or requirement constraints)}$$

and $x_{ij} \geq 0$ for all i and j.

Conveniently, the above can be represented as a tabular as follows.

		Destinations				
	To / From	D_1	D_2	....	D_m	Supply or avaliable
Origins	O_1	C_{11} $\boxed{x_{11}}$	C_{12} $\boxed{x_{12}}$	....	C_{1n} $\boxed{x_{1n}}$	a_1
	O_2	C_{21} $\boxed{x_{21}}$	C_{22} $\boxed{x_{22}}$	....	C_{2n} $\boxed{x_{2n}}$	a_2
	⋮	⋮	⋮	....	⋮	⋮
	O_m	C_{m1} $\boxed{x_{m1}}$	C_{m2} $\boxed{x_{m2}}$	....	C_{mn} $\boxed{x_{mn}}$	a_m
Demand or Requirement		b_1	b_2	....	b_n	$\sum a_i$ / $\sum b_j$

Remark :

In the above table, if total supply $\left(\sum\limits_{i=1}^{m} a_i\right)$ = Total demand $\left(\sum\limits_{j=1}^{n} b_j\right)$, the transportation problem is said to be balanced transportation, otherwise it is unbalanced transportation.

The above condition is said to be *"balance condition"* or *"rim condition"* of T.P.

5.2 Transportation Problem Formulated as an LPP

From the above section we can understand that the TP is a case in LPP. Thus the given TP can be formulated into an LPP as illustrated in the following example.

ILLUSTRATION 1 ————————————————————————————————————

Forumulate the following TP into an LPP

	D_1	D_2	D_3	Supply
O_1	5	6	7	50
O_2	3	7	4	60
O_3	8	5	4	40
Demand	30	55	65	150
				150

Solution :

In the above TP,

Let   the no. of units transported from O_1 to $D_1 = x_{11}$

the no. of units transported from O_1 to $D_2 = x_{12}$ and so on.

Summarily, the no. of units transported from O_i to $D_j = x_{ij}$

where $i = 1, 2, 3$ and $j = 1, 2, 3$

The the total cost can be obtained by summation of the products of the number of units to be transported from i^{th} origin (row) to j_{th} destination (column) with unit cost given in the corresponding cell (C_{ij}). This cost is to be minimised.

∴ *Objective function* is

Minimize $Z = 5x_{11} + 6x_{12} + 7x_{13} + 3x_{21} + 7x_{22} + 4x_{23} + 8x_{31} + 5x_{32} + 4x_{33}$

Subject to the constraints

(i) Supply constraints

$x_{11} + x_{12} + x_{13} = 50$

$x_{21} + x_{22} + x_{23} = 60$

$x_{31} + x_{32} + x_{33} = 40$

(ii) Demand constraints

$$x_{11} + x_{21} + x_{31} = 30$$

$$x_{12} + x_{22} + x_{32} = 55$$

$$x_{13} + x_{23} + x_{33} = 65 \text{ and } x_{ij} \geq 0 \text{ where } i = 1, 2 \ 3 \text{ and } j = 1, 2, 3$$

Remark :

For the above TP, we have $r \times c = 3 \times 3 = 9$ variables and $r + c = 3 + 3 = 6$ equations. Thus it is difficult or time taking process to solve by simplex technique. Further, it becomes too difficult if there are more number of row or/and columns. The other easier techniques are available in TP which will be discussed at length in this chapter.

5.3 *Transportation Algorithm: Solution Method*

The solution algorithm of transportation problem (T.P) is summarised as follows.

Step 1: *Formulate the Problem in the Form of Matrix :* The formulation of the problem is similar to that of linear programming. Here the standard form of objective function is to minimse the total transportation cost and the constraints are the supply (available) and demand (requirement) to each origin and destination respectively.

Step 2 : *Standardize the TP :* A T.P is said to be standard if it is to minimise the total costs. If the given problem is not standard, it is to be first standardized.

5.3.1 *Maximisation Case in Transportation Problem*

The T.P. (Maximisation type) is standardised (i.e., converted to minimisation type) by one of the following two methods.

(i) Identify highest element among all the elements in the cells of transportation matrix. Subtract the element of every cell from this highest element and put in their respective cells. This is now called *equivalent cost matrix*.

(ii) Multiply by (– 1) to the elements in all the cells of transportation matrix.

Ex : *Maximisation :*

	D_1	D_2	D_3	Supply
O_1	9	2	5	20
O_2	8	0	– 3	30
O_3	6	7	9	50
Demand	15	45	40	

Ex : *Minimisation :*

by method (i)			Supply
0	7	4	50
1	9	12	60
3	2	0	40
Demand 15	45	40	

by method (ii)			Supply
-9	-2	-5	20
-8	0	$+3$	30
-6	-7	-9	50
Demand 15	45	40	

Step 3 : *Check Rim Condition and Obtain Balanced Transportation Problem :*

5.3.2 *Unbalanced Transportation Problem*

The rim condition is that the total supply and total demand are equal

i.e.,   Total supply (availability) = Total demand (requirement)

If the rim condition is not satisfied in the given T.P. , it is said to be *unbalanced* *T.P.* and is therefore to be balanced by creating a dummy row/column with the amount equal to short fall (or defficiency), of the demand / supply respectively. The respective transportation are assumed to be zeros.

Example 1: Total supply < Total demand

Unbalanced

	D_1	D_2	D_3	Supply
O_1	19	12	15	20
O_2	18	10	13	25
O_3	16	17	19	50
Demand	15	45	40	95 / 100

Balanced

	D_1	D_2	D_3	Supply
O_1	19	12	15	20
O_2	18	10	13	25
O_3	16	17	19	50
Dummy (row)	0	0	0	5
Demand	15	45	40	100 / 100

Example 2 : Total Supply > Total demand

Unbalanced

	D_1	D_2	D_3	Supply
O_1	19	12	15	20
O_2	18	10	13	35
O_3	16	17	19	50
Demand	15	45	40	105 / 100

Balanced Dummy

	D_1	D_2	D_3	(column)	Supply
O_1	19	12	15	0	20
O_2	18	10	13	0	25
O_3	16	17	19	0	50
Demand	15	45	40	5	105 / 105

Step 4: *Obtain Initial Basic Feasible Solution (IBFS) :*

After getting standard balanced TP, an initial basic feasible solution is obtained by one of the following methods.

(i) North West Corner Method (NWCM)

(ii) Vogel's Approximation Method (VAM)

(iii) Least Cost Entry Method (LCEM) or Matrix Minima Method

These methods are explained in the sections to follow in this chapter.

Step 5: *Test the presence of degeneracy and obtain non-dengerate basic feasible solution :*

5.3.3 *Degeneracy in Transportion Problem*

The IBFS obtained (in the step - 4) is said be degenerate if the number of allocated cells is less than $r + c - 1$ where r is number of rows and c is number of columns in the TP matrix.

Degeneracy : No. of allocated cells i.e., $n\ (C_{ij}) < r + c - 1$

Nondegeneracy : No. of allocated cells i.e., $n\ (C_{ij}) \geq r + c - 1$

If the problem is non-degenerate proceed to step - 6, otherwise resolve the degeneracy by assuming a minute allocation ε (read as Epsilon) to a suitable cell such that $x_{ij} + \varepsilon = x_{ij} - \varepsilon = x_{ij}$ and $\varepsilon \neq 0$. Then proceed to step - 6.

Step 6 : *Optimality test :*

Test the non degenerate basic feasible solution for optimality by one of the following methods.

(i) Stepping stone method

(ii) MOdified DIstribution (MODI) Method :

If the solution passes the test, then it is the optimal solution, otherwise go to step - 7

Step 7 : *Update the solution :*

The solution is improved by the given set of rules (discussed in the sections to follow) till the optimal solution reached.

At every improvement, step 5 and step 6 are repeated

5.4 *Methods of Finding Initial Basic Feasible Solution (IBFS)*

There are several methods available for finding initial basic feasible solution of transportation problem. In this chapter we discuss three important methods viz. North West Corner Method (NWCM), Vogel's Approximation Method (VAM) and Least Cost Entry Method (LCEM).

Flow Chart for Solution of Transportation Problem :

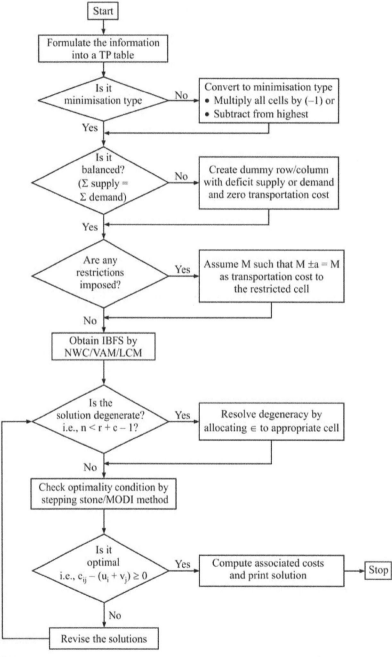

FIGURE 5.1:

5.4.1 *North West Corner Method*

This is the simplest method of all methods for finding IBFS of a TP. This method is independent of the cost or profit of transportation. The algorithm of NWCM goes as follows.

Step 1 : *Formulation :* Given information is formulated in TP matrix.

Step 2 : *Standaradising (Minimisation form):* Check whether the given TP is in standard form. A TP is said to be standard if the objective function is to minimise the total cost. However, if the TP is to maximise, it may be converted to minimisation form either of the two methods given below.

   (i) From highest among all transportation profits, subtract all profits and put back in their respective cells (or)

   (ii) Multiply profits in all cells by (-1).

Note : *As this method is independent of costs or profits this step may be ignored also. But be careful while optimising the initial solution.*

Step 3 : *Balancing :* Check whether the TP is balanced. The balanced TP is that in which total availability (supply) is equal to total requirement (demand) of origins and destinations respectively. If the TP is balanced proceed to next step, otherwise balance by adding the defict amount of availability/requirement to a dummy origin/destination respectively with unit costs as zeros.

Step 4 : *Allocation at North West Corner :*

In the given TP matrix identify the cell at north west corner i.e., top left cell. At the starting of the problem it is the cell connecting origin-1 (O_1) to diestination-1 (D_1). Now check the availability of O_1 and requirement of D_1 and allocate as many units as possible (i.e., minimum of availability of O_1 and requirements of D_1) in this cell and indicate at one of the corners, usually at the top-right of the cell.

Ex :

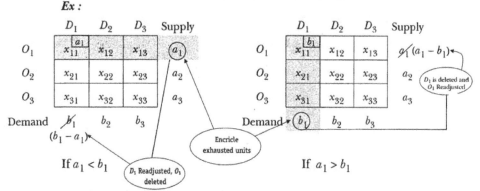

Now, if availability of O_1 i.e. a_1 is less than requirement of D_1 i.e., b_1 then allocate a_1 units to the cell (C_{11}), otherwise (if $b_1 < a_1$) allocate b_1 units to the cell (C_{11}). When an availability or requirement is exhausted encircle it and delete the corresponding row/column for the next iteration. Also readjust the availability or requirement of row or column respectively by subtracting the allocated units.

If $a_1 = b_1$ both row & column will be exhausted and deleted for next iteration. This leads to degeneracy in T.P. while optimising.

Step 5 : Repeat the above step - 4 till all the availability and requirement are exhausted.

Step 6 : Calculate the total costs by summing up the products of unit cost in the cell with number of units of all allocated cells. This is the total transportation cost given by the initial basic feasible solution.

Note : *The TP is to reverted to its original matrix before calculating the maximum profit, if it is is converted to minimisation case earlier i.e. in step - 2*

Flow Chart :

The flow chart of North West Corner Method is given below.

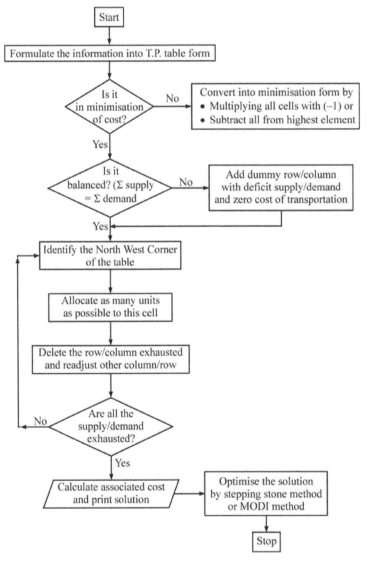

FIGURE 5.2 :

The above method is illustrated with an example here below.

ILLUSTRATION 2 ───

Consider the following transportation problem

Source	Destination				Total
	D_1	D_2	D_3	D_4	
O_1	1	2	1	4	30
O_2	3	3	2	1	50
O_3	4	2	5	9	20
Total	20	40	30	10	100

Determine the initial feasible solution   (JNTU Mech-Prod-Mechantronics-Chem-2001)

Solution :

Step 1 : *Formulation :* The given problem is already formulated, hence proceed to step 2.

Step 2 : The given problem is to minimise i.e., Standard. (If it is not mentioned, it is understood as standard TP).

Step 3 : *Balancing :* Since the row totals (100) equal to column totals (100), it is already a balanced TP.

Step 4 : Allocation by NWCM

ITERATION TABLEAU - I :

In the above TP table, the northwest corner cell is (C_{11}) or (O_1, D_1). In this cell, we allocate as many units as possible. The availability at this source (O_1) is 30, but the requirement for the destination (D_1) is only 20. Therefore we can transfer only 20 units from O_1 to D_1 (20 < 30). Therefore O_1 will have 10 more units to supply to any destination other than D_1. And since the D_1 requirement is exhausted we delete this column (D_1) for next iteration while we adjust the supply of O_1 from 30 to 10.

Verdict of I - Tableau - 1 :

Cell allocated (C_{11}) or (O_1, D_1)

No. of units allocated 20

Deletion for next iteration : D_1 column (exhausted for D_1)

Readjustment for O_1 row : 30 to 10 (As 20 units are allocated)

Cell cost of transportation : $20 \times 1 = 20$

ITERATION TABLEAU - 2 :

The above step is repeated

	D_2	D_3	D_4	Suppl
O_1	[10] 2	1	4	(10)
O_2	3	2	1	50
O_3	2	5	9	20
Demand	4̶0̶ / 30	30	10	80 / 80

Allocated cell (C_{12}) is (O_1, D_2)
No. of units allocated is 10
Deletion for next iteration is O_1
Readjustment for D_2: 40 to 30
Cost of transportation from O_1 to D_2
$= 2 \times 10 = 20$

ITERATION TABLEAU - 3 :

	D_2	D_3	D_4	Supply
O_2	[30] 3	2	1	5̶0̶ 20
O_3	2	5	9	20
Demand	30	30	10	70 / 70

Allocated cell (C_{22}) is (O_2, D_2)
No. of units allocated is 30
Deletion for next iteration is D_2
Readjustment for O_2 is 50 to 20
Cost of transportation is $3 \times 30 = 90$

ITERATION TABLEAU - IV :

	D_3	D_4	Supply
O_2	[20] 2	1	(20)
O_3	5	9	20
Demand	3̶0̶ / 10	10	40 / 40

Allocated cell (C_{23}) is (O_2, D_3)
No. of units allocated is 20
Deletion for next iteration is O_2
Readjustment for D_3 is 30 to 10
Cost of transportation is $2 \times 20 = 40$

ITERATION TABLEAU - 5 :

	D_3	D_4	Supply
O_3	10 / 5	10 / 9	20
Demand	10	10	20 / 20

Allocated cells (C_{33} and C_{34}) are (O_3, D_3) and (O_3, D_4)

No. of units allocated are 10 and 10 respectively.

Cost are 5×10 and 9×10

i.e., 50 and 90 respectively.

All the units are exhausted

$\therefore$ IBFS is obtained.

Summary :

	D_1	D_2	D_3	D_4	Supply
O_1	20 / 1	10 / 2	1	4	30
O_2	3	30 / 3	20 / 2	1	50
O_3	4	2	10 / 5	10 / 9	20
Demand	20	40	30	10	

Cost Calculations of IBFS by NWCM

Allocated Cell	From	To	Unit Cost	No. of Units	Total Cost
C_{11}	O_1	D_1	1	20	20
C_{12}	O_1	D_2	2	10	20
C_{22}	O_2	D_2	3	30	90
C_{23}	O_2	D_3	2	20	40
C_{33}	O_3	D_3	5	10	50
C_{34}	O_3	D_4	9	10	90
Grand Total Cost of Transportation					310
Total cost of transportation = Rs. 310/-					

ILLUSTRATION · 3 ──

A dealer stocks and sells four types of Bicycles namely Atlas, Bharath, Champion, Duncan which he may procure from three different suppliers namely Priyanshu, Qureshi and Raju. His anticipated sales for the bicycles for the coming seasons are 410, 680, 310 and 550 nos. respectively. He can obtain 900 bicycles from Priyanshu, 600 from Qureshi and 560 from Raju at suitable prices. The profit per bicycle in rupees for each supplier is tabulated below.

Type \ Supplier	Atlas (A)	Bharath (B)	Champion (C)	Duncan (D)
Priyanshu (P)	21.50	26.00	19.50	21.00
Qureshi (Q)	20.50	24.00	20.00	21.00
Raju (R)	18.00	19.50	19.00	19.50

Formulate the above information as transportation model and obtain initial solution by North West Corner Rule.   [JNTU - ECE/EEE - 95/0T]

Solution :

Step 1 : Formulation

Type \ Supplier	Atlas (A)	Bharath (B)	Champion (C)	Duncan (D)	Availability
Priyanshu (P)	21.50	26	19.5	21	900
Qureshi (Q)	20.50	24	20	21	600
Raju (R)	18	19.5	19	19.5	560
Requirement	410	680	310	550	2060 / 1950

Step 2 : *Standardisation :* The above problem contains profit matrix and therefore it is to be maximised. This is non-standard form. As standard T.P. is to be minimised an equivalent cost matrix is to be obtained by any of the two methods (i) Multiplying with (– 1) in all cells (ii) subtracting each cell value from highest among the cell values. Here (i) is used. In fact, there is no need for standardising here, since NWCR is independent of cost or profit. However, for having a common practice this is carried out.

Type \ Supplier	Atlas (A)	Bharath (B)	Champion (C)	Duncan (D)	Dummy (E)	Availability
Priyanshu (P)	– 21.5	– 26	– 19.5	– 21	0	900
Qureshi (Q)	– 20.5	– 24	– 20	– 21	0	600
Raju (R)	– 18	– 19.5	– 19	– 19.5	0	560
Requirement	410	680	310	550	110	2060

Step 2 : *Balancing :* The above problem is unbalanced since the requirement (1950) is not equal to availability (2060). As the availability is excess, we create a Dummy column (Say E) with no profit for any supplier and with a requirement of the deficit i.e, 110.

Type Supplier	Atlas (A)	Bharath (B)	Champion (C)	Duncan (D)	Dummy (E)	Availability
Priyanshu (P)	21.5	26	19.5	21	0	900
Qureshi (Q)	20.5	24	20	21	0	600
Raju (R)	18	19.5	19	19.5	0	560
Requirement	410	680	310	550	110	2060 2060

Step 4 : With usual rules we iterate as follows to get initial solution.

ITERATION TABLEAU 1 :

	A	B	C	D	E	Avl.
P	410 – 21.5	– 26	– 19.5	– 21	0	900 490
Q	– 20.5	– 24	– 20	– 21	0	600
R	– 18	– 19.5	– 19	– 19.5	0	560
Req.	(410)	680	310	550	110	2060 1650

Verdict :
Allocation to (P, A)
No. of units 410
Deletion col. A
Adjustment to P 900 to 490
Cost is $– 21.5 \times 410 = – 8815$
(Negative cost means profit here)

ITERATION TABLEAU - 2 :

	A	B	C	D	E	Avl.
P	410 – 21.5	490 – 26	– 19.5	– 21	0	900 (490)
Q	– 20.5	– 24	– 20	– 21	0	600
R	– 18	– 19.5	– 19	– 19.5	0	560
Req.	410	680 190	310	550	110	1650 1160

Verdict :
Allocation to (P, B)
No. of units 490
Deletion : row P
Adjustment to B 680 to 190
Cost : $– 26 \times 490 = – 12740$

ITERATION TABLEAU - 3 :

	A	B	C	D	E	Avl.
P	410 – 21.5	490 – 26	– 19.5	– 21	0	900
Q	– 20.5	190 – 24	– 20	– 21	0	600 410
R	– 18	– 19.5	– 19	– 19.5	0	560
Req.	410	680 190	310	550	110	1160 970

Verdict :
Allocation to (Q, B)
No. of units : 190
Deletion : col . B
Adjustment to Q as 600 to 410
Cost : $– 24 \times 190 = – 4560$

ITERATION TABLEAU - 4 :

	A	B	C	D	E	Avl.
P	410 / −21.5	490 / −26	−19.5	−21	0	900
Q	−20.5	190 / −24	310 / −20	−21	0	600 / ~~410~~ 100
R	−18	−19.5	−19	−19.5	0	560
Req.	410	680	(310)	550	110	~~970~~ 660

Verdict :
Allocation to (Q, C)
No. of units : 310
Deletion col : C
Adjustment to Q as 410 to 100
Cost : $-20 \times 310 = -6200$

ITERATION TABLEAU 5 & 6 :

	A	B	C	D	E	Avl.
P	410 / −21.5	490 / −26	−19.5	−21	0	900
Q	−20.5	190 / −24	390 / −20	100 / −21	0	600 / ~~410~~ 100
R	−18	−19.5	−19	450 / −19.5	110 / 0	~~560~~ 110
Req.	410	680	310	~~550~~ (450)	(110)	~~660~~ ~~560~~ 450

Verdict for Tableau - 5 :
Allocation to (Q, D)
No. of units : 100
Deletion : row Q
Adjustment to D as 550 to 450
Cost : $-21 \times 100 = -2100$

Verdict for Tableau - 6 :
Allocation to (R, D) and (R, E)
No. of units 450 and 110
(all units exhausted)
Cost : $= -19.5 \times 450$ and
$0 \times 100 = 8775$ and 0

ITERATION TABLEAU - 7 :

Final Tableau - 7 for IBFS by NWCM :

Equivalent Cost Matrix

	A	B	C	D	E	Avl.
P	410 / −21.5	490 / −26	−19.5	−21	0	900
Q	−20.5	190 / −24	310 / −20	100 / −21	0	600
R	−18	−19.5	−19	450 / −19.5	110 / 0	560
Req.	410	680	310	550	110	2060

Profit Matrix

	A	B	C	D	E	Avl.
P	410 / 21.5	490 / 26	19.5	21	0	900
Q	20.5	190 / 24	310 / 20	100 / 21	0	600
R	18	19.5	19	100 / 19.5	110 / 0	560
Req.	410	680	310	550	110	2060

Profit Calculation :

Allocated Cell	From	To	Unit Cost	No. of Units	Total Cost
C_{11}	P	A	21.5	410	8815
C_{12}	P	B	26	490	12740
C_{22}	Q	B	24	190	4560
C_{23}	Q	C	20	310	6200
C_{33}	Q	D	21	100	2100
C_{34}	R	D	19.5	450	8775
C_{35}	R	E	0	110	00
				Total Profit	43190

5.4.2 *Vogel's Approximation Method (VAM) or Penalty Method*

This method is a heuristic method and concentrates to minimise penalty cost. In this method, the allocation is made on the basis of the opportunity or penalty (extra cost) that would have incurred if the allocation in certain cell with minimum unit transportation cost is missed. The major advantage of this method is that the initial solution obtained by this method is very close to the optimal solution or most times, it is the optimal solution itself.

The Algorithm of VAM :

The algorithm of VAM is as follows :

The first three steps will be common as those of NWCM.

Step 1 : *Formulation :* The given information is formulated in tabular form with unit costs, supply and demand.

Step 2 : *Standardisation :* The T.P is said to be standard if it is in the form of minimisation of total transportation 'cost'. If the T.P has a profit matrix (i.e., to maximise total profit), it will be converted to equivalent cost matrix by one of the two methods given below.

  (i) Multiply by (-1) to all cell profits.

  (ii) Subtract every cell profit from highest profit among all cells.

Step 3 : *Balancing :* The T.P is said to be balanced if total supply is equal to total demand, otherwise, it is balanced by creating a dummy row/column (whichever is necessary) with unit costs as 'zeros' to each cell and with the deficit availability/requirement (or supply/demand).

Step 4 : *Calculation of Penalties :* Calculate penalty for each row as well as each column by taking difference between least and next least in the row/column. If in a row/column, there are two least costs (equal and minimum), then penalty is zero. This penalty indicates the minimum extra cost or penalty that has to be paid if the allocation is not made at the cell with least cost.

Step 5 : *Allocation :* Select a row or column that has highest penalty and allocate as much as possible to the cell with least cost in the selected row or column. If there is a tie in selection of highest penalty, selection can be made arbitrary.

Step 6 : *Adjustments :* Delete the row or column for next iteration whose availability or demand is exhausted/satisfied. Adjust the supply/demand of row/column after subtracting the allocated units. If a row and column satisfy their supply and demand simultaneously both will be removed for the next iteration.

Step 7 : *Repetition :* Repeat the step 4 to 6 , till the entire units of supply for origins and entire units of demand for destinations are satisfied.

Step 8 : *Cost Calculation :* Calculate the cost by summing up the products of number of units with unit cost of transportation for all allocated cells. If the final table is equivalent cost matrix converted from profit matrix in step - 3, it is reverted to profit matrix before calculating the profits.

Flow Chart for VAM :

Flow chart for Vogel's Approximation Method for finding IBFS.

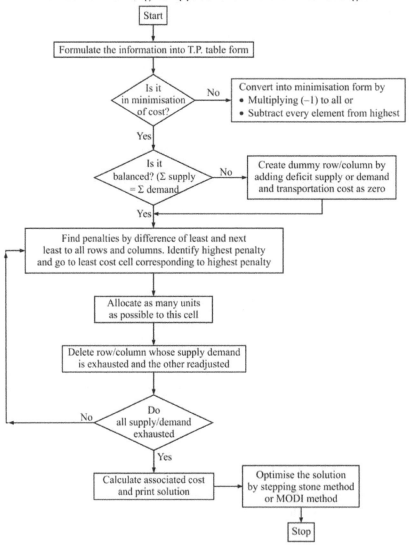

FIGURE 5.3 :

ILLUSTRATION 4

> *(Standard and balanced T.P) [Consider problem given illustration-2]*
> *Consider the transportation problem*
>
> | Source | Destination | | | | Total |
> |--------|-----|-----|-----|-----|-------|
> | | D_1 | D_2 | D_3 | D_4 | |
> | O_1 | 1 | 2 | 1 | 4 | 30 |
> | O_2 | 3 | 3 | 2 | 1 | 50 |
> | O_3 | 4 | 2 | 5 | 9 | 20 |
> | Total | 20 | 40 | 30 | 10 | 100 |
>
> *Determine the initial feasible solution.*
>
> [OU MSc - 84, IAS - 88,
> JNTU B.Tech Mech, Prod, Mechatronics, Chem - 2001]

Solution :

Step 1 : *Formulation :* The given problem is already formulated, therefore go to step - 2.

Step 2 : *Standardisation :* TP is in standard form i.e., to minimise the total transportation costs.

Step 3 : *Balancing :* The given TP is already balanced since the total supply (30 + 50 + 20 = 100) is equal total demand (20 + 40 + 30 + 10 = 100). Therefore go to step - 3.

Step 4 : Calculation of penalties for rows and columns.

ITERATION TABLEAU 1 :

	D_1	D_2	D_3	D_4	Supply	Penalty
O_1	1	2	1	4	30	0 (1-1)
O_2	3	3	2	1	50	1 (2-1)
O_3	4	2	5	9	20	2 (4-2)
Demand	20	40	30	10	100	
Penalty	2 (3-1)	0 (2-2)	1 (2-1)	3 (4-1)		

Penalty for O_1, row : least is 1, and next least is also 1, therefore difference is zero.

Penalty for O_2 row : least is 1, next least is 2 therefore penalty is 1.

Similarly for O_3 : 4 - 2 = 2.

Also for D_1 : 3 - 1 = 2, for D_2 : 2 - 2 = 0 for D_3 : 2 - 1 = 1 and D_4 : 4 - 1 = 3.

Step 5 : *Allocation :*

ITERATION TABLEAU 2 :

	D_1	D_2	D_3	D_4	Supply	Penalty
O_1	1	2	1	4	30	0
O_2	3	3	2	[10] 1	5̶0̶ 40	1
O_3	4	2	5	9	20	2
Demand	20	40	30	(10)	100	
Penalty	2	0	1	(3) ↑		

Verdict :

Allocation at C_{24} i.e., (O_2, D_4)

No. of units 10, deletion D_4 col.

Adjustment for O_2 from 50 to 40

Cost : $1 \times 10 = 10$

Step - 6 : *Adjustments :* Highest penalty among {0, 1, 2 and 2, 0 ,1, 3} is 3. Therefore we choose D_4 column and we allocate in the cell with least cost i.e. (O_2, D_4) cell i.e., C_{24}. The maximum number of units we can allocate for this cell is 10 units, since its demand is only 10 though supply is 50. Thus, the demand for D_4 is satisfied, therefore we delete this column for next iteration. And for row O_2, as 10 units are sent to D_4, we adjust it to 40 as further capacity to supply as shown in iteration tableau - 2.

Step - 7 : Repeat steps 4 through 6, till all the centres are satisfied.

ITERATION TABLEAU - 3 :

	D_1	D_2	D_3	Supply	Penalty
O_1	1	2	1	30	0
O_2	3	3	2	40	1
O_3	4	[20] 2	(20) 5	20	(2) ←
Demand	20	4̶0̶ 20	30	90	
Penalty	2	0	1		

Verdict :

Allocation at C_{32} i.e., (O_3, D_2)

No. of units 20, deletion O_3

Adjustment for O_2 from 40 to 20

Cost : $2 \times 20 = 40$

Here, highest penalties (2) are in tie. therefore we select arbitrarily as O_3.

ITERATION TABLEAU - 4 :

	D_1	D_2	D_3	Supply	Penalty
O_1	[20] 1	2	1	3̶0̶ 10	0
O_2	3	3	2	40	1
Demand	(20)	20	30	70	
Penalty	(2) ↑	1	1		

Verdict :

Allocation at C_{11} i.e., (O_1, D_1)

No. of units 20, deletion D_1

Adjustment for O_1 from 30 to 10

Cost : $1 \times 20 = 20$

ITERATION TABLEAU - 5 :

	D_2	D_3	Supply	Penalty
O_1	2	10 / 1	⑩	(1) ←
O_2	3	2	40	1
Demand	20	3̶0̶ 20	50	
Penalty	1	1		

Verdict :

Tie for penalty is resolved by arbitration and selected O_1

Allocation (C_{13}) i.e., (O_1 , D_3)

No. of units : 10, Deletion : O_1

Adjustment for D_3 from 30 to 20

Cost = $1 \times 10 = 10$

ITERATION TABLEAU - 6 :

	D_2	D_3	Supply	Penalty
O_2	20 / 3	20 / 2	40	1
Demand	20	20	40	

Here no calculation is required and directly we can allocate as 20 units each to (O_2, D_2) and (O_2, D_3) to satisfy all supply and demand with costs as 3×20 and 2×20 i.e., 60 and 40

Summary :

	D_1	D_2	D_3	D_4	Supply
O_1	20 / 1	2	10 / 1	4	30
O_2	3	20 / 3	20 / 2	10 / 1	50
O_3	4	20 / 2	5	9	20
Demand	20	40	30	10	

(Check whether sums of the allocated units are same as their respective demand/supply)

Step 8 : *Cost calculation :*

Allocation cell	From	To	No. of units	Unit cost	Total
C_{11}	O_1	D_1	20	1	20
C_{13}	O_1	D_3	10	1	10
C_{22}	O_2	D_2	20	3	60
C_{23}	O_2	D_3	20	2	40
C_{24}	O_2	D_4	10	1	10
C_{31}	O_3	D_2	20	2	40
				Total	180

ILLUSTRATION · 5 ——————————————————————————————————

(Maximisation Case and Balanced)
Obtain IBFS of transporation problem whose profit matrix is given below.

		Markets			
		M_1	M_2	M_3	**Stock**
Godowns	G_1	4	4	9	25
	G_2	3	5	8	20
	Sales	18	16	11	45

Solution :

Step 1 : *Formulation :* The given TP is already formulated.

Step 2 : *Equivalent Cost Matrix :* As the given TP is in maximisation form, it is to be converted to equivalent cost matrix. This can be done in two methods [one method is discussed in illustration - 3, other method is taken here].

ITERATION TABLEAU · 1 :

Equivalent Cost Matrix :

		M_1	M_2	M_3	Stock
Godowns	G_1	5	5	0	25
	G_2	6	4	1	20
	Sales	18	16	11	45

Highest among all unit profits is 9. All other unit cost are calculated by subtracting from 9.

Step 3 : *Check Balancing :* As total stock (25 + 20 = 45) is equal to total sales (18 + 16 + 11 = 45), the TP is balanced.

Step 4 to 6 : Calculation of penalties

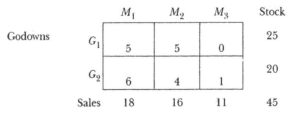

	M_1	M_2	M_3	Stock	Penalty
G_1	5	5	0 ⟶ 11	25 / 14	5
G_2	6	4	1	20	3
Sales	18	16	(11)	45	
Penalty	1	1	1		

Verdict :

Highest penalty is 5 among 5, 3, 1, 1, 1

∴ G_1 is chosen to allocated at C_{13} i.e., (G_1, M_3)

No. of units : 11
Deletion : M_3
Adjustment to G_1 as 25 to 14
Cost = $0 \times 11 = 0$
Profit = $9 \times 11 = 99$

Step 7 : Repetition of steps 4 to 6

	M_1	M_2	Stock	Penalty
G_1	5	5	14	0
G_2	6	16 / 4	2̶0̶ / 4	2 ←
Sales	18	(16)	34	
Penalty	1	1		

Verdict :

Highest penalty is 2

∴ G_2 is selected; allocation to C_{22}

i.e., (G_2 , M_2)

No. of units : 16

Deletion : M_2

Adjustment to G_2 from 20 to 4

Cost = $4 \times 16 = 64$

Profit = $5 \times 16 = 80$

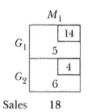

	M_1	Stock
G_1	14 / 5	14
G_2	4 / 6	4
Sales	18	

Summary :

Equivalent Cost Matrix

	M_1	M_2	M_3	Stock
G_1	14 / 5	5	11 / 0	25
G_2	4 / 6	16 / 4	1	20
Sales	18	16	11	45

Profit Matrix (Original)

	M_1	M_2	M_3	Stock
G_1	14 / 4	4	11 / 9	25
G_2	4 / 3	16 / 5	8	20
Sales	18	16	11	45

Step - 8 : *Profit calculations :*

Allocation cell	From	To	No. of units	Unit profit	Total profit
C_{11}	G_1	M_1	14	4	56
C_{13}	G_1	M_3	11	9	99
C_{21}	G_2	M_1	4	3	12
C_{22}	G_2	M_2	16	5	80
Grand total profit					247

Note : *This VAM can also be applied to the maximisation problem directly i.e., not converting into minimisation by taking penalty as highest and next highest and allocating at high profit cell. However this method may lead to some confusion, therefore students are advised to follow the above method only.*

■■■ ILLUSTRATION - 6 ————————————————————————————

(Minimisation case with unbalanced TP)

Priyanshu Enterprise has three factories at locations A, B and C which supplies three warehouses located at D, E and F. Monthly factory capacities are 10, 80 and 50 units respectively. Monthly warehouse requirements are 75, 20 and 50 units respectively. Unit shipping cost (in Rs.) are given as

<table>
<tr><td></td><td colspan="3" align="center">*Warehouse*</td></tr>
<tr><td>*Factory*</td><td>*D*</td><td>*E*</td><td>*F*</td></tr>
<tr><td>*A*</td><td>*5*</td><td>*1*</td><td>*7*</td></tr>
<tr><td>*B*</td><td>*6*</td><td>*4*</td><td>*6*</td></tr>
<tr><td>*C*</td><td>*3*</td><td>*2*</td><td>*5*</td></tr>
</table>

The penalty costs for not satisfying demand warehouses D, E and F are Rs. 5, Rs. 3 and Rs. 2 per unit respectively. Determine the optimal distribution for Priyanshu using transportation technique. [JNTU (CSE) 2000]

Solution :

Note : *Only IBFS is illustrated here and it may not be optimal solution. You have to optimise after finding IBFS by using stepping stone or MODI method.*

Step 1 : *Formulation :* The given information is formulated as follows :

ITERATION TABLEAU - 1 :

<table>
<tr><td></td><td colspan="3" align="center">Warehouse</td><td></td></tr>
<tr><td>Factory</td><td>D</td><td>E</td><td>F</td><td>Capacity</td></tr>
<tr><td>A</td><td>5</td><td>1</td><td>7</td><td>10</td></tr>
<tr><td>B</td><td>6</td><td>4</td><td>6</td><td>80</td></tr>
<tr><td>C</td><td>3</td><td>2</td><td>5</td><td>50</td></tr>
<tr><td>Requirement</td><td>75</td><td>20</td><td>50</td><td>140
145</td></tr>
</table>

Step 2 : *Standardisation :* Since the given TP has cost matrix, it is in the standard form i.e., to minimise of total transportation costs.

Step 3 : *Check balancing :*

In the above TP, total capacity (140) is less than total requirement (145), Therefore, this TP is unbalanced. To balance this TP we create another dummy factory say K, with the deficit capacity, i.e., 145 - 140 = 5.

The transportation costs usually we assume zeros, but here these are taken as 5, 3 and 2 instead of zeros since these are penalty costs for not satisfying the demand.

Thus the reformulated TP tableau is

ITERATION TABLEAU - II :

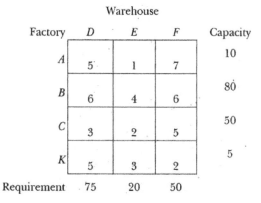

Warehouse

Factory	D	E	F	Capacity
A	5	1	7	10
B	6	4	6	80
C	3	2	5	50
K	5	3	2	5
Requirement	75	20	50	

Step 4 to 6 : Penalty calculation and allocation and adjustments :

ITERATION TABLEAU - III :

Warehouse

Factory	D	E	F	Capacity	Penalty
A	5	1 [10]	7	(10)	(4) ◄-----
B	6	4	6	80	2
C	3	2	5	50	1
K	5	3	2	5	1
Requirement	75	20 / 10	50	145	

Verdict :

Highest penalty for A i.e., 4
Allocation to C_{12} i.e., A to E
No. of units 10
Deletion A (exhausted)
Adjustment to E as 20 to 10
Cost : $1 \times 10 = 10$

Step 7: Repeat the above steps :

ITERATION TABLEAU - IV :

Warehouse

Factory	D	E	F	Capacity	Penalty
B	6	4	6	80	2
C	3	2	5	5	1
K	5	3	2 [5]	(5)	1
Requirement	75	10	50 / 45	135	
Penalty	2	1	3 ↑		

Verdict :

Highest penalty is for F i.e., 3.
Allocation to C_{43} i.e., K to F
No. of units : 5
Deletion : K (exhausted)
Adjustment : for F from 50 to 45
Cost : $2 \times 5 = 10$

ITERATION TABLEAU - V :

Warehouse

Factory	D	E	F	Capacity	Penalty
B	6	4	6	80	2
C	3 [5]	2	5	(50)	1
Requirement	~~75~~ 25	10	45	130	
Penalty	3 ↑	2	1		

Verdict :

Highest penalty is for f i.e., 3.
Allocation to C_{43} i.e., K to F
No. of units : 5
Deletion : K (exhausted)
Adjustment : for F from 50 to 45
Cost : $2 \times 5 = 10$

ITERATION TABLEAU - VI :

Warehouse

Factory	D	E	F	Capacity	Penalty
B	6 [25]	4 [10]	6 [45]	80	2
Requirement	25	10	45	80	
Penalty	3 ↑	2	1		

Verdict :

Penalty calculation is not required
Allocations B to D : 25, B to E : 10
and B to F : 45

Summary :

Factory	D	E	F	Capacity
A	5	1 [10]	7	10
B	6 [25]	4 [10]	6 [45]	80
C	3 [50]	2	5	50
K	5	3	2 [5]	5
Requirement	75	20	50	145

Step 8 : *Cost calculation :*

Allocation	From	To	No. of units	Unit cost	Total cost
C_{12}	A	E	10	1	10
C_{21}	B	D	25	6	150
C_{22}	B	E	10	4	40
C_{23}	B	F	45	6	270
C_{31}	C	D	50	3	150
C_{43}	K	F	5	2	*10
Grand total transport cost					620 + 10 = 630

* Total transportation cost is Rs. 620 and penalty for not satisfying the demand of 5 units to F is 10. Thus total is Rs. 630/-

5.4.3 *Least Cost Entry Method (Matrix Minima Method)*

This method concentrates on the cells having least cost. Thus the allocation entry is to be made in the cell with least cost among all the cells. The algorithm is as follows:

Step 1 : *Formulation :* Formulate the information into TP matrix.

Step 2 : *Standard TP :* If the TP is non-standard i.e., maximisation, convert it to standard i.e., minimisation form by either multiplying all profit cells with -1 or by subtracting from highest profit cell to make equivalent cost matrix.

Step 3 : *Balancing :* Check whether the TP is balanced by total supply = total demand and add dummy row/column with deficit and zero costs in the case of unbalanced.

Step 4 : *Allocation :* Select the cell with least cost* among all the cost cells and allocate as many units as possible, and delete the row/column whose supply/demand is exhausted while the demand/supply of column/row is readjusted by subtracting the allocated units.

If there is tie in selection least cost, choose arbitrarily or select the one where maximum allocation can be made.

Step 5 : *Repeat :* Step - 4 is repeated till all the units of supply/demand are exhausted.

Step 6 : *Cost Calculation :* Calculate the total transport cost by summing up the products of allocated units with unit cost.

---

* Least cost entry method is further divided into three types, such as row minima method, column minima method and overall minima or matrix minima method. In row minima method we choose least row-wise to allocate while in column minima method we prefer to take least of each column to allocate. However, these method are not much popularly used since these are not close to optimal solution. Therefore overall minima or matrix minima method only is discussed here.

Flow chart for LCEM or Matrix Minima Method :

The flow chart for LCEM is given below :

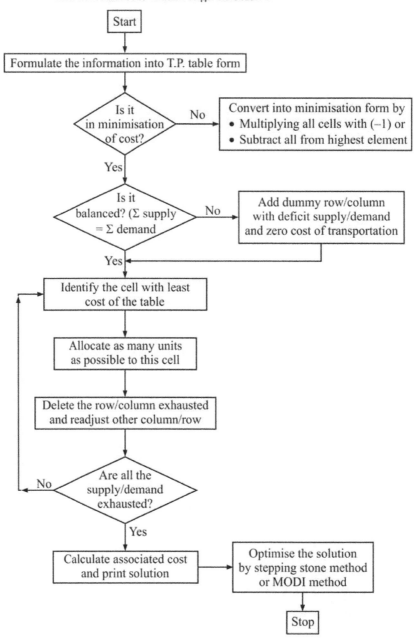

FIGURE 5.4 :

ILLUSTRATION 7 ────────────────────────────────

(Consider the TP given in illustration - 2 and illustration - 4).

Source	Destination				
	D_1	D_2	D_3	D_4	Total
O_1	1	2	1	4	30
O_2	3	3	2	1	50
O_3	4	2	5	9	20
Total	20	40	30	10	100

Determine the initial feasible solution.

[OU Msc 84, IAS - 88 JNTU Mech, Prod/Mechatronics, chem - 2001]

Solution :

Step 1 to 3 are already done i.e., the given TP is formulated, balanced and standard.

Step 4 : *Allocation :* The least cost among all the costs is '1' appearing at three places viz. at O_1 to D_1, O_1 to D_3 and O_2 to D_4 though we can choose arbitrarily the maximum units can be allocated to (O_1, D_3) cell i.e., 30. Here, the cell has supply of 30 and requirement also as 30. Hence both row and column are deleted for further iteration.

ITERATION TABLEAU - I :

Verdict :

Chosen at (O_1, D_3)

Allocation to C_{13} i.e., O_1 to D_3

No. of units = 30

Deletion O_1 & D_3

Adjustments : Nil

Cost : $1 \times 30 = 30$

Note : *We can notice degeneracy while optimisation of the above TP if both row and column get deleted simultaneously. This degeneracy is explained in the next section under optimisation of TP with degeneracy.*

Step 5 : Repeat step - 4, till all the units of supply/demand are exhausted.

ITERATION TABLEAU - II :

Verdict :

Least cost is '1' at (O_2, D_4) i.e., C_{24}

No. of units = 10

Deletion = D_4

Adjustment for O_2 from 50 to 40

Cost = $1 \times 10 = 10$

ITERATION TABLEU - 3 :

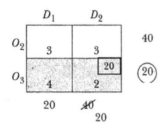

ITERATION TABLEU - 4 :

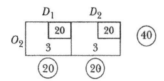

Verdict :

   Least cost is '2'

   Allocated cell C_{32} i.e., O_3 to D_2

   No. of units : 20

   Deletion O_3

   Adjustment to D_2 is from 40 to 20

   Cost $= 2 \times 20 = 40$

Verdict :

   It can be directly allocated as :

   C_{21} or O_2 to D_1 : 20

   C_{22} or O_2 to D_2 : 20

   Costs $3 \times 20 + 3 \times 20 = 60 + 60 = 120$

Summary :

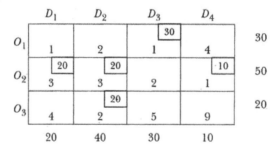

Step 6 : Cost calculation :

Allocated	From	To	Unit cost	No. of units	Total cost
C_{13}	O_1	D_3	1	30	30
C_{21}	O_2	D_1	3	20	60
C_{22}	O_2	D_2	3	20	60
C_{24}	O_2	D_4		10	10
C_{32}	O_3	D_2	2	20	40
				G. Total cost	200

ILLUSTRATION 8

Consider the problem in illustration- 6 which is unbalanced standard T.P. The same is illustrated with this method here.

Solution :

Since first three steps are same, the solution is given from step 4.

ITERATION TABLEAU - 1 :

	D	E	F	
A	5	[10] 1	7	(10)
B	6	4	6	80
C	3	2	5	50
K	5	3	2	5
	75	~~20~~ 10	50	145

Verdict :
Least cost = '1'
Allocated cell C_{12} i.e., A to E
No. of units = 10
Deletion A
Adjustment to E is 20 to 10
Cost = $1 \times 10 = 10$

ITERATION TABLEAU - 2 :

	D	E	F	
B	6	4	6	80
C	3	[10] 2	5	~~50~~ 40
K	5	3	2	5
	75	(10)	50	135

Verdict :
Least cost is 2 at (C, E) and (K, F)
Selection made arbitrarily at (C, E)
i.e., C_{32}
No. of units : 10
Deletion : E
Adjustment to C as 50 to 40
Cost = $2 \times 10 = 20$

ITERATION TABLEAU - 3 :

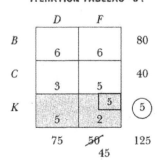

	D	F	
B	6	6	80
C	3	5	40
K	5	[5] 2	(5)
	75	~~50~~ 45	125

Verdict :
Least cost : 2
Allocation cell (C_{43}) or (K, F)
No. of units : 5
Deletion : K
Adjustment to F : 50 to 45
Cost = $2 \times 5 = 10$

ITERATION TABLEAU - 4 & 5 :

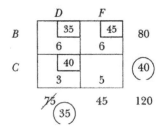

Verdict :

Verdict :

Least cost : 3 at (C, D)

∴ 40 units are allocated and hence C is deleted, then other units allocated directly at (B, D) and (B, F) as 35 and 45 respectively.

Costs : $3 \times 40 + 6 \times 35 + 6 \times 45$
     $= 120 + 210 + 270 = 600$

Summary :

	D	E	F	
A		10		10
	5	1	7	
B	35		45	80
	6	4	6	
C	40	10		50
	3	2	5	
K			5	5
	5	3	2	
	75	20	50	145

Cost calculation :

Allocated	From	To	Unit cost	No. of units	Total cost
C_{12}	A	E	1	10	10
C_{21}	B	D	6	35	210
C_{23}	B	F	6	45	270
C_{31}	C	D	3	40	120
C_{32}	C	E	2	10	20
C_{43}	K	F	2	5	10*
Total transportation cost + penalty cost					630 + 10* = 640

* Penalty for not satisfying the demand (5) of F is Rs. 2×5 = Rs. 10

Now, let us take up another problem with non-standard i.e., maximisation model.

Consider the problem under in illustration 5, to illustrate in this method

ILLUSTRATION 9 ───

Obtain IBFS for the profit matrix given below.

	M_1	M_2	M_3	Stock
G_1	4	4	9	25
G_2	3	5	8	20
Demand	18	16	11	45

Solution :

Step 1 : *Formulation :* It is already formulated.

Step 2 : *Standard TP :* The given matrix is profit matrix, therefore it is to be maximised. As standard TP is to be minimised, an equivalent cost matrix is to obtained by one of the following two methods.

(i) Multiplying by (-1) to all cells.

(ii) Subtracting all profits from highest value among all profit cells to make equivalent costs. Both methods are simultaneously discussed here below.

Equivalent Cost Matrix :

ITERATION TABLEAU - 1 (a) :   ITERATION TABLEAU - 1 (b) :

	M_1	M_2	M_3	
G_1	5	5	0	25
G_2	6	4	1	20
	18	16	11	

OR

	M_1	M_2	M_3	
G_1	-4	-4	-9	25
G_2	-3	-5	-8	20
	18	16	11	

Step 3 : *Balance :* It is already balanced since total stock $(25+20)$ = total sales $(18+16+11)$

Step 4 : *Allocation :* The least cost among the cells is 'zero' in 1 (a) [or -9 in 1 (b)] which appears in C_{13} cell i.e., from G_1 to M_3. A maximum of 11 units can be allocated here. Subsequently, M_3 gets exhausted and deleted for further iteration while stock G_1 is adjusted to 14.

ITERATION TABLEAU - 2 (a) :

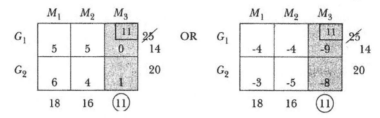

ITERATION TABLEAU - 2 (b) :

Step 5 : *Repeat :* Step 4 till all the units are exhausted.

ITERATION TABLEAU - 3(a) :

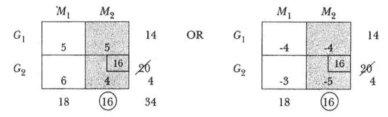

ITERATION TABLEAU - 3(b) :

Leat cost is 4 in iteration tableau - 3(a) or -5 in iteration 3(b), which appears in C_{22} cell i.e., G_2 to M_2. So, 16 units can be allocated here since M_2 requires 16 only though G_2 can supply 20 units.

ITERATION TABLEAU - 4(a) :

ITERATION TABLEAU - 4(b) :

IBFS : $G_1 - M_3$: 11 units, $G_1 - M_1$: 14 units; $G_2 - M_1$: 4; $G_2 - M_2$: 16

Summary :

The allocated units are now entered in the respective cells of original TP (profit matrix).

	14	4	11	25
	4		9	
	4	16	8	20
	3	5		
	18	16	11	

Profit Calculation :

Allocated cell	From	To	No. of units	Unit profit	Total profit
C_{11}	G_1	M_1	14	4	56
C_{13}	G_1	M_3	11	9	99
C_{21}	G_2	M_1	4	3	12
C_{22}	G_2	M_2	16	5	80
		Total units	45	Total profit	247

Aliter : *The maximisation case may also be solved by allocating as many units as possible in the cell with highest profit. This is a direct method called "highest profit entry method". Without converting the matrix into equivalent cost matrix. However, this method may lead to certain confusion and hence not discussed at length.*

Note : *Answer of IBFS with the three methods (NWCM, VAM and LCEM) need not be same.*

5.5 *Optimisation of Transportation Problem*

After obtaining IBFS by one of the three method viz., NWCM, VAM or LCEM, we move towards optimality. To find the optimal solution we have two methods

1. Stepping Stone method.

2. **MO**dified **DI**stribution (MODI) method.

These are discussed in the following sections.

5.5.1 *Stepping Stone Method*

This method is developed by *A*. Charnes. In this method we use the *improvement index*. The algorithm is as follows.

Algorithm :

Obtain IBFS :

Step 1 : *Obtained Initial solution* by any suitable method such as NWCM, VAM, LCEM etc.

Step 2 : *Check degeneracy :*

Check whether there is any degeneracy. This is checked by the degeneracy condition that the number of allocated cells is less than sum of number of rows and number of columns minus one.

Non-degenerate : $n\ (C_{ij}) \geq n\ (r) + n\ (c) - 1$ [or simply $n\ (C_{ij}) \geq r + c - 1$]

Degenerate : $n\ (C_{ij}) < n\ (r) + n\ (c) - 1$ [or simply $n\ (C_{ij}) < r + c - 1$]

Where $n\ (C_{ij})$ is no. of allocated cells in IBFS, $n\ (r)$ is no. of rows and $n\ (c)$ is no. of columns of TP. If there is degeneracy, resolve it, else move to step-3. (*Degeneracy case is explained later*).

Step 3 : *Calculate Improvement Index :* IBFS may or may not be optimum. To check whether the IBFS is optimum or not, we calculate Improvement Index, which indicates possibility of cost reduction by assigning one unit to unallocated cell.

   This is done as follows :

   (i) Connect one un-allocated cell at a time with all other allocated cells as corners of a loop. The loop need not always be a square. It may be any shape as given below.

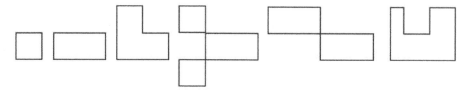

FIGURE 5.5 :

   Further, the loop should always be closed and should contain either horizontal or vertical lines only, provided all corners are allocated cells except one whose index is to be found. Also the loop lines may intersect, but the point of intersection is not a corner and has no importance. The loop line may go over an unallocated or allocated cell and such cells have no interference in calculation of index.

   Thus after drawing the loop, give positive sign to the un-allocated cell cost and alternatively negative and positive to the corners of the cells. Sum up these costs of cells to find the index.

Step 4 : *Revision to Find Improved Matrix :* If all the indices are positive, the optimal solution is reached (IBFS itself is OBFS). But, if any value is negative, select unallocated cell with most negative then transfer possible number of units to the selected cell along the loop and readjust the supply/demand of allocated cells.

Step 5 : *Repeat :* Repeat step 3 & 4 untill all the indices are positive.

Flow Chart for Stepping Stone Method :

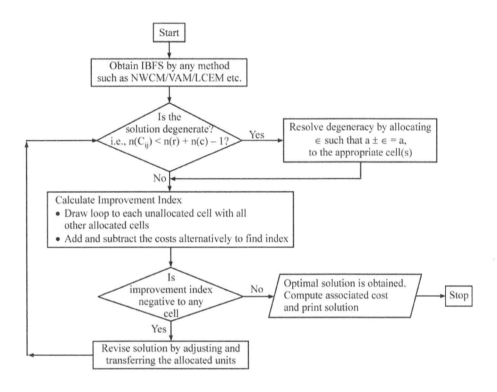

FIGURE 5.6 :

Consider the problem given in illustration - 2

ILLUSTRATION 10 ....

	D_1	D_2	D_3	D_4	Supply
O_1	1	2	1	4	30
O_2	3	3	2	1	50
O_3	4	2	5	9	20
Demand	20	40	30	10	

Solution :

Now let optimise this by stepping stone method.

Step 1 : *Obtain IBFS :* (Refer Illustration - 2) the IBFS by North west corner rule is.

	D_1	D_2	D_3	D_4	Supply
O_1	20 / 1	10 / 2	1	4	30
O_2	3	30 / 3	20 / 2	1	50
O_3	4	2	10 / 5	10 / 9	20
Demand	20	40	30	10	100

Step 2 : Degeneracy test by $n \cdot (C_{ij}) \geq r + c - 1$

No. of allocated cells = 6

$[(O_1, D_1), (O_1, D_2), (O_2, D_2), (O_2, D_3), (O_3, D_3), (O_3, D_4)]$

No. of rows = 3

No. of columns = 4

As no. of allocated cells i.e., $n (C_{ij})$ is (6) equal to $n (r) + n (c) - 1$ i.e., $3 + 4 - 1 = 6$, there is no degeneracy in the TP.

Step 3 : *Optimality Test By Improvement Index :* Loops are drawn for unallocated cells as given below

Loop for (O_1, D_3) Cell.   Loop for (O_1, D_4)

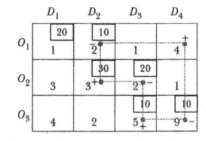

Index : $+ 1 - 2 + 3 - 2 = 0$ *Index* : $+ 4 - 9 + 5 - 2 + 3 - 2 = - 1$

Loop for (O_2, D_1) :

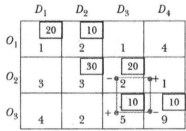

	D_1	D_2	D_3	D_4
O_1	20	10		
	⌐ 2 ● + 1		1	4
O_2	+ 30	20		
	● 3	3 −	2	1
O_3		10	10	
	4	2	5	9

Index : $+ 3 - 1 + 2 - 3 = + 1$

Loop for (O_2, D_4) Cell.

	D_1	D_2	D_3	D_4
O_1	20	10		
	1	2	1	4
O_2		30	20	
	3	3	− 2 ● + 1	
O_3			10	10
	4	2	+ 5 −	9

Index : $+ 1 - 2 + 5 - 9 = - 5$

Loop for (O_3, D_1) :

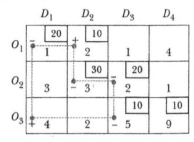

	D_1	D_2	D_3	D_4
O_1	20	10		
	− 1 + 2		1	4
O_2		30	20	
	3	− 3	2 −	1
O_3			10	10
	+ 4	2	− 5	9

Index : $+ 4 - 5 + 2 - 3 + 2 - 1 = - 1$

Loop for (O_3, D_2) :

	D_1	D_2	D_3	D_4
O_1	20	10		
	1	2	1	4
O_2		30	20	
	3	− 3	+ 2	1
O_3			10	10
	4	+ 2	− 5	9

Index : $+ 2 - 5 + 2 - 3 = - 4$

Of all the above the most negative index is for (O_2, D_4) cell i.e., $- 5$. Therefore we transfer max. possible number of units (i.e., 10) to this cell from (O_3, D_4).

Step 4 : *Revision :* When 10 units are transferred from (O_3, D_4) to (O_2, D_4), we have to reduce 10 units from (O_2, D_3) and add at (O_3, D_3). This addition or subtraction can be easily understood by sign on the corner of the loop in the cell. Thus the revised matrix is.

	D_1	D_2	D_3	D_4	
O_1	20	10			30
	1	2	1	4	
O_2		30	10	10	50
	3	3	2	1	
O_3			20		20
	4	2	5	9	
	20	40	30	10	

Step 5 : *Repeating Steps 2 to 4 :* No degeneracy is found and the indices for unallocated cells are as follows.

Unallocated	Loop	Index
(O_1, D_3)	$(O_1, D_3) \rightarrow (O_2, D_3) \rightarrow (O_2, D_2) \rightarrow (O_1, D_2)$	$+1 - 2 + 3 - 2 = 0$
(O_1, D_4)	$(O_1, D_4) \rightarrow (O_2, D_4) \rightarrow (O_2, D_2) \rightarrow (O_1, D_2)$	$+4 - 1 + 3 - 2 = +4$
(O_2, D_1)	$(O_2, D_1) \rightarrow (O_1, D_1) \rightarrow (O_1, D_2) \rightarrow (O_2, D_2)$	$+3 - 1 + 2 - 3 = +1$
(O_3, D_1)	$(O_3, D_1) \rightarrow (O_1, D_1) \rightarrow (O_1, D_2) \rightarrow (O_2, D_2)$ $\rightarrow (O_2, D_3) \rightarrow (O_3, D_3)$	$+4 - 1 + 2 - 3 + 2 - 5 = -1$
(O_3, D_2)	$(O_3, D_2) \rightarrow (O_2, D_2) \rightarrow (O_2, D_3) \rightarrow (O_3\, D_3)$	$+2 - 3 + 2 - 5 = \boxed{-4} \leftarrow$
(O_3, D_4)	$(O_3, D_4) \rightarrow (O_2, D_4) \rightarrow (O_2\, D_3) \rightarrow (O_3\,, D_3)$	$+9 - 1 + 2 - 5 = +5$

As (O_3, D_2) shows most negative index we have to transfer 20 units, (this max possible number of units to be transferred can be found by least allocated units among the negative cornered cells of the loop). The revised solution is

20	10			30
1	2	1	4	
	10	30	10	50
3	3	2	1	
	20			20
4	2	5	9	
20	40	30	10	

Again, the indices in the revised matrix are

for (O_1, D_3): 0 ; (O_1, D_4) : +4; (O_2, D_1) : + 1 remain same as above tableau.

for (O_3, D_1) : Loop is $(O_3, D_1) \rightarrow (O_3, D_2) \rightarrow (O_1, D_2) \rightarrow (O_1, D_1)$ and index is
$$+4 - 2 + 2 - 1 = +3$$

for (O_3, D_3) : Loop is $(O_3, D_3) \rightarrow (O_2, D_3) \rightarrow (O_2, D_2) \rightarrow (O_3, D_2)$ and index is
$$+5 - 2 + 3 - 2 = +4$$

for (O_3, D_4) : Loop is $(O_3, D_4) \rightarrow (O_2, D_4) \rightarrow (O_2, D_2) \rightarrow (O_3, D_2)$ index is
$$+9 - 1 + 3 - 2 = +9$$

Thus all the indices are positive, therefore optimal solution is obtained. Solution is

Allocated	From	To	No. of units	Unit cost	Total cost
C_{11} or (O_1, D_1)	O_1	D_1	20	1	20
C_{12} or (O_1, D_2)	O_1	D_2	10	2	20
C_{22} or (O_2, D_2)	O_2	D_2	10	3	30
C_{23} or (O_2, D_3)	O_2	D_3	30	2	60
C_{24} or (O_2, D_4)	O_2	D_4	10	1	10
C_{32} or (O_3, D_2)	O_3	D_2	20	2	40
				G. Total cost	180

Remark : There is an alternate optimal solution the above problem, which is indicated by zero index value for unallocated cells. Here (O_1, D_3) shows zero. Thus the solution is

	D_1	D_2	D_3	D_4	
O_1	20 1	2	10 1	4	30
O_2	3	20 3	20 2	10 1	50
O_3	4	20 2	5	9	20
	20	40	30	10	

Optimal (Min) cost $= 1 \times 20 + 1 \times 10 + 3 \times 20 + 2 \times 20 + 1 \times 10 + 2 \times 20$

$= 180$ (Ref. Illustration - 20)

Consider the TP of maximisation case given in illustration - 5 and 9.

ILLUSTRATION 11 ————————————————————————————

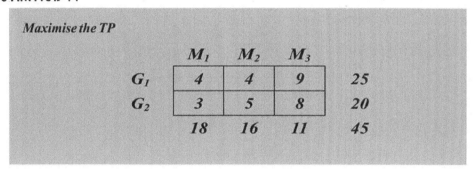

Maximise the TP

	M_1	M_2	M_3	
G_1	4	4	9	25
G_2	3	5	8	20
	18	16	11	45

Solution :

Step 1 : The IBFS by VAM or LCEM is

Equivalent Cost Matrix

14 5	5	11 0	25
4 6	16 4	1	20
18	16	11	

Profit Matrix (Original)

14 4	4	11 9	25
4 3	16 5	8	20
18	16	11	

Step 2 : *Degeneracy test :*

No. of rows (2) + No. of col. (3) – 1 = 4

No. of allocated cells = 4; i.e., $n(C_{ij}) = n(r) + n(c) - 1$

∴ No degeneracy

Step 3 : *Optimality test :*

For (G_1, M_2) For $(G_2 - M_3)$

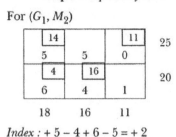

14		11	25
5	5	0	
4	16		20
6	4	1	
18	16	11	

14		11	25
5	5	0	
4	16		20
6	4	1	
18	16	11	

Index : $+ 5 - 4 + 6 - 5 = + 2$ *Index :* $+ 1 - 0 + 5 - 6 = 0$

Since indices for all unallocated cells is positive (or zero), the optimal solution is attained. And the solution is

Max profit $= 4 \times 14 + 9 \times 11 + 3 \times 4 + 5 \times 16 = 247$ (*For detailed calculation refer illustration 5 or 9*)

> **Remark:** *The above problem yields multiple optimal solution. This can be identified by zero index value for non-allocated cells.*

The solution is

	M_1	M_2	M_3	
G_1	18		7	25
	4	4	0	
G_2		16	4	20
	3	5	8	
	18	16	11	

Profit $= 4 \times 18 + 9 \times 7 + 5 \times 16 + 8 \times 4$

$= 72 + 63 + 80 + 32 = 247$

Note : *For unbalanced case also, the stepping stone method is applied in the similar way after balancing and obtaining IBFS.*

5.5.2 *MODI Method*

Modified Distribution (MODI) method is based on the properties of PRIMAL and DUAL and is simplified or modified version of stepping stone method. The algorithm is given below.

Algorithm :

Step 1 : *Initial Solution :* Develop IBFS by any method such as NWCM, VAM, LCEM etc.

Step 2 : *Degeneracy Test :* Check the presence of degeneracy by $n(C_{ij}) < n(r) + n(c) - 1$ where $n(C_{ij})$ is number of allocated cells, $n(r)$ is number of rows and $n(c)$ is number of columns. If $n(C_{ij}) \geq n(r) + n(c) - 1$, then there is no degeneracy, otherwise resolve degeneracy *(case of degeneracy is explained in the section to follow)*.

Step 3 : *Shadow Prices Calculation :* Find shadow prices of each row and each column with the formula

$$C_{ij} = u_i + v_j$$

where $\qquad\qquad C_{ij}$ = Cost in allocated cell

u_i = Shadow price of ith row

v_j = Shadow price of jth column.

As the number of equations will be equal to number of allocated cells and number of variables are sum of no. of rows and no. of columns, value of one of the variables is to assumed as zero while calculating the values of u_i and v_j.

Step 4 : *Optimality Test :* Find the cost difference $(C_{ij} - Z_{ij})$ for unallocated cells by using the formula.

$$C_{ij}' - (u_i + v_j)$$

where C_{ij}' is cost in unallocated cell and u_i & v_j are the corresponding values (shadow prices) found in step - 3.

Now, if all the values of $C_{ij}' - (u_i + v_j) \geq 0$ i.e., positive, then the solution is optimal, otherwise improve the solution by step 5.

Step 5 : Identify most negative value of $C_{ij}' - (u_i + v_j)$ calculated in step - 4 among all unoccupied cells and draw a closed loop with (only vertical/ horizontal lines) with this cell as one of its corners while all other corners as occupied (allocated) cells. [Refer stepping stone method for further details about drawing loop].Starting at the selected cell as *+ve*, give alternate *+ve* and *– ve* signs to all corners. Among all the negative sign corners choose the least allocation and transfer this amount along the loop. [Add at positive corner and subtract at negative corner].

Step 6 : Repeat steps 2 to 5 till the $C_{ij}' - (u_i + v_j) \geq 0$ for all the unoccupied cells.

Step 7 : Calculate the cost [Profit in the case of maximisation TP]

Flow Chart for MODI Method :

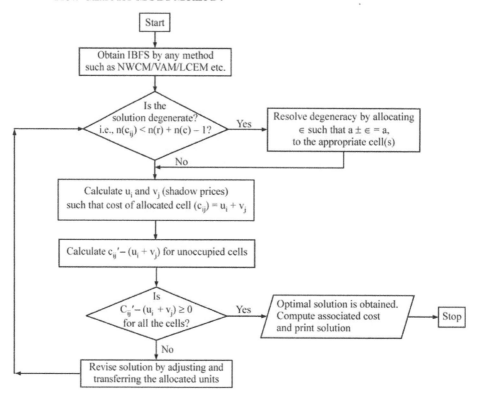

FIGURE 5.7 :

Consider the TP of illustration - 2. Let us optimise this by MODI method now.

ILLUSTRATION 12

Optimise the T.P

	D_1	D_2	D_3	D_4	*Supply*
O_1	1	2	1	4	30
O_2	3	3	2	1	50
O_3	4	2	5	9	20
Demand	20	40	30	10	45

Solution :

Step 1 : *Obtain Inital Solution :*

Refering to illustration - 2, the IBFS for the above TP by North West Corner Method is as follows.

ITERATION TABLEAU 1 :

	D_1	D_2	D_3	D_4	Supply
O_1	20 / 1	10 / 2	1	4	30
O_2	3	30 / 3	20 / 2	1	50
O_3	4	2	10 / 5	10 / 9	20
Demand	20	40	30	10	100

Step 2 : *Degeneracy test :*

No. of occupied cells $n\ (C_{ij}) = 6$

No. of rows i.e., $n\ (r) = 3$

No. of columns i.e., $n\ (c) = 4$

$n\ (C_{ij}) = n\ (r) + n\ (c) - 1$,

There is no degeneracy.

ITERATION TABLEAU 2 :

	D_1	D_2	D_3	D_4	Supply	Shadow price (u_i)
O_1	20 / 1	10 / 2	1	4	30	$u_1 = 0$ (assumed)
O_2	3	30 / 3	20 / 2	1	50	$u_2 = 1$
O_3	4	2	10 / 5	10 / 9	20	$u_3 = 4$
Demand	20	40	30	10		

Shadow Price (v_j) $v_1 = 1$ $v_2 = 2$ $v_3 = 1$ $v_4 = 5$

(Assumed) For Occupied Cells Assume $C_{ij} = u_i + v_j$

$$u_1 + v_1 = 1, u_1 + v_2 = 2, u_2 + v_2 = 3,$$
$$u_2 + v_3 = 2, u_3 + v_3 = 5, u_3 + v_4 = 9$$

We get above 6 equations with seven variables u_1, u_2, u_3 and v_1, v_2, v_3 and v_4. Thus one of the variables (say u_1) is assumed to be zero.

Then all other values can be found as mentioned in the above tableau.

Step 4 : *Optimality test :* To check $C_{ij}' - (u_i + v_j) \geq 0$

ITERATION TABLEAU 3 :

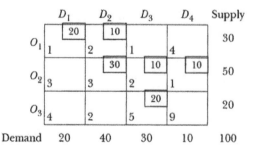

	D_1	D_2	D_3	D_4	Supply	
O_1	20 / 1	10 / 2	0 / 1	– 1 / 4	30	$u_1 = 0$
O_2	+ 1 / 3	30 / 3	20 / 2	– 5 / 1	50	$u_2 = 1$
O_3	– 1 / 4	– 4 / 2	10 / 5	10 / 9	20	$u_3 = 4$
Demand	20	40	30	10		

$v_j \quad v_1 = 1 \quad v_2 = 2 \quad v_3 = 1 \quad v_4 = 5$

Sample Calculations :

for (O_1, D_3) cell $C_{ij}' = 1$, $u_i = 0$, $v_j = 1$

$\therefore \quad C_{ij}' - (u_i + v_j) = 1 - (0 + 1) = 0$

for (O_2, D_4) cell $C_{24} = 1$, $u_2 = 1$, $v_4 = 5$

$\therefore \quad C_{24} - (u_2 + v_4) = 1 - (1 + 5) = -5$

For (O_3, D_1) $4 - (4 + 1) = -1$

and similarly for other unoccupied cells

Step - 5 : *Revising the Solution :* From the above calculations, we find most negative value of $c_{ij}' - (u_i + v_j)$ for (O_2, D_4) cell as – 5. We start constructing closed loop from this cell with other occupied cells i.e.; $(O_2, D_4) \to (O_2, D_3) \to (O_3, D_3) \to (O_3, D_4)$ and on the corners of this loop we give + , – alternatively from (O_2, D_4) onwards. Thus (O_2, D_4) and (O_3, D_3) get addition (+) while (O_2, D_3) and (O_3, D_4) get subtraction (–). Among these negative marked cells, (O_3, D_4) has least allocation as 10 and thus direction is form this cell. (10 units are transferred upwards). Revising the allocations in this way, we get.

ITERATION TABLEAU 4 :

	D_1	D_2	D_3	D_4	Supply
O_1	20 / 1	10 / 2	1	4	30
O_2	3	30 / 3	10 / 2	10 / 1	50
O_3	4	2	20 / 5	9	20
Demand	20	40	30	10	100

Step 6 : *Repeat steps 2 to 5 :*

ITERATION TABLEAU 5 :

	D_1	D_2	D_3	D_4	Supply	u_i
O_1	20 / 1	10 / 2	0 / 1	+4 / 4	30	0
O_2	+1 / 3	30 / 3	+10 / 2	10 / 1	50	1
O_3	−1 / 4	−4 / 2	20 / 5	+5 / 9	20	4
Demand	20	40	30	10		
v_j	1	2	1	0		

ITERATION TABLEAU - 6 :

	D_1	D_2	D_3	D_4	Supply	u_i
O_1	20 / 1	20 / 2	0 / 1	+4 / 4	30	0
O_2	+1 / 3	10 / 3	30 / 2	10 / 1	50	1
O_3	+3 / 4	20 / 2	+4 / 5	+9 / 9	20	0
Demand	20	40	30	10		
v_j	1	2	1	0		

In the above solution, the values of $C'_{ij} - (u_i + v_j) \geq 0$ for all unoccupied cells. Therefore the optimality is reached.

Step 7 : *Cost calculations :*

The solution is tabulated as follows.

Cell	From	To	Unit Cost	No. of Units	Total Cost
C_{11} or $(O_1 D_1)$	O_1	D_1	1	20	20
C_{12} or (O_1, D_2)	O_1	D_2	2	10	20
C_{22} or (O_2, D_2)	O_1	D_2	3	10	30
C_{23} or (O_2, D_3)	O_2	D_3	2	30	60
C_{24} or (O_2, D_4)	O_2	D_4	1	10	10
C_{32} or (D_3, D_2)	O_3	D_2	2	20	40
				Total Cost	180

Note : *This problem yields alternate optimal solution. Refer illustration 20 under section 5.11.1*

ILLUSTRATION 13 ————————————————————————————————

Describe the transportation problem. Use Vogel's approximation method to obtain an initial feasible solution to the following product distribution problem.

Transportation Costs
Distribution Centres

Factory	Mumbai	Bangalore	Delhi	Chennai	Factory Capacity
Kolkata	6	5	8	8	30
Ranchi	5	11	9	7	40
Ahmedabad	8	9	7	13	50
Centre Demand	35	28	32	25	

Solution :   [JNTU (EEE/OT) 98]

The problem has been solved using Vogel's approximation method for IBFS.

Step 1 : *Standardisation :* Since this problem is a minimization case, it is in standard from.

Step 2 : *Balance Checking :* The TP is balanced since the total capacity is equal to total demand.

Step 3 : The next step is to determine penalties for each row and each column. A penalty is the difference beween the least cost and the next least cost in corresponding row (or column), for e.g. The penalty for the first row i.e., corresponding to Kolkata is $6 - 5 = 1$.

Step 4 : Choose the maximum penalty among all the penalties which have been determined in the previous step.

Factory	Mumbai	Bangalore	Delhi	Chennai	Capacity	Penalty
Kolkata	6	28 / 5	8	8	3̶0̶ / 2	1
Ranchi	5	11	9	7	40	2
Ahmedabad	8	9	7	13	50	1
Demand	35	(28)	32	25		
Penalty	1	4 ↑	1	1		

(Here 4 is the maximum penalty)

Allocate maximum possible units to the cell whose cost is minimum in the row or column selected (corresponding to max. penalty). If the capacity or demand is exhausted, delete the row or column for next iteration. Prepare the next iteration table with revised capacities and demand.

After the first 4 steps, we can derive that the allocated cell is Kolkata to Banglore. The number of units is 28 units. The deleted column is Banglore column and revised capacity of Kolkata is 2.

Repeat the above 4 steps and prepare the next iteration tables.

ITERATION TABLE 3 :

Factory	Mumbai	Delhi	Chennai	Capacity	Penalty
Kolkata	6	8	8	2	2
Ranchi	35 / 5	9	7	4̶0̶ / 5	(2)
Ahmedabad	8	7	13	50	1
Demand	35	32	25		
Penalty	1	1	1		

Inference: Allocation to "Ranchi to Mumbai" cell–35 units. Delete – Mumbai column, Revise capacity of Ranchi Row to 5.

ITERATION TABLE 3 :

Factory	Delhi	Chennai	Capacity	Penalty
Kolkata	8	8	2	0
Ranchi	9	7	5	2
Ahmedabad	32 / 7	13	5̶0̶ / 18	6
Demand	(32)	25		
Penalty	1	1		

Inference: Allocation to "Ahmedabad to Delhi" cell –32 units, Deletion Delhi column and capacity Revised to Ahmedabad Row as 18.

ITERATION TABLE 4 :

In this case since there is only one column, penalties need not be calculated.

	Chennai	Capacity
Kolkata	2 8	2
Ranchi	5 7	5
Ahmedabad	18 13	18
Demand	25	

The obvious allocations are

Kolkata to Chennai cell	→ 2
Ranchi to Chennai cell	→ 5
and Ahmedabad to Chennai cell	→ 18

Now the IBFS is

	Mumbai	Bangalore	Delhi	Chennai	Capacity
Kolkata	6	28 5	8	2 8	30
Ranchi	35 5	11	9	5 7	40
Ahmedabad	8	9	32 7	18 13	50
Demand	35	28	32	25	

Moving Towards Optimality :

First we have to check the non-degeneracy ($n \geq r + c - 1$)

No. of allocated cells (say n) = 6

No. of rows (r) + No. of columns (c) − 1

$$= 4 + 3 - 1 = 6$$

Hence there is no degeneracy as $n = r + c - 1 = 6$.

Now calculating u_i and v_j values using allocated cell cost $(c_{ij}) = u_i + v_j$.

Where u_i and v_j are shadow cost variables of ith row and j^{th} column respectively. And then calculate $c_{ij}' - (u_i + u_j)$ where C_{ij}' is cost in nonallocated cell.

ITERATION TABLE 5 :

	Mumbai	Bangalore	Delhi	Chennai	Capacity	u_i
Kolkata	+0 / 6	28 / 5	+6 / 8	8	2 / 30	0
Ranchi	35 / 5	+7 / 11	+8 / 9	7	5 / 40	−1
Ahmedabad	−3 / 8	−1 / 9	32 / 7	18 / 13	5 / 50	5
Demand	35	28	32	25		
v_j	6	5	2	8		

$u_1 + v_2 = 5$; $u_2 + v_1 = 5$

$u_2 + v_4 = 7$; $u_3 + v_3 = 7$

$v_3 + v_4 = 13$; $v_1 + v_4 = 8$

Note : *As there are 6 equations, but seven variables, one of them is assumed to be zero ($u_1 = 0$.)*

As there is more negativeness existing at Ahmedabad to Mumbai cell, the table is to be revised by transfering maximum possible units along the loop line.

ITERATION TABLE 6 :

Repeating the step as same as the above in Iteration Tableu 5. (No degeneracy as $n = r + c − 1 = 6$).

	Mumbai	Bangalore	Delhi	Chennai	Capacity	u_i
Kolkata	6	28 / 5	+3 / 8	2 / 8	30	0
Ranchi	17 / 5	+7 / 11	+5 / 9	23 / 7	40	−1
Ahmedabad	18 / 8	+2 / 9	32 / 7	+3 / 13	50	2
Demand	35	28	32	25		
v_j	+0 / 6	5	5	8		

As $C_{ij}' \geq u_i + v_j$ *for all non allocated cells, the optimal solution is reached. The solution is as follows :*

From	To	No. of Units	Unit Cost	Total Cost
Kolkata	Bangalore	28	5	140
Kolkata	Chennai	2	8	16
Ranchi	Mumbai	17	5	85
Ranchi	Chennai	23	7	161
Ahmedabad	Mumbai	18	8	144
Ahmedabad	Delhi	32	7	224
	G. Total Units	120	G. Total Cost	770

ILLUSTRATION 14 --

Solve the following transportation problem, i.e., Find the optimal solution, where the entries are cost coefficients.

From	To	Destination				Availability
		1	2	3	4	
Origins	1	15	0	20	10	50
	2	12	8	11	20	50
	3	0	16	14	18	100
Requirement		30	40	60	70	200

Solution :   [JNTU (EEE/OT) 97/S]

The TP is solved by using Vogel's approximation method for Initial Basic Feasible Solution.

Step 1 : Since the TP is a minimization case, it is in standard form.

Step 2 : Given TP is balanced since the total availability is equal to total requirement.

Step 3 : The penalties corresponding to each row and each column are determined.

(Penalty is difference between the lowest number and the next lowest number of a row or column.)

Ex : The penalty corresponding to 1st row is calculated as $10 - 0 = 10$

ITERATION TABLEAU - 1 :

To From	D_1	D_2	D_3	D_4	Avail	Penalty
O_1	15	0	20	10	50	10
O_2	12	8	11	20	50	3
O_3	[30] 0	16	14	18	~~100~~ 70	14 ←
Required	~~30~~	40	60	70	200	
Penalty	12	8	3	8		

In this TP the max. penalty is 14, hence 30 units are allocated to the cell having minimum cost i.e., 0 of row O_3 and consequently 30 units are allocated (in this case) hence the 1st column is deleted and 3rd row availability is revised to 100–30=70.

The above steps are repeated for successive iterations.

ITERATION TABLEAU - 2 :

To From	D_2	D_3	D_4	Avail	Penalty
O_1	0	20	10	~~50~~ 10	10 ←
O_2	40 8	11	20	50	3
O_3	16	14	18	70	2
Required	(40)	60	70	170	
Penalty	8	3	8		

The allocation is from O_1 to D_2 and no. of units allocated are 40 units. Revised availability for O_1 row is 10.

ITERATION TABLEAU - 3 :

To From	D_3	D_4	Avail	Penalty
O_1	20	10 10	(10)	10 ←
O_2	11	20	50	9
O_3	14	18	70	4
Required	60	~~70~~ 60	130	
Penalty	3	2		

Allocation is in cell $(O_1$ to $D_4)$ and equal to 10 units, delete O_1 and revise requirement of D_4 as 60

ITERATION TABLEAU - 4 :

	D_3	D_4	Avial	Penalty
O_2	50 11	20	~~50~~ 0	9 ←
O_3	14	18	70	4
Required	~~60~~ 10	60	120	
Penalty	3	2		

Allocation is from O_2 to D_3 and is equal to 50 units; Delete O_2 and revise requirement of D_3 as 10

ITERATION TABLEAU - 5 :

	D_3	D_4
O_3	10 14	60 18

Summary :

	D_1	D_2	D_3	D_4
O_1		40 0		10 10
	15	0	20	10
O_2			50 11	
	12	18	11	20
O_3	30 0		10 14	60 18
	0	16	14	18

The total cost incurred is,

$$10 \times 10 + 11 \times 50 + 14 \times 10 + 18 \times 60 = 1870$$

Step 5 : *Checking for Optimality :* The given problem is non–degenerate (no degeneracy) as number of allocated cells $n\,(c_{ij}) = 6$

No. of rows $(r) = 3$, No. of columns $(c) = 4$; $r + c - 1 = 3 + 4 - 1 = 6$

$$[\; \therefore \; n\,(c_{ij}) = r + c - 1\,]$$

Now, equations with allocated cells as $c_{ij} = u_i + v_j$

We get,

$$u_1 + v_2 = 0; \qquad u_1 + v_4 = 10$$
$$u_2 + v_3 = 11\;; \qquad u_3 + v_1 = 0$$
$$u_3 + v_3 = 14; \qquad u_3 + v_4 = 18$$

Assuming $u_1 = 0$, we have $u_2 = 5$, $u_3 = 8$,

$$v_1 = -8 \;, v_2 = 0 \;, v_3 = 6 \text{ and } v_4 = 10$$

Now checking $c_{ij}' - (u_i + v_j) \geq 0$ for non allocated cells,

for c_{11} $15 - (-8+0) = +23$;   for c_{13} $20 - (0+6) = +14$

for c_{21} $12 - (-8+5) = 15$;   for c_{22} $8 - (5+0) = +3$

for c_{24} $20 - (5+10) = +5$, and   for c_{32} $16 - (8+0) = +8$.

Since all $c_{ij}' - (u_i + v_j)$ are positive, the obtained solution is optimal.

Hence the minimum cost = **Rs. 1870/-**

and allocations are

$$\textbf{\textit{O}}_1 \textbf{ to } \textbf{\textit{D}}_2 = \textbf{40 units}\;; \qquad \textbf{\textit{O}}_1 \textbf{ to } \textbf{\textit{D}}_4 = \textbf{10 units}$$
$$\textbf{\textit{O}}_2 \textbf{ to } \textbf{\textit{D}}_3 = \textbf{50 units}\;; \qquad \textbf{\textit{O}}_3 \textbf{ to } \textbf{\textit{D}}_1 = \textbf{30 units}$$
$$\textbf{\textit{O}}_3 \textbf{ to } \textbf{\textit{D}}_3 = \textbf{10 units} \textbf{ and } \textbf{\textit{O}}_3 \textbf{ to } \textbf{\textit{D}}_4 = \textbf{60 units}$$

ILLUSTRATION 15 ────────────────────────────────────

> *An oil company has got three refineries P, Q and R and it has to send petrol to four different depots A, B, C and D. The cost of shipping 1 gallon of petrol at the refineries are given in the table. The requirement of the depots and the available petrol are also given. Find the minimum cost of shipping after obtaining an initial solution by VAM.s*

Refineries	Deposit				Available
	A	B	C	D	
P	10	12	15	8	130
Q	11	11	9	10	150
R	20	9	7	18	170
Required	90	100	140	120	450

[JNTU (Mech) 99]

Solution :

The given problem is

Step 1 : Minimisation case of TP → cost minimisation (standard form).

Step 2 : Balanced TP model → Requirements = Availability (450 = 450)

Step 3 : We find the penalities and start allocating the row or column which has maximum penalty.

ITERATION TABLEAU - 1 :

Refineries		Depot				Available	Penalty
		A	B	C	D		
P		10	12	15	8 [120]	130/10	2
Q		11	11	9	10	150	1
R		20	9	7	18	170	2
Required		90	100	140	(120)		
Penalty		1	2	2	2 ↑		

Step 4 : Remove the column *D* as it is filled to its requirement and start allocating to other columns & rows.

ITERATION TABLEAU - 2 :

	A	B	C	Availability	Penalty
P	10	12	15	10	2
Q	11	11	9	150	2
R	20	9	7 [140]	170/30	2
Required	90	100	(140)		
Penalty	1	2	2 ↑		

Step 5 : Column C is neglected as it is filled to its requirement.

ITERATION TABLEAU - 3 :

	A	B	Availability	Penalty
P	10	12	10	2
Q	11	11	150	0
R	20	30 / 9	(30)	11
Required	90	100/70		
Penalty	1	2		

Step 6 : Row *R* is neglected

ITERATION TABLEAU - 4 :

	A	B		Penalty
P	10 / 10	12	10	2 ←
Q	80 / 11	70 / 11	150	0
	90	70		
Penalty	1	1		

Step 7 : Finding the Costs of an Allocated Cells :

	A	B	C	D	
P	10 / 10	+11 / 12	+7 / 15	120 / 8	$u_1 = 10$
Q	80 / 11	70 / 11	0 / 9	+1 / 10	$u_2 = 11$
R	+11 / 20	30 / 9	140 / 7	+11 / 18	$u_3 = 9$
	$v_1 = 0$	$v_2 = 0$	$v_3 = -2$	$v_4 = -2$	

∴   Initial solution by VAM itself is optimal.

Since the values $c_{ij}' - (u_i + v_j) \geq 0$.

$= 10 \times 10 + 11 \times 80 + 11 \times 70 + 9 \times 3 + 8 \times 120 + 7 \times 14 =$ Rs. 3960.

∴   Minimum cost = Rs. 3960.

Note : *This problem yields an alternate optimal solution.*
For hints, Refer illustration - 20 under section 5. 11.1 and try for another solution

ILLUSTRATION 16 ——————————————————————————————

Solve the Transportation Problem whose unit cost matrix, supply and demand are given below:

	D_1	D_2	D_3	D_4	D_5	Supply
O_1	7	7	10	5	11	45
O_2	4	3	8	6	13	90
O_3	9	8	6	7	5	96
O_4	12	13	10	6	3	75
O_5	5	4	5	6	12	105
Demand	120	80	50	75	85	

[JNTU (Mech) 98/s]

Solution :

As the matrix is cost matrix, the TP is to be minimised i.e., in the standard form, hence we procede directly.

	D_1	D_2	D_3	D_4	D_5	Supply
O_1	7	7	10	5	11	45
O_2	4	3	8	6	13	90
O_3	9	8	6	7	5	96
O_4	12	13	10	6	3	75
O_5	5	4	5	6	12	105
Demand	120	80	50	75	85	411 / 410

As the Demand (410) $\neq$ Supply (411), the given TP is unbalanced. Hence we create a dummy destination D_6 with a demand of 1 unit so as to make demand = supply.

Obtaining IBFS by Vogel's approximation :

ı ITERATION TABLEAU 1 :

	D_1	D_2	D_3	D_4	D_5	D_6	Supply	Penalty
O_1	7	7	10	5	11	0 [1]	45/44	5
O_2	4	3	8	6	13	0	90	3
O_3	9	8	6	7	5	0	96	5
O_4	12	13	10	6	3	0	75	3
O_5	5	4	5	6	12	0	105	4
Demand	120	80	50	75	85	(1)	411	
Penalty	1	1	1	1	2	0	410	

* (Penalty is difference of least and next least of respective row/column) Highest penalty is for O_1 therefore one unit is allocated to the cell (O_1, D_6) and all the units in D_6 are exhausted. Hence D_6 is eliminated for next iteration while supply of O_1, reduces to 44 units.

ITERATION TABLEAU 2 :

	D_1	D_2	D_3	D_4	D_5	Supply	Penalty
O_1	7	7	10	5	11	44	2
O_2	4	3	8	6	13	90	1
O_3	9	8	6	7	5	96	1
O_4	12	13	10	6	3 [75]	(75)	3 ←
O_5	5	4	5	6	12	105	1
Demand	120	80	50	75	85 / 10	410 / 335	
Penalty	1	1	1	1	2		

Allocation to (O_4, D_5) – 75 units

Demand for D_5 becomes 10, O_4 is deleted.

ITERATION TABLEAU 3 :

	D_1	D_2	D_3	D_4	D_5	Supply	Penalty
O_1	7	7	10	5	11	44	2
O_2	4	3	8	6	13	90	1
O_3	9	8	6	7	5 [10]	96 / 86	1
O_5	5	4	5	6	12	105	1
Demand	120	80	50	75	10 / 0	335 / 325	
Penalty	1	1	1	1	6		

Allocation (O_3, D_5) – 10 units

Supply of O_3 reduces to 86, D_5 is removed.

ITERATION TABLEAU 4 :

	D_1	D_2	D_3	D_4	Supply	Penalty
O_1	7	7	10	5 [44]	(44) 0	2 ←
O_2	4	3	8	6	90	1
O_3	9	8	6	7	86	1
O_5	5	4	5	6	105	1
Demand	120	80	50	7̶5̶ 31	3̶2̶5̶ 281	
Penalty	1	1	1	1		

Allocation (O_1 , D_4) – 44 units, deletion – O_1.
Demand reduction for D_4 to 31.

ITERATION TABELEAU 5 :

	D_1	D_2	D_3	D_4	Supply	Penalty
O_2	4	3 [80]	8	6	9̶0̶ 10	1 ←
O_3	9	8	6	7	86	1
O_5	5	4	5	6	105	1
Demand	120	8̶0̶	50	31	2̶8̶1̶ 201	
Penalty	1	1	1	0		

Note : Highest penalty is choosen orbitrarily.
Allocation (O_2 , D_2) — 80 units, Deletion – D_2
Supply reduction of O_2 to 10.

ITERATION TABLEAU 6 :

	D_1	D_3	D_4	Supply	Penalty
O_2	4 [10]	8	6	(10)	2
O_3	9	6	7	86	1
O_5	5	5	6	105	0
Demand	1̶2̶0̶ 110	50	31	2̶0̶1̶ 191	
Penalty	1	1	0		

Allocation (O_2 , D_1) — 10 units, Deletions → O_2;
Demand reduction on D_1 to 110.

	D_1	D_3	D_4	Supply
O_3	9	6	7	86
O_5	[105] 5	5	6	(1̶0̶5̶)
Demand	1̶1̶0̶ 5	50	31	1̶9̶1̶ 86
Penalty	4	1	1	

Allocation (O_5, D_1) Deletion — O_5

Demand reduction of D_1 to O_5.

	D_1	D_3	D_4	
O_3	[5] 9	[50] 6	[31] 7	86
	5	50	31	8̶6̶ 0

Allocation (O_3, D_1) — 5; (O_3, D_3) — 50,

(O_3, D_4) — 31 units.

	D_1	D_2	D_3	D_4	D_5	D_6	
O_1	7	7	10	[44] 5	11	[1] 0	45
O_2	[10] 4	[80] 3	8	6	13	0	90
O_3	[5] 9	8	[50] 6	[31] 7	[10] 5	0	96
O_4	12	13	10	6	[75] 3	0	75
O_5	[105] 5	4	5	6	12	0	105
	120	80	50	75	85	1	411

Transportation cost by VAM is : $5 \times 44 + 0 \times 1 + 4 \times 1 + 3 \times 80 + 9 \times 5$

$+ 6 \times 50 + 7 + 31 + 5 \times 10$

$+ 3 \times 75 + 5 \times 105 = 1862$

Optimality Test :

Degeneracy test :   No. of allocated cells = 10 ;

No. of rows (r) = 5 ; No. of columns (c) = 6; $r + c - 1 = 10$

No. of allocated cells (10) $\geq r + c - 1$ (*i.e* 10) ; so no degeneracy.

	D_1	D_2	D_3	D_4	D_5	D_6		u_i
O_1	0 / 7	+1 / 7	+6 / 10	44 / 5	+8 / 11	1 / 0	45	0
O_2	10 / 4	80 / 3	+7 / 8	+4 / 6	+13 / 13	+3 / 0	90	−3
O_3	5 / 9	0 / 8	50 / 6	31 / 7	10 / 5	−2 / 0+	96	2
O_4	+5 / 12	+7 / 13	+6 / 10	+1 / 6	75 / 3	0 / 0	75	0
O_5	105 / 5	0 / 4	+3 / 5	+3 / 6	+11 / 12	+2 / 0	105	−2
	120	80	50	75	85	1		
v_j	7	6	4	5	3	0		

$u_1 + v_6 = 0$; $u_1 + v_4 = 5$; $u_2 + v_1 = 4$

$u_2 + v_2 = 3$; $u_3 + v_1 = 9$; $u_3 + v_3 = 6$

$u_3 + v_4 = 7$; $u_3 + v_5 = 5$; $u_4 + v_5 = 3$; $u_5 + v_1 = 7$

	D_1	D_2	D_3	D_4	D_5	D_6		u_i
O_1	0 / 7	+1 / 7	+6 / 10	45 / 5	+8 / 11	+2 / 0	45	0
O_2	10 / 4	80 / 3	+7 / 8	/ 6	+3 / 3	+5 / 0	90	−3
O_3	5 / 9	0 / 8	50 / 6	30 / 7	10 / 5	1 / 0	96	2
O_4	+5 / 12	+7 / 13	+6 / 10	+1 / 6	75 / 3	+2 / 0	75	0
O_5	105 / 5	0 / 4	+3 / 5	+3 / 6	+11 / 12	+4 / 0	105	−2
	120	80	50	75	85	1		
v_j	7	6	4	5	3	−2		

From the above tableau $C_{ij}' - (u_i + v_j) \geq 0$ i.e., Non−negative for all non−allocated cells, the optimal solution is obtained as follows,

Cell	Unit cost (Rs.)	Allocated units	Total cost (Rs.)
(O_1, D_4)	5	45	225
(O_2, D_1)	4	10	40
(O_2, D_2)	3	80	240
(O_3, D_1)	9	5	45
(O_3, D_3)	6	50	300

(O_3, D_4)	7	30	210
(O_3, D_5)	5	10	50
(O_3, D_6)	0	1	0
(O_4, D_5)	3	75	225
(O_5, D_1)	5	105	525
		Grand Total	Rs. 1860/-

The total minimum transporation cost (optimum) is Rs. 1860/- and one unit in the supply of origin –3 is not transported to any destination (unutilized resource).

Note : *The above TP yields multiple optimal solutions which will be explained in later sections. Students may try.*

ILLUSTRATION 17

Find the optimal solution to the Transportation Problem given below:

		Warehouses				
		W_1	W_2	W_3	W_4	*Supply*
Factory	F_1	14	25	45	5	6
	F_2	65	25	35	55	8
	F_3	35	3	65	15	16
	Demand	4	7	6	13	*30 (total)*

[JNTU (CSE) 97]

Solution :

	Ware House				
Factory	W_1	W_2	W_3	W_4	Supply
F_1	14	25	45	5	6
F_2	65	25	35	55	8
F_3	35	3	65	15	16
Demand	4	7	6	13	30

Step 1 : As it is minimization case, the TP is in standard form.

Step 2 : The given TP is balanced as total supply = Total demand.

Step 3 : Finding IBFS by Vogels approximation method.

ITERATION TABLEAU - 1 :

	W_1	W_2	W_3	W_4	Supply	Penalties
F_1	14	25	45	5	6	9
F_2	65	25	35	55	8	10
F_3	35	3 [7]	65	15	16̸ 9	12
Demand	4	(7)	6	13	30	
Penalties	21	**22**	10	10		

ITERATION TABLEAU 2 :

	W_1	W_3	W_4	Supply	Penalties
F_1	[4] 14	45	5	6̸ 2	9
F_2	65	35	55	8	20
F_3	35	65	15	9	20
Demand	(4)	6	13	30	
Penalties	21 ↑	10	10		

ITERATION TABLEAU 3 :

	W_3	W_4	Supply	Penalties
F_1	45	5	2	40
F_2	35	55	8	20
F_3	65	15 [9]	(9)	50 ←
Demand	6	13̸ 4		
Penalties	10	10		

ITERATION TABLEAU 4 :

	W_3	W_4	Supply	Penalties
F_1	45	5 [2]	(2)	40
F_2	35 [6]	55 [2]	8	20
Demand	6	4̸ 2		
Penalties	10	50		

IBFS TABLEAU :

	W_1	W_2	W_3	W_4	Supply
F_1	[4] 14	25	45	[2] 5	6
F_2	65	25	[6] 35	[2] 55	8
F_3	35	[7] 3	65	[9] 15	16
Demand	4	7	6	13	

Initial basic feasible solution $= 14 \times 4 + 3 \times 7 + 6 \times 35 + 2 \times 5 + 2 \times 55 + 9 \times 15 = 542$

$n\,(C_{ij}) = n\,(r) + n\,(c) - 1$ i.e., $6 = 4 + 3 - 1$, Hence no degeneracy

Optimisation :

	W_1	W_2	W_3	W_4	
F_1	[4] 14	+32 25	+60 45	[2] 5	$u_1 = 0$
F_2	+1 65	−18 25	[6] 35	[2] 55	$u_2 = 50$
F_3	+70 35	[7] 3	+70 65	[9] 15	$u_3 = 10$
	$v_1 = 14$	$v_2 = -7$	$v_3 = -15$	$v_4 = +5$	

$$u_1 + v_1 = 14\,; \qquad u_1 + v_4 = 5$$
$$u_2 + v_3 = 35\,; \qquad u_2 + v_4 = 55$$
$$u_3 + v_2 = 3\,; \qquad u_3 + v_4 = 15$$

	W_1	W_2	W_3	W_4		
F_1	+19 [4] 14	+32 25	+42 45	[2] 5	6	$u_1 = 0$
F_2	+11 65	[2] 25	+52 [6] 35	55	8	$u_2 = 32$
F_3	35	[5] 3	65	[11] 15	16	$u_3 = 10$
	4	7	6	13		
	$v_1 = 14$	$v_2 = -7$	$v_3 = 3$	$v_4 = 5$		

$$u_1 + v_1 = 14\ ; \qquad u_1 + v_4 = 5$$
$$u_2 + v_2 = 25\,; \qquad u_2 + v_3 = 35$$
$$u_3 + v_2 = 3\,; \qquad u_3 + v_4 = 15$$

$\therefore$ Optimal solution

$$= 4 \times 14 + 2 \times 25 + 3 \times 5 + 6 \times 35 + 2 \times 5 + 11 \times 15 = 506$$

5.6 *Degeneracy in Transportation Problem*

A transportation problem is said to be degenerate if its initial solution violates the condition that the number of allocated (occupied) cells n (C_{ij}) is greater than or equal to sum of number of rows n (r) and number of columns n (c) minus one. In other words if n $(C_{ij}) < n$ $(r) + n$ $(c) - 1$, the TP is degenerate.

It becomes difficult while computing optimal solution if the TP has degeneracy. Therefore the degeneracy is to be first removed before going to computation of IBFS.

How to Resolve Degeneracy :

Degeneracy in TP can be resolved by allocating a minute quantity ε (read as epsilon) to an appropriate cell. This ε is assumed so negligible quantity such that $a + \varepsilon = a - \varepsilon = a$ and $\varepsilon \neq 0$ (also $\varepsilon + \varepsilon = \varepsilon$, $\varepsilon - \varepsilon = 0$).

Where a is any quantity.

Thus ε is a very minute and negligible non zero quantity that does not alter any value when added or subtracted any number of times.

***e.g.* :** ε is like one Rupee tip given to worker for loading/unloading in a business where some lakhs or some thousands are spent in transporting the goods. Thus ε does not affect that total transportation cost.

The degeneracy may be experienced in the initial stage of optimisation or at the subsequent iterations.

In general the degeneracy can be identified in transportation problem while finding IBFS itself by the following points.

1. If the partial sum of availability is equal to partial sum of requirements, there is a chance of getting degeneracy.

2. In any method (of NWCM, VAM, LCEM etc.), if a row and a column are simultaneously satisfied/exhausted and hence both deleted for next iteration, then degeneracy will occur while optimising.

5.6.1 *Degeneracy in the Initial Solution*

ILLUSTRATION 18 ────────────────────────────────────

Determine the Optimum basic feasible solution to the following Transporation problem.

		To			
		A	**B**	**C**	**Available**
From	**1**	50	30	220	1
	2	90	45	170	3
	3	250	200	50	4
Required		4	2	2	

[JNTU (Mech.) 99/CCC, (ECE) 99/CCC]

Solution :

Step 1: The given Transportation problem (TP) is in the standard from (i.e., minimization case). Hence we can proceed directly.

Step 2 : In the given TP, the sum of availability (8) is equal to the sum of required (8). Therefore it is balanced.

Step 3 : Preparation of IBFS by North West Corner Method.

To

		A	B	C	Available
From	1	50 [1]	30	220	1
	2	90 [3]	45	170	3
	3	250	200 [2]	50 [2]	4
Required		4	2	2	

Step 4 : *Optimization :* The number of allocated cells say (*n*) = 4; No. of rows + No. of columns – 1 say $r + c - 1 = 3+3-1=5$

As $n < r + c - 1$, there is a degeneracy. *Hence assuming ϵ as an allocation at the cell (1, B) such that $\epsilon \neq 0$ and $a + \epsilon = a - \epsilon = a$.

Then, the costs u_i and v_j are computed.

	A	B	C	u_i
1	50 [1]	30 [ϵ]	220	$u_1 = 0$ (assumed)
2	90 [3]	45	170	$u_2 = 40$
3	250	200 [2]	50 [2]	$u_3 = 170$
v_j	$v_1 = 50$	$v_2 = 30$	$v_3 = -120$	

$u_i + v_j = C_{ij}$ for allocated cells
$u_1 + v_1 = 50$
$u_1 + v_2 = 30$
$u_2 + v_1 = 90$
$u_3 + v_2 = 200$
$u_3 + v_3 = 50$

(As we have five equations and six variables, we have to assume one of these as equal to zero. Hence $u_1 = 0$ is assumed)

Now, calculating the value of $c_{ij}' - (u_i + v_j)$ for non–allocated cells, to check the optimality, we get

	A	B	C	u_i
1	$50^{+\delta}$ [1] $^{-\delta}$ 30 [ϵ]		+340 220	0
2	$90^{-\delta}$ [3] $^{+\delta}$ -25 45		+250 170	40
3	+30 250	200 [2]	50 [2]	170
v_j	4 50	2 30	2 -120	

* The degeneracy can be detected in initial stages while finding IBFS, since the partial sum of availability of 1st & 2nd row i.e., (1 + 3) is equal to partial required of A i.e., 4. Also, when cell $(2, A)$ is allocated with 3 units both row and column get satisfied.

As $c_{ij}' - (u_i + v_j) < 0$ for cell $(2, B)$, we revise it by transfering some units given by least figure among the corners marked by –ve on the loop i.e. ε here and then the above step is repeated.

	A	B	C	u_i
1	1 50	+25 30	+365 220	0
2	3 90	ε 45	+275 170	40
3	+5 250	2 200	2 50	195
Required	4	2	−2	
v_j	50	5	−145	

As the values of $c_{ij}' - (u_i + v_j) \geq 0$ (non negative) for all non–allocated cells, the optimal basic feasible solution is obtained.

The solution is,

From	To	Unit Cost	No. of Units	Total Cost
1	A	50	1	50
2	A	90	3	270
2	B	45	ε	45ε
3	B	200	2	400
4	C	50	2	100
		Total	$8 + \varepsilon$	$320 + 45\varepsilon$

Hence the total tranportation cost is 320 units (or Rs.)

5.6.2 *Degeneracy in the Subsequent Iterations*

ILLUSTRATION 19 —————————————————————————

Optimise the following TP

	D_1	D_2	D_3	Supply
S_1	8	5	6	120
S_2	15	10	12	80
S_3	3	9	10	80
	150	80	50	

Solution :

With usual steps (as explained) the initial solution for the above TP (minimisation & balanced) by North-west corner method is given below.

ITERATION TABLEAU 1 :

	D_1	D_2	D_3	Supply
S_1	120 \ 8	5	6	120
S_2	30 \ 15	50 \ 10	12	80
S_3	3	30 \ 9	50 \ 10	80
	150	80	50	

Since the number of occupied cells n (C_{ij}) is 5 and No. of rows n (r) + No. of columns i.e., $n\,(c) - 1$ is $3 + 3 - 1 = 5$, there is no degeneracy. Thus u_i and v_j are computed with usual set of rules, Also, values $C_{ij} - (u_i + v_j)$ for unoccupied cells are calcaulted. The loop is identified for most negative.

ITERATION TABLEAU 2 :

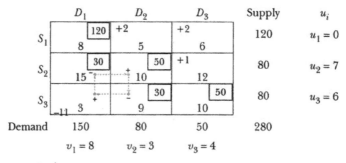

	D_1	D_2	D_3	Supply	u_i
S_1	120 \ 8	+2 \ 5	+2 \ 6	120	$u_1 = 0$
S_2	30 \ 15	50 \ 10	+1 \ 12	80	$u_2 = 7$
S_3	3	30 \ 9	50 \ 10	80	$u_3 = 6$
Demand	150	80	50	280	
	$v_1 = 8$	$v_2 = 3$	$v_3 = 4$		

ITERATION TABLEAU 3 :

	D_1	D_2	D_3	Supply
S_1	120 \ 8	5	6	120
S_2	15	80 \ 10	12	80
S_3	30 \ 3	9	50 \ 10	80
Demand	150	80	50	280

In IT - 3 we notice degeneracy* since occupied cells are only four while n $(r) + n$ $(c) - 1 = 5$. Thus n $(C_{ij}) < n$ $(r) + n$ $(c) - 1$.

Therefore we have to resolve degeneracy first, by allocating ε to (S_2, D_1) or (S_3, D_2) such that $a + \varepsilon = a - \varepsilon = a$ and $\varepsilon \neq 0$ and proceed with usual rules.

Note : *Here degeneracy occured since both cells have turned out to be unoccupied when 30 units have been transferred.*

---

* The chance of occuring degeneracy can be predicted in the beginning since partial sum of availability of S_1 & S_3 (120 + 80) is equal to partial sum of requirement of D_1 and D_3 (150 + 50).

ITERATION TABLEAU 4 :

	D_1	D_2	D_3	Supply	u_i
S_1	120 −9 −⋈ 8	−9 5	−9 ↑ +6	120	0
S_2	+11 15	80 10	+1 12	80	−4
S_3	↑ 30 + 3	ε 9 ⋈	50 − 10	80	−5
Demand	150	80	50	280	
v_j	8	14	15		

ITERATION TABLEAU 5 :

	D_1	D_2	D_3	Supply	u_i
S_1	70 −9 − 8	+ 5	50 6	120	0
S_2	+11 15	80 +10 10	+10 12	80	−4
S_3	80 3 +	ε 9 −	+9 10	80	−5
Demand	150	80	50	280	
v_j	8	14	6		

ITERATION TABLEAU 6 :

	D_1	D_2	D_3	Supply	u_i
S_1	70 8	ε 5	50 6	120	0
S_2	+2 15	80 +1 10	+1 12	80	+5
S_3	80 3	+9 9	+9 10	80	−5
Demand	150	80	50	280	
v_j	8	5	6		

Since $C_{ij}' \geq (u_i + v_j)$ for all the unoccupied cells, the optimal solution is arrived. The optimal cost calculations are shown below.

Cells	From	To	No. of Units	Units Cost	Total Cost
C_{11}	S_1	D_1	70	8	560
C_{12}	S_1	D_2	ε	5	5ε
C_{13}	S_1	D_3	50	6	300

C_{22}	S_2	D_2	80	10	800
C_{31}	S_3	D_1	80	3	240
		G. Total	280	–	1900 + 5ε = 1900

Since value of ε is negligible, the total transportation cost = 1900 units.

5.7 *Special Cases in TP*

There are some special cases in TP and are listed below. [We may have a combination of two or more of these specialities]

1. Maximisation cases

2. Unbalanced cases

3. Multiple or alternate optimal solution case

4. Restricted transportation case

5. Conditional T.P.

6. Trans-shipment problem.

Of the above the first two cases have been explained in the earlier illustrations including them as first two steps of IBFS of the T.P. However, these cases are explained again in combination with restricted and conditional transportation problems under illustration - 21 and 22 to follow. The third case has been explained in illustration 10, 11 & 12. However, the alternate optimal solution in illustration - 12 is shown here.

5.7.1 *Multiple Optimal Solution or Alternate Optimal Solution*

ILLUSTRATION 20

The optimal solution of TP given in illustration - 10 or 12 is considered here to illustrate the alternate optimal solution.

Solution :

After arriving the optimal solution by MODI or stepping stone method, we now identify alternate optimal solution as follows

With usual calculations of u_j & u_j find $C_{ij}' - (u_i + v_j)$ for non-allocated cells. As all are + ve we obtain optimal solution equal to 180.

Since $C_{ij} - (u_i + v_j)$ value for (O_1, D_3) cell is zero (or $C_{ij}' = u_i + v_j$), this cell is as good as any allocated cell (whose $C_{ij} = u_i + v_j$) and thus can enter the basis (allocation). We take up this cell and construct a loop as shown in the above table. From the loop $(O_1, D_3) \rightarrow (O_2, D_3) \rightarrow (O_2, D_2) \rightarrow (O_1, D_2)$, we transfer the units from (O_1, D_2) or (O_2, D_3) and the maximum possible units to transfer is only 10. Therefore we choose the direction from (O_1, D_2) to (O_1, D_3). Thus the revised matrix is given below.

[Add 10 to (O_1, D_3), subtract 10 to (O_2, D_3), Add 10 to (O_2, D_2) and subtract 10 to (O_1, D_2). Thus (O_1, D_3) becomes occupied i.e., basic while (O_1, D_2) becomes unoccupied i.e., non basic.]

Note : *This resembles the same $Z_j - C_j = 0$ values in simplex table. Compare for your better understanding.*

	D_1	D_2	D_3	D_4	Supply	u_i
O_1	20 | 0 1	2	10 1	+4 4	30	0
O_2	+1 3	20 3	20 2	10 1	50	1
O_3	+3 4	20 2	+4 5	+9 9	20	0
Demand	20	40	30	10	100	
v_j	1	2	1	0		

From the above table the values of $C_{ij}' - (u_i + v_j) \geq 0$ for all unoccupied cells, thence it is optimal solution.

The cost calculation are given below.

Cells	From	To	Units Cost	No. of Units	Total Cost
C_{11}	O_1	D_1	1	20	20
C_{13}	O_1	D_3	1	10	10
C_{22}	O_2	D_2	3	20	60
C_{23}	O_2	D_3	2	20	40
C_{24}	O_2	D_4	1	10	10
C_{32}	O_3	D_2	2	20	40
Grand Total transportation cost					180

5.7.2 *Restricted TP*

Here, we will consider a problem with a combination of maximisation and restriction (or prohibition) in transportation.

In solving the problems, the rectricted cell is assumed to have an unaffordable cost M such that $M + a_{ij} - M - a_{ij} = M$.

ILLUSTRATION 21

ABC agency transports material from one place to the other on commission basis. The following are the estimated commissions per unit of materials to be transported from the plants P_1, P_2 and P_3 to market centres M_1, M_2 and M_3. Optimise the commissions to be earned by the agency (note that there is not route available to transport from P_2 to M_1).   [MBA-OU-97]

	M_1	M_2	M_3	*Supply*
P_1	6	9	8	120
P_2	–	4	2	80
P_3	11	5	4	80
Demand	150	70	60	280

Solution :

As the given transportation problem (TP) is in non standard from (since commission is to be maximised), the TP is first converted to standard from by subtracting every element from the greatest among all the elements. The equivalent cost matrix is,

	M_1	M_2	M_3	Supply
P_1	5	2	3	120
P_2	–	7	9	80
P_3	0	6	7	80
Demand	150	70	60	280

As there is no route available from P_2 to M_1, let us assign an unaffordable cost M such that $M + a_{ij} = M - a_{ij} = M$.

And the given TP is balanced as total demand = total supply = 280.

Now finding IBFS by Vogel's Approximations,

	M_1	M_2	M_3	Supply	Penalty
P_1	5	2	3	120	1
P_2	M	7	9	80	2
P_3	0 [80]	6	7	(80) 0	6 ←
Demand	150 / 70	70	60	280	
Penalty	5	4	4		

Allocation to $(P_3, M_1) \rightarrow 80$, Deletion $\rightarrow P_3$

Revision for M_1 from 150 to 70.

	M_1	M_2	M_3	Supply	Penalty
P_1	10 5	2	3	$\cancel{120}$ 50	1
P_2	M	7	9	80	2
Demand	$\cancel{70}$ 0	70	60		
Penalty	$M - 5 = M$ ↑	5	6		

Allocation : $(P_1, M_1) \rightarrow 70$, Deletion $\rightarrow M_1$

Revision : P_1 from 120 to 50.

	M_2	M_3	Supply	Penalty
P_1	2	50 3	50	1
P_2	70 7	10 9	80	2
Demand	70	60		
Penalty	5	6 ↑		

Now, we place the allocations in the equivalent cost matrix and check for optimality. [The TP is not degenarate as the number of allocated cells (5) is equal to no. of rows (3) plus no. of columns (3) minus 1 i.e (5)].

	M_1	M_2	M_3	u_i
P_1	70 +1 5	2	50 3	0
P_2	+M M	70 7	10 9	6
P_3	80 +10 0	+9 6	7	– 5
v_j	5	1	3	

Note : *Assumed $u_1 = 0$ and other u_i and v_j are calculated using $C_{ij} = u_i + v_j$ where C_{ij} is allocated cell cost) and then checked for $C_{ij}' - (u_i + v_j) \geq 0$ for non - allocated cells.*

Since $C_{ij}' - (u_i + v_j) \geq 0$ for all non allocated cells, the solution is optimal.

Replacing the allocations in the original matrix.

	M_1	M_2	M_3	Supply
P_1	70 / 6	9	50 / 8	120
P_2	—	70 / 4	10 / 2	80
P_3	80 / 11	5	4	80
Demand	150	70	60	280

Calculations of optimum (maximum) commission.

From	To	Per Unit Commission	Units Transported	Total Commission
P_1	M_1	6	70	420
P_1	M_3	8	50	400
P_2	M_2	4	70	280
P_2	M_3	2	10	20
P_3	M_1	11	80	880
			Grand Total	2000

Optimum Commission = Rs. 2000.

5.7.3 Conditional TP

We now illustrate a TP with maximisation, unbalanced, multiple optimal solution and other typical conditions.

ILLUSTRATION 22

A company has 3 factories and 3 ware houses with the following shipping cost and other information.

Ware House	Factories			Sales Price (Rs.)	Ware House Capacity
	F_1	F_2	F_3		
W_1	3	9	5	34	80
W_2	1	7	4	32	120
W_3	5	8	3	31	150
R/M cost	15	18	14		
Labour cost	10	9	12		
Factory capacity	150	100	130		

Due to prior commitment F_1 must supply 50 units of W_2 and for every unit sent to W_1 an octroi of Rs. 2/- per unit is imposed, also only at F_3 left out inventory costs Re. 1/- per unit left out.
Find the optimal solution for the company so as to make maximum profits or minimum cost/loss.

Solution :

Total of warehouse capacity (i.e., requirement here)
$$= 80 + 120 + 150 = 350$$

Total of factory capacities
$$= 150 + 100 + 130 = 380$$

Since $\sum$ requirement $\neq$ $\sum$ capacities, the TP is unbalanced. Therefore we introduce a dummy warehouse (row) whose capacity requirement is 30 (deficit) units.

In the above TP, there is sale price, raw material cost, labour cost apart from the shipping cost. First we have to calculate the net profit or loss to each cell with the following formula.

Profit/loss to each cell :

C_{ij} = sales price of corresponding warehouse - shipping cost of the cell - cost of R/M at corresponding factory - cost of labour at each factory.

An exemplary calculation for cell C_{11} i .e., W_1 to F_1 is shown here below.

Profit / loss for C_{11} = 34 − 3 − 15 − 10
$$= 34 − 28 = 6 \text{ (profit)}$$
for C_{12} = 34 − 9 − 18 − 9 = −2 (loss) and so on.

The profit cost calculation for each cell in the dummy row will be taken as follows.

Capacity of Dummy row = 30.

Sales price at Dummy row = zero (0)

Cost of R/M with F_1, F_2 and F_3 = 15, 18 and 14 respectively.

Cost of labour with F_1, F_2 and F_3 = 10, 9 and 12 respectively.

Cost of shipping at each cell in Dummy row = 0

An example for dummy cell C_{41} i.e., from D to F_1 is

$$C_{41} = 0 − 0 − 15 − 10 = −25 \text{ (loss)}$$

	F_1	F_2	F_3	Sales price	Requirement
W_1	3	9	5	34	80
W_2	1	7	4	32	120
W_3	5	8	3	31	150
D	0	0	0	0	30
R/M cost	15	18	14	Total	380
Labour cost	10	9	12		
	150	100	130	Total	380

Revised TP Matrix :

	F_1	F_2	F_3	
W_1	$34 - 3 - 15 - 10 =$ 6	$34 - 9 - 18 - 9 =$ $- 2$	$34 - 5 - 14 - 12 =$ 3	80
W_2	$32 - 1 - 15 - 10 =$ 6	$32 - 7 - 18 - 9 =$ -2	$32 - 4 - 14 - 12 =$ 2	120
W_3	$31 - 5 - 15 - 10 =$ 1	$31 - 8 - 18 - 9 =$ -4	$31 - 3 - 14 - 12 =$ 2	150
D	$0 - 0 - 15 - 10 =$ -25	$0 - 0 - 18 - 9 =$ -27	$0 - 0 - 14 - 12 =$ -26	30
Cap	150	100	130	380

Now, let us fulfill the conditions imposed.

Condition 1 :

50 units must be transported from F_1 to W_2. Let us first allocate this quantity irrespective of profit/loss due to the prior commitment. Here, we get a profit @ Rs. 6 per unit on transportation of these 50 units from F_1 to W_2 i.e., Rs. 300/-. And when these 50 units are transported, further capacities at F_1 and W_2 will be readjusted as 100 and 70 and the totals will be 330. The profit of the Rs. 300/- we keep aside for a while and add at the last.

Condition -2 :

Octroi @ Rs. 2 per unit is imposed on the units sent to W_1, therefore every cell in the first row, a cost of Rs. 2 per unit is to be deducted.

Condition -3 :

At F_3 if inventory is leftout an inventory cost of Re. 1/- per unit is levied. Therefore, to the cell (D, F_3) i.e., C_{43} we deduct Re 1/- since, this is the only cell representing left out inventory at F_3.

Thus with the above conditions, the TP matrix is re-written as follows.

	F_1	F_2	F_3	Requirement
W_1	$(6 - 2)$ 4	$(- 2 - 2)$ $- 4$	$(3 - 2)$ 1	80
W_2	6	$- 2$	2	70 (120 – 50)
W_3	1	$- 4$	2	150
D	$- 25$	$- 27$	$(- 26 - 1)$ $- 27$	30
Capacities	100 (150 – 50)	100	130	330 (380 – 50)

[Allocation made to (W_2, F_1) is 50 units @ Rs. 6/- profit = 300.]

Now, the figures in the above TP represent the profits, which is a non-standard case. These are to be converted to equivalent cost matrix. This can be done by two methods (multiply by -1 to all cells or subtract from highest value).

The equivalent matrix is given below by multiplying with (-1) to all cells.

	F_1	F_2	F_3	Requirement
W_1	– 4	4	– 1	80
W_2	– 6	2	– 2	70
W_3	– 1	4	– 2	150
D	25	27	27	30
Capacities	100	100	130	330

For the above TP, we now find IBFS by North West Corner method. The initial solution is given below:

	F_1	F_2	F_3	Requirement
W_1	80 / – 4	4	– 1	80
W_2	20 / – 6	50 / 2	– 2	70
W_3	– 1	50 / 4	100 / – 2	150
D	25	27	30 / 27	30
Capacities	100	100	130	330

In the above TP, there is no degeneracy since $n(C_{ij}) = n(r) + n(c) - 1$

i.e., $6 = 4 + 3 - 1$

We proceed to find u_i and v_j values with occupied cells using $C_{ij} = u_i + v_j$ and then calculate $C_{ij}' - (u_i + v_j)$ for unoccupied cells. A loop is constructed for cell with most negative value of $C_{ij}' - (u_i + v_j)$.

	F_1	F_2	F_3	Requirement	u_i
W_1	80 ; 0 ; −4	4	+1 ; −1	80	0
W_2	20 ; −6	50 ; 2	+2 ; −2	70	−2
W_3	+3 ; −1	50 ; 4	+100 ; −2	150	0
D	0 ; 25	−6 ; 27	30 ; 27	30	29
Capacities	100	100	130	330	
v_j	−4	+4	−2		

	F_1	F_2	F_3	Requirement	u_i
W_1	80 ; 0 ; −4	4	+1 ; −1	80	0
W_2	20 ; −6	50 ; 2	+2 ; −2	70	−2
W_3	+3 ; −1	20 ; 4	130 ; −2	150	0
D	+6 ; 25	30 ; 27	+6 ; 27	30	23
Capacities	100	100	130	330	
v_j	−4	+4	−2		

In the above table all the values $C_{ij}' - (u_i - v_j) \geq 0$, therefore optimal solution is attained.

The solution is to be transferred to profit matrix by changing the sign for each unit cost.

- As we have $C_{ij}' - (u_i - v_j) = 0$ for (W_1, F_2) cell, we get an alternative optimal solution, which can be obtained by transferring 50 units from (W_2, F_2) and doing necessary adjustments.

Solution - I :

	F_1	F_2	F_3	Req.
W_1	80 ; 4	−4	1	80
W_2	20 ; 6	50 ; −2	2	70
W_3	1	20 ; −4	130 ; 2	150
D	−25	30 ; −27	−27	30
Cap.	100	100	130	330

Solution - II :

	F_1	F_2	F_3	Req.
W_1	30 ; 4	50 ; −4	1	80
W_2	70 ; 6	−2	2	70
W_3	+1	20 ; −4	130 ; 2	150
D	−25	30 ; −27	−27	30
Cap.	100	100	130	330

Solution - I :

Allocated Cell	From	To	Unit Loss or Profit	No. of Unit	Total Profit or Loss	Remark
C_{11}	F_1	W_1	4	80	320	Profit
C_{21}	F_1	W_2	6	20	120	Profit
C_{22}	F_2	W_2	-2	50	-100	Loss
C_{32}	F_2	W_3	-4	20	-80	Loss
C_{33}	F_3	W_3	2	130	260	Profit
C_{42}	F_2	(Dummy)	-27*	30*	(-810*)	*Cost that can be minimised by not producing
				Total	700-180 = 520	

Solution - II :

Allocate Cell	From	To	Unit Loss or Profit	No. of Unit	Total Profit or Loss	Remark
C_{11}	F_1	W_1	4	30	120	Profit
C_{12}	F_2	W_1	-4	50	-200	Loss
C_{21}	F_1	W_2	6	70	420	Profit
C_{32}	F_2	W_3	-4	20	-80	Loss
C_{33}	F_3	W_3	2	130	260	Profit
C_{42}	F_2	(Dummy)	-27*	30*	(-810*)	Not to produce
				Total	800-280 = 520	

The over all profit Rs. 520 can be obtained provided the production at F_2 is restricted to 70 units only instead of 100, i.e., 30 units excess produced from F_2 is to be reduced (*otherwise it will cost Rs. 810 which turns the profit of Rs. 520 to loss of Rs. 290).

Further the profit of Rs. 300 on allocating 50 units at (W_2, F_1) adds to the answer:

∴ Optimal profit = 520 + 300 = Rs. 820/-.

Practice Problems

Obtain IBFS by NWCM for the following transportation problems whose unit costs of transportation, availabilities and requirements are given in the matrices, and then optimize

1. [JNTU (CSE) - 2000/S]

Source	Destination				Supply
	1	2	3	4	
1	3	7	6	4	5
2	2	4	3	2	2
3	4	3	8	5	3
Demand	3	3	2	2	10

2.

Party \ Consumer	A	B	C	Supply
1	6	8	4	14
2	4	9	3	12
3	1	2	6	5
Demand	6	10	15	

3. [IAS -89, OU - MBA - 92]

	D_1	D_2	D_3	D_4	Supply
O_1	6	4	1	5	14
O_2	8	9	2	7	16
O_3	4	3	6	2	5
Demand	6	10	15	4	

4. [IAS - 89 , AU - B.Tech -90]

	I	II	III	IV	Supply
O_1	13	11	15	20	200
O_2	17	14	12	13	600
O_3	18	18	15	12	700
Demand	300	300	400	500	

5. [Madras B.Sc. 81]

	D_1	D_2	D_3	D_4	D_5	D_6	Supply
S_1	9	12	9	8	4	3	5
S_2	7	3	6	8	9	4	8
S_3	4	5	6	8	10	14	6
S_4	7	3	5	7	10	9	7
S_5	2	3	8	10	2	4	3
Demand	3	4	5	7	6	4	

There is degeneracy in this problem

6. [Delhi (stat) - 73]

	1	2	3	4	5	Supply
A	2	11	10	3	7	4
B	1	4	7	2	1	8
C	3	9	4	8	12	9
Demand	3	3	4	5	6	

7.

	D_1	D_2	D_3	D_4	Supply
A	11	13	17	14	250
B	16	18	14	10	300
C	21	24	13	10	400
Demand	200	225	275	250	950

8.   [OU-B.Tech (Mech) - 91, JNTU - (FDH) - 92, OU - MBA -Sep. 99]

	D_1	D_2	D_3	D_4	Supply
S_1	19	30	50	10	7
S_2	70	30	40	60	9
S_3	40	8	70	20	18
Demand	5	8	7	14	34

Answer : $19 \times 5 + 30 \times 2 + 30 \times 6 + 40 \times 3 + 70 \times 4 + 20 \times 14 = 1015$

9. Find the optimum solution to the following transportation problem in which the cells contain the transportation cost in rupees :

	W_1	W_2	W_3	W_4	W_5	Available
f_1	7	6	4	5	9	40
f_2	8	5	6	7	8	30
f_3	6	8	9	6	5	20
f_4	5	7	7	8	6	10
Required	30	30	15	20	5	

Answer : $7 \times 30 + 6 \times 10 + 5 \times 20 + 6 \times 10 + 9 \times 5 + 6 \times 15 + 8 \times 5 + 6 \times 5 = 635$

10. Find IBFS by matrix minimum method to the following TP, each cells value being the unit profit, and find the corresponding profit. Also compare your results by any other method.   [OU - MBA 90]

	D_1	D_2	D_3	D_4	D_5	Avail.
O_1	34	55	47	24	32	250
O_2	41	28	32	46	51	150
O_3	20	31	35	47	50	175
O_4	53	35	26	28	39	125
Req.	130	280	110	60	120	700

Answer : $34 \times 130 + 55 \times 120 + 28 \times 150 + 31 \times 10 + 35 \times 110$
$$+ 47 \times 155 + 28 \times 5 + 39 \times 120 = 26785$$

11. Find IBFS to the following TP using WWC, VAM and LCEM, each cell value being cost, and find cost.   [OU - MBA - 91]

	D_1	D_2	D_3	D_4	D_5	Avail.
O_1	17	11	12	20	8	300
O_2	16	9	10	15	14	180
O_3	18	19	9	11	12	145
O_4	13	15	14	16	15	130
Req.	200	250	150	80	120	

Answer : $17 \times 200 + 11 \times 100 + 9 \times 150 + 10 \times 30 + 9 \times 120 + 11 \times 25 + 16 \times 55 = 8385$

12. Find the initial basic feasible solution to the following T.P using VAM given the cost matrix   **[OU - MBA - Sep 2001, May 92]**

	D_1	D_2	D_3	D_4	Supply
F_1	20	25	28	31	300
F_2	32	28	32	41	180
F_3	18	35	24	32	110
Demand	150	40	180	170	

13. Solve the following transportation problem using matrix minimum methods each cell value being the unit profit.   **[OU - MBA - Sep. 2000]**

	D_1	D_2	D_3	D_4	D_5	Supply
F_1	4	1	3	4	4	60
F_2	2	3	2	2	3	35
F_3	3	5	2	4	4	40
Demand	22	45	20	18	30	135

14. A company has four factories situated in different locations and five ware houses in different cities. The matrix of transportation cost (Rs. per unit) is given below with capacity (unities) as factories and the requirement (units) of the ware houses.   **[OU - MBA - Apr. 94]**

Ware House	I	II	III	IV	Req. (Units)
A	4	8	7	6	150
B	9	5	8	8	50
C	6	5	8	7	40
D	5	8	6	3	60
E	7	6	5	8	200
Capacity Units	100	80	120	100	

Find initial basic feasible solution by matrix method.

15. Find the initial basic feasible solution to the following transportation problem given the unit cost matrix, using least cost entry method.   **[OU - MBA - Feb. 93, Sep. 99]**

	D_1	D_2	D_3	D_4	D_5	Supply
O_1	20	28	32	55	70	50
O_2	48	36	40	44	25	100
O_3	35	55	22	45	48	150
Demand	100	70	50	40	80	

16. Find the IBFS to the following transportation problem. Fiven the unit profit Matrix. Using Vogel's approximation method. **[OU - MBA - May 95]**

Factory	A	B	C	Supply
F_1	10	8	8	8
F_2	10	7	10	7
F_3	11	9	7	9
F_4	12	14	10	8
Demand	10	10	8	

17. Find an IBFS for the following transportation problem using Vogel's approximation method. Compare with the result of NWC.

[OU - MBA Apr. 98, Sep. 99, May 95]

	D_1	D_2	D_3	D_4	Avail.
F_1	190	300	500	100	50
F_2	700	300	400	600	100
F_3	400	100	600	200	150
Req.	50	80	70	140	

18. A garment manufacturing company has two centers at Nizamabad and Hyderabad where it manufactures special skirts of same specifications, but at varying costs. Nizamabad centre has a capacity of 2500 skirts at cost of production as Rs. 23/- per skirt while that at Hyderabad it can produce 2100 skirts @ Rs. 25/- per skirt. Four boutiques are willing to purchase these skirts with the maximum requirements as 1800, 2300, 550 and 1750 and offer the prices as 39, 37, 40 and 36 respectively. The cost (in rupees) of the shipping one skirt from Nizamabad center to the four boutiques is 6, 8, 11 and 9 while that from Hyderabad is 12, 6, 8 and 5 respectively. Determine the optimal distribution arrangement for the garment manufacturer. **[IGNOU (MCA) DEC 97, JULY 99]**

Hint : The above information can be formulated into a TP matrix as follows.

	Boutiques				Cap.	Mfg. Cost
	1	2	3	4		
Nizamabad @ Rs. 23/-	$39 - 23 - 6 =$ 10	$37 - 23 - 8 =$ 6	$40 - 23 - 11 =$ 6	$36 - 23 - 9 =$ 4	2500	23
Hyderabad @ Rs. 25	$39 - 25 - 12 =$ 2	$37 - 25 - 6 =$ 6	$40 - 25 - 8 =$ 7	$36 - 25 - 5 =$ 6	2100	25
Req.	1800	2300	550	1750	6400	4600
Sales Price	39	37	40	36		

[Then proceed with creating a dummy row of (production) capacity 1800 and zero as cell cots. Notice that the given problem is non-standard, hence proceed by assigning negative sign or subtracting from highest.]

19. *ABC* Enterprises is having three plants manufacturing dry-cells, located at different locations. Production cost differs from plant to plant. There are five sales offices of the company located in different parts of the country. The sales prices can differ from place to place. The shipping cost from each plant to each sales office and other related data are given in the following tables :

Production Data table :

Production Cost Per Unit	Maximum Capacity in Number of Units	Plant Number
20	150	1
22	200	2
18	125	3

Shipping Cost and Demand and Sales Prices Table :
Shipping Cost :

Sales Office	1	2	3	4	5
Plant 1	1	1	5	9	4
Plant 2	9	7	8	3	6
Plant 3	4	5	3	2	7

Demand and Sales Prices :

Demand	80	100	75	45	125
Sales price	30	32	31	34	29

Find the production and distribution schedule most profitable to the company.

[IGNOU - MCA - June 1997, Dec. 99]

Hint : Refer the hint of its previous problem.

20. Raju and Co is having three manufacturing centres at Hyderabad, Vishakapatnam and Nizamabad with the production capacities as 200, 300 and 500 units respectively. Its markets are situated at Secunderabad, Vijayawada, Tirupathi and Warangal with the demands as 250 each. The costs of transportation are found to be Rs. 20, 40, 25 and 30 per unit from Hyderabad to various stations respectively. From Vizag (Vishaka) these costs are 70, 40, 80 and 60 and from Nizamabad Rs. 50, 70, 30 and 40 per unit respectively. When there is a raise in demand of the product, the markets require 50% extra. For urgent delivery cases Nizamabad centre can produce 25% extra and can send at Rs. 10 per unit extra load carried to various places. Hyderabad centre can

sub-contract 50% of its extra capacity and can supply at Rs. 20 hike in transportation. Formulate TP and suggest the optimal policy for the case of raised demand.

Hint : Create two more production centres at Nizamabad and Hyderaad respectively with 25% and 50% increased capacity and Rs. 10/- and Rs. 20/- hiked cost. Also add increased demand to the markets. The following is the formulation.

	Secunderabad	Vijayawada	Tirupathi	Warangal	Capacity
Hyderabad	20	40	25	30	200
Vishakapatnam	70	40	80	60	300
Nizamabad	50	70	30	40	500
Nizamabad-2 (25% extra)	60	80	40	50	125
Hyderabad-2 (50% extra)	40	60	45	50	100
Demand	375	375	375	375	1500 / 1225

21. A product is manufactured by four factories, A, B, C and D. The unit production costs in them are Rs. 2, Rs. 3, Re. 1 and Rs. 5 respectively. Their production capacities are 50, 70, 30 and 50 units respectively. These factories supply the product to four stores, demands of which are 25, 35, 105 and 20 units respectively. Unit transportations cost in rupees from each factory to each store is given in the following table :

Stores

Factories	1	2	3	4
A	2	4	6	11
B	10	8	7	5
C	13	3	9	12
D	4	6	8	3

Determine the extent of deliveries from each of the factories to each of the stores so that the total production and transportation cost is minimum.

[JNTU (Mech) 98]

22. A manufacturer has distribution at Agra, Allahabad and Kolkata. These centres have available 40, 20 and 40 units of his product. His retail outlets require the following number of units : $A - 25$, $B - 10$, $C - 20$, $D - 30$, $E - 15$.

The shipping cost per unit in rupees between each centre and outlet is given in the table.

Retail Outlets

	A	B	C	D	E
Agra	2	4	6	11	50
Allahabad	10	8	7	5	60
Kolkata	13	3	9	12	30

Determine the optimal shipping cost.

[JNTU - Mech./Prod./Chem./Mechatronics 2001/S]

23. A company has three plants A, B, C which supplies to ware houses 1, 2, 3, 4, 5. Monthly plant capacities are 800, 800, 900 units respectively while the monthly requirements at the warehouses are 400, 400, 500, 400, 800 units respectively. The unit transportation costs are shown in table below.

Determine optimum distribution for the company in order to minimize the total transportation cost :

Warehouses

		1	2	3	4	5
	A	5	8	6	6	3
Plants	B	4	7	7	6	6
	C	8	4	6	6	3

[JNTU - CSE/E.Comp E - 2001]

24. A company has three plants A, B and C and four warehouses W, X, Y and Z. The number of units available at the plants are 15, 25 and 5 respectively and the demand at warehouses are 5, 15, 15 and 10 respectively. The unit transportation costs are as follows.

		W	X	Y	Z
	A	10	0	20	11
Plants	B	12	7	9	20
	C	0	14	16	13

Find the allocation that gives minimum transportation cost

25. Find the optimal solution to following TP, given unit cost

	D_1	D_2	D_3	D_4	Supply
O_1	15	19	15	22	50
O_2	19	13	18	17	60
O_3	21	12	16	15	40
Demand	20	25	16	35	

[OU - MBA - April 94]

26. Find optimal solution to the following TP given unit cost of transportation matrix.

	D_1	D_2	D_3	Supply
O_1	16	20	12	200
O_2	24	8	18	160
O_3	26	24	16	90
Demand	180	120	150	

[OU - MBA - Feb. 93]

27. Find optimal solution to the following TP using VAM - MODI method each cell value being the unit cost **[OU - MBA - July 2000, Dec. 2000]**

	D_1	D_2	D_3	D_4	D_5	Supply
C_1	35	41	28	16	20	285
C_2	14	21	28	30	15	145
C_3	45	18	17	29	26	165
Demand	125	125	100	100	175	

28. Find optimal cost the following TP **[OU - MBA 91]**

	D_1	D_2	D_3	D_4	Supply
O_1	12	18	13	20	50
O_2	17	11	16	15	60
O_3	11	10	14	13	40
Demand	20	25	10	35	

29. Find optimal cost of the following TP **[OU - MBA Nov. 94]**

	D_1	D_2	D_3	D_4	Supply
O_1	42	27	24	35	50
O_2	46	37	32	32	60
O_3	40	40	30	35	40
Demand	40	20	60	30	

30. Find OBFS of following TP. Using VAM for IBFS   [OU - MBA - Sep. 98]

	W_1	W_2	W_3	W_4	Supply (tons)
F_1	100	120	95	110	250
F_2	85	75	100	120	300
F_3	110	105	80	100	450
F_4	75	85	70	90	500
F_5	140	130	120	125	400
Demand	650	500	450	400	

31. Consider a TP with three ware houses and four markets warehouses capacities are : $W_1 = 30$, $W_2 = 70$ and $W_3 = 50$ market demands are : $M_1 = 40$, $M_2 = 30$, $M_3 = 40$, $M_4 = 40$. The unit cost movement in Rs. is given below.

	M_1	M_2	M_3	M_4	Supply
W_1	20	20	20	10	50
W_2	100	80	50	40	60
W_3	70	60	60	80	40
Demand	40	20	60	30	

Find optimal solution   [OU - MBA - Apr. 99]

32. A potato chip manufacturer has three plants and four ware houses. The shipping costs are given below. Find optimal solution after using northwest corner rule.   [OU - MBA - Apr. 99]

	Plants				
Warehouses	1	2	3	4	Avail.
1	5	8	6	6	235
2	4	7	7	6	280
3	8	4	6	6	110
Req.	125	160	110	230	625

33. A department store whisehs to purchase the following quantities of sarees.

Warehouses	A	B	C	D	E
Qty.	150	100	75	250	200

Tenders are submitted by 4 different manufactuers who undertake to supply not more than the quantities mentioned below. (All types of serees combined)

Manufacturer	W	X	Y	Z
Qty.	300	250	150	200

The store estimates that the profit per saree will vary with the manufacturer as shown in the folowing matrix.

Sarees

	A	B	C	D	E
W	275	350	425	225	150
X	300	325	450	175	100
Y	250	350	475	200	125
Z	325	275	400	250	175

How should the orders be placed **[OU - MBA - May. 92]**

5.7.4 *Trans-shipment Problem*

In transportation problem, the transportation or shipping occurs between sources to destinations. And the units cannot be transferred from one source to the other nor one destination to the other. But in practical situations, when a destination requires more than its estimated demand, it may bring the unsold (or surplus) units lying at another destination. This is even applicable to the source. Thus if the transportation is, allowed among the sources as well as among destinations, apart from the regular transportation from sources to destinations, the problem is said to be trans-shipment. Diagrammatically it is shown below.

(a) Transportation problem   (b) Trans-shipment problem

FIGURE 5.5 :   DIFFERENCE BETWEEN TRANSPORTATION AND TRANS-SHIPMENT

Thus in trans-shipment problem, sources will play the role of destinations also, and destinations may be sources too. Thus a transportation problem of 'm' sources and 'n' destinations will become a trans-shipment problem with $(m + n)$ sources and $(m + n)$ destinations. i.e., $m \times n$ matrix becomes $(m + n) \times (m + n)$ matrix.

And if total number of units transported is N, then add this N at supply and demand at each source as well as destination. Then the problem is solved as usually with the rules of TP.

ILLUSTRATION 23

A firm having two sources, S_1 and S_2 wishes to ship its products to two destinations D_1 and D_2. The number of units of units available at S_1 and S_2 are 10 and 30 respectively while the demands at D_1 and D_2 are 25 and 15 units respectively. The firm instead of shipping from source to desinations, decides to investigate the possibility of trans-shipment. The unit cost of transportation (in Rs.) is given as follows:

From S_1 to S_2, D_1 and D_2 : 3, 4, 5 respectively.

From S_2 to D_1 and D_2 : 3, 5 respectively

From D_1 to D_2 : 2

and the cost of upward shipment os same as that of downward shipment this means cost of S_1 and S_2 is same as S_2 to S_1 and so-on. (Symmetric)

Solution :

The formulation of the above information in the form of a trans-shipment problem is as follows :

From	To S_1	S_2	D_1	D_2	Supply
S_1	0	3	4	5	10 + 40*
S_2	3	0	3	5	30 + 40*
D_1	4	3	0	2	40*
D_2	5	5	2	0	40*
Demand	40*	40*	25 + 40*	15 + 40*	(40 + 160*)

* The total number of units available at S_1 and S_2 are (10 + 30 = 40) is e ual to total demands at D_1 and D_2 (25 + 15 = 40). Thus maximum number of units that can be shipped is 40 units. Since all the supply points (sources as well as destinations) and demand points are able to supply this total number units, these quantities are regarded as buffer stock and is added to all sources as well as destinations at both availability and demand. Thus the TP is revised and IBFS is then found by least cost entry method and then for OBFS, MODI method is applied as follows :

	S_1	S_2	D_1	D_2	Supply	u_i
S_1	40 0	+3 3	+1 4	10 5	50	0
S_2	+3 3	40 0	25 3	5 5	70	0
D_1	+7 4	+6 3	40 0	0 2	40	– 3
D_2	+10 5	+10 5	+4 2	40 0	40	– 5
Demand	40	40	65	55	200	
v_j	0	0	3	5		

All the values of $C_{ij}' - (u_i + v_j) \geq 0$, the above TP has reached optimal solution.

The solution is S_1 to D_2 : 10 units $\times$ 5 Rs. = Rs. 50

S_2 to D_1 : 25 units $\times$ 3 Rs. = Rs. 75

S_2 to D_2 : 5 units $\times$ 5   Rs. = Rs. 25

Total = Rs. 150

However, 5 units of (D_1, D_1) cell can be transferred to cell (D_1, D_2). Thus an alternate solution that is in accordance with trans-shipment is as follows.

	S_1	S_2	D_1	D_2	Supply
S_1	40 0	3	4	10 5	50
S_2	3	40 0	30 3	5	70
D_1	4	3	35 0	5 2	40
D_2	5	5	2	40 0	40
Demand	40	40	65	55	200

The solution to the above optimal T.P is

From	To	No. of Units	Unit Cost	Total	Remarks
S_1	D_2	10	5	50	S_1, S_2 and D_2 have each 40 units buffer while 35 units are buffer at D_1 and 5 more units are sent from D_1 to D_2
S_2	D_1	30	3	90	
D_1	D_2	5	2	10	
			Total	150	

Review Questions

1. Give the applications of transportation problem in the industries.

   [JNTU - CSE 98, OU - MBA M 91, F 93]

2. Write the algorithm for solving a TP. [OU - MBA - M 92]

3. Bring out the relation between LPP and TP. [JNTU - Mech. 97/P]

4. TP is a case of LPP — discuss. [OU - MBA - 98, JNTU EEE 91]

5. How do you formulate a TP problem as an LPP.
   Explain with an example of your choice. [JNTU - EEE 92]

6. With an example of your choice, explain the algorithm of North West corner
   method to obtain IBFS in a TP. [JNTU - EEE 93/OT]

7. What is IBFS. Give step by step approach to find IBFS in a TP using Vogel's
   approximation method. [JNTU EEE - 93 / OT]

8. Write the algorithm for finding IBFS in a TP using least cost entry method.
   Give suitable example.

9. What is non-standard TP. How do you optimise in this case.

10. What is meant by unbalanced TP. How do you find optimal solution in this case.

    [PU - MBA 92, JNTU - Mech./Prod./Chem - 2001/S]

11. What is meant by degeneracy in TP. How do you resolve it?

    [JNTU - Mech 96, 98/C, CSE 94,]

12. Give step by step rules for stepping stone method for finding OBFS in a TP.

    [OU - MBA - J 2000, S 2001]

13. Write the algorithm for MODI method. [JNTU - CSE 95/S]

14. Give flow chart to find IBFS of TP using the following methods.
    (a) NWCM (b) VAM (c) LCEM

15. Draw a flow chart to find OBFS of TP using
    (a) Stepping stone method.
    (b) MODI-method. [CSE - 92/S]

16. What is meant by a prohibited (restricted) TP. How do you proceed to find
    OBFS in this case?

17. What is meant by a conditional TP. Explain with an example.

18. Write a short note on trans-shipment problem with an example.

    [JNTU - Mech. 96, 97/P, 98/C, CSE, 96, 97/S]

19. When do you get multiple optimal solution in TP. Explain with an example,
    how you identify this situation.

    [Mech. 97/C, 99/P,OU - MBA Dec. 90, F 93, M 92, Sep. 2001]

20. Draw a flow chart for optimisation of TP including its initial solution.

21. When does degeneracy occur in TP? [OU - MBA - 91 M, 94 D, 95 D, 95 M, J 2000, S 99]

22. What are the differences between general LPP and TP?

[OU - MBA - 95 Dec, 95 M, D 2000, May 99]

23. What is balanced TP? [OU - MBA - Apr 98]

24. What are the differences between stepping stone method and MODI Method?

[OU - MBA 93]

25. What is a TP. [OU - MBA - M 95]

26. What is shadow prices in TP [OU - MBA -98]

Objective Type Questions

1. Which of the following is not false in the case of the concept of 'ϵ' introduced to resolve degeneracy in T.P.

(a) $\epsilon + a = \epsilon$

(b) $\epsilon - a = \epsilon$

(c) $\epsilon = 0$

(d) $\epsilon + \epsilon = \epsilon$

2. Which of the following is false in optimal T.P. whose $C_{ij} - (u_i + v_j)$ values of ___

(a) all unoccupied cells must be equal to zero

(b) all unoccupied cells must be less than zero

(c) all occupied cells must be equal to zero

(d) all occupied cells must be greater than or equal to zero

3. Which of the following method may yield same IBFS for both minimisation and maximisation of same data.

(a) Vogel's approximation

(b) least cost entry method

(c) north west corner rule

(d) row minima method

4. In which of the following cases degeneracy may not appear

(a) partial sum of supplies is equal to partial sum of demands

(b) both row and column are satisfied in same step in Vogel's approximation

(c) same least cost for more than one cells

(d) no. of allocated cells is less than rows + columns – 1

5. Which of the following methods uses penalties to find IBFS

(a) north west corner method

(b) Vogel's approximation method

(c) least cost entry method

(d) column minima method

6. Which of the following method to find IBFS of transportation problem is independent of costs /profit of transportation matrix

(a) north west corner rule

(b) Vogel's approximation

(c) matrix minima method

(d) row minima method

7. Multiple optimal solutions in TP are observed if

    (a) $C_{ij} = u_i + v_j$ for some occupied cells
    (b) $C_{ij} \neq u_i + v_j$ for some occupied cells
    (c) $C_{ij} = u_i + v_j$ for some unoccupied cells
    (d) $C_{ij} \neq u_i + v_j$ for some unoccupied cells

8. The solution is said to be optimal in T.P if

    (a) $C_{ij} \geq u_i + v_j$ for all occupied cells
    (b) $C_{ij} \leq u_i + v_j$ for all unoccupied cells
    (c) $C_{ij} \geq u_i + v_j$ for all unoccupied cells
    (d) $C_{ij} \leq u_i + v_j$ for all unoccupied cells

9. A T.P. is said to be balanced if

    (a) no. of rows = no. of columns
    (b) no. of allocated cells = no. of rows + no. of columns − 1
    (c) Total supply = total demand
    (d) no. of supply centres = no. of demand centres

10. In a restricted (or prohibited) T.P., we use M as the cost in corresponding cell with the concept that

    (a) $x + M = x - M = x$ (b) $M + x = M - x = x$
    (c) $M + x = M - x = M$ (d) $M + x - M - x = 0$

11. In a maximisation case of T.P, we convert it to minimisation case by _____

    (a) adding every cell value to highest among them
    (b) subtracting every cell value from highest among them
    (c) dividing cell value with lowest among them
    (d) subtracting least value of each row in the corresponding row.

12. Which of the following is false with trans-shipment problem as compared to its corresponding transportation problem.

    (a) supplies can be done by demand centres
    (b) demand can be adjusted among demand centres
    (c) transportation is made between supply centres and demand centres
    (d) a supply centre can not move its goods to other supply centre

13. In a T.P. Matrix of the order 4×3, the degeneracy if found if number of allocated cells are

    (a) 11   (b) 7   (c) 6   (d) 5

14. A non degenerate T.P. has five allocated cells. Then it may have

    (a) 6 rows   (b) 6 columns

    (c) 3 rows & 2 columns   (d) 2 rows & 4 columns

15. To find IBFS in maximising T.P, which of the following is wrong

    (a) allocation is made at cell with highest value in LCEM
    (b) allocation is made by using highest penalty at the cell with least cost
    (c) all cells are multiplied by (−1) before allocating
    (d) all values are subtracted from highest value among all cells.

16. If total supply > total demand, then add

    (a) dummy supply centre
    (b) dummy demand centre
    (c) dummy supply and dummy demand centre
    (d) difference at any demand centre

17. Degeneracy is observed if

    (a) a row and column are simultaneously filled-up in north west corner method.
    (b) no. of allocated cells plus one is less than sum of no. of rows and no. of columns in a T.P.
    (c) minimum ratio is found equal for two or more basic variables in simplex
    (d) any of the above

18. Given T.P. matrix as follows, cost by NWCM is

	P	Q	R	Supply
A	5	3	2	60
B	4	2	1	40
Dem.	20	30	50	100

    (a) 200   (b) 250   (c) 300   (d) 100

19. In the above T.P. matrix, the IBFS by LCEM gives a cost of

    (a) 200   (b) 250   (c) 300   (d) none of the above

20. In which of the following method of finding IBFS, the top-left corner cell is compulsory allocated

    (a) VAM   (b) ROW Minima   (c) NWCM   (d) LCEM

21. Which of the following is false in the case of loop drawn while optimising a T.P.

    (a) every loop must have even number of corners
    (b) a loop must have at least four corners
    (c) closed loop may or may not be a square/rectangle
    (d) the loop lines need not be horizontal and vertical lines

22. With reference to the loop draw while T.P. optimization, the minimum number of corners is

    (a) 1   (b) 2   (c) 3   (d) 4

23. A loop in T.P optimisation should have _____ at the corners,

    (a) all allocated cells   (b) all unoccupied cells

    (c) at least one unoccupied cell   (d) only one unoccupied and all other
                                                   occupied cells

24. In the case of loop drawn while optimising TP, which condition is false

    (a) each row must have only one minus and one plus sign

    (b) loop must be closed

    (c) loop must have odd number of occupied cells at its corner

    (d) shape of loop must be square

25. In IBFS of the following transportation problem, the some figures were found
    missing. Identify the missed cell C_{ij} i.e., cell of ith, row and j_{th} column

    (a) C_{12} (b) C_{23} (d) C_{31} (d) C_{32}

26. The number of units missed in the above problem is

    (a) 50   (b) 20   (c) 10   (d) 70

27. For the T.P. given in Q. No. 25, the transportation cost by NWCM is _____

    (a) 410   (b) 420   (c) 450   (d) 460

28. For the above problem given in Q. No. 25, the transportation cost by least cost
    entry method is

    (a) 410   (b) 420   (c) 460   (d) 470

29. For the problem given in Q. No. 25, which method does not develop
    degeneracy at the initial stage

    (a) VAM   (b) LCEM   (c) NWCM   (d) none of these

30. The cost of transportation in a dummy cell is _____

    (a) unity   (b) zero   (c) ∞   (d) negative

Fill in the Blanks

1. In NWCM, the allocation is independent of _____ of the transportation.

2. The top-left cell of the TP is necessarily allocated in _____ method

3. The total supply must be equal to total demand. This condition is called

   _____

4. If partial sum of supply is equal to partial sum of demands then _____ may appear

5. Penalties are used in _____ method of finding IBFS in TP

6. The criterion used in LCEM is cell units _____ of TP matrix

7. In MODI method we use u_i and v_j such that cost of cell $(C_{ij}) = u_i + v_j$ for _____

8. In MODI method, if C'_{ij} is the cost of an unoccupied cell in i^{th} row, j^{th} column and u_i and v_j are assumed costs of ith row and j^{th} column, then the optimal solution should constitute _____ for all unoccupied cells

9. A TP is said to be balanced if _____

10. A TP is said to be non-degenerate if _____

11. The criterion used in VAM (to find IBFS) is_____

12. If some unoccupied cells have the costs such that $C'_{ij} - (u_i + v_j) = 0$, then the TP may yield _____

13. In restricted TP, we use the cost 'M' to the restricted cell with the condition that

    _____

14. The epsilon (ϵ) used to resolve degeneracy is based on the conceptual conditions given by mathematical expression as _____

15. In a TP of 'm' rows and 'n' columns, there exist _____ number of basic feasible solutions

16. A loop drawn in method of optimization of TP should consist at least _____ corners

17. The costs of dummy cells are taken as _____

18. Fill up the missing numbers

50	T			100
4	6	2	1	
	60	Q		80
3	5	7	2	
		R	S	80
6	3	0	5	
50	P	50	50	

$P =$ _____ $Q =$ _____ $R =$ _____ $S =$ _____ $T =$ _____

19. If supply exceeds demand, we add _____

20. A transportation problem in which available commodity frequently moves from one source to another source or destination before reaching its actual destination is called _____

Answers

Objective Type Questions :				
1. (b)	2. (c)	3. (c)	4. (c)	5. (b)
6. (a)	7. (c)	8. (c)	9. (c)	10. (c)
11. (b)	12. (d)	13. (d)	14. (d)	15. (b)
16. (b)	17. (d)	18. (b)	19. (b)	20. (c)
21. (d)	22. (d)	23. (d)	24. (d)	25. (d)
26. (c)	27. (d)	28. (d)	29. (c)	30. (b)

Fill in the Blanks :	
1. cost/profit	2. north west corner
3. rim condition	4. degeneracy
5. Vogel's approximation	6. lowest cost
7. all occupied cells	8. $C'_{ij} - (u_i + v_j) \geq 0$ or $C'_{ij} \geq u_i + v_j$
9. total supply = total demand	10. $n (C_{ij}) = n (r) + n (c) - 1$
11. Penalty	12. Alternate or multiple optimal solution
13. $M \pm a_{ij} = M$	14. $a_{ij} \pm \epsilon = a_{ij} ; \epsilon \neq 0$
15. $m + n - 1$	16. four
17. zero	18. 110, 20, 30, 50, 50

6

Chapter

Assignment Problem

CHAPTER AT A GLANCE

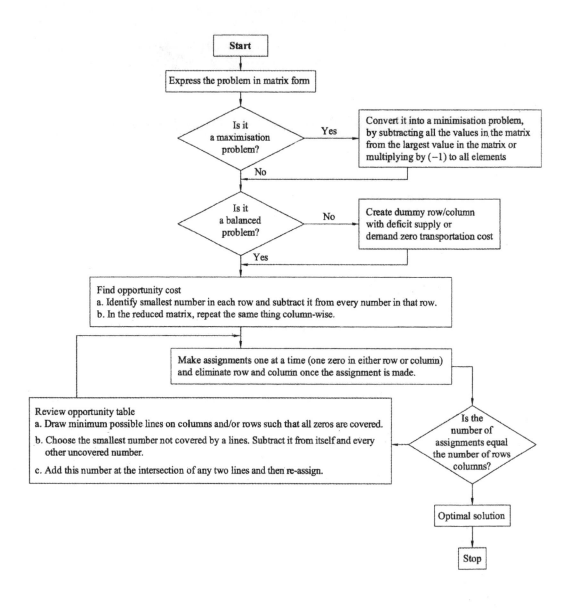

6.0 *Introduction*

In most cases the departments will have the specialists for operating certain critical or sophisticated machinery/equipment while in some other cases any body can operate any machine. Whatever the type of set up may it be, it is essential for a manager to see that maximum work is to be derived from his man power resources, which are precious and scarce. This can be done when and only when right job is given to right man.

Generally, though any body can do any job, all the men will not have same efficiency and knowledge on all the jobs. The same job, one may do fast while the other may do it slowly or a person may do one job fast and other slow. Depending on the variations of their efficiency on individual jobs, if the jobs are assigned according to their efficiencies, the maximum output can be derived in minimum time. Often, the assignment of these jobs will be done based on the experience and commonsense of the plant engineer, which may result in success in some cases but not always. The engineer will face the hardship if it leads to any loss and becomes so difficult to escape from the situations because there will be neither any authentic record nor a scientific/proven methodology to safeguard him in such situations. Moreover, the organisation may experience the irreparable loss.

The scientific and proven method to assignment is Hungarian method. This most suitable method for job shop production (Intermittent production), is developed by Hungarian Mathematician D. Konig. It is also known as 'Floods' technique or 'Fast food technique' It works on the principle of reducing the cost matrix to opportunity costs which shows the relative penalties associated with assigning resources to an activity as opposed to making the best or least cost assignment. If we can reduce the cost matrix to the extent of having at least one zero in each row and column, then it will be possible to make optimal assignment. The algorithm is as follows.

6.1 *The Assignment Algorithm*

Step 1 : *Standard Form :* Develop the cost table from the given information. *The minimisation case is supposed to be the standard form in assignment problems.* The cost or time or wastage etc., is included in this type. If profit, production quantity are given they are to be maximised. In this case the given problem is to be first converted to the minimisation form by using one of the following rules.

(a) Identify maximum cell value and subtract all the cell values from it to get corresponding cost values. (or)

(b) Multiply all the values with minus one (-1).

Step 2 : *Balance Checking : An assignment problem (AP) is said to be balanced if the matrix is a square i.e., the number of rows is equal to the number of columns.* If the number of rows is not equal to the number of columns then a dummy row or dummy column must be added by putting the assignment costs for the entire dummy cells as zeros.

Step 3 : *Opportunity Cost Table :* Prepare the opportunity cost table by following rules.

(a) Locate the smallest element in each row of the given cost table and then subtract from each element of that row.

(b) In the reduced matrix obtained from 3(a), locate the smallest element in each column and then subtract from each element of that column. Each row and column now must have at least one zero value (*This may be used as a check*).

Step 4 : *Assignment :* Make assignments in the opportunity cost matrix using the following procedure.

(a) Examine rows successively until a row with exactly one unmarked zero is obtained. Make an assignment to the single zero by making a square (□) around it.

(b) For each zero value that becomes assigned, cross out (×) all other zeros in the same row and/or column.

(c) Repeat steps 4(a) and 4(b) for each column also with exactly single zero value cell that has not been eliminated.

(d) If a row/or column has two or more unmarked zeros, then choose the zero arbitrarily.

(e) Continue this process until all zeros in rows/columns are either enclosed by square (assigned) or crossed out by ×.

Step 5 : *Optimality Criterion :* If the number of assigned cells is equal to the number of rows/columns, then it is an optimal solution. The total cost (or profit in the case of maximisation) associated with this solution is obtained by adding original cost (or profit) figures in the occupied cells. If no optimal solution is found then go to step - 6.

Step 6 : *Marking :* Draw a set of horizontal and/or vertical lines to cover all the zeros in the revised cost table obtained from step -5 using the following procedure.

(a) Mark (* or ✓) the row(s) in which no assignment was made.

(b) Examine the marked rows. If any zeros occur in those rows, mark a tick or star to the respective columns that contain those zeros.

(c) Examine marked columns. If any assigned zero occurs in those columns, then tick the respective rows that contain those assigned zeros.

(d) Repeat this process until no more rows/column can be marked.

(e) Draw a straight line through each marked column and each unmarked row.

Remember :

Draw lines on 'URMC' i.e., Unmarked Rows and Marked Columns. If the number of lines drawn is equal to the number of rows or columns, then the current solution is the optimal solution, otherwise go to step-7. (*This checking of number of lines can be used as a check point here*).

Step 7 : *Revising :* Develop the new revised opportunity cost table.

(a) Among the cells not covered by any line, choose the smallest element say, k, the key element.

(b) Subtract k from every element in the cell not covered by a line.

(c) Add k to every element in the cell covered by the two lines, i.e., intersection of two lines.

(d) Elements in cells covered by one line remain unchanged.

Step 8 : *Repeat :* Repeat steps - 4 to 7 until an optimal solution is obtained.

This algorithm is shown through a flow chart as given below :

Flow chart :

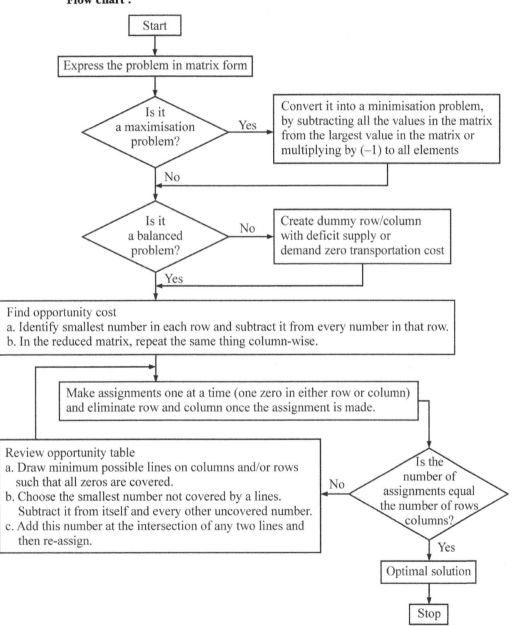

FIGURE 6.1 : FLOW CHART OF STEPS IN THE ASSIGNMENT METHOD (HUNGARIAN METHOD OR FLOODS TECHNIQUES)

The method is explained with the following illustrations.

ILLUSTRATION 1

Four machines, namely lathe, milling, grinding and drilling machines are to be repaired by four maintenance operators who can perform all the jobs but differ in their efficiency and the tasks differ in their intrinsic difficulty. The estimates of the times each man would take to perform each is given below in the matrix:

Machine Operator	Lathe	Milling	Grinding	Drilling
Venkat	8	24	17	11
Shyam	13	28	4	26
Sundar	38	19	18	15
Raju	19	26	24	10

How should the repair tasks be allocated to the operators to minimise the total man-hours.

Solution :

Step 1 : *Standard Form :* As the given matrix is to minimise (man-hours) the assignment is in the standard form. Therefore we can proceed directly.

Step 2 : *Balance Checking :* The given Assignment Problem (AP) is balanced as number of rows (4) = number of columns (4).

Step 3 : (a) *Row Iteration :* Select the least of each row and subtract from the corresponding row elements.

Machine Operator	Lathe	Milling	Grinding	Drilling
Venkat	0	16	9	3
Shyam	9	24	0	22
Sundar	23	4	3	0
Raju	9	16	14	0

(b) *Column Iteration :* Now, in the reduced matrix given above, select least number of each column and subtract from its column members of row iterated matrix.

Machine Operator	Lathe	Milling	Grinding	Drilling
Venkat	0	12	9	3
Shyam	9	20	0	22
Sundar	23	0	3	0
Raju	9	12	14	0

Step 4 : *Allocation :* Assign the repair jobs to the men by putting the box or rectangle (☐) over the zeros. While allocating in this way, see that each row/column get one and only one assignment. When a zero is enrectangled, the other zeros of its row and column are to be crossed out. (Normally, we give preference to allocate those zeros which are only one available in a row/column).

Machine Operator	Lathe	Milling	Grinding	Drilling
Venkat	☐ 0	12	9	3
Shyam	9	20	☐ 0	22
Sundar	23	☐ 0	3	⊠
Raju	9	12	14	☐ 0

Thus the optimal assignment for the minimum man-hours is

Venkat to repair Lathe machine	— 8 hours
Shyam to repair Grinding machine	— 4 hours
Sundar to repair Milling machine	— 19 hours
Raju to repair Drilling machine	— 10 hours.
Total	— 41 hours

ILLUSTRATION · 2

There are five jobs to be assigned on 5 machines and associated cost matrix is as follows:

Machines

		1	2	3	4	5
	A	11	17	8	16	20
	B	9	7	12	6	15
Jobs	C	13	16	15	12	16
	D	21	24	17	28	26
	E	14	10	12	11	15

Find the optimum assignment and associated cost using the assignment technique.

Solution :

Step 1 : *Standard Form :* As the given matrix is to minimise (cost), the assignment is in the standard form. Therefore we can proceed directly.

Step 2 : *Balance Checking :* The given assignment problem (AP) is balanced as number of row (5) = number of columns (5).

Step 3 : (a) *Row Iteration :* Select the least of each row and subtract from the corresponding row elements.

Machines

	1	2	3	4	5
A	3	9	0	8	12
B	3	1	6	0	9
C	1	4	3	0	4
D	4	7	0	11	9
E	4	0	2	1	5

Jobs

(b) *Column Iteration :* Select least number of each column and subtract from its column members.

Machines

	1	2	3	4	5
A	2	9	0	8	8
B	2	1	6	0	5
C	0	4	3	0	0
D	3	7	0	11	5
E	3	0	2	1	1

Jobs

Step 4 : *Allocation :* Assign the jobs to the machines by putting the box or rectangle (☐) over the zeros. While allocating in this way, see that each row/column get one and only one assignment. When a zero is enrectangled, the other zeros of its row and column are to be crossed out (✗) (Normally, give preference to allocate those zeros which are only one available in a row/column).

Step 5 : *Optimality Criterion :* As the number of assigned cells is not equal to the number of rows/columns, the optimal solution is not obtained. (Job *D* is not assigned to any machine and machine 5 is left unassigned). Hence we move to the next step.

Machines

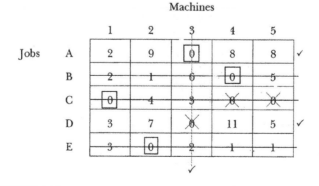

Step 6 : *Marking :* Draw a set of horizontal and vertical lines to cover all the zeros in the revised cost table obtained from step-5 using the following procedure.

(a) Mark (✓) the row in which there is no assignment i.e., Row 'D'

(b) If any zeros occur in Marked row, mark the respective columns that contain those zeros, i.e., row '*D*' has zero in col. 3: Therefore mark col. 3.

(c) If any assigned zero occurs in marked column, then tick the respective rows that contain those assigned zeros. Here marked column i.e., col. 3. has assigned zero in first row i.e., Row *A*. So mark it.

(d) Repeat this process until no more rows/columns can be marked (no more).

(e) Draw a straight line through each ***Marked Column*** i.e., col. 3 and each ***Unmarked Rows***. i.e., col. *B, C* and *E*

(If the number of lines drawn is equal to the number of rows or columns then the current solution is the optimal solution, otherwise go to step-7).

As the number of minimum lines connecting all the zeros (4) is less than the number of rows/columns (5) we move to step - 7 to revise the matrix.

Step 7 : *Revising :* Develop the new revised opportunity cost table.

(a) Among the cells not covered by any line, choose the smallest element. Here, This value is '2' (key element).

(b) Subtract '2' from every element in the cell not covered by a line.

(c) Add 2 to every element in the cell covered by the two lines, i.e., intersection of two lines.

(d) Elements in cells covered by one line remain unchanged.

Then allocate again using step - 4:

Machines

Jobs		1	2	3	4	5
	A	0	7	⦻	6	6
	B	2	1	8	0	5
	C	⦻	4	5	⦻	0
	D	1	5	0	9	3
	E	3	0	4	1	1

Since all the jobs are assigned and all the machines are assigned, the optimal solution is arrived. The optimal solution is shown in the tabular form.

Job	Machine	Cost
A	1	11
B	4	6
C	5	16
D	3	17
E	2	10
	Total	60

ILLUSTRATION 3

Find the minimum cost assignment for the following problem, explaining each step.

Workers	I	II	III	IV	V
A	6	5	8	11	16
B	1	13	16	1	10
C	16	11	8	8	8
D	9	14	12	10	16
E	10	13	11	8	16

Solution :

Steps 1 & 2: As the given A.P has cost matrix i.e to minimise, it is the standard form. Also as no. of row=no. of columns, given A.P is balanced.

Jobs

Workers	I	II	III	IV	V
A	6	5	8	11	16
B	1	13	16	1	10
C	16	11	8	8	8
D	9	14	12	10	16
E	10	13	11	8	16

Step 3 : *(a) Row Iteration :* Select least member of each row and subtract in its corresponding row members.

	I	II	III	IV	V
A	1	0	3	6	11
B	0	12	15	0	9
C	8	3	0	0	0
D	0	5	3	i	7
E	2	5	3	0	8

(b) Column Iteration : Select least no. of each column and subtract from members of corresponding column.

→ Since every column is having a zero, this step results in the same matrix as that Step 3 (a).

Step 4 : *Allocation :* Allocate by enrectangling (☐) on zeros such that each row/column will have one and only one allocation. When a zero is allocated, the other zeros in its row and column will be crossed out. (✕). (In doing the above, preference will be given to the zero if it is only one present in the row/column).

	I	II	III	IV	V
A	1	[0]	3	6	1
B	✕0	12	15	0	9
C	8	3	[0]	✕0	0
D	[0]	5	3	1	7
E	2	5	3	[0]	8

Step 5 : *Marking :* As worker B is not assigned & Job V is unassigned, we move towards optimization by marking by following rules.

(a) Mark (✓) the unassigned row i.e., Row '*B*'

(b) Mark the columns in which unassigned marked row has zero. i.e., col. I & IV

(c) Mark the row in which marked columns have the assignment i.e., rows D & E and strike off unmarked rows and marked column by a line.

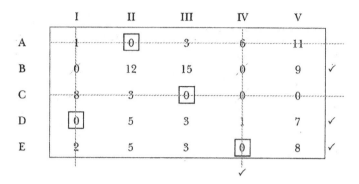

All the zeros are connected with zeros and the number of lines (4) is less than the no. of rows/columns. (5) Hence move towards step 6.

Step 6 : *Optimisation :* For finding optimal solution we iterate the above table as follows.

    (a) Idenfity least digit of unlined numbers. i.e., 3.

    (b) Subtract this number in all unlined numbers,

    (c) Add this member at the intersected numbers (junctions of two lines).

    (d) Keep all others (lined) unaltered.

Step 7 : Repeat step 4 i.e., allocation

	I	II	III	IV	V
A	4	0	3	9	11
B	0	9	12	0	6
C	11	3	0	3	0
D	0	2	0	1	4
E	2	2	0	0	5

The optimal assignment is

Worker	Job	Cost.
A	II	5
B	I	1
C	V	8
D	III	12
E	IV	8
Total		34

Total cost (min.) = 34 units.

The above assignment yields another (alternate) optimal solutions as follows.
(Refer see 6.2.2, multiple optimal solution)

A	II	5
B	IV	1
C	V	8
D	I	9
E	III	11

Total 34

ILLUSTRATION 4

There are five jobs to be assigned on each to 5 machines and the associated cost matrix as follows.

Jobs		I	II	III	IV	V
	A	11	17	8	16	20
	B	9	7	12	6	15
	C	13	16	15	12	16
	D	21	24	17	28	26
	E	14	10	2	11	15

Find the optimum assignment and the associated cost using the Assignment Technique.

Solution :

Step 1 & Step 2 : Already standard and balanced, hence proceed to step 3:

Step 3 (a): Performing row Iteration we get.

Jobs		I	II	III	IV	V
	A	3	9	0	8	12
	B	3	1	6	0	9
	C	1	4	3	0	4
	D	4	7	0	11	9
	E	4	0	2	1	5

Step 3 (b): *Column Iteration :*

		I	II	III	IV	V
	A	2	9	0	8	8
	B	2	1	6	0	5
	C	0	4	3	0	0
	D	3	7	0	11	5
	E	3	0	2	1	1

Step 4 : Allocation, marking and optimality check.

	I	II	III	IV	V	
A	2	9	0	8	8	✓
B	2	1	6	0	5	
C	0	4	3	X	X	
D	3	7	X	11	5	✓
E	3	0	2	1	1	
		✓	✓			

Step 5 : Revising and improvin

	I	II	III	IV	V
A	0	7	X	6	6
B	2	1	8	0	5
C	X	4	5	X	0
D	1	5	0	9	3
E	3	0	4	1	1

Solution : $A \to 1$ $B \to 4$ $C \to 5$ $D \to 3$ $E \to 2$; Total min cost = 60

6.2 *Variations in Assignment Problem*

The following special cases are common to occur in assignment problem.

1. Maximisation case.
2. Unbalanced AP.
3. AP with multiple optimal solutions.
4. Restricted (or prohibited) assignment.
5. Conditional assignment.
6. Crew assignment.
7. Travelling salesman problem.
8. TP as AP.

These are discussed, here below.

6.2.1 *Maximisation Case in AP*

In the algorithm of AP, it is already mentioned that the standard form is to minimise the costs or time and when it is in the non-standard form i.e., to maximise (profits or revenue), the AP is converted to standard form first before assigning.

In business practices, there may be some cases, where the profits may be obtained by allocating the given jobs to men or machines. In such cases, the AP is said to be a maximisation case, which is supposed as non-standard here. This matrix can be converted into standard form i.e., minimisation by obtaining the equivalent cost matrix. To obtain this equivalent cost matrix we choose one of the following methods (same as those in TP).

1. Identify the highest value among all the cell values and subtract each value from this highest. Put them in their respective positions.

   (This means that the lost profit as compared to the highest.)

2. Multiply by minus one (-1) to all the cell values.

   (This implies that a negative profit is loss.)

   After obtaining the equivalent cost matrix, we proceed as usually by the given set of rules (algorithm) and obtain the solution. The final solution is computed with the original matrix i.e., profit matrix.

   This is illustrated through following example.

ILLUSTRATION 5

A company has 4 territories and four salesmen for assignment. The territories are not equally rich in their sales potential. It is estimated that a typical salesman operating in each territory would bring the following annual sales.

Territory	I	II	III	IV
Annual Sales (rs)	60000	50000	40000	30000

The four salesmen are also considered to differ in ability, it is estimated that working under same condition their yearly sales could be proportionately as follows.

Sales man	A	B	C	D
Proportion	0.7	0.5	0.5	0.4

If criteria is to maximise expected sales, the intuitive answer is to assign the best salesman to the richest territory and next best to second richest and so on. Verify this answer by assignment technique.

[Dr. BRAOU - MBA 98, JNTU (Mech) - FDH-94]

Solution :

Formulation :

The sales performance of each sales man in various territories can be calculated as sample calculation shown below.

$$\text{Sales man } A \text{ in territory } I = \frac{0.7}{2.1} \times 60000 = \frac{42000}{2.1}$$

$$\text{Sales man } A \text{ in territory } II = \frac{0.7}{2.1} \times 50000 = \frac{35000}{2.1}$$

$$\text{Sales man } B \text{ in territory } I = \frac{0.5}{2.1} \times 60000 = \frac{30000}{2.1} \text{ and so on.}$$

In order to provide ease in calculation let us calculate for 2.1 years instead of 1 year and consider the figures in thousands. Thus we remove three zeros in numerator and 2.1 in the denominator. Thus the formulation of the assignment is as follows.

Territory \ Salesman	A	B	C	D
I	42	30	30	24
II	35	25	25	20
III	28	20	20	16
IV	21	15	15	12

Standardization :

As the given figures represent the sales (per thousands of Rs. in 2.1 years) and sales are to be maximised, it is in non-standard form. To standardise this, (i) we change the signs of all the cell values or (ii) Subtract all values from highest. We use (ii) here to find equivalent cost matrix.

	A	B	C	D
I	0	12	12	18
II	7	17	17	22
III	14	22	22	26
IV	21	27	27	30

Balance Checking :

In the above AP, we have four territories (rows) and four salesman (columns), therefore it is balanced. We proceed to the next step.

Obtaining opportunity cost matrix.

(a) Row Iteration :

	A	B	C	D
I	0	12	12	18
II	0	10	10	15
III	0	8	8	12
IV	0	6	6	9

(b) Column Iteration :

	A	B	C	D
I	0	6	6	9
II	0	4	4	6
III	0	2	2	3
IV	0	0	0	0

Allocation and Marking :

	A	B	C	D
I	☐0	6	6	9	✓
II	0	4	4	6	✓
III	0	2	2	3	✓
IV	0	☐0	8	9

Revised matrix and reallocation (repeating above step)

	A	B	C	D
I	☐0	4	4	7	✓
II	0	2	2	4	✓
III	0	☐0	3	1
IV	2	0	☐0	0

Re-revised matrix (repeating the above steps)

	A	B	C	D
I	☐0	2	2	5
II	0	☐0	0	2
III	2	0	☐0	1
IV	4	0	0	☐0

	A	B	C	D
I	☐0	2	2	5
II	0	0	☐0	2
III	2	☐0	0	1
IV	4	0	0	☐0

There are two solutions possible in the above AP. The solutions are I-A, II-B, III-C and IV-D or I-A, II-C, III-B and IV-D. The profits will be as follows :

Territory	Solution - I		Solution - II	
	Sales man	Profit	Sales man	Profit
I	A	$\frac{42000}{2.1} = 20000$	A	$\frac{42000}{2.1} = 20000$
II	B	$\frac{25000}{2.1} = 11904.76$	B	$\frac{20000}{2.1} = 9523.81$
III	C	$\frac{20000}{2.1} = 9523.81$	C	$\frac{25000}{2.1} = 11904.76$
IV	D	$\frac{12000}{2.1} = 5714.28$	D	$\frac{12000}{2.1} = 5714.28$
Total sales in (Rs.)		47142.82		47142.82

6.2.2 *Unbalanced Assignment Problem*

An assignment problem is said to be unbalanced if the number of rows is not equal to the number of columns. Number of row n (r) = Number of columns n (c) balanced AP. Number of rows n (r) ≠ number of column n (c) unbalanced AP.

In other words the AP is balanced if the AP matrix is a square and unbalanced if it is not a square matrix.

In unbalanced AP, we notice that either number of jobs are greater or lesser than number of men or machine. In such case, a dummy row/column will be created with zero costs to each cell. Observe the following example to convert unbalanced AP to balanced AP.

	M_1	M_2	M_3
J_1	x_{11}	x_{12}	x_{13}
J_2	x_{21}	x_{22}	x_{23}
J_3	x_{31}	x_{32}	x_{33}
J_4	x_{41}	x_{42}	x_{43}

Unbalanced : Machines (3) ≠ jobs (4)

	M_1	M_2	M_3	Dummy
J_1	x_{11}	x_{12}	x_{13}	0
J_2	x_{21}	x_{22}	x_{23}	0
J_3	x_{31}	x_{32}	x_{33}	0
J_4	x_{41}	x_{42}	x_{43}	0

Balanced : machines (4) = jobs (4) here D is dummy machine.

Similarly, if jobs are less, we create a dummy row with zero costs.

After converting into balanced AP, the solution can be obtained by usual steps given in the algorithm 6.1.

It is explained through following illustrative example.

ILLUSTRATION 6 ————————————————————————

Raju and Co has four lathe machines on which four workers operate. Any worker can operate any machine but due to the difference in skill and machine complexity the time of operation varies. The average times in hours when same job done on each machine by each worker is given below:

	L_1	L_2	L_3	L_4
W_1	7	6	4	9
W_2	5	5	8	8
W_3	4	5	4	6
W_4	7	8	5	8

> *(a) Find optimal allocation.*
> *(b) The company wants to replace the less efficient lathe with a new machine. The probable times (in hrs) that each worker can operate is estimated as 4, 5, 6 and 6 respectively. Verify whether the company has to replace any machine. If so, which machine is to be replaced.*

Solution :

Here, we have two sub-problems. One is to find optimal solution for existing machines and workers and second is to verify if any machine is to be replaced. The first one is balanced while the second is unbalanced AP.

(a) The given problem is standard since we have to minimise the time of operation. Also, it is balanced since it is a square matrix. The number of lathes (4) = number of workers (4). Therefore we find the opportunity cost matrix by rc v and column iterations as shown below.

Row Iteration :   *Column Iteration :*

	L_1	L_2	L_3	L_4
W_1	3	2	0	5
W_2	0	0	3	3
W_3	0	1	0	2
W_4	2	3	0	3

	L_1	L_2	L_3	L_4
W_1	3	2	0	3
W_2	0	0	3	1
W_3	0	1	0	0
W_4	2	3	0	1

Now, we assign as follows. Also marking is done since W_4 and L_4 are not assigned.

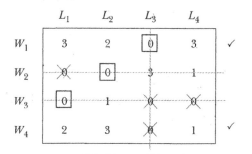

(Marking and lining)   *(Assignment in revised matrix)*

The optimal solution is

Worker	Lathe	Time
W_1	L_3	4
W_2	L_2	5
W_3	L_1	4

W_4	L_4	8
	Total cost	21

(b) Now, if a new machine say N is considered for replacement. It is unbalanced. so let us assume a dummy worker (D) with zero times on any machine which means that this fictitious worker allocated to any machine does not work i.e., this machine can to be removed being less efficient.

	L_1	L_2	L_3	L_4	N
W_1	7	6	4	9	4
W_2	5	5	8	8	5
W_3	4	5	4	6	6
W_4	7	8	5	8	6
D	0	0	0	0	0

Since No. of rows (4) $\neq$ No. of col. (5)

∴ we add a dummy worker (D) with zero times.

Now, we find the opportunity cost matrix by row iteration. [Column iteration is not necessary since we have zero in every column].

	L_1	L_2	L_3	L_4	N
W_1	3	2	0	5	0
W_2	0	0	3	3	0
W_3	0	1	0	2	2
W_4	2	3	0	3	1
D	0	0	0	0	0

Allocation :

	L_1	L_2	L_3	L_4	N
W_1	3	2	⊠	5	⬚0
W_2	⊠	⬚0	3	3	⊠
W_3	⬚0	1	⊠	2	2
W_4	2	3	⬚0	3	1
D	⊠	⊠	⊠	⬚0	⊠

As there is a unique zero in fourth column (5th row), we assign this. Obviously; other zeros in fifth row will be crossed out which makes zero at (W_2, L_2) unique in the column. Then on assigning this, zeros at (W_2, L_1) and (W_2, N) will be crossed out. Now zeros at (W_3, L_1) and (W_1, N) are unique in their columns. Assigning them, we cross out zeros at (W_3, L_3) and (W_1, L_3). Obviously zero at (W_4, L_3) is the remaining option to allocate. Thus we get the optimal solution as follows. [You can also procede starting with (W_4, L_3), which is unique in its row].

Optimal Solution :

Worker	Lathe	Time
W_1	New lathe	4
W_2	L_2	5
W_3	L_1	4
W_4	L_3	5
	Total	18

From the above, we conclude that L_4 is to be replaced by the new machine which can minimise the optimal time by 3 hours i.e., total time = 18 hours.

6.2.3 *Multiple Optimal Solution*

There may be more than one optimal solution possible for AP. In the above illustrations - 3 and 5, we have two solutions each. This can be identified by the zero positions.

If we can connect all the zeros as the corners of a rectangle or a closed loop provided the alternate corners occupy assigned, or allocated zeros, then we get alternative optimal solution by shufling assignment from one corner to the next.

In illustration - 3, we can connect zeros of cells (B, I), (B, IV), (E, IV), (E, III), (D, III). By shifting the assignment rectangle by one unit along the loop we get alternative solution. Thus the other possible solution is A - II, B - IV, C - V, D - I, E - III get another solution shifting assignment by one cell i.e., (B, I) to (B, IV); (E, IV) to (E, III) and (D, III) to (D, I). Similarly you can find in illustration - 5 as shown.

Yet other example is gien here below in illustration - 7

Maximisation of Unbalanced Assignment Problem :

ILLUSTRATION 7

There machinists are to be assigned for five jobs that will result in maximum profit.
                                                                    [JNTU (CSE) 1998]

		A	*B*	*C*	*D*	*E*
				Job		
Machinist	*1*	*6*	*8*	*5*	*10*	*7.5*
	2	*7*	*8.5*	*6*	*7*	*6.5*
	3	*5*	*6.5*	*9*	*8*	*8.5*

Solution :

1. Given Matrix is unbalanced. Hence we introduce two dummy rows with zero profit.

2. Given assignment is to maximise the profit. Hence it is to be converted to equivalent cost matrix by multiplying with (– 1) to all the cells and rewrite.

	A	B	C	D	E	
1	– 6	– 8	– 5	– 10	– 7.5	Multiplied by (-1) in all
2	– 7	– 8.5	– 6	– 7	– 6.5	the cells to convert into
3	– 5	– 6.5	– 9	– 8	8.5	equivalent cost matrix.
D_1	0	0	0	0	0	
D_2	0	0	0	0	0	

Row Iteration:

4	2	5	0	2.5	– 10 is subtracted
1.5	0	2.5	1.5	2	– 8.5 is subtracted
4	2.5	0	1	0.5	– 9 is subtracted
0	0	0	0	0	—
0	0	0	0	0	—

Note : *Column iterations not necessary as every column has zero.*

Allocation :

4	2	5	[0]	2.5
1.5	[0]	2.5	1.5	2
4	2.5	[0]	1	0.5
[0]	⊠	⊠	⊠	⊠
⊠	⊠	⊠	⊠	[0]

1 — D — 10
2 — B — 8.5
3 — C — 9

Total Profit Units   27.5

6.2.4 *Restricted (or Prohibited) Assignment*

If there is any restriction or prohibition on the assignment, of a particular cell, it is called restricted or prohibited assignment. As we should not have to allocate in this cell, we assume a highest penalty to this cell say M such that $M + C_{ij} = M - C_{ij} = M$ (where C_{ij} is cost of cell in *ith* row and *jth* column). This means that we never allocate in this cell due to its highest cost. Except for this assumption of M to the prohibited cell, the solution method remains same. An illustration is given below :

ILLUSTRATION 8

Five lecturers have to deal a subject common to five different branches of engineering. Due to understanding level of the branches and the efficiency of lecturer, the probable number of periods of 45 minutes required for each lecturer is given in the following matrix. A particular lecturer, namely, Mrs. Aparna refuses to go to CSE branch and CSIT students do not accept Mr. Prasad. Obtain an optimal assignment.

	ECE	*EEE*	*CSE*	*CSIT*	*MECH*
Mrs. Aparna	*50*	*50*	*-*	*20*	*60*
Dr. Raju	*70*	*40*	*20*	*30*	*40*
Mr. Prasad	*90*	*30*	*50*	*-*	*30*
Mr. Rajesh	*70*	*20*	*60*	*70*	*20*
Mr. Laxman	*60*	*50*	*70*	*90*	*10*

Solution :

In the above assignment matrix, it is to minimise the time i.e., in standard form. Further it is balanced since the number of lecturers (5) is equal to the number of branches.

But as Mrs. Aparna refuses to deal CSE and CSIT's do not accept Mr. Prasad, these two cells will be assigned an unaffordable penalty, say M such that $M + a = M - a = M$.

Now we have the matrix as

	ECE	EEE	CSE	CSIT	MECH
Mrs. Aparna	50	50	M	20	60
Dr. Raju	70	40	20	30	40
Mr. Prasad	90	30	50	M	30
Mr. Rajesh	70	20	60	70	20
Mr. Laxman	60	50	70	90	10

We now calculate opportunity cost (in terms of time) matrix by row iteration and subsequent column iteration.

Row Iteration :

	ECE	EEE	CSE	CSIT	MECH
Mrs. Aparna	30	30	M	0	40
Dr. Raju	50	20	0	10	20
Mr. Prasad	60	0	20	M	0
Mr. Rajesh	50	0	40	50	0
Mr. Laxman	50	40	60	80	0

Column Iteration :

	ECE	EEE	CSE	CSIT	MECH
Mrs. Aparna	0	30	M	0	40
Dr. Raju	20	20	0	10	20
Mr. Prasad	30	0	20	M	0
Mr. Rajesh	20	0	40	50	0
Mr. Laxman	20	40	60	80	0

We allocate now, with usual rules. And then marking is done.

	ECE	EEE	CSE	CSIT	MECH	
Mrs. Aparna	[0]	30	M	0	40	
Dr. Raju	20	20	[0]	10	20	
Mr. Prasad	30	[0]	20	M	0	✓
Mr. Rajesh	20	0	40	50	0	✓
Mr. Laxman	20	40	60	80	[0]	✓

Note that in the unassigned row (Rajesh), we have two zeros, therefore we mark two columns. EEE and Mech and thence mark the row Prasad and Laxman which have assignments in the marked columns.

On revising and reallocating the above matrix, we get.

	ECE	EEE	CSE	CSIT	MECH
Mrs. Aparna	0	50	M	[0]	60
Dr. Raju	20	40	[0]	10	40
Mr. Prasad	10	0	0	M	[0]
Mr. Rajesh	0	[0]	20	30	0
Mr. Laxman	[0]	40	40	60	0

Least of unlined is 20, added at intersection, subtracted in unlied and others unchanged

The above matrix yields multiple optimal solutions as follows :

Lecturer	Solution I		Solution II		Solution III	
	Branch	Periods	Branch	Periods	Branch	Period
Mrs. Aparna	CSIT	20	CSIT	20	CSIT	20
Dr. Raju	CSE	20	CSE	20	CSE	20
Mr. Prasad	Mech	30	EEE	30	EEE	30
Mr. Rajesh	EEE	20	Mech	20	ECE	70
Mr. Laxman	ECE	60	ECE	60	Mech	10
Total		150		150		150

Total number of periods = 150 :   Total time = $150 \times 45 = $ **6750** min.

Note : Observe the rectangular loops that can be constructed with zeros loop-1 : $C_{32} \to C_{35} \to C_{45} \to C_{42}$; Loop 2: $C_{41} \to C_{45} \to C_{55} \to C_{51}$. Thus we get one alternate solution to each loop. (Loop 2 is otherwise $C_{41} \to C_{42} \to C_{32} \to C_{35} \to C_{55} \to C_{51}$).

6.2.5 *Conditional Assignment*

When a condition is imposed in the assignment, it is to be satisfied before proceeding to calculate the solution. One such problem is illustrated here below.

ILLUSTRATION 9

Sai Nath Institute of Science and Technology (SNIST) is providing transport to its students in four routes. There are four parties made their bids as given below:

Party	Bids in routes (in '000 Rs/month)			
	Koti	Secunderabad	Mehdipatnam	Charminar
Pradeep travels	4	5	7	6
Raju & Co	10	5	4	4
Harika Bus	3	6	2	5
Lavanya transport	6	4	4	5

The institute wishes to allocate one route to each party. The Pradeep travels has offered a discount of Rs. 1000/- on additional route if they are allocated more than one routes. Find the optimal assignment to minimise their monthly costs. Also check whether the SNIT has to consider the offer given by Prodeep Travels.

Solution :

Here, along with the regular assignment problem, we have a condition to be checked. Let us consider the condition as fifth alternative for SNIST with the discounted prices and create a dummy route. Now, if the discounted row is allocated any route, we consider the offer otherwise we reject. Here, we have to remember that the condition is implied for additional route. Thus Pradeep travels must be allocated to at least one route and can not occupy the dummy route in first route. Thus we take M for Pradeep at dummy.

	K	S	M	C	Dummy
P	4	5	7	6	M
R	10	5	4	4	0
H	3	6	2	5	0
L	6	4	4	5	0
P' (Discounted)	3	4	6	5	0

Row Iteration :   **Column Iteration :**

	K	S	M	C	D			K	S	M	C	D
P	0	1	3	2	M		P	0	0	1	0	M
R	10	5	4	4	0		R	10	4	2	2	0
H	3	6	2	5	0		H	3	5	0	3	0
L	6	4	4	5	0		L	6	3	2	3	0
P'	3	4	6	5	0		P'	3	3	4	3	0

Allocation and Marking :   **Revised Matrix, its Allocation and Marking :**

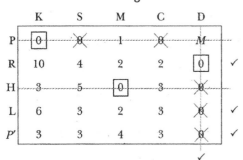 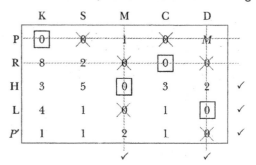

Note : In above matrix (right side) observe the marking. First, we marked P' for not having assignment, and then marked the fifth column as the marked row has zero in this column. This marked column (5th column) contains assigned zero at 4th row (of L) and hence 4th row (L) is marked. But here, this row L has one more zero at third column (i.e., col. M) and we have to consider this also. Therefore we mark third column (M) and subsequently the third row (H) as the assigned zero of marked column, appears at (H, M) i.e., in row H. Thus marking is finished and then draw the lines on unmarked rows and marked columns. As we have four lines only, we revise the matrix again. The allocation in revised matrix is shown below.

	K	S	M	C	D
P	[0]	0̸	2	0̸	M
R	8	2	1	[0]	1
H	2	4	[0]	2	2
L	3	0̸	0̸	0̸	[0]
P'	0̸	[0]	2	0̸	0̸

The institution has two options now, as understood from the above table.

Option - 1 : Allocate routes Koti and Secunderabad to Pradeep Travels and delete Lavanya Transport.

Option - 2 : Allocate Pradeep Travels to Koti route and Lavanya transport to Secunderabad route denying second route to Pradeep Travels.

Cost Calculation :

Party	Option - I		Option - II	
	Route	Cost (Rs)	Route	Cost (Rs)
Pradeep Travels	Koti*	4000	Koti	4000
	Secunderabad*	4000	–	–
Raju and Co	Charminar	4000	Charminar	4000
Harika Bus	Mehdipatnam	2000	Mehdipatnam	2000
Lavanya transport	-	-	Secunderabad	4000
Total cost		14000		14000

* If Koti route is treated as second route and Secunderabad as first route to Pradeep the cost break up will be 5000 and 3000 respectively which may not be called an alternate solution to this problem.

ILLUSTRATION 10 ——————————————————————————

Engineering College has constructed the third floor of a multistoreyed building. There are five rooms in this floor meant for five classrooms. Each room has its own advantages and disadvantages. Some have windows, some are close to washrooms or to the canteen or good ventilation and air flow etc. The rooms are of all different sizes and shapes. Five student leaders of different branches were asked to rank their room preferences amongst the rooms 301, 302, 303, 304 and 305. Their preferences were recorded in a table as indicated below:

	S_1	S_2	S_3	S_4	S_5
	302	302	303	302	301
	303	304	301	305	302
	304	305	304	304	304
		301	305	303	
			302		

Most of the students did not list all the five rooms since they were not satisfied with some of these rooms and they have left off these from the list. Assuming that their preferences can be quantified, find out the best possible arrangement so that their total preference ranking is minimum.

Solution :

The formulation for the above problem in the assignment form is as follows.

The figures in the matrix are the ranks given by each student leader to each lecture hall.

	S_1	S_2	S_3	S_4	S_5
301	-	4	2	-	1
302	1	1	5	1	2
303	2	-	1	4	-
304	3	2	3	3	3
305	-	3	4	2	-

We assign penalty M for the halls not preferred by the students.

	S_1	S_2	S_3	S_4	S_5
301	M	4	2	M	1
302	1	1	5	1	2
303	2	M	1	4	M
304	3	2	3	3	3
305	M	3	4	2	M

Row Iteration :

	S_1	S_2	S_3	S_4	S_5
301	M	3	1	M	0
302	0	0	4	0	1
303	1	M	0	3	M
304	1	0	1	1	1
305	M	1	2	0	M

The column iteration is not required since every column has at least one zero. Therefore we proceed to allocate.

	S_1	S_2	S_3	S_4	S_5
301	M	3	1	M	[0]
302	[0]	X	4	X	1
303	1	M	[0]	3	M
304	1	[0]	1	1	1
305	M	1	2	[0]	M

From the above table, we have arrived the optimal solution as follows.

301 to S_5;   302 to S_1;   303 to S_3;   304 to S_2 and   305 to S_4

* *A typical problem, with maximisation, unbalanced and restricted / conditional assignment problem is explained in the following illustrative example.*

ILLUSTRATION 11

The Indian Cricket captain Rahul Dravid has to decide first five batting positions from the following five batsman whose average runs in various positions, estimated from the past records is given below. Help him in assigning batting order.

Batsmen	Position				
	I	II	III	IV	V
Mahendra S. Dhoni	40	40	35	25	50
Sachin Tendulkar	42	30	16	25	27
Rahul Dravid	50	48	50	60	50
V. Sehwag	20	19	20	18	25
VVS. Laxman	58	60	59	55	53

If another batsman Yuvraj Singh with the following batting order positions and corresponding record of average runs, is also considered, find the optimum assignment (He never played in the V position and treat that it is risky). If Yuvraj is to be retained in the team then which of the current five batsmen will be out? What is the change in number of runs now?

Batting Position	I	II	III	IV	V
Average runs	45	52	38	50	-

Solution :

Step 1 : *Standard Form :* As the given matrix is to maximise (runs) the assignment is not in the standard form. Therefore we make it as standard form (equivalent cost matrix) by subtracting every element from highest amongst all.

	I	II	III	IV	V
M.S. Dhoni	20	20	25	35	10
Sachin Tendulkar	18	30	44	35	33
Rahul Dravid	10	12	10	0	10
V. Sehwag	40	41	40	42	35
VVS Laxman	2	0	1	5	7

Step 2 : *Balance Checking :* The given assignment problem (AP) is balanced as number of rows (5) = number of columns (5).

Step 3 : *Preparing Opportunity Cost Matrix :*

(a) *Row Iteration :* Select the least of each row and subtract from the corresponding row elements.

	I	II	III	IV	V
M.S. Dhoni	10	10	15	25	0
Sachin Tendulkar	0	12	26	17	15
Rahul Dravid	10	12	10	0	10
V. Sehwag	5	6	5	7	0
VVS Laxman	2	0	1	5	7

(b) *Column Iteration :* Select least number of each column and subtract from its column members.

	I	II	III	IV	V
M.S. Dhoni	10	10	14	25	0
Sachin Tendulkar	0	12	25	17	15
Rahul Dravid	10	12	9	0	10
V. Sehwag	5	6	4	7	0
VVS Laxman	2	0	0	5	7

Step 4 : *Allocation :* Assign the batting positions to the batsmen by putting the box or rectangle (☐) over the zeros. While allocating in this way see that each row/column get one and only one assignment. When a zero is enrectangled, the other zeros of its row and column are to be crossed out. (Normally we give preference to allocate those zeros which are only one available in a row/column).

	I	II	III	IV	V
M.S. Dhoni	10	10	14	14	$\boxed{0}$
Sachin Tendulkar	$\boxed{0}$	12	25	17	15
Rahul Dravid	10	12	9	$\boxed{0}$	10
V. Sehwag	5	6	4	7	⨯
VVS Laxman	2	$\boxed{0}$	⨯	5	7

Step 5 : *Optimality Criterion :* As the number of assigned cells not equal to the number of rows/columns, the optimal solutions not obtained. (V. Sehwag is not assigned to any position and position III is left unassigned). Hence we move to marking.

Step 6 : Marking : Draw a set of horizontal and vertical lines to cover all the zeros in the revised cost table obtained from step - 5 using the following procedure.

Mark (✓) V.Sehwag's row since there is no assignment. In this row, there is a zero in col. V and so mark col. V. Again in col. V there is assignment with first row (S. Ganguly). Therefore mark first row.

Draw a straight line through each marked column and each unmarked row, i.e., on 2nd, 3rd and 5th rows and column V.

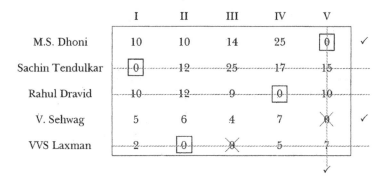

As the number of minimum lines connecting all the zeros (4) is less than the number of rows/columns (5) we move to step - 7 to revise the matrix.

Step 7 : *Revising :* Develop the new revised opportunity cost table.

Among the cells not covered by any line, choose the smallest element, this value is 4. Subtract 4 from every element in the cell not covered by a line. Add 4 to every element in the cell covered by the two lines, i.e., intersection of two lines. Elements in cells covered by one line remain unchanged.

The allocate again using step - 4.

	I	II	III	IV	V
M.S. Dhoni	6	6	10	21	[0]
Sachin Tendulkar	[0]	12	25	17	19
Rahul Dravid	10	12	9	[0]	14
V. Sehwag	1	2	[0]	3	⊠
VVS Laxman	2	[0]	⊠	5	11

The number of runs is now to be read from the original matrix and the total runs are calculated as follows :

Position	Name of the Batsman	No. of Runs
I	Sachin Tendulkar	42
II	VVS Laxman	60
III	V. Sehwag	20
IV	Rahul Dravid	60
V	M.S. Dhoni	50
TOTAL	5 Batsmen	232

Now, if Yuvraj Singh (with the averages 45, 52, 38 and 50 runs in first four positions and never went in fifth position) is considered, the given assignment becomes unbalanced. To balance we create a dummy position VI in which no batsman will make any runs (i.e., assigned zeros to all the batsmen). {step - 1}

	I	II	III	IV	V	VI
M.S. Dhoni	40	40	35	25	50	0
Sachin Tendulkar	42	30	16	25	27	0
Rahul Dravid	50	48	50	60	50	0
V. Sehwag	20	19	20	18	25	0
VVS Laxman	58	60	59	55	53	0
Yuvraj Singh	45	52	38	50	–	0

Now, the problem is converted into the standard form i.e., equivalent cost matrix.

	I	II	III	IV	V	VI
M.S. Dhoni	20	20	25	35	10	60
Sachin Tendulkar	18	30	44	35	33	60
Rahul Dravid	10	12	10	0	10	60
V. Sehwag	40	41	40	42	35	60
VVS Laxman	2	0	1	5	7	60
Yuvraj Singh	15	8	22	10	M	60

As it is risky to allocate the batsman Yuvraj Singh to fifth position and to avoid this, we put an unaffordable penalty cost M to this cell in the standard form i.e., in equivalent cost matrix.

Preparation of Opportunity Cost Matrix : Row iteration :

	I	II	III	IV	V	VI
M.S. Dhoni	10	10	15	25	0	50
Sachin Tendulkar	0	12	26	17	15	42
Rahul Dravid	10	12	10	0	10	60
V. Sehwag	5	6	5	7	0	25
VVS Laxman	2	0	1	5	7	60
Yuvraj Singh	7	0	14	2	M	52

Column iteration :

	I	II	III	IV	V	VI
M.S. Dhoni	10	10	14	25	0	25
Sachin Tendulkar	0	12	25	17	15	17
Rahul Dravid	10	12	9	0	10	35
V. Sehwag	5	6	4	7	0	0
VVS Laxman	2	0	0	5	7	35
Yuvraj Singh	7	0	13	2	M	27

Allocation :

	I	II	III	IV	V	VI
M.S. Dhoni	10	10	14	25	[0]	25
Sachin Tendulkar	[0]	12	25	17	15	17
Rahul Dravid	10	12	9	[0]	10	35
V. Sehwag	5	6	4	7	✗	[0]
VVS Laxman	2	✗	[0]	5	7	35
Yuvraj Singh	7	[0]	13	2	M	27

As every row and every column are assigned (the total number of assignments i.e., 6 = the number of rows or columns i.e., 6) the optimal solution is obtained. The allocation is as follows.

Position	Name of the Batsman	No. of Runs
I	Sachin Tendulkar	42
II	Yuvraj Singh	52
III	VVS Laxman	59
IV	Rahul Dravid	60
V	M.S. Dhoni	50
Total	5 Batsmen	263

With the entry of Yuvraj Singh in second position, the other positions are readjusted and V. Sehwag (allocated to dummy) is deleted. The increase in total runs is 31. (i.e., 232 to 263).

Remarks :

1. In the above problem, if it so comes that Yuvraj Singh is allocated to the dummy position then he should not be taken into the team.

2. If it is insisted that Yuvraj Singh must be included in the team in the problem (compulsory conditional assignment) the problem can be solved by assigning unaffordable highest penalty cost M to dummy cell against Yuvraj Singh.

6.2.6 *Crew Assignment Problem*

ILLUSTRATION 12

An airline operates seven days a week has time table shown below. Crews must have a minimum lay over (rest) time of 5 hours between flights. Obtain the pair of flights that minimises lay over time away from home for any given pair the crew will be based at the city that results in the smaller lay over.

Delhi-Jaipur			Jaipur-Delhi		
Flight No.	*Departure*	*Arrival*	*Flight No.*	*Departure*	*Arrival*
1	7.00 A.M	8.00 A.M	101	8.00 A.M	9.15 A.M
2	8.00 A.M	9.00 A.M	102	8.30 A.M	9.45 A.M
3	1.30 P.M	2.30 P.M	103	12.00 Noon	1.15 P.M
4	6.30 P.M	7.30 P.M	104	5.30 P.M	6.45 P.M

For each pair, mention the town where the crews should be based ?

[JNTU-(Mech) 99/CCC, (ECE) 99/CCC]

Solution :

As the service time is constant, it does not affect the stationing the crew. If all crew members are asked to reside at Delhi (so that they start from Delhi and come back to Delhi with minimum halt at Jaipur), then the waiting times at Jaipur for different service time connections is shown in the following table (the figures shown are hours). If the difference is less then 5, then it is calculated for the next day.

ITERATION TABLEAU 1(a) :

	101	102	103	104
1	24	24.5	28	9.5
2	23	23.5	27	8.5
3	17.5	18	21.5	27
4	12.5	13	16.5	22

Similarly if the crew is assumed to reside at Jaipur (so that they start from Jaipur and come back with minimum waiting time at Delhi), the table showing the waiting times of service line connections is :

ITERATION TABLEAU 1(b) :

	1	2	3	4
101	21.75	22.75	28.25	9.25
102	21.25	22.25	27.75	8.75
103	17.75	18.75	24.25	5.25
104	12.25	13.25	18.75	23.75

As the crew may be asked to either reside at Delhi or Jaipur, the minimum waiting time from Table 1(a) and 1(b) can be obtained for different flight connections by choosing the minimum value of the two corresponding waiting times, provided the value is more than 5 hours (the desired lay over time or rest). These are illustrated by encircling in the following tableaus

Note : *The figures of matrix 1(b) are to be transposed to compare the matrices before selecting the minimal values.*

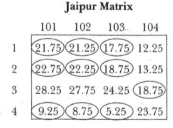

Delhi Matrix					**Jaipur Matrix**			
	101	102	103	104		101	102	103	104
1	24	24.5	28	(9.5)	1	(21.75)	(21.25)	(17.75)	12.25
2	23	23.5	27	(8.5)	2	(22.75)	(22.25)	(18.75)	13.25
3	(17.5)	(18)	(21.5)	27	3	28.25	27.75	24.25	(18.75)
4	12.5	13	16.5	(22)	4	(9.25)	(8.75)	(5.25)	23.75

ITERATION TABLEAU 2 :

	101	102	103	104
1	21.75	21.25	17.75	9.5*
2	22.75	22.25	18.75	8.5*
3	17.5*	18*	21.5*	18.75
4	9.25	8.75	5.25	22*

Star (*) marked are residing at Delhi and others at Jaipur.

Now, applying Hungarian Method.

ROW ITERATION

	101	102	103	104
1	12.25	11.75	8.25	0
2	14.25	13.75	10.25	0
3	0*	0.75	4	1.25
4	4	1.25	4.75	16.75

Star (*) marked are residing at Delhi.

ASSIGNING & MARKING

	101	102	103	104
1	12.25	11.25	8.25	[0*] ✓
2	14.5	13.25	10.25	✗* ✓
3	[0*]	✗*	4*	1.25
4	4	3	[0]	16.75

REVISION AND RE-ALLOCATION

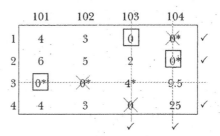

	101	102	103	104
1	4	3	[0]	✗* ✓
2	6	5	2	[0*] ✓
3	[0*]	✗*	4*	9.5
4	4	3	✗	25 ✓

Important Note : *Observe typical marking in the above table (right side).*

Marking : We start with marking 4th row as it has no assignment. In the marked (4th) row, there is a zero in 3rd (i.e., F. No. 103) column. So we mark this. We have assignment at (1, 103) in the marked column (103) and so 1st row is marked.

Now, sine 1st row (marked) contains one more zero at (1, 104) we have to mark co. 104. Subsequently we mark 2nd row as the marked column (104) has assigned zero in 2nd row. With this marking is complete and then put line over unmarked row (3rd) and marked columns (103 & 104)

RE-REVISING AND ALLOCATING

	101	102	103	104
1	1	[0]	✗	✗*
2	3	2	2	[0*]
3	[0*]	✗*	7*	12.5
4	1	✗	[0]	25

There are two solutions

Solution - 1:   1 - 102, 2* - 104, 3* - 101 & 4 - 103

Solution - 2:   1 - 103, 2* - 104, 3* - 101, 4 - 102

Total layover = 52.5 hrs.

The crew with 2-104 & 3 - 101 will stay at Delhi and other at jaipur.

The pattern of optimal assignment among the routes with respect to the waiting time (in hours) is given in the following table.

Crew	Residing at	Solution - 1		Solution - 2	
		Route No.	Layover	Route No.	Layover
1	Jaipur	1 to 102	21.25	1 to 103	17.75
2	Delhi	2 to 104	8.5	2 to 104	8.5
3	Delhi	3 to 101	17.5	3 to 101	17.5
4	Jaipur	4 to 103	5.25	4 to 102	8.75
	Total Layover		52.5		52.5

6.2.7 *Travelling Salesman Problem (or Routing Problem)*

Travelling Salesman Problem : The travelling salesman problem can be solved using assignment technique with an additional conditions on the choice of assignment. i.e. how should a travelling salesman travel starting from his home city, visiting each city once and returning to his home city, so that the total distance is minimum? For e.g. given n cities with distances d_{ij} (cost c_{ij} or time t_{ij}) from city 'i' to city 'j' the salesman starts from city 1, then any permutation of 2, 3. ... n represents the number of possible ways for his tour. Thus there are $(n-1)!$ possible ways of his tour and optimal value is to be selected. From the solution of assignment, cyclic assignment is identified by manipulating with minimum difference so as to get the solution for travelling salesman problem.

ILLUSTRATION 13

A machine operator processes five types of items on his machine each week and must choose sequence for them. The set-up cost per change depends on the items presently on the machine and item to be made, according to the table.

From Item	To Item				
	A	B	C	D	E
A	α	4	7	3	4
B	4	α	6	3	4
C	7	6	α	7	5
D	3	3	7	α	7
E	4	4	5	7	α

If he produces each type of item once and only once each week, how should he sequence the item on his machine in order to minimise the total set up cost.

[JNTU (Mech) 99/CCC, (ECE) 99/CCC]

Solution :

With usual algorithm of assignment (Hungarian method), Find the solution. Here, the elements along principal diagonals (α's) are ignored for calculations.

ROW ITERATION

	A	B	C	D	E
A	α	1	4	0	1
B	1	α	3	0	1
C	2	1	α	2	0
D	0	0	4	α	4
E	0	0	1	3	α

COLUMN ITERATION

	A	B	C	D	E
A	α	1	3	0	1
B	1	α	2	0	1
C	2	1	α	2	0
D	0	0	3	α	4
E	0	0	0	3	α

Now allocate as per the given set of rules.

	A	B	C	D	E	
A	α	1	3	[0]	1	✓
B	1	α	2	0̸	1	✓
C	2	1	α	2	[0]	
D	0̸	[0]	3	α	4	
E	[0]	0̸	0̸	3	α	

Modify the above table by subtracting the lowest unlined element i.e. 1 from all the elements not covered by lines and adding the same at the intersection of two lines.

	A	B	C	D	E
A	α	0̸	2	[0]	0̸
B	[0]	α	1	0̸	0̸
C	2	1	α	3	[0]
D	0̸	[0]	3	α	4
E	0̸	0̸	[0]	4	α

The optimum assignment is $A \to D$, $B \to A$, $C \to E$, $D \to B$ and $E \to C$ with a minimum cost of 20.

This assignment schedule does not provide us the solution of travelling salesman problem as it gives $A \to D$, $D \to B$, $B \to A$ while B is not allowed to follow A unless C and E are processed. This means that the assignment should be cyclic.

Therefore, now, we try to find the next best solution, which satisfies this extra–restriction. The next minimum (non–zero) element in the matrix is 1. So we try to bring 1 into the solution. But the element '1' occurs at two places. We shall consider all the cases seperately until the acceptable solution is reached.

We start with making an assignment at (B, C) instead of zero assignments at (B, A). The resulting feasible solution then will be

$$A \to D, D \to B, B \to C, C \to E, E \to A$$

When an assignment is made at (C, B) instead of assignment at (C, E) the resulting feasible solution will be

$$A \to E, E \to C, C \to B, B \to D, D \to A$$

The total set–up in both the programmes comes out to be 21.

Therefore both solution hold good with an extra cost of one unit.

6.2.8 *A Transportation Problem as Assignment Problem*

A transportation problem can be converted to assignment problem if the supply/demand is in less quantity. This is illustrated through an example given below:

ILLUSTRATION 14

APSRTC can provide 6 special buses to operate for four university colleges as required by them.
The distances in k.m., to be travelled are given in the following matrix. Solve the transportation problem as assignment problem (with Hungarian method) so as to minimise total distance travelled.

College	BKP	MP	KP	KG	Buses required
Dr. DRAOU	10	12	15	11	1
OU	8	9	7	12	2
JNTU	6	9	8	10	2
KU	6	7	6	6	1
Buses available	2	1	2	1	6 6

Solution :

The above TP is first converted into assignment model. For this we consider each bus in each row and each column. This is done as follows.

Univ College	BKP		MP	KP		KG	Required
Dr. BRAOU	10	10	12	15	15	11	1
OU	8	8	9	7	7	12	1
	8	8	9	7	7	12	1
JNTU	6	6	9	8	8	10	1
	6	6	9	8	8	10	1
KU	6	6	7	6	6	6	1
Available	1	1	1	1	1	1	

Now the above assignment problem is solved by usual procedure as follows :

Since the given AP is standard (minimising distance) and balanced (row = 6 = columns), we proceed to row/column iteration to find opportunity matrix.

ROW ITERATION

Univ. College	BKP1	BKP2	MP	KP1	KP2	KG
BRAOU	0	0	2	5	5	1
OU_1	1	1	2	0	0	5
OU_2	1	1	2	0	0	5
$JNTU_1$	0	0	3	2	2	4
$JNTU_2$	0	0	3	2	2	4
KU	0	0	1	0	0	0

COLUMN ITERATION

Univ. College	BKP1	BKP2	MP	KP1	KP2	KG
BRAOU	0	0	1	5	5	1
OU$_1$	1	1	1	0	0	5
OU$_2$	1	1	1	0	0	5
JNTU$_1$	0	0	2	2	2	4
JNTU$_2$	0	0	2	2	2	4
KU	0	0	0	0	0	0

ALLOCATION AND MARKING

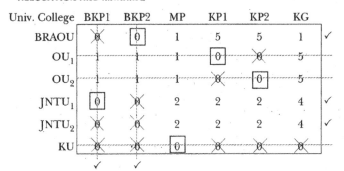

Univ. College	BKP1	BKP2	MP	KP1	KP2	KG	
BRAOU	⊠	[0]	1	5	5	1	✓
OU$_1$	1	1	1	[0]	⊠	5	
OU$_2$	1	1	1	⊠	[0]	5	
JNTU$_1$	[0]	⊠	2	2	2	4	✓
JNTU$_2$	⊠	⊠	2	2	2	4	✓
KU	⊠	⊠	[0]	⊠	⊠	⊠	
	✓	✓					

Important Note :

Observe marking which is typical here. Similar marking is shown in illustration 12, under crew assignment problem pleaser refer.

REVISION AND RE-ALLOCATION

Univ. College	BKP1	BKP2	MP	KP1	KP2	KG
BRAOU	⊠	⊠	[0]	4	4	⊠
OU$_1$	2	2	1	[0]	⊠	5
OU$_2$	2	2	1	⊠	[0]	5
JNTU$_1$	[0]	⊠	1	1	1	3
JNTU$_2$	⊠	[0]	1	1	1	3
KU	1	1	⊠	⊠	⊠	[0]

From the above table, we can derive two solutions with a total 44 km distance and detailed calculations are shown in the following table.

Institute	No. of buses	Bus depot		Distance (km)	
		Option I	Option II	Option I	Option II
Dr. BRAOU	1	MP	KG	12	11
OU	2	KP	KP	14	14
JNTU	2	BKP	BKP	12	12
KU	1	KG	MP	6	7
			Total	44	44

Distinction Between Assignment and Transportation Problem :

Transportation Problem is sometimes referred to as Allocation Problem. Both these problems are the cases of linear programming problems with some distinctions. These are distinguished as follows :

	Assignment Problems	Allocation or Transportation Problem
1.	This problem is used to assign the jobs to machines or machines to men etc.	This problem is used in transporting material from origins (like plant) to destinations (such as godown/market) etc.,
2.	No availability (supply) and requirement (demand) are needed to solve the problem	Supply & demand are needed.
3.	If number of rows is equal to no. of column the AP is said to be balanced otherwise unbalanced	If supply is equal to demand the TP is said be balanced otherwise unbalanced
4.	There will not be any degeneracy in this problem	A possibility of degeneracy either at initial stage or subsequent stages may occur.
5.	This problem is solved by converting into oportunity costs (Hungarian method)	This can be solved without converting the given costs (North west corner or Vogel's approximation or (least cost). In VAM we use penalities in LCM we use least cell cost in NWC we use position of top–left in matrix.
6.	Travelling salesmen problem and crew Assignment are its extensions.	Trans shipment problem is its extension.
Similarties		
1.	Standard AP is to minimise the cost or time.	Standard TP is to minimise the cost
2.	Max. problem is converted to min. problem by subtracting from highest among all or by multiplying all by (–1).	Max. problem is converted to min. problem by subtracting from highest among all or by multiplying all by (–1).
3.	Restricted Assignment uses 'M', an unaffordable cost.	Restricted TP uses 'M', an unaffordable cost.

Practice Problems

1. Find the assignment that minimises total machining time. The data is

Tasks	A	B	C	D
M/C1	4	9	8	5
M/C2	5	4	3	6
M/C3	7	8	6	12
M/C4	10	9	6	7

[JNTU - (Mech) 92]

Answer : $M_1 - A$, $M_2 - B$, $M_3 - C$, $M_4 - D$, Min time = 21 units

2. Jawaharlal Nehru Technological University has decided to give their four departmental buildings for construction to four contractors. They have called for tenders from approved contractors for construction of buildings. The tender finalisation committee felt that the quotations of contractors namely Philip & Co., Qutub & Bros., Raju & Sons and Sai Consultants are reasonable and they decided to reject the remaining tenders. What decision the committee should take so that amount to be spent by the university on construction is the minimum? The following table gives quotations of four contractors in lakhs for constructing the respective buildings.

Buildings Contractors	VC Building	Exams Centre	Class Room Block	Admin Building
Philip & Co	10	24	30	15
Qutub & Bros	16	22	28	12
Raju & Sons	12	20	32	10
Sai Consultants	9	26	34	16

Answer :   Philip & Co — VC Building, Qutub & Bros. — Classroom block, Raju & Sons — Admin. Buildings and Sai Consultants — Exams centre : Minimum cost is Rs. 74 lakhs.

3. A company has six maintenance groups to repair six machines. The following table gives the return in rupees when the ith job is assigned to the jth mechanic $(i, j = 1, 2, 3, 4, 5, 6)$. How should the jobs be assigned to the mechanics so as to maximise the overall return?

Machine Mechanic	I	II	III	IV	V	VI
1	9	22	58	11	19	27
2	43	78	72	50	63	48
3	41	28	91	37	45	33
4	74	42	27	49	39	32
5	36	11	57	22	25	18
6	13	56	53	31	17	28

[JNTU - (CSE) 93]

Answer :   1 — VI,   2 — V,   3 — III,   4 — 1

5 — IV and 6— II; Maximum return = Rs. 333).

4. A department has four subordinates and four tasks to be performed. The sub-ordinates differ in efficiency and the tasks differ in their intrinsic difficulties. The estimate of man-hours, each man would take to perform the task is given below :

Sub-ordinate \ Task	I	II	III	IV
1	18	26	17	11
2	13	28	14	26
3	38	19	18	15
4	19	26	24	10

How should the tasks be allotted to men to optimise the total man-hours?

[JNTU Mech/Prodn/Chem & Mechatronics 2000,
CSE/ECE-2001]

Answer : 1 - III, 2 - I , 3 - II, 4 - IV, Man hrs = 59

5. Solve the following assignment model. The assignment costs in rupees are given below :

	P	Q	R	S	T
A	3	9	2	3	7
B	6	1	5	6	6
C	9	4	7	10	3
D	2	5	4	2	1
E	9	6	2	4	6

[JNTU - (Mech.) 96P/S]

Answer : $A - P, B - Q, C - T, D - S, E - R$ or
$A - S, B - Q, C - T, D - P, E - R$ Min cost = Rs 11

6. In a textile showroom, four salesmen A, B, C and D are available to four counters P, Q, R and S. Each salesman can handle any counter. The service (in hour) of each counter when manned by each salesman is given below :

Counter	Salesman			
	A	B	C	D
P	41	72	39	52
Q	22	29	49	65
R	27	39	60	51
S	45	50	48	52

How should the salesman be allocated appropriate counters so as to minimise the service time? Each salesman must handle only one counter.

(IGNOU - MCA/ADCA - Dec 99)

Answer : $P - C, Q - B, R - A , S - D$, Min time = 147 hrs

7. Solve the following cost minimising Assignment.

Jobs

	I	II	III	IV	V
A	45	30	65	40	55
B	50	30	25	60	30
C	25	20	15	20	40
D	35	25	30	30	20
E	80	60	60	70	50

Answer :   **[JNTU - (Mech.) 96/P/S]**

$A \rightarrow III,$ $B \rightarrow V,$ $C \rightarrow I,$ $D \rightarrow IV,$ $E \rightarrow II$ or

$A \rightarrow III,$ $B \rightarrow V,$ $C \rightarrow IV,$ $D \rightarrow I.,$ $E \rightarrow II$ or min cost = 210.

8. Solve the following cost minimising problems.

	I	II	III	IV	V
A	2	9	2	7	1
B	6	8	7	6	1
C	4	6	5	3	1
D	4	2	7	3	1
E	5	3	9	5	1

Answer :   **[JNTU - (Mech.) - 93]**

$A \rightarrow II,$ $B \rightarrow III,$ $C \rightarrow I,$ $D \rightarrow IV,$ $E \rightarrow V$ or

$A \rightarrow II,$ $B \rightarrow III,$ $C \rightarrow IV,$ $D \rightarrow I,$ $E \rightarrow V$ min cost = 24.

9. A company has six jobs to be done on six machines; any job can be done on any machine. The time in hours taken by the machines for the different jobs are as given below. Assign the machines to jobs so as to minimise the total machine hours.

Jobs

	1	2	3	4	5	6
1	2	6	7	3	8	7
2	6	1	3	9	7	3
3	3	6	5	7	3	5
4	2	2	7	8	4	8
5	4	9	6	8	7	6
6	7	5	5	7	7	5

Answer :   [JNTU - (Mech.) 91]

$1 \to 4$,	$2 \to 2$,	$3 \to 5$,	$4 \to 1$,	$5 \to 3$,	$6 \to 6$ or
$1 \to 4$,	$2 \to 2$,	$3 \to 5$,	$4 \to 1$,	$5 \to 6$,	$6 \to 3$ or
$1 \to 4$,	$2 \to 2$,	$3 \to 5$,	$4 \to 3$,	$5 \to 1$,	$6 \to 6$ or

Min time = 20 hours.

10. Solve the following minimal assignment problem :

Job / machine	I	II	III	IV
1	2	3	4	5
2	4	5	6	7
3	7	8	9	8
4	3	5	8	4

[JNTU - (Mech.) -92/S]

Answer : $1 \to I, 2 \to II, 3 \to III, 4 \to IV$ or Min : 20
$1 \to II, 2 \to I,\ 3 \to IV, 4 \to III$

11. Solve the following minimising assignment problem :

Job/Machine	1	2	3	4	5
I	12	8	7	15	4
II	7	9	17	14	10
III	9	6	12	6	7
IV	7	6	14	6	10
V	9	6	12	10	6

[JNTU - (Mech) - 89]

Answer : $I \to 3, II \to 1, III \to 2, IV \to 4, sV \to 5$ or
$1 \to 3, II \to 1, III \to 4, IV \to 2, V \to 5$ Min : 32

12. Minimise the following assignment problem :

	(1)	(2)	(3)	(4)	(5)	(6)
1	9	22	58	11	19	27
2	43	78	72	50	63	48
3	41	28	91	37	45	33
4	74	42	27	49	39	32
5	36	11	57	22	25	18
6	3	56	53	31	17	28

[JNTU - (Mech.) 90]

Answer : $1 \to (4), 2 \to (1), 3 \to (6), 4 \to (3), 5 \to (2), 6 \to (5)$ or
$1 \to (4), 2 \to (6), 3 \to (2), 4 \to (3), 5 \to (5), 6 \to (1)$ Min cost = 142.

13. Four salesmen are to be assigned to four districts. Estimates of the sales revenue is hundred of Rs. for each salesmen are as under :

<table>
<tr><td></td><td colspan="4" align="center">Districts</td></tr>
<tr><td>Salesmen</td><td align="center">A</td><td align="center">B</td><td align="center">C</td><td align="center">D</td></tr>
<tr><td>1</td><td align="center">320</td><td align="center">350</td><td align="center">400</td><td align="center">280</td></tr>
<tr><td>2</td><td align="center">400</td><td align="center">250</td><td align="center">300</td><td align="center">220</td></tr>
<tr><td>3</td><td align="center">420</td><td align="center">270</td><td align="center">340</td><td align="center">300</td></tr>
<tr><td>4</td><td align="center">250</td><td align="center">390</td><td align="center">410</td><td align="center">350</td></tr>
</table>

Give the assignment pattern that maximises the sales revenue. **[IGNOU - 1991]**

Answer : $1 \to C, 2 \to A, 3 \to D, 4 \to B$ Max Rs. 1490/-

14. Find the assignment that minimises the total time in hours.

<table>
<tr><td></td><td align="center">A</td><td align="center">B</td><td align="center">C</td><td align="center">D</td><td align="center">E</td></tr>
<tr><td>M1</td><td align="center">3</td><td align="center">8</td><td align="center">7</td><td align="center">5</td><td align="center">2</td></tr>
<tr><td>M2</td><td align="center">3</td><td align="center">4</td><td align="center">2</td><td align="center">1</td><td align="center">5</td></tr>
<tr><td>M3</td><td align="center">5</td><td align="center">8</td><td align="center">10</td><td align="center">7</td><td align="center">6</td></tr>
</table>

Answer : $M_1 - E, M_2 - D, M_3 - A$, Min time = 8

15. The owner of a small machine shop has four machinists available to assign jobs for the day. Five jobs are offered with expected profit for each machinist on each job are as follows.

<table>
<tr><td></td><td></td><td colspan="5" align="center">Jobs</td></tr>
<tr><td></td><td></td><td align="center">A</td><td align="center">B</td><td align="center">C</td><td align="center">D</td><td align="center">E</td></tr>
<tr><td></td><td>I</td><td align="center">60</td><td align="center">76</td><td align="center">48</td><td align="center">99</td><td align="center">80</td></tr>
<tr><td></td><td>II</td><td align="center">69</td><td align="center">82</td><td align="center">59</td><td align="center">71</td><td align="center">57</td></tr>
<tr><td>Machinists</td><td>III</td><td align="center">85</td><td align="center">90</td><td align="center">109</td><td align="center">69</td><td align="center">79</td></tr>
<tr><td></td><td>IV</td><td align="center">46</td><td align="center">62</td><td align="center">85</td><td align="center">75</td><td align="center">78</td></tr>
</table>

Which jobs is to be declined with determining the optimum schedule ?

[JNTU - (Mech.) 95]

Answer : $I - D, II - B, III - C, IV - E$ Dummy $- A$ (decline); Profit : 368

16. The owner of a small machine shop has four machinists available to assign to jobs for the day. Five jobs are offered with expected profit for each machinist on each job as follows :

<table>
<tr><td></td><td align="center">A</td><td align="center">B</td><td align="center">C</td><td align="center">D</td><td align="center">E</td></tr>
<tr><td>1</td><td align="center">62</td><td align="center">78</td><td align="center">50</td><td align="center">101</td><td align="center">82</td></tr>
<tr><td>2</td><td align="center">71</td><td align="center">84</td><td align="center">61</td><td align="center">73</td><td align="center">59</td></tr>
<tr><td>3</td><td align="center">87</td><td align="center">92</td><td align="center">111</td><td align="center">71</td><td align="center">81</td></tr>
<tr><td>4</td><td align="center">48</td><td align="center">64</td><td align="center">87</td><td align="center">77</td><td align="center">80</td></tr>
</table>

Find by using the assignment method, the assignment of machinists to jobs that will result in a maximum profit. Which job should be declined.

[JNTU - (ECE/Mech) - 93/CCC]

Answer : $1 \rightarrow D, 2 \rightarrow B, 3 \rightarrow C, 4 \rightarrow E$: Max. profit = Rs. 376.

and Job A is declined

17. In a machine shop, a supervisor wishes to assign five jobs among six machines. Any one of the jobs can be processed completely by any one of the machines as given below :

		\multicolumn{6}{c}{Machine}
		A	B	C	D	E	F
Job	1	13	13	16	23	19	9
	2	11	19	26	16	17	18
	3	12	11	4	9	6	10
	4	7	15	9	14	14	13
	5	9	13	12	8	14	11

The assignment of jobs to machines be on a one-to-one basis. Assign the jobs to machines so that the total cost is minimum. Find the minimum total cost.

[JNTU - (Mech.) 95]

Answer : $1 \rightarrow F, 2 \rightarrow A, 3 \rightarrow E, 4 \rightarrow C, 5 \rightarrow D$ Min cost = 43.

18. A department head has six jobs and five subordinates. The subordinates differ in their efficiency and the tasks differ in their intrinsic difficulty. The department head estimates the time each man would take to perform each task as given in the effectiveness matrix below :

		\multicolumn{6}{c}{Task}
		A	B	C	D	E	F
Man	1	20	15	26	40	32	12
	2	15	32	46	26	28	20
	3	11	15	2	12	6	14
	4	8	24	12	22	22	20
	5	12	20	18	10	22	15

Only one task can be assigned to one man. Determine how should the jobs are allocated so as to minimise the total man hours. Find the minimum total man hours.

[JNTU - (Mech.) 88]

Answer : $1 \rightarrow F, 2 \rightarrow A, 3 \rightarrow E, 4 \rightarrow C, 5 \rightarrow D$, or

$1 \rightarrow B, 2 \rightarrow F, 3 \rightarrow C, 4 \rightarrow A, 5 \rightarrow D$ min time = 55 hours.

19. A truck company on a particular day has 4 trucks for sending material to 6 terminals. The cost of sending material from some destination to different trucks will be different as given by the cost matrix below. Find the assignment of 4 trucks to 4 terminals out of six at the minimum cost.

Trucks

	A	B	C	D	E
Terminals 1	3	6	2	6	5
2	7	1	4	4	7
3	3	8	5	8	3
4	6	4	3	7	4
5	5	2	4	3	2
6	5	7	6	2	5

[JNTU - 95/S]

Answer : $1 \to C, 2 \to B, 3 \to A, 6 \to D$, Min cost = 8.

20. Solve the following unbalanced assignment problem of minimising total time for doing all the jobs.

Jobs

	1	2	3	4	5
Operator 1	6	2	5	2	6
2	2	5	8	7	7
3	7	8	6	9	8
4	6	2	3	4	5
5	9	3	8	9	7
6	4	7	4	6	8

[JNTU - (CSE) 95, (CA) May 85]

Answer : $1 \to 4, 2 \to 1, 3 \to$ dummy i.e., $6, 4 \to 5, 5 \to 2, 6 \to 3$

Min time = 16 units.

21. A company has 4 machines of which to do 3 jobs. Each job can be assigned to one only and only one machine. The cost of each job on each machine is given in the following table :

Machine

	W	X	Y	Z
Job A	18	24	28	32
B	8	13	17	18
C	10	15	19	22

What are job assignments which will minimise the cost? [JNTU 96/P]

Answer : $A \to W, B \to X, C \to Y$, or $A \to W, B \to Y, C \to X, Z* = 50$.

22. Five operators have to be assigned to five machines. The assignment costs are given in the table below.

Machine

Operator		I	II	III	IV	V
	A	5	5	-	2	6
	B	7	4	2	3	4
	C	9	3	5	-	3
	D	7	2	6	7	2
	E	6	5	7	9	1

Operator A cannot operate machine III and operator C cannot operate machine IV. Find the optimal assignment schedule. **[JNTU - (Mech.) 88]**

Answer : $A - IV, B - III, C - II, D -I , E - V$ or

$A - IV, B - III, C - V, D - II, E- I$, min cost $= 15$

23. Four new machines M_1, M_2, M_3 and M_4 to be installed in a machine shop. There are five vacant places, A, B, C, D and E available. Because of limited waiting space machine M_2 cannot be placed at C and M_3 cost C_{ij} of machine i to place j in rupees, is shown below.

	A	B	C	D	E
M_1	9	11	15	10	11
M_2	12	9	-	10	9
M_3	-	11	14	11	7
M_4	14	8	12	7	8

Obtain the optimal solution using best starting solution. **[JNTU - (Mech.) 88/S]**

Answer : $M_1 - A, M_2 - B, M_3 - E , M_4 - D$; cost $= 32$

24. The secretary of a school is taking bids on the four school bus routes. Four companies have made the bids as detailed in the following table.

Bids

Company	Route 1	Route 2	Route 3	Route 4
1	4000	5000	-	-
2	-	4000	-	4000
3	3000	-	2000	-
4	-	-	4000	5000

Suppose each bidder can be assigned only one route. Use the assignment model to minimise the school's cost at running the four bus routes.

Hint : Since some of the companies have not made job for certain routes, assign a very high bids M for all such routes. Then apply assignment algorithm.

Answer : Company 1 → route 1,   Company 2 → route 2,

               Company 3 → route 3,   Company 4 → route 4

               Minimum cost = Rs. (4000 + 4000 + 2000 + 5000 = Rs. 15000.

25. Five workmen have to be assigned to repair five machines. The assignment costs are given in the following table below :

Machines Workmen	Lathe	Milling	Jig Boring	Shearing	SPM
Ramu	5	5	-	2	6
Kishore	7	4	2	3	4
Prasad	9	3	5	-	3
Sandeep	7	2	6	7	2
Pradeep	6	5	7	9	1

Ramu cannot repair Jig boring machine and Prasad cannot repair Shearing machine. Find the optimal assignment schedule. If Pradeep, the specialist in SPM repairs is exclusively used for SPM and will not be sent to repair any other machine what will be the effect on the above assignment? In case Pradeep refuses to repair SPM, do you find any effect on the minimum repair cost?

Answer :   *Solution set - 1 :* Ramu - Shearing , Kishore - Jig boring,

                      Prasad - Milling, Sandeep - Lathe, Pradeep - SPM or

       Solution set - 2 : Ramu - Shearing, Kishore - Jig boring,

                      Prasad - SPM, Sandeep - Milling and Pradeep - Lathe.

                      Minimum cost = 15.

If Pradeep is exclusively kept for repair of SPM then the solution will be unique i.e., the solution set 1. There will be no effect on the minimum cost if Pradeep refuses to repair SPM since we can use the second solution set.

26. Four operators O_1, O_2, O_3 and O_4 available to a manager has to get four jobs j_1, j_2, j_3 and j_4 done by assigning one job to each operator, given the times needed by different operators for different jobs in the matrix below.

	j_1	j_2	j_3	j_4
O_1	12	10	10	8
O_2	14	12	15	11
O_3	6	10	16	4
O_4	8	10	9	7

(i) How should the manager assign the jobs so that the total time needed for all four jobs is minimum?

(ii) If job j_2 is not to be assigned to operator O_2, what should be the assignment and how much additional total time will be required?

[JNTU - (CSE) 95/S]

Answer : (i) $O_1 \rightarrow j_3$, $O_2 \rightarrow j_2$, $O_3 \rightarrow j_4$, $O_4 \rightarrow j_1$, min time = 34

(ii) $O_1 \rightarrow j_2$, $O_2 \rightarrow j_4$, $O_3 \rightarrow j_1$, $O_4 \rightarrow j_3$ min time = 36

∴ Additional time required = 36 - 34 = 2 units of time.

27. Five swimmers are eligible to complete in a relay team which is to consist of four swimmers, swimming four different swimming styles; back stroke, breast stroke, free style and butterfly. The time taken for the five swimmers — Anitha, Bhavana, Chandi, Dolly and Easwari — to cover a distance of 100 meters in various swimming styles are given below in minutes, seconds.

Anitha swims the back stroke in 1 : 09, the breast stroke in 1:15 and has never competed in the free style or butterfly.

Bhavana is a free style specialist averaging 1:01 for the 100 meters but can also swim the breast stroke in 1:16 and butterfly in 1:20.

Chandi swims all styles — back 1:10 butterfly 1:12 free style 1:05 and breast stroke 1:20.

Dolly swims only the butterfly 1:11 while Easwari swims the back stroke 1:20 the breast stroke in 1:16 and the free style 1:06 and the butterfly 1:10. Which swimmer should be assigned to which swimming style? Who will not be in the relay.

Hint : The assignment matrix with time expressed in seconds and adding a dummy style to balance it is given by

	Back Stroke	Breast Stroke	Free Style	Butterfly	Dummy
Anitha	69	75	-	-	0
Bhavana	-	76	61	80	0
Chandi	70	80	65	72	0
Dolly	-	-	-	71	0
Easwari	80	76	66	70	0

Answer : Anitha will be in breast stroke (times 75 secs).

Bhavana will be in free stroke (time 61 secs).

Chandi will be in back stroke (time 70 secs)

Dolly will not participate (dummy).

Easwari will be in butterfly (time 70 secs).

Dolly will be out of the relay.

Total minimum time in the relay = 276 secs or 4 min 36 secs.

28 Alpha corporation has four plants each of which can manufacture any one of four products: Production costs differ from one plant to another as do sales revenue. Given the revenue and cost data below, obtain which product each plant should produce to maximise profit.

Sales revenue (Rs. '000) product

Plant	1	2	3	4
A	50	68	49	62
B	60	70	51	74
C	55	67	53	70
D	58	65	54	69

Production costs (Rs. '000)

Plant	1	2	3	4
A	49	60	45	61
B	55	63	45	69
C	52	62	49	68
D	55	64	48	66

[JNTU, 2002 (Set-2) & (Set-4)]

Hint : Construct the profit matrix by using the fact : Profit matrix = revenue matrix - cost matrix.

To make use of minimisation technique subtract each element of profit matrix from the maximum element which will be 8. Then apply assignment rule in usual manner.

Answer : A - 2 ,   B - 4,   C - 1,   D - 3; Max. Profit = Rs. 22000/-

29. A small aeroplane company, operating seven days a week serves three cities *A, B* and *C* according to the schedule shown in the following table. The lay over cost per stop is roughly proportional to the square of the layover time. How should planes be assigned the flights so as to minimise the total lay over cost?

Flight No. and Index	From	Departure	To	Arrival
$A_1 B$	A	09 AM	B	Noon
$A_2 B$	A	10 AM	B	01 PM
$A_3 B$	A	03 PM	B	06 PM
$A_4 C$	A	08 PM	C	Mid Night
$A_5 C$	A	10 PM	C	02 AM
$B_1 A$	B	04 PM	A	07 AM
$B_2 A$	B	11 AM	A	02 PM
$B_3 A$	B	03 AM	A	06 PM
$C_1 A$	C	07 AM	A	11 AM
$C_2 A$	C	03 AM	A	07 PM

(Bombay DMS 83; Madurai BE (Mech) 78)

Answer : Min Layover = 1066 units

Flight No.	1	2	3	4	5
Departure route	$A_1 B$	$A_2 B$	$A_3 B$	$C_1 A$	$C_2 A$
Arrival route	$B_3 A$	$B_1 A$	$B_2 A$	$A_4 C$	$A_5 C$

30. A trip from Hyderabad to Vijayawada takes 6 hours by bus. A typical time table of bus services in both directions is given below.

Hyderabad - Vijayawada			Vijayawada - Hyderabad		
Route No :	Departure	Arrival	Route No.	Departure	Arrival
a	06.00	12.00	1	05.30	11.30
b	07.30	13.30	2	9.00	15.00
c	11.30	17.30	3	15.00	21.00
d	19.00	01.00	4	18.30	00.30
e	00.30	06.30	5	00.00	06.00

The cost of providing this service by the transport company depends upon the time spent by the bus crew (driver and conductor) away from their places in addition to service times. There are five crews. There is a constraint that every crew should be provided with 4 hours of rest before return trip again and should not wait for more than 24 hours for the return trip. The company has residential facilities for the crew at Hyderabad as well as at Vijayawada. Obtain the pairing of routes so as to minimise the cost.

Answer :

Crew	1	2	3	4	5
Residence	Hyderabad	Vijaywada	Vijaywada	Hyderabad	Vijaywada
Service No.	$d1$	$2e$	$3a$	$b4$	$5c$
Waiting time (hr)	4.5	9.5	9.0	5.0	5.5

Minimum total waiting time = 33.30 hours.

31. An Airline that operates seven days a week, has the time table shown below crews must have a minimum layover of 1 hour between flights. Obtain the pairing of flights that minimises lay over time away from home. For any given pairing, the crew will be based at the city that returns in the smaller layover for each pair. Also mention the town where crew should be based.

Hyderabad - Tirupati			Tirupati - Hyderabad		
Flight No.	Depart	Arrive	Flight No.	Depart	Arrive
1	7:30 AM	9:00 AM	2	7:00 AM	10:00 AM
3	8:45 AM	9:45 AM	4	7:45 AM	10:45 AM
5	2:00 PM	3:30 PM	6	11:00 AM	2:00 PM
7	5:45 PM	7:15 PM	8	6:00 PM	9:00 PM
9	7:00 PM	8:30 PM	10	7:30 PM	10:30 PM

[JNTU - (Mech.) 95/CCC]

Answer : $1 \to 10, 3 \to 6, 5 \to 8, 7 \to 2, 9 \to 4$

Minimum layover time 28:75 hours..

32. An airline that operates seven days a week, has the time table shown below. Crews must have a minimum lay over of 6 hours between flights. Obtain the pairing of flights that minimises lay over time away from home. For any given pairing the crew will be based at the city that results in the smaller layover. For each pair also mention the town where crew should be based.

Delhi - Kolkatta			Kolkata - Delhi		
Flight No.	Depart	Arrive	Flight No.	Depart	Arrive
1	7:00 AM	9:00 AM	101	9:00 AM	11:00 AM
2	9:00 AM	11:00 AM	102	10:00 AM	12:00 Noon
3	1:30 PM	3:30 PM	103	3:30 PM	5:30 PM
4	7:30 PM	9:30 PM	104	8:00 PM	10:00 PM

(ICWA June 1990)

Answer : $1 \to 103$ (base at Delhi) $3 \to 101$ (base at Delhi)

$2 \to 104$ (base at Delhi) $4 \to 102$ (base at Kolkatta).

33. Solve the following travelling 'salesman problem' given by the following data :

$C_{12} = 16, C_{13} = 4, C_{14} = 12, C_{23} = 6, C_{34} = 5 , C_{25} = 8, 3_{35} = 6$ and $C_{45} = 20$

Where $C_{ij} = C_{ji}$ and $C_{ij} = \infty$ if $i = j$

Answer : $1 \to 2 \to 5 \to 3 \to 4 \to$ & cost = Rs. 47

34. A medical representative has to visit five stations A, B, C, D and E. He does not want to visit any station twice before completing his tour of all the stations and wishes to return to the starting station. Costs of going from one station to another are given below. Determine the optimal route.

	A	B	C	D	E
A	∞	2	4	7	1
B	5	∞	2	8	2
C	7	6	∞	4	6
D	10	3	5	∞	4
E	1	2	2	4	∞

Answer : $A \rightarrow B \rightarrow C \rightarrow D \rightarrow E \rightarrow A.$ **[JNTU - (Mech.) 96/P]**

35. Solve the travelling salesman problem in the matrix shown below :

To

From		1	2	3	4	5
	1	∞	6	12	6	4
	2	6	∞	10	5	4
	3	8	7	∞	11	3
	4	5	4	11	∞	5
	5	5	2	7	8	∞

Answer : $3 \rightarrow 5 \rightarrow 2 \rightarrow 4 \rightarrow 1 \rightarrow 3;\ Z = 27.$ **[JNTU - (Mech.) 96/S]**

36. A sales man has to visit five cities A, B, C, D and E. The distance (in hundred miles) between the five cities are as follows :

From	A	B	C	D	E
A	-	7	6	8	4
B	7	-	8	5	6
C	6	8	-	9	7
D	8	5	9	-	8
E	4	6	7	8	-

If the salesman starts from city A and has to come back to city A, which route should he select so that the total distance travelled is minimum.

[IGNOU - (MBA) Assignment 96, JNTU - 2003/S (Set - 4)]

Answer : $A \rightarrow E \rightarrow B \rightarrow D \rightarrow C \rightarrow A$ min distance = 30 hundred miles.

37. A machine operator processes five types of items on his machine each week, and must choose a sequence for them. The main costs involved are setup costs and priority. The setup costs per change depends on the item presently on the machine and the setup to be made, according to the following table :

		To item				
		A	B	C	D	E
From item	A	α	4	7	3	4
	B	4	α	6	3	4
	C	7	6	α	7	5
	D	3	3	7	α	7
	E	4	4	5	7	α

In addition, there is a priority rating among the items, A and B having a higher priority than C, D, E. This may be interpreted as adding an additional cost of 5 when any high-priority item immediately follows a low-priority item. If he processes each type of item once and only once each week, how should be sequence the items on his machine? **[JNTU - (Mech.) 96/P, OU - MBA 2001/S]**

Hint : See illustration - 13

Additional Problems

1. Find the optimal solution to the following assignment problem, each cell value being the processing time in minutes.

	I_1	I_2	I_3	I_4	I_5
A	20	15	14	19	30
B	24	18	31	28	19
C	32	21	25	20	27
D	34	20	12	16	19
E	31	26	25	21	22

2. Find the optimal assignment cost.

	F	G	H	I	J
A	74	35	28	10	31
B	24	63	21	32	40
C	16	45	51	20	38
D	60	56	46	36	28
E	35	61	70	42	25

3. Solve the following assignment problem given the cost matrix.

Workers	J_1	J_2	J_3	J_4
W_1	24	27	18	20
W_2	26	23	20	31
W_3	24	22	34	26
W_4	19	21	21	22
W_5	30	25	28	27

[OU - MBA 93]

4. The quotations (Rs. in lakhs) received for 4 projects from four contractors are given below :

Projects	C_1	C_2	C_3	C_4
A	7	3	7	8
B	5	7	4	7
C	9	5	8	5
D	10	8	9	2

If only one project is to be awarded to one contractor, find out the assignment of projects to contraction in order to minimise the total cost.   **[OU - MBA Apr. 94]**

5. Solve the following assginment problem given the profit matrix.

	J_1	J_2	J_3	J_4	J_5
P_1	12	6	9	6	10
P_2	6	12	11	10	10
P_3	10	11	10	8	4
P_4	6	4	9	8	11
P_5	12	13	8	6	5

[OU - MBA Dec. 1995]

6. A national truck rental service has a surplus of one truck in each of the cities 1, 2, 3, 4, 5 and 6 and a deficit of one truck in each of the cities 7, 8, 9, 10, 11 and 12. The distances in km between the cities with a surplus and the cities with a deficit are displayed below :

	7	8	9	10	11	12
1	31	62	29	42	15	41
2	12	19	39	55	71	40
3	17	29	50	41	22	22
4	35	40	38	42	27	33
5	19	30	29	16	2	23
6	72	30	30	50	41	20

How should the truck be dispersed so as to minimise the total distance travelled?
[OU - MBA - May 92, March 99, July 2000]

Answer : ($1 \rightarrow 11$, $2 \rightarrow 8$, $3 \rightarrow 7$, $4 \rightarrow 9$, $5 \rightarrow 10$, $6 \rightarrow 12$, Distance = 125km)

7. Check if the following assignment is optimal. If not find the optimal.

Jobs

	J_1	J_2	J_3	J_4	J_5
M_1	20	15	14	–	30
M_2	24	–	31	28	19
M_3	32	21	–	20	27
M_4	–	20	12	16	19
M_5	31	26	25	21	–

Men

[OU – MBA Dec. 2000]

8. The following cost matrix related to an assignment problem

Machines

	1	2	3	4	5
1	7	3	4	6	2
2	5	2	8	7	5
3	6	9	6	5	2
4	10	8	3	9	4

Problem

(i) Find the optimum assignment

(ii) If man 2 should not be assigned to machine 2, what is the best assignment

[OU – MBA Sep. 1998]

9. Fire workers are available to work with the machines and the repsective costs (in Rs.) associated with each worker's machine assignment is given below. A sixth machine is available to replace one of the existing and the associated costs and also given below.

(i) Determine whether the new machine can be accepted.

(ii) Determine also optimal assignment and the associated saving cost.

Machines

	M_1	M_2	M_3	M_4	M_5	M_6
W_1	12	3	6	–	5	9
W_2	4	11	–	5	–	8
W_3	8	2	10	9	7	5
W_4	–	7	8	6	12	10
W_5	5	8	9	4	6	–

Workers

10. Solve the following assignment problem given the cost matrix.

	D_1	D_2	D_3	D_4	D_5
O_1	6	2	5	2	6
O_2	2	5	8	7	7
O_3	7	8	6	9	8
O_4	6	2	3	3	5
O_5	9	3	8	9	7
O_6	4	7	4	6	8

11. A project work consists of four major jobs for which four contraction have submitted tenders. The tender amounts quoted in thousands of rupees are given in the matrix below.

Jobs

		J_1	J_2	J_3	J_4
	C_1	15	29	35	20
	C_2	21	27	33	17
Contractors	C_3	17	25	37	15
	C_4	14	31	39	21

Find the assignment which minimises total cost of the projects each contractor has to be assigned one job only. **[OU - MBA March - 1999]**

12. Suggest optimum assignment of four workers A, B, C and D to four jobs I, II, III, IV the time taken by different workers in completing the different jobs is given below.

Jobs

	I	II	III	IV
	8	10	12	16
Workers	11	11	15	8
	9	6	5	14
	15	14	9	7

The indicate the total time taken is completing the job.

[OU - MBA Aug/Sep 1999]

13. A company is facing with the problem of assigning four different salesmen to four territories for promoting its sales. Territories are not equally rich in their sales potential and the salesman also differ in their ability to promote sales. the following tables gives the expected annual sales (in thousands of Rs.) for each salesman if assigned to various territories. Find the assignment of salesmen so as to maximize the annual sales.

Territories\ Salesmen	1	2	3	4
1	60	50	40	30
2	40	30	20	15
3	40	20	35	10
4	30	30	25	20

[JNTU C.S.E 2003/S (Set - 2)]

Answer : $S_1 \to T_1$ (60), $S_2 \to T_2$ (30), $S_3 \to T_3$ (35), $S_4 \to T_4$ (20) or

$S_1 \to T_2$ (50), $S_2 \to T_1$ (40), $S_3 \to T_3$ (35)and $S_4 \to T_4$ (20), Max. Profit = 145000

14. There as five jobs to be assigned, one each to 5 machines and the associated cost matrix is as follows :

Man\ Job	I	II	III	IV	V
A	11	17	8	16	20
B	9	7	12	6	15
C	13	16	15	12	16
D	21	24	17	28	26
E	14	10	12	11	15

[AGRA M.Sc. MATHS. 1990, M.S. BARODA. B.Sc., 78, MADRAS IIT (M.TECH) 78, DELHI B.Sc. (MATHS) 77, MBA 74, M.Sc. MATHS 71, INDIAN STATISTICAL INSTITUTE 71]

Answer : $A \to I, B \to IV, C \to V, D \to III, E \to II$ Min. cost = 60

15. Solve the following minimal assignment problem.

Man\ Job	I	II	III	IV	V
A	1	3	2	3	6
B	2	4	3	1	5
C	5	6	3	4	6
D	3	1	4	2	2
E	1	5	6	5	4

[MEERUT STAT 1974, MATHS 92]

Answer : $(A \to I, B \to IV, C \to III, D \to II, E \to V)$

16. Solve the following minimal assignment problem.

Job \ Man	I	II	III	IV
A	12	30	21	15
B	18	33	9	31
C	44	25	24	21
D	23	30	28	14

[MEERUT B.Sc. 1990/S]

Answer : $(A \to I, B \to III, C \to II, D \to IV)$

17. Solve the following minimal assignment problem.

Job \ Man	I	II	III	IV
A	2	3	4	5
B	4	5	6	7
C	7	8	9	8
D	3	5	8	4

[MEERUT 1975, 91, Delhi 68]

Answer : $A \to I, B \to II, C \to III, D \to IV,$ or $A \to II, B \to I, C \to IV, D \to III)$

18.

Project \ Location	I	II	III	IV	V
A	15	21	6	4	9
B	3	40	21	10	7
C	9	6	5	8	10
D	14	8	6	9	3
E	21	18	18	7	4

[AGRA M.Sc. MATHS 1986, GAUHATI STAT. 69]

Answer : $(A \to IV, B \to I, C \to II, D \to III, E \to V)$

19. Solve the assignment problem represented by the following matrix which gives the distances from the customers A, B, C, D, E to the depots, a, b, c, d and e each depot has one car. How should the cars be assigned to the customers as to minimize the distance travelled?

	a	b	c	d	e
A	160	130	175	190	200
B	135	120	130	160	175
C	140	110	155	170	185
D	50	50	80	80	110
E	55	35	70	80	105

[MEERUT MATHS 1980, STAT. 87, CALCUTTA (APP. MATHS) M.Sc. 76,
KURUKSHETRA. M.Sc. MATHS 75, ROHITKHAND 82]

Answer :   Total distance = 570 km $(A \to e, B \to c, C \to b, D \to a, E \to d)$

20. The assignment cost of assigning any one operator to any one machine is given in the following table.

Operators Machine	I	II	III	IV
A	10	5	13	15
B	3	9	18	3
C	10	7	3	2
D	5	11	9	7

Find the optimal assignment by Hungarian method   **[JNTU Mech. 2003/S (Set - 4)]**

Answer : $A \rightarrow II$ (5), $B \rightarrow IV$ (3), $C \rightarrow III$ (3) and $D \rightarrow I$ (5), Total cost = 16 units

21. Solve the following assignment problem

Jobs Operators	1	2	3	4
A	10	12	9	11
B	5	10	7	8
C	12	14	13	11
D	8	15	11	9

[MADRAS B.Sc. MATHS 84, MADURAI B.Sc. (MATHS) 83, MADURAI M.E. 1984]

Answer : $(A \rightarrow III, B \rightarrow I, C \rightarrow II, D \rightarrow IV)$ min cost = 37

22. Solve the following assignment problem

Jobs Machines	I	II	III	IV	V
A	11	10	18	5	9
B	14	13	12	19	6
C	5	3	4	2	4
D	15	18	17	9	12
E	10	11	19	6	14

[DELHI B.Sc. (MATHS) 93, 89]

Answer : $(A \rightarrow II, B \rightarrow V, C \rightarrow III, D \rightarrow IV, E \rightarrow I)$ and min cost = 39

23. A team of 5 horses and 5 riders has entered a jumping show contest. The number of penalty points to be expected when each rider rides any horse in shown below.

Riders / Horses	R_1	R_2	R_3	R_4	R_5
H_1	5	3	4	7	1
H_2	2	3	7	6	5
H_3	4	1	5	2	4
H_4	6	8	1	2	3
H_5	4	2	5	7	1

How should the horses be alloted to the riders so as to minimize the expected loss of the team **[MEERUT M.Sc. (MATHS) 84]**

Answer : $(H_1 \rightarrow R_5, H_2 \rightarrow R_1, H_3 \rightarrow R_4, H_4 \rightarrow R_3, H_5 \rightarrow R_2)$ min loss = 8

24. Solve the following assignment problem. The elements given in the matrix are the profits in Rs. derived for such assignment.

Machines / Jobs	P	Q	R	S
A	51	53	54	50
B	47	50	48	50
C	49	50	60	61
D	63	64	60	60

Answer : $A \rightarrow R$ (54), $B \rightarrow Q$ (50), $C \rightarrow S$ (61), $D \rightarrow Q$ (63) Max. Profit = Rs. 228
[JNTU C.S.E 2003/S (Set - 2)]

25. Find the optimal solution for the assignment problem with the following cost matrix.

	I	II	III	IV
A	5	3	1	8
B	7	9	2	6
C	6	4	5	7
D	5	7	7	6

Answer : $(A \rightarrow III, B \rightarrow IV, C \rightarrow II, D \rightarrow I)$ min cost = 16 **[DELHI 71]**

26. Solve the following assignment problems.

Men / Machines	I	II	III	IV	V
A	1	3	2	8	8
B	2	4	3	1	5
C	5	6	3	4	6
D	3	1	4	2	2
E	1	5	6	5	4

Answer : $(A \rightarrow I, B \rightarrow IV, C - III, D \rightarrow II, E \rightarrow V)$ **[MEERUT STAT. 74]**

27. Solve the following assignment problems

Tasks＼Persons	I	II	III	IV
A	5	12	19	11
B	5	10	7	8
C	12	14	13	11
D	8	15	11	9

Answer : $(A \to I, \ B \to III, \ C \to II, \ D \to IV)$ min cost $= 35$

28. Solve the following assignment problems

Jobs＼Persons	I	II	III	IV	V
A	12	8	7	15	4
B	7	9	17	14	10
C	9	6	12	6	7
D	7	6	14	6	10
E	9	6	12	10	6

[MADURAI B.Sc. (MATHS) 89, OSMANIA M.Sc. (STAT.) 84, IAS (MATHS) 89, BOMBAY B.Sc. (STAT) 85, KERALA B.Sc. (ENGG) 85, CALCUTTA B.Sc. (MATHS) 84]

Answer :   (1) $(A \to III, \ B \to I, \ C \to II, \ D \to IV, \ E \to V)$

(2) $(A \to III, \ B \to I, \ C \to IV, \ D \to II, \ E \to V)$

29. Solve the following assignment problems

Machines＼Jobs	I	II	III	IV	V
A	8	4	2	6	1
B	0	9	5	5	4
C	3	8	9	2	6
D	4	3	1	0	3
E	9	5	8	9	5

[MADURAI B.Sc. (MATHS) 89, CALCUTTA B.Sc. MATHS 84]

Answer : $(A \to V, \ B \to I, \ C \to IV, \ D \to III, \ E \to II)$ min cost $= 9$

30. The jobs A, B, C are to be assigned to three machines X, Y, Z. The processing costs (Rs) are as given in the matrix shown below. Find the allocation which will minimize the overall processing cost.

Jobs＼Machines	X	Y	Z
A	19	28	31
B	11	17	16
C	12	15	13

Answer : $(A \to X, \ B \to Y, \ C \to Z)$ **[BOMBAY (DIP OPER. MAN) 74]**

31. A project work consists of four major jobs for which four contractors have submitted tenders. The tender amounts quoted in lakhs of rupees are given in the matrix below.

Contractor \ Jobs	a	b	c	d
1	10	24	30	15
2	16	22	28	12
3	12	20	32	10
4	9	26	34	16

Find the assignment which minimizes the total cost of project [each contract has to be assigned at least one job] **[MEERUT 1980, ISI (DIP) 76]**

Answer :   Three alternative assignments are

(1) $(1 \to b, \ 2 \to c, \ 3 \to d, \ 4 \to a)$

(2) $(1 \to c, \ 2 \to b, \ 3 \to d, \ 4 \to a)$

(3) $(1 \to c, \ 2 \to d, \ 3 \to b, \ 4 \to a)$ min cost = Rs. 71,00,000

32. A company producing a single product and selling it through five agencies situated in different cities. All of a sudden there is a demand for the production at another five cities not having any agency of the company. The company faced with the problem of deciding as how to assign the existing agencies to dispatch the product to needy cities in such a way that the total travelling distance is minimised. The distance between the surplus and deficit cities (in kilometers) is given by

Surplus city \ Deficit city	A'	B'	C'	D'	E'
A	10	5	9	18	11
B	13	19	6	12	14
C	3	2	4	4	5
D	18	9	12	17	15
E	11	6	14	19	10

Determine the optimum assignment schedule

[MADRAS (M.Tech.) 77, MEERUT (STAT) 76]

Answer : $(A \to A', \ B \to C', \ C \to D', \ D \to B', \ E \to E')$ optimal distance = 39 km

33. Find the minimum cost solution for the 5×5 assignment problem whose cost coefficients are as given below.

1	2	3	4	5
-2	-4	-8	-6	-1
0	-9	-5	-5	-4
-3	-8	0	-2	-6
-4	-3	-1	0	-3
-9	-5	-9	-9	-5

[ROORKEE M.E. (Mech.) 77, IIT KARAGPUR (DIP) 78]

Answer : $(1 \to 3,\ 2 \to 2,\ 3 \to 5,\ 4 \to 4,\ 5 \to 1)$ or
$(1 \to 4,\ 2 \to 2,\ 3 \to 3,\ 4 \to 5,\ 5 \to 1$ optimal cost $= 36)$

34. Six wagons are available at six stations A, B, C, D, E and F. these are required at stations, I, II, III, IV, V and VI. The mileage between various stations is given by the following table.

	I	II	III	IV	V	VI
A	20	23	18	10	16	20
B	50	20	17	16	15	11
C	60	30	40	55	8	7
D	6	7	10	20	100	9
E	18	19	28	17	60	70
F	9	10	20	30	40	55

How should the wagons be transported in order to minimize the total mileage covered   [ALIGRAH (COMP. Sc) 77]

Answer : $(A \to IV,\ B \to VI,\ C \to V,\ D \to III,\ E \to I, F \to II)$ total mileage $= 66$

Review Questions

1. Write a short note on assignment problem.

2. Give the step by step procedure for solving assignment problem by Hungarian method.

3. Discuss how you get multiple solutions in an assignment problem.

4. What is meant by unbalanced assignment problem.

[OUT - MBA 94, OU - MBA Dec. 2000, OU - MBA Apr. 94,
JNTU Mech/Prod/Mechantronics/Chem. - 2001]

5. Explain in detail assignment model of LP problem. How does it differ from transportation problem. **[JNTU Mech. 96]**

6. Distinguish between assignment and allocation problem. **[JNTU CSE - 98]**

7. Write a short note on travelling salesman problem. **[OU -MBA Sep. 98, Dec. 2000 JNTU CSE - 97, Mech. 99, Mech/Prod/Mechatronics/Chem. 2001/S]**

8. Distinguish between assignment and travelling salesman problem.

9. What is routing problem? How does it differ from assignment problem.

10. How can you convert a transportation problem into an assignment? Explain with an example.

11. Discuss the maximisation case in assignment problem.

12. What is meant by restricted (or prohibited) assignment? Explain how you can solve the AP in such case.

13. How do you apply assignment in the case of allocating crew in transport agencies/railways/airways etc.

14. Discuss the applications of assignment problems.

15. Give the solution method by Flood's technique to assignment problem in the form of flow chart.

16. What are the difference between transportation prolem and an assignmen problem? **[OU - MBA Feb. 1993, Dec. 1990, March/April 1999]**

Objective Type Questions

1. Optimality is said to be reached in AP if

   (a) $C_{ij} \geq u_i + v_j$ for occupied cells

   (b) $Z_j - C_j$ is positive for all cells

   (c) highest time of middle jobs $\geq$ least time of first or last series

   (d) no. of minimum lines connecting all the zeros of opportunity matrix is at least equal to its order

2. An assignment problem is said to be balanced if

   (a) no. of rows = no. of columns

   (b) no. of allocated cells = no. of rows + no. of columns − 1

   (c) total supply = total demand

   (d) no. of zero in the cost matrix is equal to the order

3. In a prohibited AP, a penalty cost 'M' is used such that

   (a) $a_{ij} \pm M = a_j$ (b) $M \pm a_{ij} = a_{ij}$

   (c) $M \pm a_{ij} = M$ (d) $M \pm a_{ij} = 0$

4. In a maximisation case of A.P. we convert it to minimisation case by

   (a) adding every cell value to highest among them

   (b) subtracting every cell value from highest among them

   (c) dividing every cell value with lowest among them

   (d) subtracting least value of each row in the corresponding row

5. If number of rows exceeds the number of columns by 2, then we add

   (a) a dummy row   (b) 2 dummy rows

   (c) a dummy column   (d) 2 dummy columns

6. The cost of assignment in a dummy cell is

   (a) unity   (b) zero

   (c) ∞   (d) negative

7. An opportunity cost table of AP is prepared by

   (a) subtracting every element from highest among them

   (b) multiplying by (−!) in all cells

   (c) subtracting least of each row in corresponding row and then least of every column in corresponding column

   (d) transposing the matrix

8. While revising the opportunity cost table of AP, we put the lines across

   (a) marked rows and unmarked columns

   (b) marked rows and marked columns

   (c) unmarked rows and marked columns

   (d) unmarked rows and unmarked columns

9. which of the following is not be marked while revising the AP,

   (a) a row which has no assignment

   (b) a column which has no assignment

   (c) a row in which marked column has an assigned zero

   (d) a column in which marked row has an assigned zero

10. While revising the opportunity cost table of AP, least among unlined number is taken to

    (a) add at unlined numbers

    (b) subtract at double lined numbers

    (c) add at single lined numbers

    (d) subtract at unlined numbers

11. If the number of minimum lines is less the order of matrix then it means

    (a) optimisation is already reached

    (b) one more line is to be drawn

    (c) allocation is wrong

    (d) there is a scope for further optimisation

12. In AP, we get multiple optimal solutions if

    (a) a loop can be constructed with zeros having assignment at alternate corners

    (b) minimum number of lines is greater than the order

    (c) minimum number of lines is equal to the order

    (d) minimum number of lines is less than the order

13. In a travelling salesman problem the additional condition on the choice of assignment is to find

    (a) in what direction should the salesman travel from home city

    (b) cycle in which he has to travel

    (c) a solution such that he has to return to the home city

    (d) any of the above

14. A matrix of travelling salesman problem contains

    (a) no figures along at least one row

    (b) no figures along at least one column

    (c) no figures along the diagonals

    (d) no figures along the principal diagonal

15. An opportunity cost matrix of AP should have

    (a) at least one zero in each row

    (b) at least one zero in each column

    (c) both (a) & (b) of the above

    (d) no zeros in at least one column or row

16. In an unbalanced AP of 5×4 matrix, while finding the opportunity cost matrix, we need not perform

    (a) row iteration   (b) column iteration

    (c) total matrix iteration   (d) marking 4×4 matrix

17. In an optimal solution of A.r. of 4×4 matrix the minimum number of lines we must get at least is _____

    (a) 3   (b) 15   (c) 4   (d) 7

18. The main criterion used in Hungarian method to solve an assignment problem is to calculate

    (a) operating cost   (b) maintenance cost

    (c) opportunity cost   (d) over heads costs

19. In an assignment whose cost matrix is given by

	I	II	III
A	5	10	15
B	6	12	6
C	11	7	7

    The optimal assignment is

    (a) $A \rightarrow II, B \rightarrow I, C \rightarrow III$ (b) $A \rightarrow I, B \rightarrow III, C \rightarrow II$

    (c) $B \rightarrow I, C \rightarrow II, A \rightarrow III$ (d) none of the above

20. The total cost in the above AP is

    (a) 18   (b) 20

    (c) 17   (d) none of the above

21. An assignment problem (AP) can be converted as transportation problem (TP) by assuming supply/ demand to each row/column as _____

    (a) zero   (b) unity

    (c) infinity   (d) no relation

22. An assignment problem can be solved by applying Johnson & Jackson's rules of sequencing if _____ conditions is imposed.

    (a) unity supply/demand   (b) zero supply/demand

    (c) no passing   (d) infinite supply/demand

23. Men are allocated to various routes so as to result in minimum idle time or cost and decisions is made regarding where these men have to be asked to reside. This is performed by _____ method

    (a) transportation problem   (b) travelling salesman problem

    (c) crew assignment   (d) routing or network problem

24. Which of the following is always a balanced problem

    (a) transportation   (b) assignment

    (c) travelling salesman   (d) none of the above

25. The column iteration is not necessary while preparing opportunity matrix in an assignment if

    (a) every column contains at least one zero after row iteration

    (b) an unbalanced assignment of order $m \times n$, where $m > n$

    (c) every column has at least one zero initially

    (d) any of the above.

Fill in the Blanks

1. If an assignment problem is square it is called _____

2. We can convert a maximisation AP to minimisation AP by multiplying every cell with _____

3. Costs of dummy row/column are usually taken as _____

4. We put lines across _____ rows and _____ column while revising opportunity cost table of AP

5. The optimisation is said to be reached in AP if minimum number of lines covering all zeros is equal to _____

6. If a rectangular loop constructed by all zeros with alternative assignment on the corners of the loop can be obtained, the AP will have _____

7. If an AP has condition that the solution must be cyclic, then the AP is said to be _____

8. The optimal solution of AP of $n \times n$ order should at least get _____ minimum number of lines connecting all zeros in the opportunity matrix

9. In a restricted AP with elements x_{ij}, we use 'M' as penalty with a condition that _____

10. While revising the opportunity cost matrix of AP, we identify least among unlined numbers and _____ to unlined numbers and _____ to double lined (intersecting) numbers.

Answers

Objective Type Questions :				
1. (d)	2. (a)	3. (c)	4. (b)	5. (d)
6. (b)	7. (c)	8. (c)	9. (b)	10. (d)
11. (d)	12. (a)	13. (d)	14. (d)	15. (c)
16. (a)	17. (c)	18. (c)	19. (b)	20. (a)
21. (b)	22. (c)	23. (c)	24. (c)	25. (d)

Fill in the Blanks :	
1. balanced AP	2. − 1
3. zeros	4. unmarked, marked
5. no. of rows or columns or order of matrix	6. alternate or multiple optimal solution
7. travelling salesman problem	8. n
9. $M \pm x_{ij} = M$	10. subtract, add

7

Chapter

Inventory Management

CHAPTER AT A GLANCE

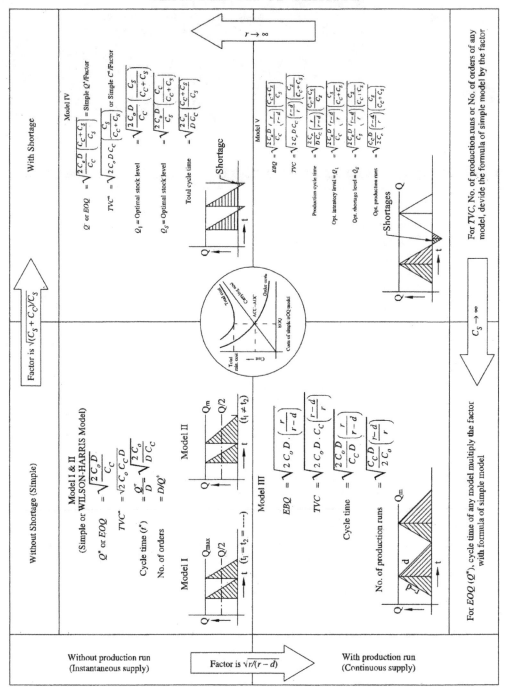

7.0 *Introduction*

It was assumed in the past that accumulation of adequate stocks was necessary and beneficial for the manufacturing concerns but, today large stocks are viewed with alarm and referred to as "graveyard" of a business and the need for control is felt. In any manufacturing concern the cost of material constitute 40% to 60% of total cost of finished product and 26% to 30% of material cost are supposed as inventory carrying cost. Therefore, large inventory can eat the profits. Thus there is more pressure for maintaining liquidity in trading and industry. But on the other hand, holding of inventories is needed in order to avoid keeping men and machines idle for the want of materials. Also, the marketing demands a wide variety and availability at any time. In this complex situation, the inventory management is really a challenging job. Moreover, inventory structure varies very widely from industry to industry, depending on nature and type of product, production technology, collaborating country, nature and extent of the imported materials, production cycle etc.

7.1 *Classification of Inventory*

Different authors have classified the inventory in different ways. These are as follows :

1. **Based on Specific Purpose of Holding :**

   (a) *Lot Size Inventory :* These are held for two reasons
      1. The large discounts on large quantities or lots.
      2. The material is available in lots easily rather than exact amounts.

   (b) *Fluctuation Inventory :* These inventories exist because of fluctuation in demand or supply. Buffer or safety stocks are of this kind.

   (c) *Anticipation Inventory :* These are built up in advance for a big selling season, a promotion program or a plant shut down period. *e.g.*, production and stocking of Diwali crackers, coolers, etc.

   (d) *Transportation Inventory :* These, also called transit or pipeline inventories arise due to transportation of inventory items to distribution centres and customers from various production centres. The amounts of transportation inventory depend on the time consumed in transportation and the nature of demand.

   (e) *De-coupling Inventory :* If various products stages operate successively, then the event of breakdown of one or any disturbance at some stage can affect the entire system. This kind of inter dependence is not only costly but also disruptive. To reduce this inter-dependence these inventories are maintained.

2. **Based on the Quantities and Measurement :**

   (a) *Single or Piece Wise :* *e.g.* pencil, chocolates.

   (b) *Unit Wise :* *e.g.* set of (6) tea cups, dinner set etc.

   (c) *By Volume :* *e.g.* oils, gases etc.

   (d) *By Weight :* *e.g.* food grains, grease etc.

3. **The Most Common Way of Classification of Inventories :**

   (a) *Production Inventories :* Items, which are used to make products, raw materials, and bought out finished components (BOF) and bought out semi finished components (BOSF).

   (b) *Maintenance, Repair and Operating (MRO) Inventories :* Items which do not form a part of final product but are consumed indirectly in the production process like spare parts, consumable items, etc.

   (c) *In Process Inventories :* Semi-finished products at various stage of production.

   (d) *Finished Goods Inventories :* completed product ready for dispatch.

   (e) *Miscellaneous Inventories :* which arise out of the above four type of inventories such as scrap, surplus and obsolete items, which are to be disposed off.

7.2 *The Operating Doctrine*

In an inventory control situation, two basic decisions are to be taken which are usually referred to as operating Doctrine of Inventory. These are :

   (i) How much to order?

   (ii) When to order?

To answer these questions one should be aware of the costs involved and other factors those influence these decisions.

7.3 *The Inventory Costs*

Various costs associated with inventory control are often classified as follows.

   (i) *Set Up Cost :* The cost associated with setting up of the machinery before starting the production, and is independent of the quantity ordered for production.

   (ii) *Ordering Cost :* This is a cost associated with ordering of raw material for production. Advertisement, consumption of stationery and postage, telephone charges, rent for space used by purchase department, travelling expenditure, etc., constitute the ordering costs.

   (iii) *Purchase or Production Costs :* This is cost of purchasing or producing a unit of item. This is very important when discounts are allowed.

   (iv) *Carrying or Holding Cost :* The costs associated with the storage costs like rent, interest on the money locked up, insurance of stored equipment, production, taxes, depreciation of equipment etc.

   (v) *Shortage or Stock Out Cost :* The penalty cost for running out of stock (i.e., when item cannot be supplied on the customer's demand). This cost includes the loss of potential profit through sales of items and loss of goodwill, in terms of permanent loss of customers and its associated lost profit in future sales.

   (vi) *Salvage Cost or Selling Price :* When the demand for certain commodity is affected by quality stocked, decision problem based on a profit maximization criterion includes the revenue from selling.

7.4 *Terminology Used in Inventory*

1. **Demand :** Demand is the number of units required per period and may be either known exactly (or deterministic) or known in terms of probabilities (probabilistic or stochastic) or completely unknown.

2. **Lead Time :** The time gap between placing of an order and its actual arrival in the inventory is termed as lead-time. The longer the lead-time, the higher is the average inventory. Lead time has two components, namely the Administrative or internal lead time (ILT) i.e., from initiation of procurement action to placing of the order, and the delivery or external lead time (ELT) from placing the order to deleviry of the material.

3. **Order Cycle :** The time period between placement of two successive orders is referred to as a reorder cycle. The reorder may be placed on the basis of two types of inventory review systems, *P*-system & *Q*-system.

4. **Stock Replenishment :** Replenishment of the stock may occur instantaneously (when purchased from outside) or uniformly (when the product is manufactured in-house).

5. **Time Horizon :** The time period over which the inventory level will be controlled is the horizon. This may be finite or infinite depending upon the nature of the demand of the commodity.

6. **Re-order Level :** The level between maximum and minimum stock, at which purchasing or manufacturing activities must start for replenishment is called the reorder level.

7. **Re-order Quantity :** This is the quantity of replenishment order. In certain cases it is known as Economic Order Quantity (EOQ).

7.5 *Inventory : Significance (Advantages of Inventory)*

Inventory is an essential requirement of every facet of business life. There is hardly any business activity that can be performed without inventory whether it is a manufacturing concern of good or trading of goods or service organisation like hospital, banks etc. Even in organisation that processes informations also, the data is supposed as the inventory and is stored for future references. In whatever form the inventory is stored, usually, we find the stock points mainly at three stages. These are at (i) input (raw material) stage (ii) conversion (work-in-process) stage and (iii) output (finished goods) stage.

In view of the uncertain situations in market, it is physically impossible and economically impractical to arrive a decision on where, when and how much quantity of each item is needed. Even if it is possible in a few occassions, it is prohibitively expensive. This is the fundamental reason in stocking and maintaining the inventory levels of each item. Thus inventories play a significant role in any organisation. Some more points to add in the list showing how inventory is advantageous and significant in an organisation are given below :

1. Inventory ensures continuous and smooth flow of production by providing right amount of stocks.

2. It avoids physical impossibility and economical impracticability of getting right amounts of stock at exact time of need.

3. Inventories enable the organisations to purcahse their materials at advantageous prices or economic times.

4. Inventory can become a reason for reasonable customer service through supply without delay. This obviously enhances customer satisfaction and thence the reputation of the organisation.

5. Carrying inventory makes it possible to maintain more stable operating or working level.

6. Selection of inventory quantity levels can provide the economy in shipping and transportation.

7. To plan overall operation strategy through decoupling of successive stages in the chain of manufacturing or transportation, inventories play a vital role to minimise the shortage cost and other related costs.

8. To facilitate economic production runs.

9. To facilitate intermittent production of several products on the same facility.

10. Inventories help in achieving favourable Return On Investment (ROI).

11. The decisions regarding effective use of capital is dependent on inventory levels.

12. The efficient usage of available storage space is based on inventory decisions.

13. Make or buy decisions can be effective and economic by proper inventory policies.

14. Inventory provides means against hedging against future price and delivery uncertainties.

7.6 *Objectives of Inventory*

The main objective of inventory control is to minimise the overall investment (or costs) or inventory carrying at lowest possible level and consistency in operating requirements. However, the following may also be considered as sub-objectives of inventory control.

- To minimise carrying cost of inventory.
- To supply the finished product/raw material/ sub-assemblies/semi-finished goods etc., to the users as per their requirements at right time and at right price.
- To minimise the inactive, surplus, waste, scrap, obsolete, spoilage materials.
- To reduce the shortage costs.
- To minimise replacement costs.
- To maximise production efficiency and distribution effectiveness.
- To maximise the reutrn on investment (ROI).

7.7 *Functions of Invenotry*

The basic function of inventory is to increase profitability through minimising the costs of procurement, storage etc.

Since zero-inventory is impractical, the stocking of materials is inevitable and has a specific objective to fulfill an intended function.

The other functions of inventory are studied under major heads as investment alternative, geographical specialisation, decoupling, balance between supply and demand and providing safety stock.

1. **Investment Alternative :** A major area of asset deployment in an industry is inventory, since inventory can provide a minimum Return on Investment (ROI). According to the concept of Marginal Efficiency of Capital (MEC), a firm should invest in those alternatives, which can provide a greater ROI than capital cost borrowed. Inventory is one of such alternatives to choose.

   In simple words, an investment on the materials can usually be realised to the extent of investments. And if there is a favourable market or value addition to the materials, it can be sold at higher price and thus can raise the profitability. Thus inventory functions as safe guarding investment and also can raise the ROI.

2. **Geographical Specialsation or Territorial Importance :** Here, the function of inventory is to permit geographical specialisation of individual operating units. Owing to various factors influencing the production and producivity such as power, water, availability of raw materials / man power, facilities, markets, means of transportation etc., it is often found economical to locate production centres considerably away from warehouses. Thus it may become necessary to collect all the products from each production centre at one place and distribute to market centres according to demand and shipping costs. Thus inventory functions to act as mediator between the production and market, on the basis of geographical specialisation.

3. **Decoupling :** This function of inventory is to enhance the efficiency of operations with in a single facility. Suppose, machines A, B and C (say milling, drilling and grinding) have to run in the order $A \rightarrow B \rightarrow C$ in a mass production shop. On any day if machine B is under breakdown, this automatically stops the machine C, though it is in very good working condition. To avoid this, an inventory can be maintained in a substore (containing the material already processed by machine B) between A and C, that can ensure smooth production flow by detaching B and coupling with the substore.

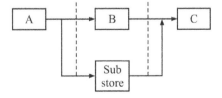

FIGURE 7.1 : DECOUPLING INVENTORY

Thus inventory acts as a machine by decoupling in the production chain. This decoupling also provides the organisaion units to plan and schedule their activities independently.

4. **Balance Between Supply and Demand :** Every one is aware of the stocking of inventory whose supply is seasonal while demand is continuous through out the year, (e.g., rice, wheat etc)., or demand is seasonal but supply is continuous (e.g., crackers, rain coats etc). In all such cases, the inventory only can function as to bring a balance between supply and demand.

5. **Safety Stock :** Due to short range variation in either demand or replacement etc., a safety stock or buffer is kept to function a smooth and uniform flow of production. Safety stock provides protection to the organisation in two types of uncertainties. First is when the sales (or consumption) is excess of forecast and second is due to delay in replenishment. Thus inventory acts as a protective shield from uncertainty of demand.

7.8 *Factors Affecting Inventory*

1. Economic parameters such as purchase price, production cost, selling price, procurement cost, carrying cost, shortage cost, operating and information processing cost etc.

2. Demand.

3. Ordering cycle and its review. (Continuous / periodic)

4. Delivery lead / lag time.

5. Time horizon.

6. Number of supply echelons.

7. Number of stages of inventory.

8. Number of items.

9. Availability and conditions.

10. Government's/Company's policy.

7.9 *Concept of Economic Order Quantity (EOQ)*

It is felt that materials are dead investments unless they are realised into money. But till then, they block the money and eat away the profits. Therefore materials managers will be asked to maintain just sufficient stocks. We know well that the higher stocks block the money while under stocks will interrupt production and makes the machines/men idle and hence affects the reputation of the industry for not meeting the demand or not fulfilling the promises. Thus there is a need to arrive a quantity to be stocked that should neither be high nor low.

Thus the operating doctrine of inventory (i.e., when and how much to order or stock) depends on how much it costs to the organisation. Thus concept of optimum order quantity (OOQ) or Economic Order Quantity (EOQ) has gained the prominent and significant role in inventory management.

The concept of EOQ was first proposed by Ford Wilson Harris in 1913. He has attempted to bring a trade off between the order costs and holding and shortage costs. Suppose you are buying some pens for which you have to spend Rs. 100/- to order. This will be more or less same whether you order 10 or 100. But when you buy 10 the unit cost   is rupees ten and when you buy 100 it is one rupee each. Thus the unit orde cost decreases on increase of quantity (but never becomes zero).

In case of carrying cost it varies linearly and the cost on storing increases with increase of quantity. Graphically, it can be observed that the ordering cost er unit decreases hyperbolically with the increase quantity while the unit carrying cost increases linearly with the increase of quantity. Thus a trade off between these two can be obtained to find a most Economic Order Quantity (EOQ) often denoted by Q* shows most minimum on total cost graph and the normal drawn at EOQ passes through the point of intersection of order cost curve and carrying cost curve indicating annual ordering cost is equal to annual carrying cost (i.e., AOC = ACC) at EOQ.

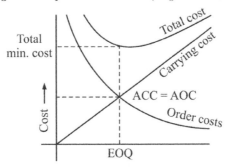

Costs of simple EOQ model

FIGURE 7.2 :

Limitations of EOQ :

1. Ordering to the nearest quantities or packing. Say, instead of ordering 11 dozens, the order may be one gross. This may not be economic.

2. Modifying an order to get a better freight rate. The saving in freight may be more than compensation the extra holding cost.

3. Simplification of routine becomes difficult : Instead of 13 times a year, one may order once a month (once in every four weeks gives 13, monthly once gives 12 order)

4. In case of perishable, or bulky items with diminishing consumption or for items whose market, prices are likely to decline, it may be better to order less than theoritical order quantity.

5. Seasonal supply factors, market conditions, availability of transport etc., may indicate larger or smaller purchase quantities. In these situations, judgement should be given more weight.

6. Liberal discounts or concessional freight rates may suggest larger quantities. The pros and cons of such purchase should be weighted carefully before taking the decision.

7.10 *Inventory Models*

There are different classifications of inventory models to obtain EOQ. The models differ based on the variations in nature of demand, allowing or not the shortages, production or replenishment instantaneously or continuously etc. However, the models considered in this book are classified as follows :

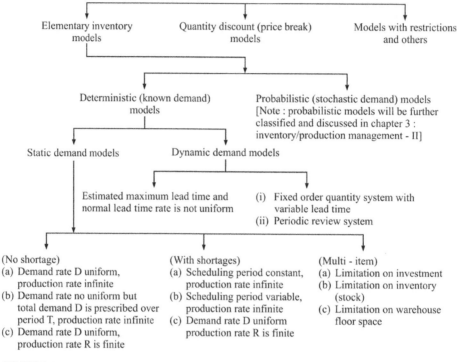

FIGURE 7.3 :

7.11 *Model -1 : Wilson Harris Simple EOQ*

Assumption In Deterministic Model :

- Demand is known exactly for a given period, usually one year
- Demand is uniform and constant over a period of time.
- No limitations are imposed by storing and clerical capacity.
- The order cost and storage cost is independent of order size.
- Orders are received instantaneously.
- Order and delivery quantities are equal.
- Buffer stock of finished product is independent of order size.
- Price of raw material is stable.
- Order costs per order are same irrespective of order size.

- Set up cost is constant.
- Rate of production is known.

Derivation of EOQ for Constant Demand :

Let C_o = ordering cost per order

C_c = carrying cost per unit per year

D = annual demand or requirement per year

p = price of item

Q = ordered quantity per order.

Now,

1. Annual Order Costs (AOC) :

Total no. of orders in the year $= \dfrac{\text{Total demand in the year}}{\text{Quantity ordered per order}} = \dfrac{D}{Q}$

∴ Total annual orders cost (AOC) = order cost per order × No. of orders per year

$$= C_o \times \dfrac{D}{Q}$$

2. Annual Carrying Costs (ACC) :

Average quantity stocked $= \dfrac{\text{Max. quantity} - \text{min. quantity}}{2} = \dfrac{Q - 0}{2} = \dfrac{Q}{2}$

Annual carrying costs = carrying cost per unit per period

$\times$ average quantity stocked $= C_c \times \dfrac{Q}{2}$

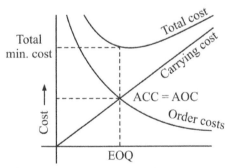

FIGURE 7.4 :   ORDER COSTS AND CARRING COSTS AT EOQ

3. Annual Purchase Cost (APC) :

Price item × demand $= p.D.$

Annual Total Cost $(ATC) = (AOC + ACC) + APC$

[$AOC + ACC$ are total variable costs while APC is total fixed costs].

$$ATC = \dfrac{C_o D}{Q} + C_c \cdot \dfrac{Q}{2} + pD$$

For, Q to be optimum, (Q_o), $\dfrac{d\,A\,T\,C}{dQ} = 0$

$$\therefore \quad -\frac{C_o\,D}{Q_o{}^2} + \frac{C_c}{2} = 0$$

$$Q_o = \sqrt{\frac{2\,C_o\,D}{C_c}}$$

Also $\qquad TVC = \sqrt{2\,C_o\,C_c\,D}$

Aliter 1 : If students do not know differentiation, they may follow the following method instead of differentiation.

After getting the expressions for AOC & ACC.

As we know that $AOC = ACC$ at EOQ,

We have $\qquad C_o \cdot \dfrac{D}{Q_o} = C_c \cdot \dfrac{Q_o}{2}$ at $Q \rightarrow Q_o$

$$\therefore \quad Q_o{}^2 = \frac{2 \cdot C_o\,D}{C_c}$$

$$\therefore \quad Q_o = \sqrt{\frac{2\,C_o\,D}{C_c}}$$

Aliter 2 : We can also get the expression by goemetric method.

AOC = No. of orders $\times$ unit cost.

$\qquad = \dfrac{\text{Annual demand}}{\text{Quantity per order}} \times$ unit order cost

$\qquad = \dfrac{D}{Q} \times C_o$

ACC = sum of areas of all triangles (i.e., inventory cycles) $\times$ carrying cost of each cycle

$$= \left\{ \frac{1}{2} \cdot Q \cdot t + \frac{1}{2} Q \cdot t + \ldots + n \text{ times} \right\} \times C_c$$

$$= \frac{1}{2} Q \left[t + t + \ldots + n \text{ cycles} \right] \cdot C_c$$

$$= \frac{1}{2} Q \left[1 \text{ year} \right] \cdot C_c = \frac{1}{2} Q \cdot C_c$$

And at EOQ $\qquad$ i.e., if $Q \rightarrow Q_o$

$\qquad AOC = ACC$

$$\therefore \quad \frac{D}{Q_o} \cdot C_o = \frac{1}{2} Q_o \cdot C_c \Rightarrow Q_o = \sqrt{\frac{2\,C_o\,D}{C_c}}$$

note : *Carrying cost will be usually given in terms of percentage of n interest on unit price of the item. In such case $C_c = i =$ interest on average inventory and $p =$ price of the item. Similarly, if the holding cost (or rent or maintenance cost etc) say H, is given along with interest (i) and price (p) then $C_c = H + ip$ is considered.*

ILLUSTRATION 1 ——————————————————————————

> *A Company uses 10000 units per year of an item. The purchase price is Rs. 1 per item. Ordering cost is Rs. 25 per order. Carrying cost per year is 12% of the inventory value. Find*
> *(i)   The EOQ*
> *(ii)   The number of orders per year.*
> *(iii) If the lead-time is 4 weeks and assuming 50 working weeks per year, find the reorder point.*
>
> [JNTU (Mech.) 98/5]

Solution :

$$\text{Annual Demand } (D) \qquad\qquad = 10{,}000 \text{ units}$$

$$\text{Order cost per order } (C_o) \qquad = \text{Rs. } 25$$

$$\text{Interest on Carrying cost per year } (i) \quad = 12\,\%$$

$$\text{Cost of item } (p) \qquad\qquad\qquad = \text{Re. } 1/\text{-}$$

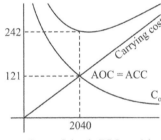

Costs of simple EOQ model

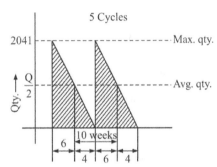

FIGURE 7.5 :

(i) $EOQ = \sqrt{\dfrac{2 \cdot C_o \cdot D}{i \cdot p}} = \sqrt{\dfrac{2 \times 25 \times 10000}{0.12 \times 1}}$

$\qquad = \sqrt{416666.6667} = 2041.24 \cong 2000 \text{ units.}$

(ii)   The number of orders $= \dfrac{D}{Q} = \dfrac{10{,}000}{2041.24} = 4.899$

$\qquad$ (or) $\dfrac{10000}{2000} = 5 \text{ orders} \cong 5 \text{ orders}$

(iii) Re order period $= \dfrac{\text{No. of working weeks/yr}}{\text{No. of orders/yr}} = \dfrac{50}{5} = 10 \text{ weeks}$

$\qquad$ Given Lead time $= 4$ weeks

$\qquad \therefore$ Re order point is at the end of 6th week of every cycle.

ILLUSTRATION 2

> *A certain item costs Rs. 235 per ton. The monthly requirement is 5 tons and each time the stock is replenished, there is a set-up cost of Rs. 1,000. The costs of carrying inventory has been estimated at 25% of the value of the stock per year. What is the optimum order quantity.* [JNTU (Mech.) 99]

Solution :

$$\text{Price } (p) = \text{Rs. 235 per ton}$$
$$\text{Annual Demand } (D) = 5 \times 12 = 60 \text{ tonnes}$$
$$\text{Order Cost/order } (C_o) = 1,000 \text{ per order}$$
$$\text{Carrying cost } (C_c) = \frac{10}{100} \times 60 \times 235$$
$$= 1410$$
$$OOQ = \sqrt{\frac{2 C_o D}{C_c}}$$
$$= \sqrt{\frac{2 \times 1000 \times 60}{1410}} = 9.2253 \text{ tonnes.}$$

FIGURE 7.6 :

ILLUSTRATION 3

> *A company uses 24000 units of a raw material which costs Rs.12.50 per unit. Placing each order costs Rs. 22.50 and the carrying cost is 5.4% per year of the average inventory. Find the economic order quantity and the total inventory cost (Including the cost of the material).* [JNTU (Mech.) 99/CC (ECE) 99/CCC]

Solution :

Given data :

Annual demand (D) = 24,000 units
cost per unit (p) = Rs. 12.5 per unit
order cost (C_o) = Rs. 22.5 per order
Inventory carrying interest
$$(i) = 5.4 \% \text{ per year of avg. inv.}$$
$$= \frac{5.4}{100} = 0.054$$

Inventory carrying cost $(C_c) = i.p = 12.5 \times 0.054$

Economic Order Quantity

$$EOQ = \sqrt{\frac{2.C_o D}{C_c}} = \sqrt{\frac{2 \times C_o \times D}{i \times p}} = \sqrt{\frac{2 \times 2.25 \times 24000}{0.054 \times 12.50}}$$
$$= \sqrt{1600000} = 1264.911 \cong 1265 \text{ units}$$

Total variable cost $= \sqrt{2 . C_o . C_c . D}$

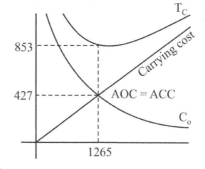

FIGURE 7.7 :

(i.e carrying costs+ordercosts) $= \sqrt{2 \times 22.5 \times 0.054 \times 12.50 \times 24000}$

$$= \sqrt{729000} = 853.82$$

Total cost including material cost

$$= \text{Total variable cost} + \text{fixed cost (i.e., material cost)}$$
$$= 853.82 + 12.5 \times 24000$$
$$= 853.82 + 300000$$
$$= \text{Rs. } 3000853.82 \text{ Ps.}$$

ILLUSTRATION 4

The production department for a company requires 3600 kg of raw materials for manufacturing a particular item per year. It has been estimated that the cost of placing an order is Rs. 36 and the cost of carrying inventory is 23% of investment on the inventories. The price is Rs. 10 per kg. The purchase manager wishes to determine the operating doctrine for raw materials.

Solution :

From the above data, we have (with usual notatios) C_0 = Rs. 36 per order D = 3600 kg/yr , C_c = 25% of investment i.e. C_c = Rs 10 × 0.25 = 2.5 per kg/yr

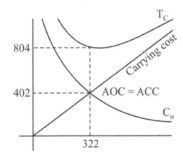

(a)

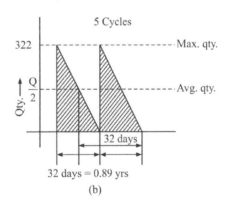

(b)

I FIGURE 7.8 :

(i) The optimal lot size $Q^* = \sqrt{\dfrac{2 C_0 D}{C_c}} = \sqrt{\dfrac{2 \times 3600 \times 36}{2.5}} \simeq 322$ kg/order

(ii) The optimal order cycle time $t^* = \dfrac{Q^*}{D} = \dfrac{322}{3600} = 0.89$ yr

Assuming 360 days in a year, t^* = 32 days

(iii) The total min. variable cost $TVC^* = \sqrt{2 C_0 C_c D} = \sqrt{2 \times 3600 \times 36 \times 2.5}$

$$= \text{Rs. } 804.98 \text{ per year.}$$

(iv) Total annual inventory cost = $TVC + TFC$

$$= 804.98 + (3600 \ kg) \ (\text{Rs. } 10/kg) = \text{Rs. } 36804.98 \text{ per yr.}$$
$$\simeq \text{Rs. } 36805 \text{ per yr.}$$

7.12 *Model - 2 : Continuous and Non-uniform Demand (Consumption)*

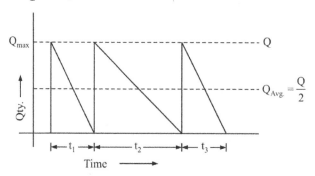

I FIGURE 7.9 :   CONTINUOUS AND NON-UNIFORM CONSUMPTION MODEL

In the above model with usual notations,

We can obtain $AOC = \dfrac{C_o \cdot D}{Q}$ (Same as in model 1)

and $ACC = C_c\left[\dfrac{1}{2}Q \times t_1 + \dfrac{1}{2}Q\,t_2 + \ldots + \dfrac{1}{2}Q \cdot t_n\right]$

$= \dfrac{1}{2} \cdot Q \cdot C_c \cdot \left[t_1 + t_2 + \ldots + t_n\right] = \dfrac{1}{2}Q \cdot C_c \,[1 \text{ year}] = \dfrac{1}{2}Q \cdot C_c$

And at *EOQ*, i.e., if $Q = Q_{opt}$

Then $AOC = ACC$

$\therefore \qquad \dfrac{C_o D}{Q_o} = \dfrac{C_c \cdot Q_o}{2}$

$\therefore \qquad Q_o = \sqrt{\dfrac{2\,C_o \cdot D}{C_c}}$

[Thus model-1 and mode-2 are one and same].

Practice Problems

1. An aircraft company uses rivets at an approximate consumption rate of 2,500 *kg* per year. The rivets cost Rs. 30 per *kg* and the company personnel estimates that it costs Rs. 130 to place an order and the inventory carrying cost is 10% per year. How frequently should orders for rivets be placed and what quantities should be ordered?   **[Maduri B.E. (Mech) 1980, JNTU (Mech.) 94/CCC]**

Answer : $Q^* = 466\,kg$ rivets ; $T^* = 0.18$ years

2. A manufacturer has to supply his customer with 600 units of his product per year. Shortages are not allowed and the storage cost amounts to Rs. 0.60 per unit per year. The set-up cost per run is Rs. 80. Find the optimum run size and the minimum average yearly cost. **[Marathwada, M.Sc. (Math), 1982, JNTU (CSE) 95/S**

Answer : $Q^* = 400$ units ; $t^* = 8$ months $TVC^* = $ Rs. 240

3. A company works 50 weeks in a year. For a certain part, included in the assembly of several parts, there is an annual demand of 10,000 units.

This part may be obtained from either an outside supplier or a subsidiary company. The following data relating to the part are given :

Description	From Outside Supplier Rs.	From Subsidiary Company Rs.
Purchase price/unit	12	13
Cost of placing an order	10	10
Cost of receiving an order	20	15
Storage and all carrying cost, including capital cost per unit per annum	2	2

(i) What purchase quantity and from which source would you recommend?

(ii) What would be the minimum total cost **[ICWA, June 1989]**

Answer : (i) $Q^* = 547.72$ units $TVC^* =$ Rs. 1,21,095.45

(ii) $Q^* = 500$ units ; $TVC^* =$ Rs. 1,31,000 Purchase must be made from an outside supplier.

4. A manufacturing company uses certain part at a constant rate of 4,000 units per year. Each unit costs Rs. 2/- and the company personnel estimates that it costs Rs. 50 to place an order, the carrying costs of inventory is estimated to be 20% per year, find the optimum size of each order and minimum yearly costs. **[Bharathidasan M.Sc. (Math) 1981, OU (Mech.) 88]**

Answer : $Q^* = 1000$ units, TVC = Rs. 400

5. Robert has to supply his customers 100 units of the product 'X' only every Friday and only then he obtains 'X' from Garry, his local supplier, at Rs. 60 per unit. The cost of ordering and transportation from Garry is Rs. 150 per order. The cost of carrying inventory is estimated at 15% per year of the cost of the product carried.

(i) Describe graphically the inventory system

(ii) Find the lot size which will minimize the cost of the system.

(iii) Determine the optimal cost. **[JNTU (Mech.) 93/MS]**

Answer : (ii) $Q^* = 416$ units; (iii) $TC^* = Rs. 6072$

6. Sainath keeps his inventory in special containers. Each container occupies 10 Sq. ft. of store space. Only 5000 sq. ft of the 9000 containers, priced Rs. 8 per container. The ordering cost is estimated at Rs. 40 per order, and the annual carrying costs amount to 25% of the inventory value.

Would you recommend Sainath to increase his storage space? If so, how much should be increased? **[Delhi Univ., M.Com. 1990, JNTU (EEE & ECE) 95/MS]**

Answer : Increase in storage space $= 1000$ sq. ft.

7. Find the economic lot size, the associated total cost and the length of time between two orders, given that the set-up cost is Rs. 100, daily holding cost per unit of inventory is 5 paise and daily demand is approximately 30 units.

[JNTU (CSE) 92]

Answer : $Q^* = 346$ units, $TVC^* = 173$; $t = 11.5$ days

8. Shivnath company buys in lots of 2000 units which is only 3 months supply. The cost per unit is Rs. 125 and the order cost is Rs. 250. The inventory carrying cost is Rs. 20% of unit value. How much money can be saved by using economic order quantity?   [JNTU (Mech. & ECE) 95/CCC]

Answer : Rs. 16000/-

9. A ship-building firm uses rivets at a constant rate of 20,000 numbers per year. Ordering costs are Rs. 30 per year. Each rivet costs Rs. 1.50 and the holding cost is estimated to be 12.5% of unit cost per unit per year. Determine EOQ.

[JNTU (Mech.) 93/MS]

Answer : $Q^* = 1005$ rivets

10. A computer manufacturing company wishes to know what EOQ should be ordered for its chips. The average daily requirement is 120 units and the company has 250 working days a year. The manufacturing cost is 50 paise per part. The sum of annual rate of interest, insurance, taxes and so forth is 20% of the unit cost and the cost of preparation is Rs. 50 per lo[JNTU (CSE) 94/S/MS]

Answer : $Q^* = 5477$ units

11. Consider the inventory system with the following data in usual notations:

$D = 1000$ units/year ; $i = 0.30$; $p =$ Rs. 0.50/unit; $C_0 =$ Rs. 10.00; $L = 2$ years (lead time).

(i) Optimal order quantity

(ii) Reorder point;

(iii) Minimum average cost.   [JNTU - CSE/ECE 2001]

7.13 *Model 3 : With Continuous Production (Supply) Rate*

This model is similar to that of model-1 except that of instantaneous replenishment. Instead of instaneous replenishment, this model assumes to replenish continuously till the maximum quantity is reached due to the fact that in many situations, the amounts ordered will not be received all at once but will be available at finite supply rate. This order may be produced internally also.

Thus in addition to the usual assumption the following points are assumed in this model.

1. Supply is continuous till Q_{max} is reached and then it stops.
2. The production rate or supply rate (say r) is greater than demand or consumption rate (d) i.e., $r > d$.
3. Production begins immediately after production set-up.
4. During production also consumption will be there and when Q_{max} reaches, production stops while consumption continues till Q_{min} is reached.

Graphically this can be shown as follows :

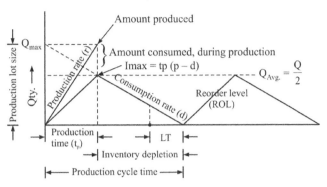

FIGURE 7.10 : CONTINUOUS SUPPLY RATE MODEL

In the above model, if t_p = production time and r = rate of production.

We have the quantity produced

$$Q = r \cdot t_p \implies t_p = \frac{Q}{r}$$

During the production time, the inventory increases at the rate of r, but due to consumption simultaneouly this inventory decreases at the rate of d.

Therefore the inventory accumulates at the rate of $(r - d)$.

Thus inventory accumulation

$$I_{\max} = \text{rate of accumulation} \times \text{production time}.$$

$$\therefore \qquad I_{\max} = (r - d)\, t_p = (r - d)\frac{Q}{r} = \left(1 - \frac{d}{r}\right) Q$$

and we have $I_{\min} = 0$.

[Since we consume till all the inventory is exhausted before starting new production cycle].

Thus average inventory $= \frac{1}{2}\left(I_{\max} + I_{\min}\right) = \frac{Q}{2}\left(1 - \frac{d}{r}\right)$

Now, annual carrying cost $(ACC) = C_c$. averrage quantity

$$= C_c \cdot \frac{Q}{2}\left(1 - \frac{d}{r}\right)$$

And we know that

Annual ordering costs $(AOC) = C_o \cdot \dfrac{D}{Q}$

$\therefore$ Total variables costs (TVC) $= C_o \dfrac{D}{Q} + C_c \cdot \dfrac{Q}{2} \left(1 - \dfrac{d}{r}\right)$

$$Q \to Q_{min} \text{ at } \dfrac{d\,(TVC)}{dQ} = 0 = -C_o \dfrac{D}{Q_o^2} + \dfrac{C_c}{2}\left(1 - \dfrac{d}{r}\right) = 0$$

$$Q_o = \sqrt{\dfrac{2\,C_o\,D}{C_c\left(1 - \dfrac{d}{r}\right)}}$$

Or otherwise, at $EOQ,\, AOC = ACC$

$\therefore$ $\dfrac{C_o\,D}{Q_o} = C_c \cdot \dfrac{Q_o}{2}\left(1 - \dfrac{d}{r}\right)$

$$Q_o = \sqrt{\dfrac{2\,C_o\,D}{C_c \cdot \left(1 - \dfrac{d}{r}\right)}} = \sqrt{\dfrac{2 \cdot C_o\,D}{C_c}\left(\dfrac{r}{r-d}\right)}$$

Other Important Formulae :

1. Total variable cost

$$TVC^* = \dfrac{D}{Q_{opt}}\,C_o + \dfrac{Q_o}{2}\left(1 - \dfrac{d}{r}\right) \cdot C_c$$

or $TVC = \sqrt{2\,C_o\,D\,C_c\left(1 - \dfrac{d}{r}\right)} = \sqrt{2\,C_o\,D \cdot C_c\left(\dfrac{r-d}{r}\right)}$

2. Optimal length of production run

$$\text{Opt.}\ t_p^* = \dfrac{Q_{opt}}{r} = \dfrac{1}{r}\sqrt{\dfrac{2 \cdot C_o\,D}{C_c}\left(\dfrac{r}{r-d}\right)} = \sqrt{\dfrac{2\,C_o\,D}{C_c \cdot r\,(r-d)}}$$

3. Optimal production cycle time

$$t^* = \dfrac{Q_{opt}}{D} = \dfrac{1}{D}\sqrt{\dfrac{2\,C_o\,D}{C_c}\left(\dfrac{r}{r-d}\right)} = \sqrt{\dfrac{2\,C_o}{C_c\,D}\left(\dfrac{r}{r-d}\right)}$$

4. Optimal number of production runs

$$N^* = \dfrac{D}{Q_{opt}} = \sqrt{\dfrac{C_c\,D}{2\,C_0}\left(\dfrac{r-d}{r}\right)}$$

Remarks :

In all the above formulae, there may be a confusion between D and d. D is annual demand or annual consumption while d is consumption rate. If there is only one cycle per year, the formula can be written as $Q_{opt} = \sqrt{\dfrac{2\,C_o\,D}{C_c}\left(\dfrac{R}{R-D}\right)}$.

Note : *It is more appropriate to use the term EBQ (Economic Batch Quantity) instead of EOQ (Economic Order Quantity) for Q_{opt} and 'set-up cost' is more appropriate term in the place of 'order cost' (C_0). Therefore these words are often found in the problems of this model.*

ILLUSTRATION 5 ───

> *A contractor has to supply 10,000 bearings per day to an automobile manufacturer. He finds that when he starts production run, he can produce 25,000 bearings per day. The cost of holding a bearing in stock for a year is Rs. 2 and the setup cost of a production run is Rs. 180. How frequently should production run be made? Also find production run time and total variable cost. (Assume 300 days in the year).*   [JNTU (Mech.) 97/p]

Solution :

According to the problem, with usual notations,

Set-up cost (C_o) = Rs. 180 per production run.

Carrying cost (C_c) = Rs. 2 per bearing per year.

Rate of production (r) = 25, 000 bearings per day.

Rate of consumption (d) = 10,000 bearings per day.

Annual demand   = $10,000 \times 300 = 30,00,000$ / year
(Assume 300 days in a year)

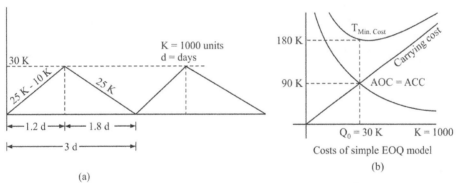

(a)

FIGURE 7.11 :

Now, knowing all the terms, we can find EBQ (Economic Batch Quantity)

$$EBQ = Q_{opt} = \sqrt{\frac{2\,C_o\,D}{C_c} \cdot \left(\frac{r}{r-d}\right)}$$

$$= \sqrt{\frac{2 \times 30,00,000 \times 180}{2} \frac{25000}{(25000 - 10000)}}$$

$$= \sqrt{54 \times 10^7 \times \frac{25}{15}} = \sqrt{90 \times 10^7} = 30000 \text{ bearings.}$$

Frequency of production run is given by $t*$.

$$t* = \frac{Q_{opt}}{D} = \frac{30000}{10000} = 3 \text{ days}$$

Production cycle time

$$t_p^* = \frac{Q_{opt}}{r} = \frac{30000}{25000} = \frac{6}{5} = 1\frac{1}{5} \text{ day.}$$

Total variable cost $= C_o \cdot \dfrac{D}{Q_o} + C_c \cdot \dfrac{Q_o}{2} \cdot \left(\dfrac{r-d}{r}\right)$

or $\sqrt{2\,C_o\,D\,C_c\left(\dfrac{r-d}{r}\right)}$

$TVC = \sqrt{2 \times 180 \times 3000000 \times 2 \times \dfrac{25000 - 10000}{10000}}$

$= \sqrt{4 \times 54 \times 10^8 \times \dfrac{15000}{10000}}$

$= \sqrt{4 \times 54 \times 10^8 \times \dfrac{15}{10}}$

$= $ Rs. 1, 80,000

ILLUSTRATION 6

The manager of a company manufacturing car parts has entered into a contract of supplying 1000 numbers per day of a particular part to a car manufacturer. He finds that his plant has a capacity of producing 2000 numbers per day of the part. The cost of the part is Rs.50. Cost of holding stock is 12% per annum and set up cost per production run is Rs. 100. What should be run size for each production run and total optimum cost/month? How frequently should production runs be made? Shortage is not permissible. [JNTU (CSE) 97/S]

Solution :

Demand rate $=$ Quantity to be supplied per day (d)$=1000$

Production rate $=$ Quantity produced per day (r) $= 2000$

cost of the part(p) $=$ Rs. 50

Interest on Holding cost (i) $= 12\%$ of annual cost.

Holding cost $(C_c{=}ip) = 0.12 \times 50$

Set up cost $(C_o) = $ Rs. 100 per production run.

No. of working days is assumed as 365 days.

$\therefore$ Demand (D) $= 365 \times 1000$

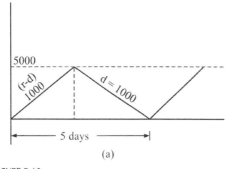

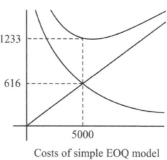

Costs of simple EOQ model

(a)   (b)

FIGURE 7.12 :

$$\text{Optimal Run Size} = Q^* = \sqrt{\frac{2 \cdot D \cdot C_o}{C_c}\left[\frac{r}{r-d}\right]}$$

$$= \sqrt{\frac{2 \times 365000 \times 100}{0.12 \times 50}\left(\frac{2000}{2000-1000}\right)}$$

$$= 4932.9 \cong 5000 \text{ parts.}$$

$$\text{Frequency of production run } (t^*) = \frac{Q^*}{d} = \frac{5000}{1000} = 5 \text{ days.}$$

$$\text{Total variable cost} = \sqrt{2 \cdot C_o \cdot C_c \cdot D\left[1-\frac{d}{r}\right]}$$

$$= \sqrt{2 \times 100 \times 0.12 \times 50 \times 356000\left[1-\frac{1000}{2000}\right]}$$

$$= \text{Rs. } 14798.65 \text{ per year}$$

$$= \text{Rs. } 1233.22 \text{ per month}$$

$$\text{Total fixed cost/month} = \frac{365000 \times 50}{12} = 15,20,833.33$$

$$\therefore \quad \text{Total optimum cost per month} = \text{Rs. } 1522066.55$$

Practice Problems

1. (a) A product 'X' is purchased by a company from outside suppliers. The consumption is 10,000 units per year. The cost of the item is Rs. 5 per unit and the ordering cost is estimated to be Rs. 100 per order. The cost of carrying inventory is 25%. If the consumption rate is uniform, determine the economic purchasing quantity.

   (b) In the above problem assume that company is going to manufacture the item with the equipment that is estimated to produce 100 units per day. The cost of the unit thus produced is Rs. 3.50 per unit. The set-up cost is Rs. 150 per set-up and the inventory carrying charge is 25%. How has the answer changed ? **[JNTU (Mech. & ECE) 92/S/CCC]**

Answer : (a) $Q^* = 1,265$ units (b) $Q^* = 2,391$ units

2. An item is produced at the rate of 50 items per day. The demand occurs at the rate of 25 items per day. If the set-up cost is Rs. 100 per setup and holding cost is 0.01 unit of item per day, find the economic lot size for one run, assuming that the shortages are not permitted. Also find the time of cycle and minimum total costs for one run. **[OU (Mech.) 91/S]**

Answer : $Q^* = 1000$ items ; $T^* = 40$ days; $TVC^* = Rs. 5$ per day.

Total cost per run = Rs. 200

3. An item is produced at the rate of 128 units per day. The annual demand is 6,400 units. The set-up cost for each production run is Rs. 24 and inventory carrying costs is Rs. 3 per unit per year. There are 250 working days for production each year. Develop an inventory policy for this item. **[IGNOU (MBA) 94]**

Answer : $Q^* = 358$ units, $t^* = 14$ working days, $T_1^* = 2.8$ days $TC =$ Rs. 858.65 per yer

4. The annual demand for a product is 1000000 units. The rate of production is 2000000 units per year. The set-up cost per production run is Rs. 5000 and the variable production cost of each item is Rs. 10. The annual holding cost per unit is 20% of its value.Find the optimum production lot size, and the length of the production run. **[IGNOU (MBA, Assignment) 95]**

Answer : $Q^* = 31600$ units ; $t^* = 115$ days

5. You have been given the following information regarding the production lot size of a particular product :

> Annual demand = 5,000 units
>
> Set-up cost = Rs. 100 per set-up
>
> Daily demand = 17 units.
>
> Production rate = 50 units per day
>
> Optimum production lot size = 275 units.

Rising interest rates have caused a 10% increase in the holding costs. Determine the new optimum production lot size for the product.

[IGNOU (MBA, Assignment) 96]

6. Assuming you are reviewing the production lot size decision associated with a production operating where production is 8000 units a year, annual demand is 2000 units, set-up costs is Rs. 300 per production run and holding cost is Rs. 1.60 per unit per year. The current production run is 500 units every 3 months.

Would you recommend a change in the production lot size ? If so, why? How much could be saved by adopting the new production run lot size?

[JNTU (Mech.) 90/S/P]

Answer : 750

7. A contractor has to supply 20,000 units per day. He can produce 30,000 units per day, the cost of holding a unit stock is Rs. 3 per year and the set-up cost per run is Rs. 50. How frequently, and of what size, the production runs be made?

[JNTU (Mech.) 92/S/P]

Answer : $Q^* = 1414$ units; $t^* = 1.68$ hrs

8. Find the most economic batch quantity of a product on a machine if the production rate of the item on the machine is 200 piece/day and the demand is uniform at the rate of 100 pieces/day. The set-up cost is Rs. 200 per batch and the cost of holding one item in inventory is Rs. 0.81 per day. How will be batch quantity vary if the machine production rate was infinite?

[JNTU (CSE) 94/S]

Answer : 157

7.14 *Model - 4 : Deterministic Model With Shortages*

So far we have seen the models with an assumption that no shortages are allowed, and the *EOQ* derivation involved a trade-off between ordering cost and carrying cost. But, there could be situations where shortages may occur or may be allowed to gain some advantages.

Advantages of Allowing Shortages :

The following advantages can be derived by allowing shortages.

1. By allowing shortages, the cycle time increases, and hence the order (or setup) cost can be spread over longer period.

2. If unit value of the inventory is high, carrying costs also will increase. In such cases the shortage cost could be less than the increased or differed carrying cost.

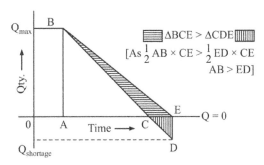

FIGURE 7.13 : MODEL WITH SHORTAGES

In the above graph, if shortage is allowed, the graph takes the form *ABCDE* and if not allowed it takes the form *ABE*. Now, we can observe that area of *BCE* ($\frac{1}{2}$ *AB* × *CE*) i.e., difference in carrying cost is greater than the area covered by Δ *CDE* ($\frac{1}{2}$ *ED* × *CE*), i.e., shortage cost.

[This is true as long as the shortage quantity is safe i.e., *ED < AB*]

Thus to encash the above advantages, managers often choose to have shortages; However, the shortage should not be permitted if it is expected to cause heavy loss of production or damage.

Now, let us derive the expression for EOQ for this model.

Expression for EOQ :

This model is based on all the assumptions of Model - 1 except that inventory can be allowed to run out of stock for a certain period of time. Usually in business situations the following two types of such running out of stock can occur.

1. Sales may be lost due to customers are not likely to buy the inventory items.

2. Customers may leave orders with the supplier and this back order is filled on stock availability.

Thus all the costs such as cost of keeping backlog reorders, loss of good will, cost of shipping to customers etc., depend on how long a customer can wait to receive the order.

In deriving the expression for EOQ of this model, the following terms are used.

t_i = time between receipt of the order and when the inventory level reaches zero.

t_s = Shortage time or stock - out time or back order time.

t = total cycle time = $t_i + t_s$

Q_s = Number of units that are back ordered or shortage quantity.

Q_i = Max. inventory quantity (when received)

Q = Total quantity = $Q_s + Q_i$

C_s = Shortage costs,   and   other terms are as usual.

C_o = Order cost per order

C_c = Carrying cost

i = Interest on average carrying inventory

p = Price of the item

D = Annual demand

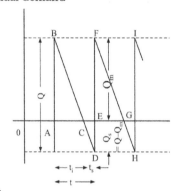

FIGURE 7.14 :   MODEL WITH SHORTAGES

In this model, except for the fixed costs i.e., procurement cost $(D.p)$, all the costs will be affected. Therefore, we need to determine the optimal values of order quantity Q_{opt} (or Q^*), optimal stock level (Q_i^*) and optimal number of shortage units that can be allowed (Q_s^*). And hence the objective function for this model would be, to minimise the total variable cost (TVC), where

TVC = Annual oder cost + Annual carrying cost + Annual shortage cost

= $AOC + ACC + ASC$

Now, we know that,

$$\text{Demand } (D) = \frac{\text{Quantity of units consumed}}{\text{Consumption period}}$$

$$\therefore \quad D = \frac{Q}{t} \quad \text{or } t = \frac{Q}{D}$$

Similarly, $t_i = \dfrac{Q_i}{D}$; $t_s = \dfrac{Q_s}{D} = \dfrac{Q - Q_i}{D}$ and $t = \dfrac{Q}{D}$

Also, from the above graph

ΔABC and ΔCDE are similar

$\therefore \qquad \dfrac{AC}{CE} = \dfrac{AB}{DE}$ $\qquad\qquad \therefore \qquad \dfrac{t_i}{Q_i} = \dfrac{t_s}{Q_s}$

also $\qquad \dfrac{t_i}{Q_i} = \dfrac{t}{Q}$

The average inentory during the period t_i, is the area of $\Delta ABC = \dfrac{1}{2} \cdot Q_i \cdot t_i$

Average carrying inventory $= \dfrac{1}{2} Q_i \cdot \dfrac{Q_i \cdot t}{Q} = \dfrac{Q_i^2}{2Q} \cdot t$

$\therefore \qquad$ Carrying cost $= \dfrac{Q_i^2}{2Q} \cdot t \cdot C_c$
per cycle

Annual carrying cost $= \dfrac{Q_i^2}{2Q} \cdot C_c$
($t = 1$ year)

Similarly, for the period t_s, the average inventory is given by the area of ΔCDE

$\therefore$ Average shortage inventory $= \dfrac{1}{2} \cdot CE \times ED$

$$= \dfrac{1}{2} Q_s \cdot t_s = \dfrac{1}{2} (Q - Q_i) \cdot t_s$$

$$= \dfrac{1}{2} (Q - Q_i) \cdot \dfrac{(Q - Q_i)}{D} \qquad \left[\because t_s = \dfrac{Q - Q_i}{D} \right]$$

$$= \dfrac{(Q - Q_i)^2}{2Q} \cdot t \qquad\qquad \left[\because D = \dfrac{Q}{t} \right]$$

Shortage cost per cycle $= \dfrac{(Q - Q_i)^2}{2Q} \cdot t \cdot C_s$

Annaual shortage cost $= \dfrac{(Q - Q_i)^2}{2Q} \cdot C_s$

($t = 1$ year)

And annual order costs $= \dfrac{C_o \cdot D}{Q}$

$\therefore \qquad TVC = \dfrac{C_o \cdot D}{Q} + \dfrac{Q_i^2}{2Q} \cdot C_c + \dfrac{(Q - Q_i)^2}{2Q} \cdot C_s$

Thus the above relation states that TVC is the function of two variables Q and Q_i. According to the principles of partial derivative, the minimum values of Q and Q_i may be calculated with the conditions that

$$\frac{\partial^2 (TVC)}{\partial Q^2} + \frac{\partial^2 (TVC)}{\partial Q_i^2} - \left\{ \frac{\partial^2 (TVC)}{\partial Q \, \partial Q_i} \right\} > 0$$

and $\quad \dfrac{\partial (TVC)}{\partial Q} = 0 \; ; \; \dfrac{\partial (TVC)}{\partial Q_i} = 0$

On solving the above equations we get, optimum order quantity.

$$EOQ = Q^* = \sqrt{\frac{2 C_o D}{C_c} \cdot \left(\frac{C_c + C_s}{C_s} \right)}$$

and   optimal stock level $= Q_i^* = \sqrt{\dfrac{2 C_o D}{C_c} \cdot \left(\dfrac{C_s}{C_c + C_s} \right)}$

and   optimal shortage level $Q_s^* = Q^* - Q_i^* = Q^* \left(\dfrac{C_c}{C_c + C_s} \right) = \sqrt{\dfrac{2 C_o D}{C_c} \cdot \dfrac{C_c}{C_c + C_s}}$

also $TVC^* = \sqrt{2 C_o D \cdot C_c \cdot \left(\dfrac{C_s}{C_c + C_s} \right)}$

and total cycle time $(t) = \dfrac{Q^*}{D} = \sqrt{\dfrac{2 C_o}{D C_c} \cdot \left(\dfrac{C_c + C_s}{C_s} \right)}$

Total cost $TVC + TFC = TVC + pD$

ILLUSTRATION 7

> *A manufacturer has to supply his customer 30,000 units of product/year. Demand is known and fixed. There is no storage space and shipping is daily. The penalty for failure to supply is Rs. 0.20 per unit per month. Inventory holding cost is Rs.0.1 per month and set up cost is Rs.350/-per production run. Find optimum lot size for the manufacturer.* [JNTU (CSE) 98]

Solution :

$D = 30000$ units/yr ('D' is fixed hence deterministic).

There is a penalty for failure to supply.

Therefore shortage cost (C_s) = Rs. 0.20 per unit per month.

$\qquad\qquad\qquad = 0.2 \times 12 =$ Rs. 2.40 per unit per year.

Holding cost = Rs. 0.1 per unit per month

Holding or carrying cost (C_C)= 0.1 ×12 = 1.2 per unit per year. Order cost or set–up cost (C_0) = Rs. 350 per prodn. run.

$$\therefore EBQ = \sqrt{\frac{2 \cdot C_0 \cdot D}{C_C} \cdot \frac{C_C + C_S}{C_S}}$$

$$= \sqrt{\frac{2 \times 350 \times 30000}{1.2} \times \frac{3.6}{2.4}}$$

$$= \sqrt{\frac{3500 \times 30000}{4}} = \frac{1}{2} \times 1000 \sqrt{105}$$

$$\cong 500 \times 10 = 5000 \text{ or } 500 \times 11 = 5500$$

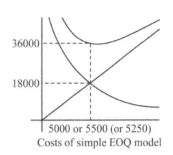

Costs of simple EOQ model

FIGURE 7.15 :

ILLUSTRATION 8

A commodity is to be supplied at a constant rate of 200 units per day. Supplies of any amount can be had at any required time, but each ordering costs Rs.50; costs of holding the commodity in inventory is Rs.2.00 per unit per day while the delay in the supply of the items induces a penalty of Rs.10 per unit per day. Find the optimal policy (Q, t), where 't' is the reorder cycle period and Q is the inventory level after re-order. Also find the optimal inventory level and shortage units. What would be the best policy, if the penalty cost becomes infinity? [JNTU (CSE) 95/S]

Solution :

From the data of problem, with usual notations, we have

D = 200 units/day $\qquad C_o = Rs.$ 50 per order

C_c = Rs. 2 per unit per day $\qquad C_s = Rs.$ 10 per unit per day

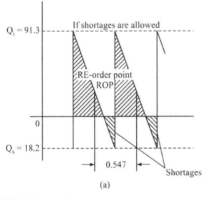

(a)

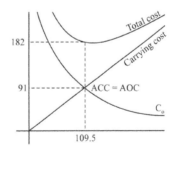

(b)

FIGURE 7.16 :

Now,

(i) Optimal order quantity $\qquad Q^* = \sqrt{\frac{2 C_o D}{C_c} \left(\frac{C_c + C_s}{C_s} \right)}$

$$= \sqrt{\frac{2 \times 50 \times 200}{2} \left(\frac{2 + 10}{10} \right)} = 109.5 \text{ units}$$

(ii) Optimal inventory level

$$Q_i^* = \sqrt{\frac{2\,C_o\,D}{C_c}\left(\frac{C_s}{C_c + C_s}\right)}$$

$$= \sqrt{\frac{2 \times 50 \times 200}{2}\left(\frac{10}{2 + 10}\right)} = 91.3 \text{ units}$$

(iii) Optimal shortage units

$$Q_s^* = Q^* - Q_i^*$$
$$= 109.5 - 91.3 = 18.2$$

(iv) Reorder cycle time

$$t^* = \frac{Q^*}{D} = \frac{109.5}{200} = 0.547 \text{ day}$$

(v) Total veriable cost

$$= \sqrt{2 \cdot C_o \cdot C_c \cdot D \cdot \left(\frac{C_s}{C_c + C_s}\right)}$$

$$= \sqrt{2 \times 50 \times 2 \times 200 \left(\frac{10}{2 + 10}\right)} = \text{Rs.}182.57$$

(vi) If penalty cost $C_s = \infty$, the shortage cost is unaffordable i.e., shortage should not be allowed, thus the expression for Q^* becomes

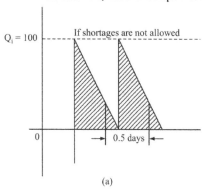

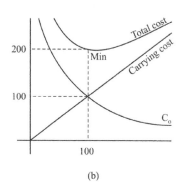

(a)   (b)

FIGURE 7.17 :

$$Q* = \sqrt{\frac{2\,C_o\,D}{C_c}} = \sqrt{\frac{2 \times 50 \times 200}{2}} = 100 \text{ units}$$

and $$t^* = \frac{Q^*}{D} = \frac{100}{200} = \frac{1}{2} \text{ day}$$

Total variable cost $= \sqrt{2 \cdot C_o \cdot C_c \cdot D} = \sqrt{2 \times 50 \times 2 \times 200} = \text{Rs. }200/\text{-}$

Practice Problems

1. The demand of an item is uniform at a rate of 25 units per month. The fixed cost is Rs. 15 each time of production run made. The production cost is Rs. 1 per item, and the inventory carrying cost is Rs. 0.30 per item per month. If the shortage cost is Rs. 1.50 per item per month determine how often to make a production run and of what size is should be?   **[Osmania M.Sc. (Stat), 1984]**

Answer : $Q^* = 54$ items ; $t^* = 2.16$ months

2. Rajiv automobiles has to supply its customers 24,000 units of its product per year. This demand is fixed and known. The customer has no storage space and so the manufacturer has to transport a day's supply each day. If the manufacturer fails to supply, the penalty is Rs. 0.20 per unit per month. The inventory holding cost amounts to Rs. 0.10 per month and the set-up cost is Rs. 350 per production run. Find the optimum lot size of the manufacturer.

[OU (M.Sc.) 1985]

Answer : $Q^* = 4,578$ units per run

3. The cost parameters and other factors for a production inventory system of automobile pistons are given below. Find : (i) Optimal lot size ; (ii) Number of shortages ; (iii) Manufacturing time and time between set-ups.

> Demand per year = 6000 units
> Set-up costs = Rs. 500
> Molding costs per year = Rs. 8
> Unit cost = Rs. 40

Production rate per year = 36,000 units, shortage cost per unit per yr - Rs. 20.

[IGNOU (MBA) 96]

Answer :   (i) $Q^* = 1,123$ units   (ii) $Q_s^* = 266.6$ units

    (iii) $t_1^* = 0.03$ per year; $t^* = 0.19$ year

4. The demand for an item is deterministic and constant over a time and it is equal to 600 units per year. The per unit cost of the item is Rs. 50 while the cost of placing an order is Rs. 5. The inventory carrying cost is 20% of the cost of inventory per annum and the cost of shortage is Re. 1 per unit per month. Find the optimal ordering quantity when stockouts are permitted. If the stockouts are not permitted, what would be the loss to the company.   [ICWA, Dec. 1985]

Answer :   (i) $Q^* = 33$ units ; $R^* = 15$ units ; $TVC^* = $ Rs. 181

    (ii) $Q^* = 24.5$ units ; $TVC^* = $ Rs. 245 loss = Rs 64

5. The demand for a purchased item is 1000 units per month, and shortages are allowed. If the unit cost is Rs. 1.50 per unit, the cost of making one purchase is Rs. 600, the holding cost for one unit is Rs. 2 per year, and the cost one shortage is Rs. 10 per year.
Determine :

  (i) The optimum purchase quantity

  (ii) The number of orders per year

  (iii) The optimal total yearly cost.

  (iv) Represent the model graphically.   **[Bangalore ME 1980]**

Answer : (i) $Q^* = 3600$ units,   (ii) 50 order/yer (iii) $TVC^* = $ Rs. 6000

7.15 *Model 5 : With Finite Replenishment and Shortages Allowed*

This model is a combination model -3 and Model - 4. The graph for this model will be as follows.

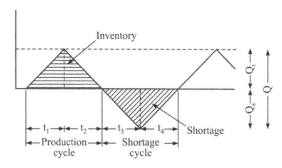

FIGURE 7.18 :   MODEL WITH SHORTAGES AND CONTINUOUS SUPPLY RATE

t_1 = period with production and consumption

t_2 = period of consumption

t_3 = period of consumption with shortage

t_4 = period of production is shortage and consumption

The expressions for various terms can be derived by clubbing the model - 3 & 4 and we get the following formulae .

1. Optimal production lot size $EBQ^* = \sqrt{\dfrac{2 \cdot C_o \cdot D}{C_c} \cdot \left(\dfrac{r}{r-d}\right)\left(\dfrac{C_c + C_s}{C_s}\right)}$

2. Optimal inventory level $Q_i^* = \left(\dfrac{r-d}{r}\right) Q^* - Q_s^*$

$$= \sqrt{\dfrac{2 C_o D}{C_c} \cdot \left(\dfrac{r-d}{r}\right)\left(\dfrac{C_s}{C_c + C_s}\right)}$$

3. Optimal level of shortage $Q_s^* = Q^* \cdot \left(\dfrac{r-d}{r}\right) \cdot \left(\dfrac{C_c}{C_c + C_s}\right)$

$$= \sqrt{\dfrac{2 C_o D}{C_c} \cdot \dfrac{C_c}{C_c + C_s} \cdot \dfrac{r-d}{r}}$$

4. Production cycle time $t^* = \dfrac{Q^*}{D} = \dfrac{1}{D} \sqrt{\dfrac{2 \cdot C_o \cdot D}{C_c} \cdot \left(\dfrac{r}{r-d}\right)\left(\dfrac{C_c + C_s}{C_s}\right)}$

$$= \sqrt{\dfrac{2 C_o}{D \cdot C_c} \cdot \left(\dfrac{r}{r-d}\right)\left(\dfrac{C_c + C_s}{C_s}\right)}$$

5. Total min. variable cost

$$TVC^* = \sqrt{2\,C_o\,D\,C_c\left(\frac{r-d}{r}\right)\left(\frac{C_s}{C_c+C_s}\right)}$$

1. *If $r = \infty$, the model becomes same as that of instant replenishment with shortage i.e., model - 4.*
2. *If $C_s = \infty$, the model becomes same as that of finite replenishment (prodn. rate) without shortage i.e., model - 3.*
3. *If $C_s = \infty$ and $r = \infty$, then the model becomes instaneous replenishment without shortages i.e., model - 1.*

Note : Students are advised to remember the formulae of this model and may deduce to any other models using the points 6 to 8 given above.

ILLUSTRATION 9

An automobile company uses 6000 pistons per year. The company can manufacture the pistons at the rate of 36000 units per year with a set-up cost of Rs. 2000. The cost of holding inventory per year is estimated to be Rs. 8/- per unit cost is Rs. 40. If the company has a provision to allow shortage at the cost of Rs. 20 per unit per year, find (a) Optimal lot size (b) No. of shortages (c) Manufacturing time and (d) time between set-ups. (e) also find total cost (including material cost) [JNTU(Mech. 92/S/P]

Solution :

The given data in the problem, with usual notations is

Demand (D)   = 6000 units/year   (consumption rate)

Production rate (r) = 3600 units/year

Set-up costs (C_o)   = Rs. 2000 per set-up

Carrying cost (C_c) = Rs. 8 per unit per year

Shortage cost (C_s) = Rs. 20 per unit per year

Price (p)   = Rs. 40

(Here $d = D$ since it is given in terms of 'per year')

(a) Economic lot size $(Q^*) = \sqrt{\dfrac{2\,C_o\,D}{C_c}\left(\dfrac{r}{r-D}\right)\left(\dfrac{C_c+C_s}{C_s}\right)}$

$\qquad = \sqrt{\dfrac{2\times2000\times6000}{8}\left(\dfrac{36000}{36000-6000}\right)\left(\dfrac{8+20}{20}\right)}$

$\qquad = 2{,}246$ units

(b) No. of shortages $(Q_s^*) = Q^*\left(\dfrac{C_c}{C_c+C_s}\right)\left(\dfrac{r-D}{r}\right)$

$\qquad = 2246 \times \dfrac{8}{8+20} \times \dfrac{36000-6000}{36000} = 533.33$ units

(c) Manufacturing time $t_1^* = \dfrac{Q^*}{r} = \dfrac{2246}{3600} = 0.06$ year

$$= 0.06 \times 365 = 21.90 \text{ days}$$

$$\simeq 22 \text{ days}$$

(d) Time between set-ups $t^* = \dfrac{Q^*}{D} = \dfrac{2246}{24000} = 0.38$ year

$$= 0.36 \times 365 = 138.7 \text{ days}$$

(e) Total cost $= \dfrac{C_o D}{Q^*} + \dfrac{C_c (Q_i^*)^2}{2Q^*} + \dfrac{C_s (Q^* - Q_i^*)^2}{2Q^*} + pD$

$$= \sqrt{2\, C_0\, C_c\, D\, \frac{r - D}{r} \cdot \frac{C_s}{C_c + C_s}} + pD$$

$$= \sqrt{2 \times 2000 \times 8 \times 6000 \times \left(\frac{36000 - 6000}{36000} \right)\left(\frac{20}{8 + 20} \right)} + 40 \times 6000$$

$$= \sqrt{1.92 \times \frac{5}{6} \times \frac{6}{7} \times 10^8} + 24000$$

$$= 10690.45 + 240000$$

$$= 250690.45$$

Practice Problems

1. The demand for an item in a company is 18,000 units per year, and the company can produce the item at the rate of 3,000 per month. the cost of one set-up is Rs. 500 and the holding cost of one unit per month is .15 rupees. The shortage cost of one unit is Rs. 240/- per year. Determine the optimum manufacturing quantity and the number of shortages. Also determine the manufacturing time and the time between set-ups.

Answer : 4,489 units, 17 units, 1.5 months, 3 months

2. The demand for a product is 25 units per month and the items are withdrawn uniformly. The set-up cost each time a production run is Rs. 15. The inventory holding cost is Rs. 0.30 per item per month.

   (i) Determine how often to make production rim, if shortages are not allowed?

   (ii) Determine how often to make production rim, if shortages cost Rs. 1.50 per item per month   **[Madras B.E. 1990]**

Answer : (i) $Q^* = 50$ (ii) 54.7 units

3. The demand of an item is uniform at a rate of 20 units per month. The fixed cost is Rs. 10 each time a production run is made. The production cost is Rs 1 per item and the inventory carrying cost is Rs. 0.25 per item per month. If the shortage cost is Rs. 1.25 per item per month, determine how often to make a production run and of what size should it be? **[Osmania M.Sc. (Stat) 1984, Mysore B.E. (Mech) 1985]**

Answer : $Q^* = 44$ items; $t^* = 2.2$ months

4. A company has a demand of 12000 units/year for an item and it can produce 2000 such items per month. The cost of one set-up is Rs. 400 and the holding cost per unit/month is Re 0.15. The shortage cost of one unit is Rs. 20 per year. find the optimum lot size and the total cost per year, assuming the cost of 1 unit is Rs. 4. Also find the maximum inventory manufacturing time and total time.

[JNTU (Mech.) 96/MS]

Answer : $Q^* = 3413$ units $TVC^* = $ Rs. 51336

5. The demand for an item in a company is 15000 units per year and the company can produce the items at a rate of 300 per month. The cost of one set-up is Rs. 500 and holding cost of 1 unit per month is 15 paise. The shortage cost of one unit is Rs. 20 per month.
Determine

   (i) Optimum production batch quantity and number of shortages

   (ii) Optimum cycle time and production time

   (iii) Maximum inventory level in the cycle

   (iv) Total associated cost per year if the cost of the items is Rs. 20 per unit.

[JNTU (Mech.) 94]

Answer : (i) $Q^* = 4,489$ $Q_s^* = 17$ units   (ii) 1.5, 3 months

   (iii) 2227 units   (iv) Rs. 334/-

7.16 *Model 6 : With Price Breaks or Quantity Discounts*

In the earlier models it is assumed that the cost of offered procurement is unaltered by ordersize. but, in some situations the discounts will be ofered to purchase more units to encourage the buyers. In such occassions, we have differential prices for different quantities ordered. Thus the decision whether the price discounts should be availed or not is another important aspect of inventory policy. The computation method is given below.

Model with 2 - Price Breaks :

Suppose the following price discounts are offered where b_1, b_2 are the limits of discounts, and c_1, c_2, c_3 are the costs respectively.

Mathematically,

Quantity	Price Per Unit
$0 < Q_1 < b_1$	c_1
$b_1 \le Q_2 < b_2$	c_2
$b_2 \le Q_3$	c_3

Notice that $c_3 < c_2 < c_1$

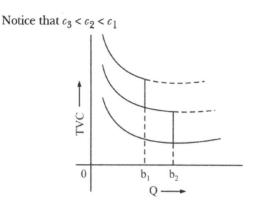

FIGURE 7.19 :   MODEL WITH PRICE BREAKS

Now, for the above problem, with usual notations, the algorithm is as follows.

The Algorithm :

Step 1 : Consider the lowest price i.e., c_3 and determine Q_3^* with usual EOQ formula

i.e., $$Q_3^* = \sqrt{\frac{2\,C_o\,D}{c_3}}$$

Now, if $Q_3^* \geq b_2$, then Q_3^* is the EOQ for the given problem i.e., $Q^* = Q_3^*$ and hence optimal TVC_3^* (for Q_3^*) can be computed. If $Q_3^* < b_2$, then go to step - 2.

Step 2 : Consider c_2 and compute Q_2^* with the EOQ formula i.e., $Q_2^* = \sqrt{\frac{2\,C_o\,D}{c_2}}$

Now if Q_2^* lies in between b_1 and b_2 i.e., if $b_1 \leq Q_2^* < b_2$ is true, then compare TVC^* (at Q_2^*) and TVC (at b_2). If TVC^* (at Q_2^*) < TVC (at b_2) then $Q^* = Q_2^*$, i.e., Q_2^* is the EOQ, otherwise $Q^* = b$ is the EOQ required.

If $Q_2^* < b_1$ then go to step 3.

Step 3 : Calculate Q_1^* based on c_1 and compare TVC (for b_1), TVC (for b_2) and TVC (for Q_1^*) to find EOQ. Obviously the quantity with the lowest cost will be the required EOQ.

Note :   Thus model can be extended to any number of price - breaks in the same pattern as above steps.

The above problem can also be computed by calculating EOQs for various conditions separately and comparing the corresponding costs.

ILLUSTRATION 10 ─────────────────────────────────────

> *Monthly demand for an item is 200 units. Ordering cost is Rs. 350, inventory carrying charge is 24% of the purchase price per year. The purchase prices are P_1 = Rs. 10 for purchasing $Q_1 < 500$; P_i = Rs. 9.25 for purchasing $500 \leq Q_2$ 750 and P_3 Rs. 8.75 for purchasing $750 \leq Q_3$.*
>
> *Determine optimum purchase quantity. If the order cost is reduced to Rs. 100 per order, compute the optimum purchase quantity.* [JNTU (Mech.) 95, IGNOU (MBA) 96]

$\mathcal{S}olution$:

   Case (i) :

$$C_o = 350$$

$$D = 200 \times 12 = 2400$$

$$C_c = 0.24 \, p_i$$

$$Q_i^* = \sqrt{\frac{2 \times C_o \times D}{C_c}} \qquad \text{where } i = 1, 2, 3 \dots$$

$$Q_3^* = \sqrt{\frac{2 \times 350 \times 2400}{0.24 \times 8.75}} = 894 \text{ units.}$$

Since 894 is greater than 750, the optimum purchase quantity is 894.

   Case (ii) :

$$C_o = 100$$

$$D = 200 \times 12 = 2400$$

and $C_c = 0.24 \, p_i$ where $i = 1, 2, 3$.

here $Q_3^* = \sqrt{\dfrac{2 \times 100 \times 2400}{0.24 \times 8.75}} = 478$ units.

since $478 < 750 = b_3$, we next compute

$$Q_2^* = \sqrt{\frac{2 \times 100 \times 2400}{0.24 \times 9.25}} = 465 \text{ units}$$

since $465 < 500 = b_2$, we next compute

$$Q_1^* = \sqrt{\frac{2 \times 100 \times 2400}{0.24 \times 10}} = 447 \text{ units} > 0.$$

We, now compare the total costs for purchasing.

$$Q_1^* = 447$$

$$b_2 = 500 \text{ and}$$

$$b_3 = 750 \text{ units respectively.}$$

$$TC_1 \text{ (for purchasing 447)} = 10 \times 2400 + \frac{100 \times 2400}{447} + \frac{1}{2}(0.24) \times 10 \times 447$$

$$= \text{Rs. } 25{,}085$$

$$TC_2 \text{ (for } Q_2 = 500) = 9.25 \times 2400 + \frac{100 \times 2400}{500} + \frac{1}{2}(0.24) \times 9.25 \times 500$$

$$= \text{Rs. } 23{,}247$$

$$TC_3 \text{ (for } Q_3 = 750) = 8.75 \times 2400 + \frac{100 \times 2400}{750} + \frac{1}{2}(0.24) \times 8.75 \times 750$$

$$= \text{Rs. } 22{,}119.50*$$

Therefore economic purchase quantity in this case is $Q_3{}^* = 750$.

Practice Problems

1. Find the optimum order quantity for a product for which the price breaks are as follows :

Quantity	Unit Cost (Rs.)
$0 \leq Q_1 \leq 800$	Re 1.00
$800 \leq Q_2$	Re 0.98

   The yearly demand for the product is 1,600 units per year, cost of placing an order is Rs 5. The cost of shortage is 10% per year.   [OU (MBA) 88]

   Answer : $Q^* = 800$ units

2. The annual demand of a product is 10,000 units. Each unit costs Rs. 100 if orders are placed in quantities below 200 units but for orders of 200 and above the price is Rs. 95. The annual inventory holding costs is 10% of the value of the item and the ordering cost is Rs. 5 per order. Find the economic lot size.

   [JNTU Mech/Chem/Prod - 2001/S, OU - (MBA) 89]

   Answer : $Q^* = 100$ *units*

3. Find the optimum order quantity for the following

   Annual demand = 3600 units ordering costs = Rs. 50
   Cost of storage = 20% of unit costs

   Price Break :

Quantity	Unit Cost (Rs.)
$0 \leq Q_1 \leq 100$	20
$100 \leq Q_2$	18

   [JNTU (CSE) 93]

   Answer : $Q^* = 316.23$

4. Find the optimal order quantity for a product for which the price - breaks are as follows :

Quantity	Unit Cost (Rs.)
Below 100	200
101 to 200	180
above 200	160

Monthly demand of the product is 400 units. The storage cost is 20% per year, of the price of the product per unit. Ordering cost is Rs. 50 per order.

[OU, B.E. (Prod)1993]

Answer : $Q^* = 200$ units

5. Find the optimum order quantity for a product for which the price breaks are as follows.

Quantity	Unit Cost (Rs.)
$0 \leq Q_1 < 100$	20
$100 \leq Q_2 < 200$	18
$200 \leq Q_3$	16

The monthly demand for the product is 400 units. The storage cost is 20% of the unit cost of the product and the cost of ordering is Rs. 25 per month.

[JNTU (ECE & EEE) 94]

Answer : $Q^* = 200$ units

6. A manufacturer of engines is required to purchase 4,800 casting per year. The requirement is assumed to be known as fixed. These castings are subject to quantity discounts. The price schedule is as follows.

Quantity	Unit Cost (Rs.)
Less than 500	150
500 to 750	138.75
750 or more	131.25

Monthly holding cost expressed as a decimal fraction of the value of the unit is 0.02. Set-up costs associated with the procurement of purchased item is Rs. 75 per procurement. Find the optimum purchase quantity per procurement.

Answer : $Q^* = 1656$ units

7. Annual demand for an item is 500 units odering costs is Rs. 18 per order. Inventory carrying cost is Rs. 15 per unit per year relationship between price and quantity ordered is as follows :

Quantity Ordered :	1 to 15	16 to 149	150 to 549	550 & above
Price Per Unit (Rs) :	10	9	8.75	8.5

Specify   optimal order quantity and the corresponding price of this item.

[I.I.I.E. Grad. 1982]

Answer : $Q^* = 16$ units $TVC^* = $ Rs. 6142.50

8. A diesel engine manufacturing company has planned its production schedule for the next year based on the forecasted demand, backorders and plant capacity. Instead of manufacturing the piston, that goes into the final product, the company has decided to buy the pistons. The pistons are required at a rate of 60 per day. Ordering costs have been estimated at Rs. 50 per order and the carrying cost fraction is 0.15. All the assumptions of the basic EOQ model are applicable. The company, however, can take advantage of one of the several quantity discounts. The pricing schedule of the pistons is :

Quantity	Unit Price (Rs)
0 - 1999	65
2000 - 4999	60
5000 - 10,000	55
above 10,000	50

(a) What is the optimal order quantity?

(b) What is the minimum inventory costs?

Answer : $Q^* = 10,000$ units $TVC^* = $ Rs. 937590

9. A company purchases a certain chemical to use in its process. No shortages are allowed. The demand for the chemical is 1,000 kg/month; the cost of one purchase is Rs. 800; and the holding cost for one unit is Rs. 10 per month. The unit cost depends on the amount purchased as given below :

Amount in kg	Cost Per kg (Rs.)
0 - 249	8
250 - 449	7.50
450 - 649	7
650 & above	6.75

Determine the optimum purchase quantity and the optimum total yearly costs.

Answer : $Q^* = 142.857$ kg $TVC^* = $ Rs. 19314

7.17 *Stochastic or Probabilistic Models of Inventory Control*

A class of inventory problems requires that the order quantity decision be made only once for the entire demand process. And this is also not certain and requires to make use to probabilities since the decision of dilemma concentrates on ordering 'enough', so that the full potential for profit may be realised, but not too much so as to avoid losses on the excess.

There are numerous examples in nature, a few are listed below.

1. How many crackers are to be manufactured and stocked for this Deewali season.

2. How many lunch plates are to be prepared for the party.

3. How many leaves of bread are to be baked for a given day

4. How many news papers are to be stocked to sell on a given day.

5. How many salwar suits of current fashions & styles are to be produced.

In all the above examples of stochastic inventory models, the major distinguishing characteristic is that the order quantity decision is a "one shot" affair and that, despite uncertainty in the demand, a proper trade–off must be achieved between the consequences of too much and too little.

Thus in the production systems also, the decision regarding optimum batch quantity becomes so critical and complicated that we have to depend on the probabilities to determine the demand first. Thus entire problem depends on the probable demand again.

Again the use of probabilities is extendable to various distributions such as Poisson, Normal and other models as applicable to the situation. The stochastic inventory control decisions also depend on the other factors such as with or with out set–up cost, instantaneous or continuous demand, descrete and continuous replenishments, and shortages allowed or not allowed and so forth.

One popular and most often occurring model of probabilistic nature is discussed here i.e., single period probabilistic model without set-up costs (with instantaneous demand with discrete replenishment unit, no set-up cost).

The retail marketers, news paper sellers, seasonal products sellers often have such problem of decision regarding how much stock is to be maintained. This decision is purely probabilistic and may depend on the past record (forecasting otherwise) and thus expected (probable) demand during the period.

In such cases, the following inventory model will be helpful.

7.18 *Model 7 : Single Period Probabilistic Model With Instantaneous Demand Without Setup Cost*

The following working rule is used :

Working Rule :

Step 1 : Calculate $\dfrac{C_s - C}{C_c + C_s}$

Where C_s = shortage cost; C = cost of item per unit (if considered) C_c = carrying cost

Step 2 : Calculate cumulative probability distribution (for each period).

Step 3 : Determine the value of Q^* and $(Q^* - 1)$ where the ratio calculated in step - 1 lies in distribution calculated in step - 2.

Step 4 : Take the higher value as optimum level

Step 5 : If I = inventory on hand before the order is placed, then order $Q^* - I$ if $Q^* > I$ and do not order if $Q^* \leq I$

ILLUSTRATION 11 ――――――――――――――――――――――――――

A newspaper boy buys papers for 3 paise and sells them for 7 paise each. He can not return unsold newspapers. Daily demand has the following distribution.

No. of customers	23	24	25	26	27	28	29	30	31	32
Probability	0.01	0.03	0.06	0.10	0.20	0.25	0.15	0.1	0.05	0.05

If each day's demand is independent of the previous day's, how many papers should the order each day? [JNTU (CSE) 97/S]

Solution :

Let Q be the number of newspapers ordered per day and let d be the demand for it, i.e., the number of newspapers actually sold per day.

Let C_1 = cost price of newspaper and

            C_2 = Gain i.e., C_1 – sales price

Given C_1 = Rs. 0.03 and C_2 = Rs. 0.07 – Rs. 0.03 = 0.04

$$\frac{C_2}{C_1 + C_2} = \frac{0.04}{0.03 + 0.04} = \frac{0.04}{0.07} = 0.57$$

The optimum solution is obtained by developing the cummutative probability distribution of daily demand as follows.

d	23	24	25	26	27	28	29	30	31	32
$P(d)$	0.01	0.03	0.06	0.10	0.20	0.25	0.15	0.10	0.05	0.05
Q	0.01	0.04	0.10	0.20	0.40	0.65	0.80	0.90	0.95	1.00

$$\sum_{r=23}$$

Since $\dfrac{C_2}{C_1 + C_2}$ = 0.57 is between 0.4 and 0.65, the condition for optimality

suggests that $Q^0 = 28$ i.e., optimum number of newspapers to be ordered is 28.

Practice Problems

1. A newspaper boy buys paper for 60 paise each and sell them for Rs. 1.40 paise He cannot return unsold papers. Daily demand has the following distribution.

No. of Customers	23	24	25	26	27	28	29	30	31	32
Probability	0.01	0.03	0.06	0.1	0.2	0.25	0.15	0.1	0.05	0.05

If each day's demand is independent of the previous day's how many papers should be ordered each day ? [JNTU (Mech.) 97]

Answer : 28

2. A newspaper boy buys papers from Rs. 1.40 and sells them for Rs. 2.45 each. He cannot return unsold newspapers. Daily demand has the following distribution:

No. of Customers	25	26	27	28	29	30	31	32	33	34	35	36
Probability	0.03	0.05	0.05	0.1	0.15	0.15	0.12	0.1	0.1	0.7	0.6	0.02

If each day's demand is independent of the previous day's, how many papers he should order each day? **[Meerut M.Sc. (Math), 1993, 1992, JNTU (CSIT) 98]**

Answer : 30

3. A contractor of second hand motor trucks uses to maintain a stock of trucks every month. Demand of the trucks occurs at a relatively constant rate but not in a constant size. The demand follows the following probability distribution

Demand	0	1	2	3	4	5	6 or above
Probability	0.4	0.24	0.2	0.1	0.05	0.01	0.00

The holding cost of an old truck in stock for one month is Rs. 100 and the penalty for truck if not supplied on demand is Rs. 1,000. Determine the optimal size of the stock for the contractor. **[Marathwada M.Sc. (App. Math), 1982]**

Answer : 3

4. A T.V dealer finds that costs of holding a television in stock for a week is Rs. 20, customers who cannot obtain new television immediately tend to go to another dealer; and he estimates that for every customer who does not get immediate delivery he loses on an average Rs. 200. For one particular model of TV, the probabilities for a demand 0, 1, 2, 3, 4, and 5 in a week are 0.05, 0.1, 0.2, 0.3, 0.2 and 0.15 respectively. How many television per week should the dealer order?

Answer : 4

Hint : $C_1 = 20$ $C_2 = 200$, $0.85 < 0.921 < 1.00$ **[JNTU (CSE) 98/S]**

5. The probability distribution of monthly sale of certain items is as follows :

Monthly Sales	0	1	2	3	4	5	6
Probability	0.01	0.06	0.25	0.35	0.2	0.03	0.1

The cost of carrying inventory is Rs. 30 per unit per month and the cost of unit shortage is Rs. 70 per month. Determine the optimum stock level which minimise the total expected cost.

Answer : 4

6. Some of the spare parts of a ship costs Rs. 1,00,000 each. These spare parts can only be ordered together with the ship. If not odered at the time when the ship is constructed, these parts cannot be available on need. Suppose of loss of Rs. 10,000,000 is suffered from each spare part that is needed when none is available in stock. Further, suppose that probabilities that the spare part will be needed as replacement during the life-time of the class of the ship discussed are

Spare part Require	0	1	2	3	4	5 or more
Probability	0.9488	0.04	0.01	0.001	0.0002	0

How many spares should be procured?

Answer : 2

7. A newspaper boy buys papers for Rs. 2.60 each and sells them for Rs. 3.60 each. He cannot return unsold newspapers. Daily demand has the following distribution.

Customers	23	24	25	26	27	28	29	30	31	32
Probability	0.01	0.03	0.06	0.1	0.2	0.25	0.15	0.1	0.05	0.05

If each day's demand is independent of the previous day's how many papers should he order each day? [JNTU (Mech. 97/S]

Answer : 27

7.19 *Inventory Control Approaches*

In the previous topics, we have discussed on *"how much to order"*, its related re-order level and its cost calculations. However, another important point of operating doctrine of inventory is "when to order?". Though, this depends on many factors and the reorder point is set basing on several aspects, there are two systems often chosen by materials managers. These are :

1. *P* - system or fixed period system

2. *Q* - system or fixed quantity system.

1. *P* - System or Fixed Period System or Periodic Review System :

In this system, the orders are placed at a predetermined fixed period irrespective of the quantities in stock. However, the re-order is placed for the quantities that are less to the target (maximum) inventory level. A review is made periodically and the period is fixed based on the estimated consumptions or past records. A graphical representation is shown below.

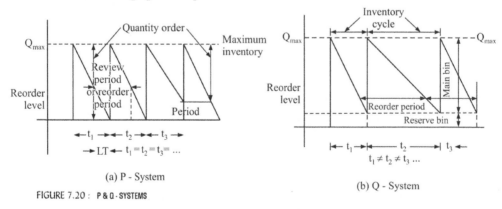

(a) P - System

(b) Q - System

FIGURE 7.20 : P & Q - SYSTEMS

2. *Q* - System or Fixed Quantity System or Two - Bin System :

In this system, a re-order level is fixed and whenever this level is reached, irrespective of time (or period), the order will be placed for procurement. Here, the reorder periods of different cycles are different and inventory cycles also differ since the re-ordering depends on 'when' the quantities reach the pre-determined minimum level. This is represented in the graph given above.

An example with a over head water tank, we can understand the above systems easily. Suppose, you depend on the municipal water supply, which gives you water only in the morning at 6 am everyday. Then you will fill your tank at 6 am everyday irrespective of the stock in the tank. This is P - system.

If you have a borewell in your campus, you will fill the tank whenever a minimum level is reached irrespective of period of consumption. This is Q - system. Now, apply the above systems for inventory, assuming tank as your inventory store and water is items procured.

Some More Examples :

P - *System :* A government employee who receives salary during the first week dumps all the grocery (inventory) on every first week of the month.

Q - *System :* A businessman may not bring his household grocery at every certain period, but brings whenever a minimum level is reached.

The petrol/diesel in the fuel tank of your scooter /bike or any vehicle has two bins (though not separate), called main bin and a reserve bin. Whenever, the stock level reaches reserve, you start your procurement action which is in support of Q - system, while you may check-up air in the tyres periodically (say weekly) which is in P - system.

Distinction Between P-system and Q - system :

S.No.	P - system	Q - system
1.	This system is based on fixed period or time	This system is based on fixed minimum quantity level.
2.	Cycle period and re-order period are constant	Cycle period and reorder period change
3.	Cycle period is equal to reorder period.	Cycle period is not equal to reorder period
4.	Only one bin is used	2-bins (main & reserve) are used.
5.	Advantageous for joint production/ transportation buying	Advantageous in situations where stock-out costs are high
6.	Useful where financial resources are avialable at fixed intervals	Useful where financial aresources are abundant or/and available at any time.

7.20 *Stores Management & Selective Inventory Control*

In general situations of materials management, all the items will not have same importance due to their cost, consumption, value, function etc.,

Thus the care to be taken obviously differs, based on the importance of materials. To have a control and modeling on the materials which are of less cost or importance will be uneconomical.

Therefore materials managers often use selective inventory controlling methods by grouping all the item in certain categories and control them with some basis.

Various techniques available for managers in this regard are *ABC* analysis, *VED, SDE, XYZ, SOs, FSND, VEIN, CIN, GOLF* etc. Of these, *ABC* analysis is the most popular and widely employed technique.

ABC **Analysis :**

   ABC analysis is a basic tool, which helps materials mangers to take a decision regarding "*how much care is to be put on each item*". This selective approach is based an "*annual consumption value*" of various items.

   For example, the items like nuts, bolts, valves etc., are though equally important as compared to engines, pistons, cylinders, gears etc., the costs differ. Also the consumption varies from item to item. The items like engines are stocked in less quantities due to their high cost and low consumption while high stocks of nuts, bolts etc., will be maintained as they are cheap and more consumed. Thus more emphasis should be given to control the stocks of engines while it is expensive and uneconomical to maintaining and recording the bolts, nuts etc.,

   Thus depending the consumption and costs of the items, the items are classified into three categories *A*, *B* and *C* in the discending order of their annual consumption cost. Thus usually, *A* - class items are costly but less consumed while *C* - class items are many cheap and more consumed. The average pattern of percentage of items and percentage of annual consumption costs are as follows.

Category	Percentage of Items	Percentage of Annual Consumption Cost
A	10% – 20%	70% – 85%
B	20% – 30%	10% – 25%
C	50% – 70%	5% – 15%

   ABC (Always Better Control) Technique is in accordance with Pareto's principle "*VITAL FEW- TRIVIAL MANY*" and thus low value items are usually consumed more and high value items are less consumed.

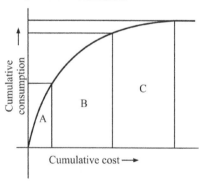

FIGURE 7.21 :   A-B-C ANALYSIS - GRAPHICAL REPRESENTATION

   The computation algorithm for *ABC* analysis is as follows :

***Step 1* :** Obtain data of annual consumption rate and unit cost of each item.

***Step 2* :** Calculate annual consumption cost of each item

   Annual cosumption cost = annual consumption × unit cost

***Step 3* :** Arrange the items in the descending order of their annual cost consumption values.

***Step 4* :** Find the cumulative value of annual consumption cost.

***Step 5* :** Express the annual consumption cost as a percentage of total value of the items.

***Step 6* :** Express percent number of items in the total number of items

Step 7 : Draw graph between the cumulative precent items to cumulative percent annual consumption cost. i.e., graph between the values of step - 5 & step 6.

Step 8 : Classify the area under graph into three class *A*, *B* and *C* according certain norms such as given below.

Class	Percentage of Items	Percentage of Annual Consumption Cost
A	10	80
B	20	15
C	70	5

ILLUSTRATION 12 ——————————————————————————————————

A company is considering a selective inventory using the following data. Classify them using ABC analysis.

Item	1	2	3	4	5	6	7	8	9	10	11	12
Consumption	6000	61,200	16800	3000	55800	22630	26640	14760	20520	9000	29940	24660
Unit Cost	4.00	0.05	2.10	6.00	0.20	0.50	0.65	0.4	0.4	0.1	0.3	0.5

[OU (MBA) 92]

Solution :

Item	Consumption	Unit Cost	Consumption Cost	Ranking
1	6000	4.00	24000	II
2	61200	0.05	3060	XII
3	16800	2.10	35280	I
4	3000	6.00	18000	III
5	55800	0.20	11160	VII
6	22680	0.50	11340	VI
7	26640	0.65	17316	IV
8	14760	0.40	5904	XI
9	20520	0.40	8208	X
10	9000	0.1	9000	VIII
11	29940	0.3	8982	IX
12	24660	0.5	12330	V

S.No.	Item No.	Annual Consumption Cost (ACC)	Comulative ACC	% Comulative ACC	% Items	Commulative % Items
	(1)	(2)	(3) = Σ (2)	(4) = (3)/164580	(5) = (1)/12	
1	3	35280	35280	21.44	8.33	A
2	1	24000	59280	36.02	16.67	A
3	4	18000	77280	46.96	25.00	B
4	7	17316	94596	57.47	33.33	B

5	12	12330	106926	64.97	41.67	B
6	6	11340	118266	71.86	50.00	B
7	5	11160	129426	78.64	58.33	C
8	10	9000	138426	84.11	66.67	C
9	11	8982	147408	89.57	75.00	C
10	9	8208	155616	94.55	83.33	C
11	8	5904	161520	98.14	91.67	C
12	2	3060	164580	100.00	100.00	C

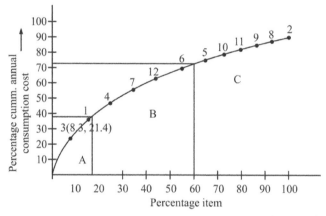

FIGURE 7.22 :   GRAPHICAL REPRESENTATION OF A-B-C ANALYSIS

F-S-N Classification : This is most widely employed control tool in marketing and in production. This classification takes into account the pattern of issues from stores. The three letters stand for *Fast moving, Slow moving* and *Non moving* (some people refer to *D* also, which stands for Dead items). This becomes very handy when it is used to control obsolescence. Items classified as S and N require very great attention, especially N items. There are several reasons for an item falling in N category like change in technology, in the specifications of a particular spare part/item might no longer be in use. However, when a fast moving item is not moved with its pace, it may go into slow moving and in turn may become non moving also. Therefore a manager has to be very careful in managing their movement or consumption we can also observe here that in a store where the items are placed according to the FSN, the 'F' items will occupy a forefront places while 'N' items occupy last or backward place. Therefore a manager has to keenly observe and act at right time to create the proper demand to these items, else they may become dead or obsolete. When an FSN classification is made all such information stands out prominently enabling managers to act on the information in the best interests of the organisation.

V-E-D Classification : This classification is based on the significance of the item in production or assembly chain. the three letters, V, E, D stand for VITAL, ESSENTIAL and DESIRABLE items. the items irrespective of their costs that are highly important without which you cannot run the production or assembly are vital. The absence of essential items may not stop the production chain but suffers production chain considerably. The desirably items are usually required to make the product good looking or aesthetic and their absence hardly has any impact on production chain. For a scooter, engine is a vital part, breaks are essential and seat cover is desirable. Note that a small value at the tyre is also vital though cheaply available.

The other selective inventory control techniques are summarised and presented in a tabular form as follows.

S.No.	Classification	Meaning	Basic Principle	Typical Area of Application
1.	ABC	A - High consumption value B - Medium consumption value C - Low consumption value	Percentage of total consumption value	Common and widely employed to any area
2.	HML	High cost , Medium cost and Low cost items	Cost of the items	Precious/costly items
3.	VED	Vital, Essential and Desirable	Item significance in the operation or machine	Spare parts/ MRO
4.	VEIN	Vital, Essential, Important and Normal	– do –	– do –
5.	GOLF	Government, Open market, Local market, Foreign market	Availability & Restrictions	Items under restrictions or constraints
6.	SOs	Seasonal, Off seasonal	Season of availability	Anticipatory inventory
7.	FSN/FSND	Fast moving, Slow moving, Non moving, Dead	Market trend	Marketing sales inventory, finished goods inventory, store issue patterns.
8.	XYZ	X - High consumption Y - Medium consumption Z - Least consumption	Consumption	General items and Regularly used items
9.	CIN	Critical, Important & Normal	Criticality	Machinery, Equipment, Components etc.,
10.	SDE	Scarece, Difficult, Easy available	Availability of the item in the market	Market demand items.

Review Questions

1. Derive EOQ formula, stating the assumption.   **[JNTU. B.Tech Mech 99]**

2. Derive a formula for optimum batch quantity if the demand is continuous at a constant rate. Assume appropriate data.   **[JNTU B.Tech CSE 1997]**

3. What is *ABC* analysis in inventory control?
   Describe it in detail with the help of graph.   **[JNTU B.Tech Mech 1998/S]**

4. Write a short note on purchasing inventory with one price break.
                                    [JNTU B.Tech Mech. 99]

5. Discuss the significance of stochastic models in inventory control of production system   **[JNTU B.Tech CSE 1997]**

6. What are the assumptions used in deriving EOQ formula in Wilson - Haris model.

7. What assumption do you take for obtaining EOQ for deterministic inventory models
   (a) With shortage
   (b) With a production rate

8. Derive the expression for EOQ in deterministic model with back orders.

9. Derive the formula to find EOQ with the production rate (r) and demand (d) without shortages. How does the formula be interpreted if the same model is assumed to have shortages. Interprete these with graphically.

10. What are the functions of inventory.   [JNTU Mech. 2000/S , CSE - 2000]

11. What are the advantages of having inventory.   [JNTU Mech, Mechatronics, prodn. & Cívil. 2001]

12. Discuss various costs involved in inventory control in an industry.
                                                                 [JNTU Mech. 96/MS]

13. Discuss the role of inventory in a large scale industry.   [JNTU Mech 96/S/MS]

14. What are the consequences of over- inventory and under - inventory situations.

15. Discuss the advantages and disadvantages of allowing shortages in the inventories.   [JNTU CSE 95/S]

16. What is selective inventory control. Why do you optimise this in large industries.
                                                                 [JNTU Mech. 96/PPC]

17. What are P - system & Q - system of inventories ? Explain.
                                                                 [JNTU Mech. 95/S, 98, 99/PPC]

18. Distinguish between P - system & Q - system of inventories.
                                                                 [JNTU Mech. 95, 96, 96/S, 97/PPC]

19. Discuss the working rule of probabilistic model without set-up cost and instantaneous demand.

20. Give the algorithm for price break models

21. Write the algorithm for ABC analysis.   [JNTU - Mech. 98/S/PPC]

22. Discuss the factors affecting inventory in the industries.   [JNTU Mech. 93/MS]

23. Write short notes on
    (a) Decoupling inventory
    (b) Transit inventory
    (c) Buffer inventory   [JNTU Mech. 93/OR

24. Give the classification of inventories.

25. Classify and discuss various inventory models briefly.   [JNTU - ECE/EEE - 93/MS]

26. Discuss various selective inventory control techniques and compare them.

27. Derive the EOQ formula for continuous & non uniform demand with instantaneous procurement.

28. Write short notes on
    (a) Lead time   (b) Inventory cycle
    (c) Re-order cycle   (d) Re-order period.   **[JNTU Met & CSE - 93/S/MS]**

29. What do you understand by the term operating doctrine in the inventory modeling.

30. What are the limitations of EOQ.

Objective Type Questions

1. At EOQ

   (a) annual purchase cost = annual ordering cost

   (b) annual ordering cost = annual carrying cost

   (c) annual carrying cost = annual shortage cost

   (d) annual shortage cost = annual purchase cost

2. If shortage cost is infinity

   (a) order cost is zero   (b) no shortages are allowed

   (c) no inventory carrying is allowed (d) purchase cost = carrying cost

3. The stock maintained to withstand unknown demand changes is known as

   (a) De-coupling inventory   (b) pipeline inventory

   (c) fluctuatory inventory   (d) anticipatory inventory

4. The inventory policy followed for water consumption from a water tank operated with municipal water spply connection is usually

   (a) P - system   (b) Q - system

   (c) 2 - bin system   (d) Stochastic model

5. The inventory system designed for a petrol tank of a two-wheeler is

   (a) P - system   (b) Q - system

   (c) Stochastic system   (d) Poisson model

6. The most suitable analysis for a retail marketer is

   (a) F - S - N analysis   (b) A - B - C analysis

   (c) G - O - L - F analysis (d) V - E - D analysis

7. Which of the following inventory is maintained to meet expected demand fluctuations

   (a) buffer stock   (b) de-coupling inventory

   (c) fluctuatory inventory   (d) anticipators inventory

8. With usual notations the formula for EOQ when shortage is allowed and finite and uniform supply rate

(a) $\sqrt{\dfrac{2\,C_o\,D}{C_c}}$

(b) $\sqrt{\dfrac{2\,C_o\,D}{C_c}\cdot\left(\dfrac{r}{r-d}\right)}$

(c) $\sqrt{\dfrac{2\,C_o\,D}{C_c}\cdot\left(\dfrac{C_c+C_s}{C_s}\right)}$

(d) $\sqrt{\dfrac{2\,C_o\,D}{C_c}\cdot\left(\dfrac{r}{r-d}\right)\left(\dfrac{C_c+C_s}{C_s}\right)}$

9. Expression for total variable cost in the case of instant replenishment without shortage, with usual notations.

(a) $\sqrt{C_o\,C_c\,D}$ (b) $\sqrt{C_o\,C_c\,D/2}$ (c) $\sqrt{2\,C_o\,C_c\,D}$ (d) none of these

10. If EOQ = demand (D), then the number of orders placed in the inventory horizon (i.e., 1 year) is _____

(a) zero   (b) one   (c) two   (d) infinite

11. If average carrying cost per unit per year are twice to that of ordering cost/order then EOQ varies with

(a) D^2 (d) D (c) $\sqrt{D}$ (d) $\sqrt{2D}$

12. Being ordering cost/order and average unit carrying costs constant, if demand suddenly falls by 75% then EOQ _____

(a) increases by 50%   (b) decreases by 50%

(c) decreases by 37.5%   (d) none of these

13. Which of the following increases with quantity ordered per order

(a) order cost   (b) carrying cost

(c) demand   (d) purchase cost

14. The time difference between ordering point and replenishment point is called _____

(a) re-order period   (b) lead time

(c) inventory cycle time   (d) inventory horizon

15. Which of the following is assumed to be constant in deterministic inventory models

(a) annual demand   (b) annual ordering cost

(c) annual carrying cost   (d) annual shortage cost

16. When shortages are allowed with constant demand and instaneous replenishment, EOQ = _____ (with usual notations)

(a) $\sqrt{\dfrac{2\,C_o\,D}{C_c}}$

(b) $\sqrt{\dfrac{2\,C_o\,D}{C_c}\cdot\dfrac{C_s}{C_c+C_s}}$

(c) $\sqrt{\dfrac{2\,C_o\,D}{C_c}\cdot\dfrac{C_c+C_s}{C_s}}$

(d) $\sqrt{\dfrac{2\,C_o\,D}{C_c}\cdot\dfrac{C_c-C_s}{C_s}}$

17. Shortages are allowed in an industry to derive the advantage of

    (a) spreading the ordering cost over a longer period

    (b) less number of orders in the inventory horizon that reduces ordering cost

    (c) increased carrying cost does not effect considerably

    (d) all the above

18. In the case of shortages, the optimal amount of back order units = _____

    (a) $\dfrac{EOQ}{2}$

    (b) 10% of EOQ

    (c) $EOQ \cdot \dfrac{C_s}{C_c + C_s}$

    (d) $EOQ \cdot \dfrac{C_c}{C_c + C_s}$

19. Total cycle time in case of model with shortages is _____

    (a) $\sqrt{\dfrac{2\,C_o\,D}{C_c} \cdot \left(\dfrac{C_c + C_s}{C_s}\right)}$

    (b) $\sqrt{\dfrac{2\,C_o\,D}{C_c} \cdot \dfrac{C_s}{C_c + C_s}}$

    (c) $\sqrt{\dfrac{2\,C_o}{D \cdot C_c} \cdot \dfrac{C_s}{C_c + C_s}}$

    (d) $\sqrt{\dfrac{2\,C_o}{D \cdot C_c} \cdot \left(\dfrac{C_c + C_s}{C_s}\right)}$

20. A continuous and finite supply rate model becomes simple EOQ [i.e., instant supply] model for the condition that production rate = _____

    (a) zero

    (b) unity

    (c) infinity

    (d) consumption rate

21. Total variable cost becomes zero if

    (a) rate of production is infinity

    (b) rate of consumption is infinity

    (c) rate of production = rate of consumption

    (d) never

22. The optimal number of shortages in the inventory model with finite and constant supply with shortages allowed is _____ (with usual notations)

    (a) $Q^* \left(\dfrac{C_c - C_s}{C_s} \cdot \dfrac{r}{r+d}\right)$

    (b) $\left(\dfrac{C_c}{C_c + C_S} \cdot \dfrac{r-d}{s}\right) \times Q^*$

    (c) $\left[\dfrac{C_s}{C_c + C_s} \cdot \dfrac{r-d}{r}\right] \times Q^*$

    (d) $\left(\dfrac{C_c + C_s}{C_c} \cdot \dfrac{r}{r-d}\right) \times Q^*$

23. In the model with continuous production with shortages, the production time is _____

    (a) EOQ/rate of produce   (b) EOQ/rate of consumption

    (c) EOQ/annual demand   (d) none of the above

24. No. of orders in year = _____

    (a) annual demand/total cost   (b) annual demand/EOQ

    (c) EOQ/demand   (d) EOQ/rate of production

25. The optimal stock level of a shortage model (with instantaneous procurement) is

    (a) $\sqrt{\dfrac{2\,C_o\,D}{C_c} \cdot \dfrac{C_c + C_s}{C_s}}$ (b) $\sqrt{\dfrac{2\,C_o\,D}{C_c} \cdot \dfrac{C_s}{C_c + C_s}}$

    (c) $\sqrt{\dfrac{2\,C_o\,D}{C_c} \cdot \dfrac{C_c - C_s}{C_s}}$ (d) $\sqrt{\dfrac{2\,C_o\,D}{C_c} \cdot \dfrac{C_c - C_s}{C_c}}$

26. The annual demand of an item is 1200 units and placing the order costs Rs. 12 per order and costs 20% of price to carry. The price of the item is Rs. 10/- The EOQ =

    (a) 12   (b) 120

    (c) 1200   (d) 20

27. The number of orders to be placed in the year in the above problem is _____

    (a) 12   (b) 10

    (c) 6   (d) 100

28. The re-order period for the above problem (Q. No. 26) is _____

    (a) monthly   (b) quarterly

    (c) half yearly   (d) yearly

29. Which of the following is the inventory purchased to get the advantage of price discounts or to reduce transportation costs_____

    (a) fluctuatory inventory   (b) lot wise inventory

    (c) anticipatory inventory   (d) transit inventory

30. The material stored to safeguard the production chain in the event of machine breakdown is termed as _____

    (a) transit inventory   (b) lot wise inventory

    (c) buffer stock   (d) decoupling inventory

31 Operating Doctrine of inventory is concerned with_____

(a)when to order and how to order

(b) when to order and who to order

(c) when to order and how much to order

(d) who to order and how to order

32. The purchase cost is assumed to be varying with quantity ordered in the inventory model with _____

(a) shortages   (b) finite production rate

(c) price breaks   (d) none

33. In which case of the following, probabilities inventory models are not employed_____

(a) stocks of retailers   (b) news papers seller

(c) dealer of single commodity   (d) a manufacturer of crackers

34. The optimum stock is determined in probabilities model with instant demand, no set-up cost, where the cumulative probability just exceeds the ratio of_____

(a) shortage cost to sum of shortage and holding costs

(b) holding cost to sum of shortage and holding cost

(c) sum of shortage and holding cost to shortage cost

(d) shortage cost to difference of holding and shortage cost.

35. Stochastic model means _____

(a) deterministic models   (b) probabilistic models

(c) fixed period models   (d) fixed quantity models

36. Which inventory costs are assumed to be zero in JIT inventory system

(a) ordering cost   (b) transportation cost

(c) carrying cost   (d) purchase cost

37. Taxes and insurance are included in

(a) ordering cost   (b) shortage costs

(c) purchase cost   (d) carrying costs

38. Which of the following is false in the concept of EOQ and its assumptions

    (a) lead time is zero
    (b) annual order cost = annual carrying cost
    (c) ordering cost remains same through out the year
    (d) none of the above

39. The criteria used in $A - B - C$ Analysis is _____

    (a) cost
    (b) value
    (c) consumption
    (d) consumption cost

40. Which of the following analysis neither considers cost nor value

    (a) $A - B - C$
    (b) $X - Y - Z$
    (c) $H - M - L$
    (d) $V - E - D$

41. The inventory controlling technique that uses significance of the item as the basis is _____

    (a) $A - B - C$
    (b) $F - S - N$
    (c) $V - E - D$
    (d) $H - M - L$

42. Items sold under the restrictions such as government policies, FERA, MRTP act etc, are classified by _____

    (a) ABC
    (b) FSND
    (c) VEIN
    (d) GOLF

43. The availability of item in the market is basis for _____ analysis

    (a) ABC
    (b) SDE
    (c) SOs
    (d) HML

44. S-D-E analysis means

    (a) source - destination - EOQ
    (b) single-double-empty
    (c) scarce - difficult-easy
    (d) specific - desirable-essential

45. If average unit carrying cost is equal to order cost per order and demand is 200 units/year then EOQ is

    (a) 20
    (b) 40
    (c) $10 \sqrt{2}$
    (d) insufficient information

46. When C_c is twice C_0 to make EOQ = 20, demand must be _____

    (a) 200
    (b) 400
    (c) 100
    (d) 20

47. Simple EOQ is independent of

    (a) holding cost

    (b) order cost

    (c) shortage cost

    (d) purchase cost

48. If demand is equal to EOQ, then

    (a) single order will be placed per year

    (b) order cost per order is equal to annual carrying cost

    (c) order cost per order is equal to half of total variable cost

    (d) all the above

49. If annual demand is square of optimal order quantity, then

    (a) $C_0 = C_c^2$

    (b) $C_0 = 2\,C_c$

    (c) $C_c = C_0^2$

    (d) $C_c = 2\,C_0$

50. In F-S-N analysis of inventory control, 'S' represents

    (a) scarcely available items

    (b) supply rate uniform

    (c) slow moving items

    (d) seasonal items

51. Which of the following product is an example for seasonal consumption but continuous production

    (a) sugar

    (b) electricity

    (c) rain coats

    (d) news paper

52. Which of the following product is not an example of seasonal production but continuous consumption

    (a) sugar

    (b) mango

    (c) rain coasts

    (d) soaps

53. The pair of inventory models with same EOQ formula, when total inventory period is one year is _____

    (a) instant procurement with shortage & without shortage

    (b) finite & continuous supply rate and instant supply

    (c) simple EOQ with uniform demand rate and different rates of demand in different cycles

    (d) shortage model and no set-up cost and probabilistic model

54. The total inventory period is referred to as _____

    (a) cycle   (b) horizon

    (c) ROP   (d) Lead time

55. Costs for tenders and auctions are included in _____

    (a) order cost   (b) holding cost

    (c) shortage cost   (d) purchase cost

56. Which of the following is considered as as a demand independent item

    (a) a wrist watch   (b) a glass on wrist watch

    (c) a jewel in watch   (d) none of the above

Fill in the Blanks

1. An inventory model with shortages will become simple EOQ model if shortage cost is _____

2. The EOQ formula was first proposed by _____

3. The stock maintained to encounter the production disturbance due to machine breakdown is _____

4. A-B-C analysis is based on _____

5. If a single order is placed in the total inventory period, then EOQ = _____

6. The total period for which inventory decision are made (usually one year) is called _____

7. The decisions such as when and how much to procure are said to be _____ of inventory

8. For production rate = _____, the model with finite production rate becomes simple EOQ model

9. The order cost/order _____ with the increase in ordered quantity

10. The time difference between any two consecutive orders is called _____

11. The time difference between any two successive replenishments is called _____

12. Keeping C_0 and C_c constant, if demand doubles suddenly then EOQ _____

13. In case of shortages, the ratio of optimal amount of back order to optimal order quantity is equal to _____

14. Inventory cycle time in the case of finite and uniform supply rate is _____

15. The optimum number of orders placed = _____

16. If C_0 is half of C_c, then annual demand is _____ of EOQ

17. F-S-N-D analysis is abbreviated for F = _____ S = _____
    N = _____ D = _____

18. The inventory controlling technique used on the basis of cost of annual consumption is _____

19. Formula for total costs in simple EOQ model is _____.

20. The time gap between the ordering point and its immediate actual stock completion point is called _____

21. Fill in the boxes :

    $$ATC = \boxed{} \times \frac{D}{Q} + \boxed{} \times \frac{Q}{2} + \boxed{} \times D$$

22. The average quantity consumed when quantity 'Q' is replenished in a cycle of time 't' is _____

23. When an inventory model is shown graphically as time on X - axis, and quantity on Y - axis, the area under the graph represents _____

24. Given commulative probability is 0.25, 0.5, 0.75 for sale of 2, 3 and 4 units of stock respectively with storage cost Rs. 2.50 ps and shortage cost Re. 1/-, the optimum number of units to keep in stock is _____

25. Given order cost Rs. 20 per order, annual demand = 1000 units, carrying cost 10% of unit price Rs. 10/- of an item, the EOQ = _____

26. The no. of orders in the above question = _____

27. If there are 300 working days, the ROP in the above question is = _____

28. The operating doctrine of inventory models is to find _____ and _____ to order.

29. The quantities kept in stock to safe guard the product in the event of excess consumption or unusual demand fluctuations etc. is called _____

30. The items whose demand is estimated based on the demand of the final product to which it is assembled is called _____

Answers

Objective Type Questions :				
1. (b)	2. (b)	3. (c)	4. (a)	5. (b)
6. (a)	7. (d)	8. (d)	9. (c)	10. (b)
11. (c)	12. (b)	13. (a)	14. (b)	15. (a)
16. (c)	17. (d)	18. (d)	19. (d)	20. (c)
21. (c)	22. (b)	23. (a)	24. (b)	25. (b)
26. (b)	27. (b)	28. (a)	29. (b)	30. (d)
31. (c)	32. (c)	33. (c)	34. (a)	35. (b)
36. (c)	37. (d)	38. (d)	39. (d)	40. (d)
41. (c)	42. (d)	43. (b)	44. (c)	45. (a)
46. (b)	47. (d)	48. (d)	49. (d)	50. (c)
51. (c)	52. (c)	53. (c)	54. (b)	55. (a)
56. (a)				

Fill in the Blanks :	
1. infinity	2. Wilson & Harris
3. decoupling inventory	4. consumption cost
5. demand	6. inventory horizon
7. operating doctrine	8. infinity
9. decreases	10. re-order period or ROP
11. inventory cycle period	12. increases by $\sqrt{2}$ times
13. $\dfrac{C_c}{C_c + C_s}$	14. EOQ/consumption rate
15. $\dfrac{\text{Demand}}{\text{EOQ}}$	16. square
17. fast moving, slow moving, non moving, dead	18. ABC analysis
19. $\sqrt{2\,C_0,\ C_c\,D} + pD$	20. Lead time
21. order cost/order (C_0), carrying cost on avg. qty. (C_c) unit price (p) quantity 'Q'	22. $\dfrac{Q.t}{2}$
23. consumption quantity in the period	24. 3 units
25. 200 units	26. 5 orders
27. 60 days	28. when and how much
29. buffer or reserve stock	30. demand dependent item

8 Chapter

Queue Models (Waiting Lines)

CHAPTER AT A GLANCE

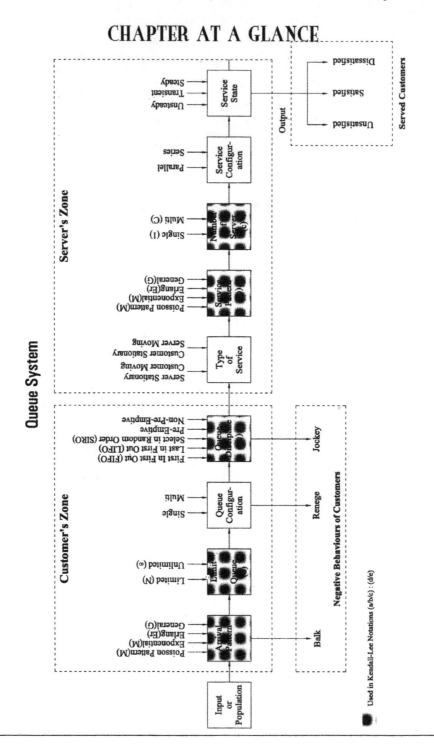

8.0 *Introduction*

One of the common situations in our daily life occurs with waiting or queue system. Rather, there will not be any person on this earth who did not experience the troubles due to waiting such as at bus stops, ticket booths, doctor's clinic, petrol bunks, bank counters, traffic lights and so on. Queues are very common in industries also, in shops where the machines wait for repair/maintenance, operators wait at tool cribs, material waiting for operations etc. These queues ruin the time of customers and it may even cost hugely in some occasions. The minimisation of waiting time and thus costs are the primary concerns of these queue models in Operation Research.

8.1 *Terminology*

The following terms are commonly used in queue models.

1. **Customers :** The persons or objects that require certain service are called customers.

2. **Server :** The person or an object or a machine that provides certain defined service is known as server.

3. **Service :** The activity between server and customer is called service that consumes some time.

4. **Queue or Waiting Line :** A systematic or a disciplined arrangement of a group of persons or objects that wait for a service is called queue or waiting line.

5. **Arrival :** The process of customers coming towards service facility or server to receive certain service is Arrival.

6. **Mean Rate of Arrival :** The average number of customers arriving per unit time is called mean rate of arrival and is denoted by 'λ' (read as LAMBDA).

   λ = Total no. of customers arriving/total time taken.

7. **Mean Inter-arrival Time :** It is the average time gap between two consecutive arrivals of customers. It is the inverse of mean arrival rate i.e.,
   $$1/\lambda = \frac{\text{Total arrival time}}{\text{Number of arrivals}}.$$

8. **Rate of Service :** It is the average number of customers served per unit time and is denoted by μ (read as MUE).
   $$\mu = \frac{\text{Total number of customers served}}{\text{Total service time}}$$

9. **Mean Service Time :** It is the average time taken by the server to serve a customers and is equal to inverse of service rate i.e., $1/\mu = \dfrac{\text{Total service time}}{\text{Number of customers served}}.$

8.2 *When Does A Queue Result?*

Suppose a queue system whose arrival rate is λ and service rate is μ.

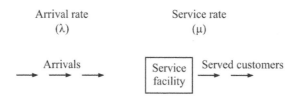

FIGURE 8.1 :   QUEUE OR WAITING LINE

Now, in this system any of the following three relations must occur (by law of trichotomy).

(i) $\lambda > \mu$ (ii) $\lambda = \mu$ (iii) $\lambda < \mu$

(i) When $\lambda > \mu$, i.e., the rate of arrival is greater than rate of service, i.e., number of customers arriving is greater than number of customers served per unit time (time kept constant), the system results in piling up of arrived customers waiting for service and thus a queue results.

(ii) If $\lambda = \mu$, i.e., number of customers arrived is equal to customers served, there will not be any waiting, and this queue system has cent percent efficiency.

(iii) If $\lambda < \mu$, i.e., number of customers arriving is less than number of customers served per unit time. This means that server will be free, waiting for customers to arrive. Thus the queue results in server's point of view.

Thus queue results where there is an imperfect matching between the rate of arrival and the rate of service.

8.3 *Costs Associated with Waiting Lines (Queuing)*

Queuing theory, can be applied to variety of operational situations where there is an imperfect matching between the rate of arrivals and the rate of service. The rate of arrival and service however, depend on the probability studies and the imperfect matching usually occurs due to inability of server to predict accurately the arrival and service time. These may cost on customer due to waiting or on server to operate the service facility. Thus the level of service (either service rate or the number of service facilities) is determined by cost analysis of the following two conflicting costs.

1. Cost of offering service.

2. Cost incurred due to delay in offering service or due to keeping customer waiting.

Now, we can analyse these as follows.

1. **Cost of Service :** If existing service level in terms of rate of service or number of service facilities (channels) is reduced, the cost of operation comes down.

In other words, increase in level of service increases the cost of operating service facilities.

2. **Waiting Costs :** In contrast to the above, the reduction in existing service level will hike the customers waiting time or may result in long queues. Thus the increase in service level will decrease the waiting costs and decrease in service level will increase the waiting costs.

Thus the optimum level of service can be obtained by achieving a balance between the above two costs or at the level where the total cost shows minimum. This is interpreted through a graph shown below.

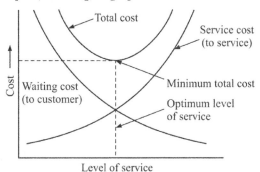

FIGURE 8.2 : COST ANALYSIS OF QUEUE SYSTEM

8.4 *Queue System : Salient Features*

A queue system may be studied by diving into two subsystems viz arrival system and service system. The arrival system is based on customer's behaviours and their process of arrival while the service system is dependent on the service mechanism and server's behaviour.

These are diagrammatically shown in the block diagram given at the beginning of this chapter.

The essential features shown in the block diagram are discussed here below in detail.

1. **Input Source or Population :** The input source or population is a set that contains the probable potential customers to come out for service. Thus the input source need not be homogenous and may consist of sub populations. The input source may contain the subset of patients for the queue system at a doctor's clinic, a sub set of vehicle owners for queue system of petrol bunk, a subset of passengers who may enter a waiting line for a bus and so on. Another point to be noted here is that all the population need not be customers, but as they come out of population with an intention to get the defined service are considered as the customers to that service system.

2. **Arrival Pattern :** The customers are expected to arrive at their own convenience and conditions. However, the time gaps between any two consecutive arrivals, if noticed, will be fit in certain fashion or pattern. These patterns of arrival times often follow one of the following probability distribution.

   (i) Poisson distribution (represented by M)

   (ii) Exponential distribution (represented by M)

(iii) Erlang distribution (represented by E_r)

(iv) General pattern (represented by G)

3. **Limit of Queue : (Restricted/unrestricted queues**) The limit of queue is some times restricted by server's behaviour. If the limit is imposed by the server, the arrivals will be limited for a given period. For example, if a doctor has condition that he would treat 50 out-patients only per day, the queue will be limited or finite and 51st out-patient will have to be in queue to enter the queue i.e., he will be in secondary queue while the first 50 will be in primary queue. However, if there is no restriction, the queue is said to be unlimited or infinite. The limited queue is denoted by N while unlimited queue by ∞.

Limited queue may be adopted by space restrictions or other conditions also.

For instance, a service facility such as a cinema hall, has restricted accommodation which can not accommodate more than its space. In such cases the server will not allow further arrivals unless some space is vacated to accommodate the new entry. In some other cases, the limited queue system may be adopted to prevent a formal queue and to avoid nuisances. These queues are often referred to as 'take-a-number' or 'take-appointment' or 'reservation' policy. However, the queue may be unlimited for getting into this 'reservation'.

4. **Queue Discipline :** The queue discipline is the order in which the customer is selected from the queue for service. There are numerous ways in which customers in queue can be served of which some are listed below.

   (a) *First In First Out (FIFO) or First Come First Serve (FCFS)* : It is the discipline in which the customers are served in the chronological order of their arrivals.

   e.g. *Tickets at a cinema hall; sales at a grocery shop, trains on a (single line) platform etc.*

   (b) *Last In First Out (LIFO) or Last Come First Serve (LCFS)* : If the service is made in opposite order of arrivals of customers, i.e., who ever comes last is served first and first obviously goes to last, it is called LIFO or LCFS system.

   e.g. *Stack of plates; loading and unloading a truck or go-down; office filing of papers in chronological orders; wearing socks and shoes; dressing a shirt and coat over it, packing systems etc.*

   (c) *Service In Random Order (SIRO)* : By this rule, the customer for service is picked up at random, irrespective of their arrivals.

   e.g. *Lottery system from which one is picked up, the dresses waiting in a ward robe from which one is to be chosen, food stuffs in a buffet, sales counter of commodities or vegetables etc.*

   (d) *Priority Service* : Under this rule, the server gives priority to certain customer(s) due to some importance or prestigeous or high cost group of the customers.

   e.g. *A telephone urgent call given to a customer is charged at higher price; a separate counter for cheques at a electricity bill payment counter; priority given at (APSRTC) reservation counter who buys a CAT(Concessinional Annual Ticket) card etc.*

(e) ***Pre-emptive Priority Rule :*** Under this rule, highest priority is given to certain customer(s) irrespective of their arrival and costs. The customer is allowed to enter into service immediately after entering into queue system, even if another (lower priority) customer is already in service. i.e., the lower priority customer who is in service will be interrupted (pre-empted) to facilitate the special customer.

  e.g. *An emergency case arriving at a doctor's clinic who is attending to a regular out-patient. (The doctor will stop his service to the regular patient and immediately rushes to emergency case,)*

  A minister or VIP coming to receive a service at counter is given highest priority,

  An expediting in production shop.

(f) ***Non Pre-emptive Priority Rule :*** This is also a rule by priority to the special customer but the priority will not emptivate the current service. The service to the special customer starts immediately after the completion of current service.

  e.g. *A medical representative will be given appointment immediately after the current service to an out-patient at a doctor's clinic.*

  In a production shop, backlog planning, chase planning is in accordance with non pre-emptive in many occasions.

Note : In the above disciplines (a) and (b) are called static queue disciplines while the rest are dynamic queue disciplines.

5. **Customer's Behaviour :** A customer coming out of population with an intention to receive certain service may be prepared to wait till he gets service. If he waits hoping to get the service, he is called 'patient customer' or 'positive customer' or 'optimistic customer'. Where as a customer may walk out of the queue without getting service due to various reasons such as long queue in front of him or not having patience to wait etc., he is said to be an 'impatient customer' or 'negative customer' or 'pessimistic customer'. Various types of negative behaviours are defined here below :

  (a) ***Balking :*** A customer who gets discouraged by seeing the length of the queue before him and thinks that he may not get service, may walk out or may not join the queue. He is said to be *'Balking'*.

  (b) ***Reneging :*** A customer who joins the queue and waits for some time but leaves the queue due to intolerable delay or impatient to wait any longer is said to be *'Reneging'*.

  (c) ***Jockeying :*** A customer who moves from one queue to another hoping to receive a more quick service is said to be *'Jockeying'*.

  Thus, the customers who come out of their original queue showing a negative behaviour that he may not get service. However, a customer who receives service also may have negative bahaviour as given below :

(d) *Unsatisfied Customer :* A customer who is not satisfied by the service by the quantity is said to be unsatisfied. For example, a customer of an hotel may not be satisfied with the quantity of food supplied to him though it is tasty.

(e) *Dissatisfied Customer :* A customer who is not satisfied by the quality of service is said to be dissatisfied. For instance, a customer getting tasteless food at an hotel though supplied in huge quantities is dissatisfied.

The customers unsatisfied or dissatisfied usually expect certain level of service and when they do not have the actual, equivalent to their expectation they will not have satisfaction. However, if a customer has the match between his expectation and the actual receipt, he will be *satisfied* and if the actual is more than his expectation, he will be '*delighted*'.

6. **Type of Service :** The service system may be in two ways as given below :

   (a) Customer stationary and server moving.

   e.g. *(i) Arranged meal at which server brings what customer desires.*
   (ii) A machine waiting for repair.

   (b) Server stationary and customer moving.

   e.g. *(i) A buffet meal where customer goes to buffet table and gets what he wants.*
   (ii) An aeroplane waiting for a run way for landing.

7. **Service Patterns :** Similar to that of inter arrival times, the service time taken by the server to serve each customers varies. These variations depend on the rate of arrival and server's behaviour. However, these times can be fit into one of the following probability distribution patterns.

   (a) Poisson distribution (denoted by M)

   (b) Exponential distribution (denoted by M)

   (c) Erlang distribution (denoted by E_r)

   (d) General fashion (denoted by G)

8. **Number of Servers :** We have already learnt that a queue system and its associated costs depend on the level of service which includes both the rate of service and number of service facilities. On one hand the rate of service is dependent on the service time distribution patterns while on the other hand the level of service can be manipulated by arranging number of servers. There are two ways, of assumption to the queue problem with regard to number of service facilities.

   (a) Single server system represented by 1.

   (b) Multiserver system represented by S.

9. **Service Configurations :** The efficiency and capacity of a queue system can be increased by arranging the service systems in an effective way, particularly when multi-service facilities are available. Various arrangements of service facilities can be classified into following three ways.

(a) Series Configuration :

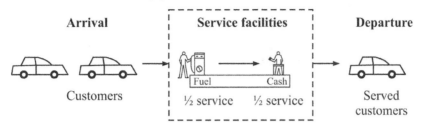

FIGURE 8.3 : SERIES CONFIGURATION

Here service is divided into certain parts served sequentially

(b) Parallel Configuration :

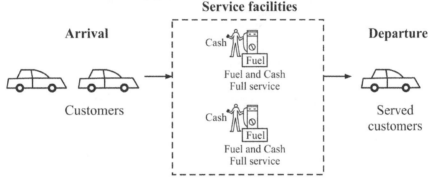

FIGURE 8.4 : PARALLEL CONFIGURATION

Here arrivals are divided into parts and separated in to different queues.

(c) Combined Configuration : This configuration is a combination of the above two configuration.

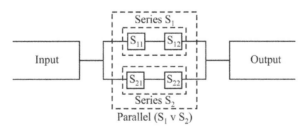

FIGURE 8.5 : COMBINATION CONFIGURATION

10. **State of Service :** We have learnt that the service system is influenced by service time distribution and on server's behaviour. These are inter dependent on arrival patterns and customer behaviour. On close observation of the changes that occur in service system, we can find three states of service viz.,

(a) Unsteady state.   (b) Transient state.   (c) Steady state.

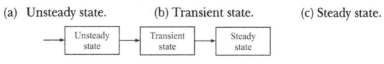

FIGURE 8.6 : QUEUE STATES

When the service system is just started, the server will not be able estimate how much time he has to take per customer. Also, he may be in a bit of confusion and non-uniformity due to initial conditions and hence takes more time for service. This state in which the server does non-uniform service is *unsteady state.*

From this state he slowly moves towards attaining uniformity. This consumes some time and the server during this time is said to be in *transient state.*

After getting enough hands on experience, the server picks up and acquires complete knowledge on the service system and takes almost same times at uniform rate to serve each customer. This state is called *steady state.* Further, this steady state can exist when the ratio of arrival rate to service rate is less than unity.

$$\text{Steady state condition : } \quad \frac{\lambda}{\mu} < 1$$

11. **Departure and Output :** A served customer goes out of the service zone and emerges as output. This activity is often known as departure in queuing terminology. A customer departed from the service zone may have been satisfied or unsatisfied or dissatisfied. This has already been explained in fifth point of this section. In the above eleven features, first five are included in customer zone while the rest in server zone.

8.5 *Queuing Models : Kendall - Lee Notations*

Queuing models can be defined by five most significant features of the above explained eleven points. D.G. Kendall in 1953 has noticed three of them in the form $(a/b/c)$ and later A.M Lee in 1966 added two more in the form (d/e) to describe a queue model. These are known as Kendall-Lee notations in the standard format as $(a/b/c) : (d/e)$ where

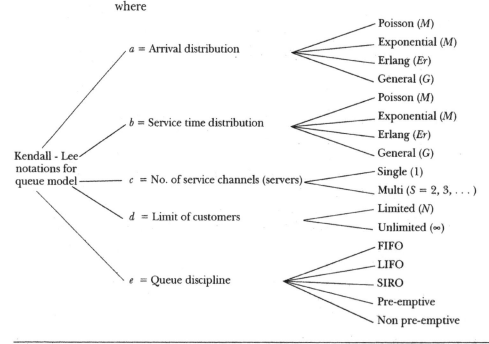

A queue model is blend of the above five points, which is diagramatically shown below.

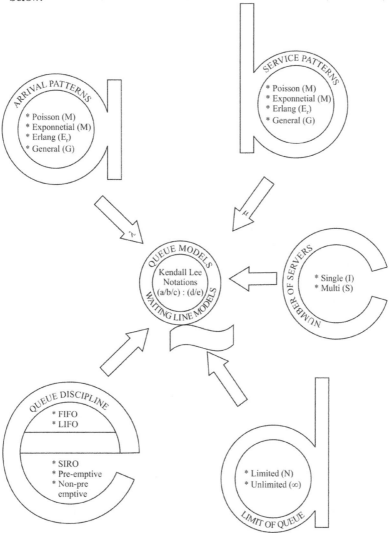

FIGURE 8.7 : QUEUE MODEL : KENDALL - LEE NOTATIONS

Thus if queue system is composed of Poisson arrival distribution, exponential service time distribution (both of these follow 'Markovian' property and hence represented by M), single channel, unlimited number of customers in queue can enter and follow a discipline, First Come First Serve, the queue is described as

$$(M/M/1) : (\infty/\text{FIFO})$$

Similarly $(M/E_r/s) : (N/\text{SIRO})$ means Poisson arrivals, Erlangian service time distribution, s number of channels, limited number of customers are allowed, selected in random order.

8.6 *Parameters of Queue System*

Every queue system has two parameters upon which entire queue system depends. These are

λ = mean arrival rate (or $1/\lambda$ = mean inter arrival time)

μ = mean service rate (or $1/\mu$ = mean service time)

8.7 *Axioms of Poisson Process*

The arrivals in a Poisson's process are characterised by the following three axioms.

Axiom 1: In a non-overlapping interval, the number of arrivals are statistically independent. This means Poisson has independent increments.

Axiom 2: The probability of occuring more than one arrival between the time interval, t and $(t + \Delta t)$, is $0\,(\Delta t)$. This means that the probability of occuring two or more arrivals during the small time interval Δt is negligible. Hence we have

$$P_0\,(\Delta t) + P_1\,(\Delta T) + 0\,(\Delta t) = 1$$

Axiom 3: The probability that an arrival occurs between time, t, and time $(t + \Delta t)$, is equal to $\lambda\,(\Delta t) + 0(\Delta t)$.

Hence, we have, $\qquad P_1(\Delta t) = \lambda\,(\Delta t) + 0\,(\Delta t)$

where $\qquad\qquad\qquad \lambda$ = a constant and independent of $N(t)$

$\qquad\qquad\qquad N(t)$ = total number of arrivals upto time t

$\qquad\qquad\qquad N(0) \quad = 0$

$$0(\Delta t) \quad = \underset{\Delta t \to 0}{Lt}\ \frac{0(\Delta t)}{\Delta t} = 0$$

8.8 *Operating Characteristics of Queue System*

Some of the operational characteristics, which are of general interest to estimate the performance of the existing queue system or/and to design a new queue system are listed below:

1. **Expected number of customers in queue or size (or length) of queue (L_q) :** This is the average number of customers waiting for service and is denoted by L_q.

2. **Expected number of customers in system or size (or length) of the system (L_s) :** It is the average number of customers expected to be waiting in queue and being in service. It is denoted by L_s.

3. **Expected waiting time in queue (W_q):** This is average time spent by a customer in queue before the commencement of the service. This is denoted by W_q.

4. **Expected Waiting time in the System (W_s) :** It is the total time spent by a customer in the system, whcih includes the waiting time in queue and the service time. It is represented with W_s.

$$W_s = W_q + \text{mean service time} = W_q + \frac{1}{\mu}.$$

5. **The server utilisation factor or fraction of busy period or traffic intensity :**
It is the proportion of the time that a server actually spends with the customers. In other words it is the probable period of server being busy. It is designated with a greek letter ρ (rho).

$$\rho = \frac{\text{Average service completion time} (1/\mu)}{\text{Average inter arrival time} (1/\lambda)} = \frac{\lambda}{\mu}$$

General Relationship Between Operating Characteristics :

1. Between L_q and L_s

   L_s = No. of customers in the system

   = No. of customers in queue + number of customers in service.

   $$\boxed{L_s = L_q + \frac{\lambda}{\mu}}$$

2. Between W_q and W_s

   W_s = Waiting time in system.

   = Waiting time in queue + waiting time in service.

   $$\boxed{W_s = W_q + \frac{1}{\mu}}$$

3. Between L_q and W_q

   $$\frac{L_q}{W_q} = \frac{\text{No. of customers in queue}}{\text{Waiting time in queue}}$$

   $$= \lambda \quad \text{i.e.,} \left(\frac{\text{No. of arrivals into queue}}{\text{Total time of arrivals into queue}} \right)$$

   $\therefore$ $\boxed{L_q = \lambda W_q}$

4. Between L_s and W_s :Similar to the above

   $$L_s = \lambda W_s$$

5. Expected number of customers served per busy period.

   $$L_b = \frac{L_s}{P (n \geq s)} = \frac{\mu}{\mu - \lambda}$$

   $\therefore$ $$L_s = L_b \cdot P (n \geq s) = \frac{\mu}{\mu - \lambda} \cdot P (n \geq s)$$

   $$= \frac{\mu}{\mu - \lambda} \cdot \rho \text{ for } (M/M/1) : (\infty/\text{FIFO})$$

6. Expected waiting time during busy period,

   $$W_b = \frac{W_q}{P_b} = \frac{1}{\mu - \lambda}$$

   $\therefore$ $$W_q = W_b \cdot P_b = \frac{1}{\mu - \lambda} \cdot P_b$$

   $$W_q = \frac{\rho}{\mu - \lambda} = \frac{\lambda}{\mu (\mu - \lambda)} \text{ for } (M/M/1) : (\infty/\text{FIFO}).$$

8.9 *Model - 1 :(M/M/1) : (∞/FIFO)*

The following expressions are used characterise this model.

1. Probability Density Function $P_n = \left(\dfrac{\lambda}{\mu}\right)^n \cdot \left(1 - \dfrac{\lambda}{\mu}\right)$

   $= \rho^n (1 - \rho)$ where $\rho < 1$; $n = 0, 1, 2 \ldots$.

2. Probability of system being empty (i.e., no customer waiting)

   $\therefore \quad P_0 = 1 - \dfrac{\lambda}{\mu} = 1 - \rho$

3. Probability of server being busy $(P_b) = 1 - P_0 = \lambda/\mu$.

4. Expected size of system

   $\therefore \quad L_s = \dfrac{\mu}{\mu - \lambda} \cdot P_b = \dfrac{\mu \left(\dfrac{\lambda}{\mu}\right)}{\mu - \lambda} = \dfrac{\lambda}{\mu - \lambda} = \dfrac{\rho}{1 - \rho}$

5. Expected length of queue

   $\therefore \quad L_q = \dfrac{\lambda^2}{\mu (\mu - \lambda)} = \dfrac{\rho^2}{1 - \rho}$

6. Expected waiting time in queue

   $W_q = \dfrac{L_q}{\lambda} = \dfrac{\lambda}{\mu (\mu - \lambda)}$

7. Expected waiting time in system

   $W_s = W_q + \dfrac{1}{\mu} = \dfrac{L_s}{\lambda} = \dfrac{1}{\mu - \lambda}$

8. Expected waiting time in the queue for busy system

   $W_b = \dfrac{W_q}{P_b} = \dfrac{W_q}{1 - P_0} = \dfrac{1}{\mu - \lambda} = W_s$

9. Expected number of customers served per busy period.

   $L_b = \dfrac{L_s}{P_b} = \dfrac{L_s}{1 - P_0} = \dfrac{\mu}{\mu - \lambda} = \dfrac{1}{1 - \rho}$

10. Probability that k or more customers waiting in the system.

    i.e., $\quad P (n \geq k) = \left(\dfrac{\lambda}{\mu}\right)^k \quad$ and $\quad P (n > k) = \left(\dfrac{\lambda}{\mu}\right)^{k+1}$

11. Probability of waiting 't' minutes or more in queue

    $P (W \geq t) = \displaystyle\int_{t}^{\infty} \dfrac{\lambda}{\mu} (\mu - \lambda) \, e^{-(\mu - \lambda) t} \, dt$

ILLUSTRATION 1 ────────────────────────────────

In a bank, cheques are cashed at a single 'Teller' counter. Customers arrive, at the counter in a Poisson manner at an average rate of 30 customers per hour. The Teller takes, on an average a minute and a half to cash cheque. The service time has been shown to be exponentially distributed.

(i) Calculate the % of time the Teller is busy and

(ii) Also calculate the average time a customer is expected to wait.

[JNTU (Mech.) 98/S]

Solution :

In the given problem, arrival process is in Poisson manner (M), service time distribution is in exponential process (M), No. of servers is 1, No. of customers allowed in queue process is ∞, and the service discipline is FCFS.

Hence the queue model is (M/M/1) : (∞ / FCFS) (according to Kendall – Lee notation).

Arrival rate $\quad (\lambda) = 30$ customers/hr.

Serivce time $\left(\dfrac{1}{\mu}\right) = 1\dfrac{1}{2}$ min $= \dfrac{3}{2}$ min

$$= \dfrac{3}{2} \times \dfrac{1}{60} \text{ hrs.} = \dfrac{1}{40} \text{ hours.}$$

Service rate $\quad (\mu) = 40$ customers/hr.

$\lambda < \mu$, hence the model–I implies at steady state

(i) Traffic intensity $(\rho) = \dfrac{\lambda}{\mu}$

(or) factor of busy period $= \dfrac{30}{40} = \dfrac{3}{4} = 0.75$

$\therefore \quad$ The percentage of the time the Teller is busy is
$0.75 \times 100 = 75\%$

(ii) Average length of queue $\quad (L_q) = \dfrac{\lambda}{\mu - \lambda}$ or $\dfrac{\rho}{1 - \rho} = \dfrac{30}{40 - 30} = 3$

$\therefore \quad$ Average waiting time that a customer is expected to wait

$$W_q = \dfrac{L_q}{\lambda} = \dfrac{3}{30} = \dfrac{1}{10} = 0.1 \text{ hours} = 6 \text{ min.}$$

ILLUSTRATION 2

The mean arrival rate to a service centre is 3 per hour. The mean service time is 10 minutes. Assuming Poisson arrival rate and exponential servicing time,

(i) Utilisation factor.

(ii) Probability of two units in the system.

(iii) Expected number of units in the system.

(iv) Expected number of units in the queue.

(v) Expected time in minutes, customer has to wait in the system.

Solution :

In the above problem let us first identify the queue system.

Arrival pattern is Poisson distribution i.e., M

Service pattern is exponentially distributed i.e., M

Number of service centres is 1.

Limit of customers arrival (No limit is imposed) $= \infty$.

Queue discipline = FIFO (When there is no mention of disciple, it may assumed to be FIFO)

Thus the above system can be described as $(M/M/1) : (\infty/FIFO)$ according to Kendall-Lee.

Next, parameters of the system are to be identified.

Mean rate of arrival $= \lambda = 3$ per hour.

Mean service time $\left(\dfrac{1}{\mu}\right) = 10 \; Min = \dfrac{10}{60} = \dfrac{1}{6} \; hrs.$

$\therefore \qquad \mu = 6$ per hour.

Traffic intensity $(\rho) = \dfrac{\lambda}{\mu} = \dfrac{3}{6} = \dfrac{1}{2}$

As $\lambda/\mu < 1$, steady state can exist.

Now, we can find all the operating characteristics that govern the queue system.

(i) Utilisation factor $= \rho = \dfrac{\lambda}{\mu} = \dfrac{3}{6} = \dfrac{1}{2}$

(ii) Probability of two units in the system $P \; (n = 2) = \rho^2 = \left(\dfrac{\lambda}{\mu}\right)^2 = \left(\dfrac{3}{6}\right)^2 = \left(\dfrac{1}{2}\right)^2 = 0.25$

(iii) Expected number of units in the system $L_s = \dfrac{\lambda}{\mu - \lambda} = \dfrac{3}{6 - 3} = \dfrac{3}{3} = 1$

(iv) Expected number of units in the queue $L_q = \dfrac{\lambda^2}{\mu \; (\mu - \lambda)} = \dfrac{9}{6 \; (6 - 3)} = \dfrac{9}{6 \times 3} = \dfrac{1}{2}$

(v) Expected time in minutes customer has to wait in the system

$$W_s = \dfrac{1}{\mu - \lambda} \times 60 = \dfrac{60}{6 - 3} = \dfrac{60}{3} = 20 \text{ minutes.}$$

ILLUSTRATIONS 3 ───────────────────────────────────

Customers arrive at a box office window, being managed by a single individual, according to a Poisson input process with mean rate of 30 per hour. The time required to serve a customer has an exponential distribution with a mean of 90 seconds. Find the average waiting time of a customer. Also determine the average number of customers in the system and average queue length.

[JNTU (Mech., ECE) 1999/CCC)]

Solution :

Arrival process is Poisson i.e., Markovian represented by '*M*'.

Service distribution is Exponential which can be represetned as '*M*' (Markovian)

Number of servers $= 1$

No. of people (customers) allowed to enter queue $= \infty$ and queue descipline is understood as *FCFS* (First Come First Serve).

Hence, the given queue model is represented as

$(M/M/1) : \qquad (\infty/FCFS)$

Also given that,

$$\text{Mean arrival rate} = \lambda = 30 \text{ per hour}$$

$$= \frac{30}{60 \times 60} \text{ per second}$$

$$\lambda = \frac{1}{120} \text{ per second}$$

$$\text{and service time} \left(\frac{1}{\mu}\right) = 90 \text{ seconds}$$

$$\therefore \text{ mean service rate } (\mu) = \frac{1}{90} \text{ per second}$$

Fraction of Busy period or Traffic intensity $(\rho) = \dfrac{\lambda}{\mu} = \dfrac{\frac{1}{20}}{\frac{1}{90}} = \dfrac{90}{120} = \dfrac{3}{4} = 0.75$

Average numbe of customers in the system (L_s)

$$= \frac{\lambda}{\mu - \lambda} \text{ or } \frac{\rho}{1 - \rho} = \frac{0.75}{1 - 0.75} = \frac{0.75}{0.25} = 3 \text{ customers}$$

Average number of customers in the system $(L_s) = 3$ customers

Average queue length $(L_q) = \dfrac{\lambda^2}{\mu(\mu - \lambda)}$ or $\dfrac{\rho^2}{1 - \rho} = 0.75 \times 3 = 2.25$

Average waiting time of a customer in queue (W_q)

$$= \frac{L_q}{\lambda} = \frac{2.25}{1/120} = 4.5 \text{ minutes or } 270 \text{ seconds.}$$

Average waiting time of a customer in the system $W_s = \dfrac{L_s}{\lambda}$ or

W_q + service time = 6 minutes or 360 seconds

ILLUSTRATION 4

At a certain petrol pump, customers arrive according to a poisson process with an average time of 5 minutes between arrivals. The service time is exponentially distributed with mean time of 2 minutes. Find (i) the average queue length (ii) the average number of customers in the queueing system (iii) average time spent by a customer in the petrol pump and (iv) average waiting time of a customer before receiving the service. (v) if the waiting time in queue is 4 min, a second pump will be opened. What should be the arrival rate for opening second pump.

[Dr. BRAOU (MBA) 97]

Solution :

This is an (M/M/1) : (∞/FCFS) queue model.

$$\text{Average inter arrival time} = \frac{1}{\lambda} = 5 \text{ min} = \frac{1}{12} hr$$

$$\text{or} \quad \lambda = 12 / \text{hr}$$

$$\text{Average service time} = \frac{1}{\mu} = 2 \text{ min} = \frac{1}{30} hr$$

$$\text{or} \quad \mu = 30 / \text{hr}$$

As $\lambda < \mu$, the steady state solution exists,

(i) Average length of queue (L_q) $= \dfrac{\lambda^2}{\mu(\mu - \lambda)} = \dfrac{144}{30 \times 18} = \dfrac{4}{15}$

(ii) Average length of system $(L_s) = \dfrac{\lambda}{\mu - \lambda} = \dfrac{12}{18} = \dfrac{2}{3}$

(iii) Average time spent in the petrol pump (system)

$$W_s = \dfrac{1}{\mu - \lambda} = \dfrac{1}{18} \text{ hr} = 3.33 \text{ min}$$

(iv) Average waiting time of customer in queue i.e., before receiving the service

$$(W_q) = \dfrac{\lambda}{\mu(\mu - \lambda)} = \dfrac{12}{30} \times \dfrac{1}{18} = \dfrac{1}{45} \text{ hr} = 1.33 \text{ min}$$

(or $W_q = W_s - \dfrac{1}{\mu} = 3.33 - 2 = 1.33$ min)

(v) To open second pump $W_q' = 4$ min

$$W_q' = \dfrac{\lambda'}{\mu(\mu - \lambda')} = 4 \text{ min} = \dfrac{1}{15} \text{ hrs.}$$

$$\dfrac{\lambda'}{30(30 - \lambda')} = \dfrac{1}{15}$$

$\lambda' = 20$/hr and inter arrival time = 3 min.

ILLUSTRATION 5 ⸻

A television repairman finds that the time spent on his jobs has a exponential distribution with mean 30 minutes. If he repairs sets in the order in which they came in, and if the arrival of sets follow a Poisson distribution approximately with an average rate of 10 per 8 hour day, what is the repairman's expected idle time each day? How many jobs are ahead of the average set just brought in?

[JNTU (Mech.) 94/P]

Solution :

Arrival fashion is Poisson distribution (M); service pattern is exponential distribution (μ), number of servers is 1, no limit is imposed, repair discipline is FIFO.

Therefore, according to Kendall-Lee, the queue system $(M/M/1) : (\infty/\text{FIFO})$

and mean arrival rate $(\lambda) = \dfrac{10}{8} = \dfrac{5}{4}$ sets per hour.

and mean service rate $(\mu) = \dfrac{1}{30} \times 60 = 2$ sets per hour

$$\dfrac{\lambda}{\mu} = \dfrac{5}{4}/2 = \dfrac{5}{8}$$

$\dfrac{\lambda}{\mu} < 1$, hence steady state can exist.

(i) Expected idle time of repairman each day

number of hours for which the repairer is busy in 8 hour day is

$$8 \times \dfrac{\lambda}{\mu} = 8 \times \dfrac{5}{8} = 5 \text{ hours}$$

$\therefore$ Idle time for repairman in an 8 hour day = $8 - 5 = 3$ hours

(ii) Expected number of TV sets in the system

$$L_s = \frac{\lambda}{\mu - \lambda} \text{ or } \frac{\rho}{1 - \rho} = \frac{\frac{5}{8}}{1 - \frac{5}{8}} = \frac{5}{3} = 2 \text{ sets approx..}$$

ILLUSTRATION 6

In a railway marshalling yard, goods trains arrive at a rate of 30 trains per day. Assuming that the inter-arrival time follows an exponential distribution and the service time (the time taken to dump a train) distribution is also exponential with an average of 36 minutes, calculate (i) expected queue size (line length), (ii) probability that the queue size exceeds 10. If the input of trains increase to an average of 33 per day, what will be the change in (i) and (ii).

[JNTU (Mech.) 97, OU 83]

Solution :

From the data, the given queue system has exponential arrival (*M*), exponential service pattern (*M*,) single track (service facility), unlimited queue size and FCFS discipline.

∴ According to Kendall-Lee, the system is $(M/M/1) : (\infty/FCFS)$

We have the parameters as

$$\lambda = \frac{30}{60 \times 24} = \frac{1}{48} \text{ trains per minute}$$

and $\mu = \dfrac{1}{36}$ trains per minute.

Then traffic intensity $(\rho) = \dfrac{\lambda}{\mu} = \dfrac{36}{48} = 0.75$ as $\rho < 1$, steady state can exist.

∴ We can proceed with the problem as follows :

(i) Expected queue size (line length)

$$L_s = \frac{\rho}{1 - \rho} = \frac{0.75}{1 - 0.75} = 3 \text{ trains.}$$

(ii) Probability that the queue size exceeds 10,

$$P(n \geq 10) = \rho^{10} = (0.75)^{10} = 0.056$$

Now, if input increase to 33 trains per day, then we have

$$\lambda = \frac{33}{60 \times 24} = \frac{11}{480} \text{ trains per minute and } \mu = \frac{1}{36} \text{ trains / min.}$$

Traffic intensity $\rho = \dfrac{\lambda}{\mu} = \dfrac{11}{480} \Big/ \dfrac{1}{36} = 0.83$

as $\rho < 1$, steady state can exist.

Hence recalculating the values of (i) and (ii), we have

(i) $L_s = \dfrac{\rho}{1 - \rho} = \dfrac{0.83}{1 - 0.83} = 5 \text{ trains (approx.)}$

(ii) $P(x \geq 10) = \rho^{10} = (0.83)^{10} = 0.155 \text{ (approx.)}.$

ILLUSTRATION 7

Arrivals at a telephone booth are considered to be Poisson with an average time of 10 minutes between one arrival to the next. The length of phone call is assumed to be distributed exponentially, with mean 3 minutes.

(i) What is the probability that a person arriving at the booth will have to
(ii) The telephone department will install a second booth when convinced that an arrival would expect waiting for at least 3 minutes for a phone call. By how much the flow of arrivals should increase in order to justify a second booth.

(iii) What is the average length of queue that forms time to time.

(iv) What is the probability that it will take him more than 10 minutes altogether to wait for the phone call and complete his call?

[JNTU (Mech.) 94/CCC, (ECE) 94/CCC]

Solution :

Given problem has Poisson arrival, exponential service time distribution, single booth (service facility), unlimited queue with FCFS discipline.

Therefore in Kendall - Lee notations, it is represented as $(M/M/1) : (\infty/FCFS)$

Also given $\quad \lambda = \dfrac{1}{10}$ persons per minute

and $\quad \mu = \dfrac{1}{3}$ persons per minute

$\therefore \qquad \rho = \dfrac{\lambda}{\mu} = \dfrac{3}{10} = 0.3$

Since, $\rho < 1$ the steady state exists, hence we proceed

(i) Probability that a person has to wait at the booth

$$p\,(n > 0) = 1 - P_o = \dfrac{\lambda}{\mu} = 0.3$$

(ii) The installation of second booth can be justified if the arrival rate is more than the waiting time. Let λ' be increased arrival rate. Then expected waiting time in queue will be

$$W_q = \dfrac{\lambda'}{\mu\,(\mu - \lambda')} \quad \Rightarrow \quad 3 = \dfrac{\lambda'}{\dfrac{1}{3}\left(\dfrac{1}{3} - \lambda'\right)}$$

$$\lambda' = \dfrac{1}{3} - \lambda' \quad \Rightarrow \quad 2\,\lambda' = \dfrac{1}{3}$$

$$\lambda' = \dfrac{1}{6} = 0.16 \quad \text{or} \quad \dfrac{1}{\lambda'} = 6 \text{ minute.}$$

Here $W_q = 3$ (given)

and increase in the arrival rate, $= \dfrac{1}{6} - \dfrac{1}{10} = \dfrac{4}{60} = \dfrac{1}{15} = 0.06$ arrivals /minutes

(iii) Average length of a busy period

$$L_b = \dfrac{\mu}{\mu - \lambda} = \dfrac{\dfrac{1}{3}}{\dfrac{1}{3} - \dfrac{1}{10}} = \dfrac{0.33}{0.23} = 1.43 \text{ minutes}$$

(iv) Probability of waiting for 10 minutes or more is given by

$$P(W \geq 10) = \int_{10}^{\infty} \frac{\lambda}{\mu} (\mu - \lambda) e^{-(\mu-\lambda)t} dt = \int_{10}^{\infty} (0.3)(0.23) e^{0.23t} dt$$

$$= 0.069 \left[\frac{e^{-0.23t}}{-0.23} \right]_{10}^{\infty} = 0.03$$

This suggests that 3% of the arrivals on an average will have to wait for 10 minutes or more before they can use the phone.

Practice Problems

1. In a large computer industry, the average rate of system break down is 10 systems per hour. The idle time cost of a system is estimated to be Rs. 20 per hour. The working hours per day are 8. The manager of industry considers two mechanics for repairing. The first mechanic A takes about 5 minutes on an average to repair a system and demands wages Rs. 10 a hour. The second mechanic B takes 4 minutes in repairing and charges at rate of Rs. 15 a hour. Assuming rate of system breakdown is Poisson distributed and repair rate exponentially distributed which of the two mechanics should be appointed.

Solution :

For mechanic A :

$$\lambda = 10 \ /hr \text{ and } \mu = 12 \ /hr.$$

Total cost = total wages + cost of non productive time

= (Hourly rate × number of hours) + (Average number of systems) × (Cost of idle system / hour) × number of hours.

$$= 10 \times 8 + \frac{\lambda}{\mu - \lambda} \times 20 \times 8 = 80 + \frac{10}{12 - 10} \times 160 = 880 \text{ Rs.}.$$

For mechanic B :

$$\lambda = 10/hr : \mu = 15/hr$$

$$\text{Total cost} = 15 \times 8 + \frac{\lambda}{\mu - \lambda} \times 20 \times 8 = 120 + \frac{10}{15 - 10} \times 160 = 440 \text{ Rs.}$$

Though mechanic B charges (440 Rs.) is less, when compared to mechanic A who charges (880 Rs.) mechanic B can repair three systems more than mechanic A. i.e., 24 systems for 8 hours (8 × 3). Thus he can save (24 × 20) i.e., 480 Rs. per day, so that total cost of mechanic B becomes 920 Rs.. (440 + 480).

So, mechanic A has to be appointed.

2. Customers arrive at a one window drive in H.M.T sales counter in a Poisson fashion with mean 10 per hour. Service time per customer is exponential with mean 5 minutes. The space in front of the window including that for serviced car can accommodate a maximum of 3 cars other cars can wait outside.

(i) What is probability that a customers can drive directly to window.

(ii) What is probability arriving customer have to wait outside indicated space.

(iii) Expected waiting time of customers trucks per day.

Solution :

$$\lambda = 10/hr; \quad \mu = \frac{60}{5} = 12 \,/ \,\text{hour}.$$

(i) $P_0 + P_1 + P_2 = \left(1 - \frac{\lambda}{\mu}\right) + \frac{\lambda}{\mu}\left(1 - \frac{\lambda}{\mu}\right) + \left(\frac{\lambda}{\mu}\right)^2\left(1 - \frac{\lambda}{\mu}\right)$

$$= \left(1 - \frac{\lambda}{\mu}\right)\left[1 + \frac{\lambda}{\mu} + \left(\frac{\lambda}{\mu}\right)^2\right] = 0.42$$

(ii) $P\,(n \geq 3) = (\lambda/\mu)^3 = (10/12)^3 = 0.48$

(iii) $W_q = 5/2$ hour or 25 minutes.

3. A scooter mechanic has an average of four customers an hour. The average service time is six minutes. Arrivals are at Poisson fashion and service is done at exponential pattern. Find the

(i) Proportion of time during which shop is empty.

(ii) Probability of finding at least 1 customer.

(iii) Average time spent (service time including)

(iv) Average number of customers in system.

Solution :

$$\lambda = 4\,/\,\text{hour} \quad \mu = \frac{60}{6} = 10\,/\,\text{hour}; \qquad \rho = \frac{\lambda}{\mu} = 0.4$$

(i) $P_0 = 1 - \rho = 0.6;$ (ii) $P\,(n \geq 1) = 0.4;$

(iii) $W_s = \frac{1}{6}$ hour or 10 minutes   (iv) $W_s = \frac{2}{3}.$

4. The average rate of arrival of trains at a railway station during the busy period is 20 per hour and actual number of arrivals in any hour follows a Poisson distribution. The maximum capacity of the railway station is 60 trains per hour on an average in good weather and 30 trains per hour in bad weather but actual number of arrivals in any hour follows a Poisson distribution with these averages. When there is congestion the trains are made to wait for clearance outside the station.

(i) How many trains would be waiting outside on an average in good weather and bad weather.

(ii) How long would a train be waiting outside before getting into station in good and bad weather.

Solution :

$$\lambda = 20; \quad \mu = 60 \text{ in good weather and 30 in bad weather.}$$

(i) $L_q = \dfrac{1}{6}$ in good weather $\qquad ; \qquad \dfrac{4}{3}$ in bad weather

(ii) $\dfrac{1}{40}$ hour in good weather $\qquad ; \qquad \dfrac{1}{10}$ hour in bad weather

5. Find the (i) average queue length, (ii) average time spent, (iii) probability that there would be two customers in the queue in a sales counter manned by one sales person where the service time is about 6 minute and arrival rate is one person every 10 minutes.

Solution :

$$\lambda = \frac{60}{10} = 6 \text{ / hour}, \ \mu = \frac{60}{6} = 10 \text{ /hour}$$

(i) $L_q = 0.9$ customers

(ii) $W_s = \dfrac{1}{4}$ hour or 15 minutes

(iii) $P_2 = (\lambda/\mu)^2 \, (1 - \lambda/\mu) = 0.144 \ or \ 1.44\%$

6. A maintenance service facility has Poisson arrival rates, negative exponential service times and operates on a first come first serve queue discipline Break downs occur on an average of three per day with a range of zero to eight. The maintenance crew can service on an average six machines per day with a range from zero to seven. Find the

   (i) Utilisation factor of the service facility.

   (ii) Mean time in the system

   (iii) Mean number in the system in break down.

   (iv) Mean waiting time in the system.

   (v) Probability of finding two machines in the system.

   (vi) Expected number in the queue.

Solution :

$$\lambda = 3/\text{day}; \qquad \mu = 6/\text{day}$$

(i) $\rho = \dfrac{\lambda}{\mu} = \dfrac{3}{6}$ or 50% $\qquad\qquad$ (ii) $W_s = \dfrac{1}{3}$ day

(iii) $W_s = 1$ machine $\qquad\qquad$ (iv) $W_q = \dfrac{1}{6}$ day

(v) $P_2 = (\lambda/\mu)^2 \, (1 - \lambda/\mu) = 0.125$ $\qquad$ (vi) $L_q = \dfrac{1}{2}$ machine

(vii) Idle time $= 1 - \rho = 50\%$.

7. Customers for paying electricity bill arrive following Poisson distribution with an average time of 5 minutes between one arrival and the next. The time of service is 3 minute and it is at exponential pattern.

   (i) What is the probability that service counter is busy.

   (ii) How many service points should be established to reduce the waiting time to less than or equal to half of present waiting time.

Solution :

$$\lambda = 12/hour;\ \mu = 20/hour$$

(i) Busy period; $1 - P_0 = \lambda/\mu = 0.60$ and $W_q = \dfrac{3}{40}$ hour

(ii) $\dfrac{1}{8}$ hour; $W_s' = \dfrac{1}{\mu'} - \lambda = \dfrac{1}{16} = \dfrac{1}{\mu' - 12}$ or $\mu' = 28$ hour.

Thus number of booths required to achieve new service rate $= \dfrac{28}{20} = 1.4$ booths.

8. Customers arrive at a counter in Poisson fashion at an average rate of 15 customers per hour in a single teller counter where bank cheques all cashed. The service time is exponentially distributed where the cashier takes 3 minutes to cash the cheque.

   (i) Calculate percentage of time the teller is busy.

   (ii) Calculate average time customer is expected to wait.

Solution :

$$\lambda = 15/hour\ ;\ \mu = 60/3 = 20/hr$$

(i) Busy period $= 1 - P_0 = \dfrac{\lambda}{\mu} = \dfrac{3}{4}$ i.e., teller is busy for 75% of time.

(ii) $W_s = \dfrac{1}{10}$ hour or 6 minutes.

9. On an average 96 patients per 24 hour day require the service of an emergency clinic. Also on average, a patient requires 10 minutes of active attention. Assume facility can handle only one emergency at a time. Suppose it costs the clinic Rs. 100 per patient treated to obtain an average service time of 10 minute and each minute of decrease in this average time would cost Rs. 10 per patient treated. How much would have to be budgeted by the clinic to decrease average size of queue from 1 ⅓ patients to ½ patient.

Solution :

$$\lambda = 90/day\ ;\ \mu = \dfrac{(24 \times 60)}{10} = 144/day$$

(i) $L_q = \dfrac{4}{3}$ patients;

When L_q is changed from $\dfrac{4}{3}$ to $\dfrac{1}{3}$

Then μ changes to $\dfrac{1}{2} = \dfrac{\lambda^2}{\mu'(\mu' - \lambda)} = \dfrac{96^2}{\mu'(\mu' - 96)}$

$\mu' = 192$ patients a day.

New service time $= 24 \times 60/192 = 7.5$ min.

Reduced time $= 2.5$ min

Increased cost due to decreased service time $= 2.5 \times 10 =$ Rs. 25

Total cost $= 192 \times 100 + 25 =$ Rs. 19225

Present cost $=$ Rs. $144 \times 100 = 14400$

Increased budget $=$ Rs. 4825

10. A cashier at a bank works on exponential fashion. At what average rate must he work in order to ensure a probability of 0.9 so that a customer will not wait longer than 12 minutes. There is only one counter where customers are served at rate of 15 per hour who are arriving at Poisson fashion. **[JNTU (CSE) 2001]**

Solution :

$$\lambda = \frac{15}{60} = \frac{1}{4} \quad P\,\overline{(W \geq 12)} = 0.9 \implies P\,(W \geq 12) = 0.1$$

$$P\,[\text{(waiting)} \geq 12] = \int_{12}^{\infty} \frac{\lambda}{\mu}\left(\mu - \lambda\right) e^{-(\mu - \lambda)\,t} \times dt = 0.10\,\text{(given)}$$

Thus $\quad e^{3\,-\,12\,\mu} = 0.4\,\mu \qquad$ or $\dfrac{1}{\mu} = 2.48$ min/service.

11. Oil tankers arrive at a refinery to transport oil for distant markets. They are served on first come first serve fashion. Tankers arrive at a rate of 10 per hour whereas loading rate is 15 per hour. Arrivals are in Poisson fashion filling in exponential fashion. Transporters are complaining that their trucks have to wait for nearly 12 min at plant. Examine whether complaint is justified. Also determine probability that loaders are idle in above problem.

Solution :

$\lambda = 10/\text{hours};\ \mu = 15/\text{hour}$

(i) $W_q = \dfrac{10}{75}$ hour or 8min, hence complaint is not jsutified.

(ii) Idle time $P_0 = 1 - \rho = \dfrac{5}{15}$ or 33.33%.

12. Consider a self service store with one cashier. Assume Poisson arrivals and exponential service times suppose that nine customers arrive at an average every 5 minutes and cashier can serve 10 in 5 minutes. Find

(i) Average number of customers in queuing for service.

(ii) The probability of having more than 10 customers in system.

(iii) Probability that a customer has to want in queue for more than 2 minutes.

If the service can speed up to 12 in 5 minute by using different cash register, what will be effect on quantities (i), (ii) and (iii).

Solution :

Case I :

$$\lambda = \frac{9}{5}; \quad \mu = \frac{10}{5};$$

(i) $L_s = 9$ customers;   (ii) $P(n \geq 10) = (0.9)^{10}$

(iii) $P(\text{waiting} \geq 2) = \int\limits_{2}^{\infty} \frac{\lambda}{\mu} (\mu - \lambda) \, e^{-(\mu - \lambda)t} \, dt = 0.67.$

Case II :

$$\lambda = \frac{9}{5}; \quad \mu = \frac{12}{5};$$

(i) $L_s = 3$ customers.   (ii) $p(n \geq 10) = (0.75)^{10}$

(iii) $P(\text{waiting} \geq 2) = \int\limits_{2}^{\infty} \frac{\lambda}{\mu} (\mu - \lambda) \, e^{-\mu - \lambda t} \, dt = 0.30.$

13. In a tool crib manned by a single assistant, operators arrive at the tool crib at a rate of 10 per hour. Each operator needs 3 minutes on an average to be served. Find out loss of production due to time lost in waiting for an operator in a shift of 8 hours if the rate of production is 100 units per shift.

Solution :

$$\lambda = 10/\text{hour}; \qquad \mu = 20/\text{hour}$$

(i) $W_q = \frac{1}{20}$ hour; average waiting time per shift $= \frac{8}{20}$ hour.

Loss of production due to waiting $= \frac{2}{5} \times \frac{100}{8} = 5$ units.

14. A fertilizer company distributes its products by trucks loaded at its only loading station, Both company and contractor trucks are used for purpose. It was found that on an average every 5 minutes one truck arrived and average loading time was 3 minutes 40% of trucks belong to contractors. Make probable assumptions and find out

(i) Probability that a truck has to wait.

(ii) Waiting time of truck that waits.

(iii) Expected waiting time of customers truck per day.

Solution :

$$\lambda = \frac{60}{5} = 12/\text{hour}; \quad \mu = \frac{60}{3} = 20/\text{hour}.$$

(i) $P_0 = \frac{\lambda}{\mu} = 0.6$

(ii) $W_s = \frac{1}{8}$ hour or 7.5 minute.

(iii) Total waiting time = number of trucks per day × (% contractors truck) × expected waiting time for truck.

$$= 12 \times 24 \times \frac{40}{100} \times \frac{\lambda}{\mu\,(\mu - \lambda)} = 8.64 \text{ hours/day}.$$

15. Customers arrive at a first class ticket counter of a theatre in a poisson distributed arrival rate of 25 per hour. Service time is constant at 2 minutes. Calculate

    (i) Mean number in waiting time.

    (ii) Mean waiting line.

Answer : (i) 5 people (ii) 10 min.

16. At a public telephone booth in post office arrivals are considered to be Poisson with an average interval time of 12 minutes length of call is distributed exponentially with average time being 4 minutes. Calculate

    (i) Probability that fresh arrival need not wait.
    (ii) Arrival have to wait more than 10 minutes before he gets chance.
    (iii) Average length of queue that form from time to time.

Answer : (i) 0.67 (ii) 0.063 (iii) 1.5

17. Weavers in a textile mill arrive at a department store room to obtain spare parts needed for keeping the looms running. The store is manned by a single attendant. The average arrival rate of weavers per hour is 10 and service rate per hour is 12. Both arrival and service rate follow Poisson process.
    Determine :
    (a) Average length of waiting time.
    (b) Average time a machine spends in the system.
    (c) Percentage of idle time of department store room (attendant).

    [JNTU (CSE) 99]

Answer : (a) 25 min (b) 30 min (c) 40%

18. In the production shop of a company the breakdown of the machines is found to be poisson with an average rate of Rs. 3 machines per hour. Break down time at one machine costs Rs. 40 per hour to the company. There are two choices before the company for hiring the repairman. One of the repairman is slow but cheap, the other fast but expensive. The slow-cheap repairman demands Rs. 20 per hour and will repair the broken down machines exponentially at the rate of 4 per hour. The fast-expensive repairman demands Rs. 30 per hour and will repair machines exponentially at an average rate of 6 per hour, which repairman should be hired? **[JNTU Mech/Prod/Chem/Mechatronics 2001/S]**

8.10 *Model - 2 : (M/M/1) : (∞/SIRO)*

Single Channel - Unrestricted Queue : This model is identical with model - 1 in all respects except for the queue discipline. Since the operating characteristics of a queue model are independent of queue discipline (as the derivation for probability is independent of queue discipline), all the results remain unchanged. Therefore the problems under this model may be solved using same formulae as those of model -1.

8.11 *Model - 3: (M/M/1) : (N/FCFS)*

Single Channel - Restricted Queue Model : This model differs with model-1 in respect of limit of queue. Here the capacity of the system is limited to N. The relevant formulae are given below :

1. $P_n = \left(\dfrac{1 - \rho}{1 - \rho^{n+1}} \right) \rho^n$ for $0 \le n \le N$, $\rho \ne 1$ i.e, $\lambda \ne \mu$

   $= \dfrac{1}{N+1}$ for $\rho = 1$ or $\lambda = \mu$

   and $P_o = \dfrac{1 - \rho}{1 - \rho^{N+1}}$; $\rho \ne 1$ and $\rho < 1$

   Where $\rho = \dfrac{\lambda}{\mu}$

   N = limit of queue system.

   n = number of customers arrived.

2. Expected size of system or expected number of customers in the system

   $$L_s = \frac{\rho}{1 - \rho} - \frac{(N+1)\,\rho^{N+1}}{1 - \rho^{N+1}} \text{ for } \rho \ne 1$$

   $$= \frac{N}{2} \quad \text{for} \quad \rho = 1$$

3. Expected length (size) of queue or expected number of customers waiting in queue.

   $$L_q = L_s - \frac{\lambda}{\mu}$$

4. Expected waiting time of customer in the system $\left(W_q + \dfrac{1}{\mu} \right)$

   $$W_s = \frac{L_s}{\lambda\,(1 - \rho_N)}$$

5. Expected waiting time in queue

   $$W_q = \frac{L_q}{\lambda\,(1 - \rho_N)}$$

ILLUSTRATION 8 ─────────────────────────────────

> *Dr. Raju's out-patient clinic can accommodate six people only in the waiting hall. The patients who arrive when hall is full, balk away. The patients arrive in Poisson fashion at an average rate of 3 per hour and spend an average of 15 minutes in Doctor's chamber which is exponentially distributed. Find*
> *(i) The probability that a patient can get directly into the doctor's chamber upon his arrival*
> *(ii) Expected number of patients waiting for treatment.*
> *(iii) Effective arrival rate.*
> *(iv) The time a patient can expect to spend in the clinic.*

Solution :

From the above problem,

Arrival : Poisson (M); service : Exponential (M);

Number of servers : One; (1) discipline : FCFS;

But the queue is limited

$\therefore$ The model is $(M/M/1) : (N/FCFS)$

Now capacity of the system

$$N = \text{hall capacity} + \text{doctor's chamber capacity}$$
$$= 6 + 1 = 7$$

Arrival rate (λ) = 3 per hour.

Service rate (μ) $= \dfrac{1}{15}$ per min = 4 per hours.

$$\rho = \frac{\lambda}{\mu} = \frac{3}{4} = 0.75 \text{ i.e., } < 1 \text{ (stead state exists)}$$

(i) Patient will directly enter doctor's chamber if the hall is empty on his arrival. The probability at this situation is P_0

$$\therefore \quad P_0 = \frac{1 - \rho}{1 - \rho^{N+1}} = \frac{1 - \left(\dfrac{3}{4}\right)}{1 - \left(\dfrac{3}{4}\right)^{7+1}} = 0.2778$$

(ii) Expected number of patients waiting for treatment is (L_q)

$$\therefore \quad L_s = \left[\frac{\left(\dfrac{3}{4}\right)}{1 - \left(\dfrac{3}{4}\right)} - \frac{8\left(\dfrac{3}{4}\right)^{8}}{1 - \left(\dfrac{3}{4}\right)^{8}} \right] = 2.1$$

$$L_q = L_s - \frac{\lambda}{\mu} = 2.1 - \left(\frac{3}{4}\right) = 1.36$$

(iii) Effective arrival rate (λ_{eff}) = $\mu \, (1 - P_0)$

$$= 4 \, (1 - 0.2778) = 2.89 \cong 3 \text{ per hour.}$$

(iv) The time the patient can expect to spend in the system

$$= \frac{L_q + (1 - P_0)}{\lambda_{\text{eff}}} = \frac{1.36 + 1 - 0.2778}{2.89} = 0.72 \text{ hours}$$

$$= 43.2 \text{ minutes.}$$

Practice Problems

1. Consider a single server queuing system with Poisson input, exponential service times. Suppose the mean arrival rate is 3 calling units per hour, the expected service time is 0.25 hour and the maximum permissible number calling units in the system is two. Derive the steady-state probabiulity distribution of the number of calling units in the system and then calcualte the expected number in the system.

Answer : The expected number of calling units in the system = 0.81.

2. If for a period of 2 hours in the day (8 to 10 a.m.) trains arrive at the yard at every 20 minutes but the service time continues to remain 36 minutes, then calculate for this period.

   (a) The probability that the yard is empty, and

   (b) The average number of trains in the system, on the assumption that the line capcaity of the yard is limited to a four trains only.

Answer :

$$\lambda = 1/20; \quad \mu = 1/36; \quad \rho = 36/20 = 1.8 \, (> 1) \text{ and } n = 4$$

(a) $P_0 = \dfrac{\rho - 1}{\rho^{N+1} - 1} = 0.04$

(b) $L_s = \displaystyle\sum_{n=0}^{4} nP_n = 2.9 \equiv 3.$

8.12 *Model - 4 (M/M/S) : (∞/FIFO)*

Multi Channel - Unrestricted Queue Model : This model can be supposed as the extension to model-1. Here instead of single channel, multi channel (multiple servers) in parallel say 's' lines are assumed. With other notations as usual we use the following formulae for this model.

1. $P_n = \dfrac{\rho^n}{\lfloor n} \cdot P_0$ for $1 \le n < s$

   $= \dfrac{\rho^n}{\lfloor s \cdot s^{n-s}} \cdot P_0$ for $n \ge s$

   Where $\rho = \dfrac{\lambda}{\mu}$

$$P_0 = \left[\sum_{n=0}^{s-1} \frac{(s\,\rho)^n}{\lfloor n} + \frac{1}{\lfloor s} \frac{(s\,\rho)^s}{1 - \rho} \right]^{-1} \quad \text{or} \quad \left[\sum_{n=0}^{s-1} \frac{1}{\lfloor n} \left(\frac{\lambda}{\mu}\right)^n + \frac{1}{\lfloor s} \left(\frac{\lambda}{\mu}\right)^s \cdot \frac{s\,\mu}{s\,\mu - \lambda} \right]$$

2. Expected length of queue

$$L_q = \left[\frac{1}{\lfloor(s-1)} \cdot \left(\frac{\lambda}{\mu}\right)^s \cdot \frac{\lambda\mu}{(s\mu-\lambda)^2} \right] P_o$$

3. Expected number of customers in the system

$$L_s = L_q + \frac{\lambda}{\mu}$$

4. Expected waiting time of the customer in queue

$$W_q = \frac{L_q}{\lambda} = \left[\frac{1}{\lfloor(s-1)} \cdot \left(\frac{\lambda}{\mu}\right)^s \cdot \frac{\mu}{s(\mu-\lambda)^2} \right] P_o$$

5. Expected waiting time in the system

$$W_s = W_q + \frac{1}{\mu} \quad \text{or} \quad \frac{L_q}{\lambda} + \frac{1}{\mu}$$

6. The probability that a customer has to wait (busy period)

$$P(n \geq s) = \frac{1}{\lfloor s} \left(\frac{\lambda}{\mu}\right)^s \cdot \frac{s\mu}{s\mu-\lambda} \cdot P_o$$

ILLUSTRATION 9

A super market has two girls ringing up sales at the counters. If the service time for each customer is exponential with mean 4 minutes, and if people arrive in a Poisson fashion at the rate of 10/hour.

(a) What is the probability of having to wait for the service.

(b) What is the expected percentage of idle time for each girl?

(c) Find the average length and the average number of units in the system ?

[JNTU CSE 97/S JNTU - Mech/Prod/Chem/Mechatronics 2001/S]

Solution :

No. of servers $(s) = 2$, Arival rate $(\lambda) = \frac{1}{6}$ and service rate $(\mu) = \frac{1}{4}$ per minute],

$$\frac{\lambda}{s\mu} = \frac{1}{6} \cdot \frac{1}{2 \cdot \frac{1}{4}} = \frac{1}{3}$$

Therefore,

$$P_0 = \left[\sum_{n=0}^{2-1} \frac{2}{n!}\left(\frac{4}{6}\right)^n + \frac{1}{2!}\left(\frac{4}{6}\right)^2 \frac{2 \cdot \frac{1}{4}}{2 \cdot \frac{1}{4} - \frac{1}{6}} \right]^{-1} = \left[1 + \frac{2}{3} + \frac{1}{3}\right]^{-1} = \frac{1}{2}$$

$$P_1 = \frac{\lambda}{\mu} P_0 = \frac{2}{3} \cdot \frac{1}{2} = \frac{1}{3}$$

(a) $P(n \geq 2) = P(W > 0)$

$$= \sum_{n=2}^{\infty} P_n = \sum_{n=2}^{\infty} \frac{1}{2! \, 2^{n-2}} \left(\frac{2}{3}\right)^n P_0 = \sum_{n=2}^{\infty} \left(\frac{1}{3}\right)^n = \sum_{n=0}^{\infty} \left(\frac{1}{3}\right)^n - \frac{1}{3} - 1$$

$$= \frac{1}{6} \text{ (or) } 0.167$$

(b) The expected idle time for each girl is $1 - \dfrac{\lambda}{s\,\mu}$ i.e., $1 - \dfrac{1}{3}$ (or) $\dfrac{2}{3}$ ($= 0.67$).

      Hence, the expected percentage of idle time for each girl is 67%.

(c) Expected length of customers waiting time $= \dfrac{1}{s\,\mu - \lambda}$

$$= \frac{1}{0.50 - 0.17} \text{ or } 3 \text{ minutes.}$$

ILLUSTRATION 10 ────────────────────────

> *A bank has two tellers working on savings accounts. The first teller handles withdrawals only. The second teller handles deposits only. It has been found that the service time distribution for deposits and withdrawals both are exponential with mean service time 3 minutes per customer. Depositors are found to arrive in a Poisson fashion through out the day with mean arrival rate 16 per hour. Withdrawals also arrive in a Poisson fashion with mean arrival rate 14 per hour. What would be the effect on the average waiting time for depositors and with drawers, if each teller could handle both withdrawals and deposits? What would be the effect of this could only be accomplished by increasing the service time to 3.5 minutes?*

Solution :

Initially, we can assume two independent queueing systems as follows.

Withdrawers System	**Depositors System**
Arrivals : Poisson (*M*)	Poisson (*M*)
Service : Eponential (*M*)	Exponential (*M*)
Number of servers : One (1)	One (1)
Limit of queue : Unlimited (∞)	Unlimited (∞)
Discipline : FCFS	FCFS
Model : (M/M/1) : (∞ / FCFS)	(M/M/1) : (∞ / FCFS)
Mean arrival rate : $(\lambda_1) = 14$/hours	$\lambda_2 = 16$/hours

Mean service rate : $\mu_1 = \dfrac{1}{3}$ per min 20/hours $\mu_2 = \dfrac{1}{3}$ per min 20/hour

$\left(\rho_1 = \dfrac{\lambda_1}{\mu_1} = 0.7 < 1, \; \because \text{ service time } \dfrac{1}{\mu_1} = 3 \text{ min } \right)$ (Mean service time $\dfrac{1}{\mu_2} = 3$ min)

$$\rho_2 = \frac{\lambda_2}{\mu_2} = \frac{16}{20} = 0.8 < 1$$

Average waiting time   Average waiting time

$$W_q = \frac{\lambda_1}{\mu_1 (\mu_1 - \lambda_1)} = \frac{14}{20 (20 - 14)} \qquad\qquad W_q = \frac{\lambda_2}{\mu_2 (\mu_2 - \lambda_2)} = \frac{16}{20 (20 - 16)}$$

$$= \frac{7}{60} = 7 \text{ min} \qquad\qquad\qquad\qquad\qquad\qquad = \frac{1}{5} \text{ hrs} = 12 \text{ min}$$

Now Consider Combined System :

> Arrivals : Poisson (*M*)
>
> Service : Exponential (*M*)
>
> No. of servers : two (2)
>
> Limit of queue : Unlimited (∞)
>
> Discipline : FCFS

∴ Model : (M/M/2) : (∞ / FCFS)

Mean arrival rate : $\lambda = 14 + 16 = 30$/hours

Mean service rate : $\mu = 20$/hours

Number of servers (*s*) = 2 and $\rho = \dfrac{\lambda}{s\,\mu} = \dfrac{30}{2 \times 20} = \dfrac{3}{4}$

($\therefore \rho < 1$)

Also $\quad \dfrac{\lambda}{\mu} = \dfrac{30}{20} = \dfrac{3}{2}$

Now $\quad P_0 = \left[\displaystyle\sum_{n=0}^{s-1} \frac{1}{n!} \left(\frac{\lambda}{\mu}\right)^n = \frac{1}{s!} \left(\frac{\lambda}{\mu}\right)^s \left(\frac{s\,\mu}{s\,\mu - \lambda}\right) \right]^{-1}$

$\qquad\qquad = \left[\displaystyle\sum_{n=0}^{1} \frac{1}{n!} \left(\frac{3}{2}\right)^n + \frac{1}{2!} \left(\frac{3}{2}\right)^2 \left(\frac{40}{40 - 30}\right) \right]^{-1}$

$\qquad\qquad = \left[\frac{1}{0} \left(\frac{3}{2}\right)^0 + \frac{1}{1} \left(\frac{3}{2}\right)^1 + \frac{1}{2 \times 1} \times \frac{9}{4} \times \frac{40}{10} \right]^{-1}$

$\qquad\qquad = \left[1 + \frac{3}{2} + \frac{9}{2} \right]^{-1} = \frac{1}{7}$

Average waiting time of arrivals in the queue

$$W_q = \frac{L_q}{\lambda} = \frac{1}{(s-1)!} \cdot \left(\frac{\lambda}{\mu}\right)^s \cdot \frac{\mu}{(s\,\mu - \lambda)^2} \cdot P_0$$

$$= \frac{1}{(2-1)!} \cdot \left(\frac{3}{2}\right)^2 \cdot \frac{20}{(40-30)^2} \cdot \frac{1}{7}$$

$$= \frac{9}{4} \times \frac{20}{100} \cdot \frac{1}{7} \qquad = \frac{9}{140} \text{ hours.}$$

$$= 3.86 \text{ min}$$

Combined waiting time with increased service time when

$$\lambda' = 30 \text{ /hour and } \frac{1}{\mu'} = 3.5 \text{ min}$$

or $\qquad \mu = \dfrac{60}{3.5} = \dfrac{120}{7}$ per hour

$$\frac{\lambda'}{s\,\mu'} = \frac{30}{2\,(120/7)} = \frac{7}{8} \text{ i.e., } < 1 \quad \text{and} \quad \frac{\lambda'}{\mu'} = \frac{30}{120/7} = \frac{7}{4}$$

$$P_0 = \left[\sum_{n=0}^{1} \frac{1}{n!} \left(\frac{7}{4}\right)^n + \frac{1}{2!}\left(\frac{7}{4}\right)^2 \frac{2 \times (120/7)}{\left\{2\,(120/7) - 30\right\}} \right]^{-1}$$

$$= \left[\frac{1}{0!}\left(\frac{7}{9}\right)^0 + \frac{1}{1!}\left(\frac{7}{4}\right)^1 + \frac{1}{2 \times 1} \times \frac{49}{16} \times \frac{2\,(120/7)}{(30/7)} \right]^{-1}$$

$$= \left[1 + \frac{7}{4} + \frac{49}{4} \right]^{-1} \qquad = [15]^{-1} \qquad = \frac{1}{15}$$

Average waiting time of arrivals in the queue

$$W_q = \frac{1}{(s-1)!}\left(\frac{\lambda'}{\mu'}\right) \frac{\mu'}{(s\,\mu' - \lambda')^2}\,P_0$$

$$= \frac{1}{(2-1)!}\left(\frac{7}{4}\right)^2 \frac{120/7}{[2\,(120/7) - 30]^2} \times \frac{1}{15}$$

$$= \frac{49}{16} \times \frac{120/7}{(30/7)^2} \times \frac{1}{15} = \frac{49}{16} \times \frac{120}{7} \times \left(\frac{7}{30}\right)^2 \times \frac{1}{15}$$

$$= \frac{343}{30 \times 60} \text{ hours}$$

$$= \frac{343}{30} \text{ min} \quad = 11.433 \text{ min.}$$

Practice Problems

1. A post office has 3 windows providing the same services. It recieves on an average 30 customers/hours. Arrivals are Poisson distributed and service time exponentially. The post office serves on average 12 customers/hours.

    (i) What is the probability that a customers will be served immediately?

    (ii) What is the probability that a customer will have to wait?

    (iii) What is the average no. of customers in the system.

    (iv) What is the average total time that customer must spend in the post office?

Answer : (i) 0.29   (ii) 0.70   (iii) 6 customers (approx)   (iv) 12 min.

2. Two repairmen are attending five machines in a workshop. Each machine breaks down according to a poisson distribution with a mean 3 per hour. The repair time per machine is exponential with mean 15 min.

    (i) Find the probability that the two repairmen are idle and that one repairman is idle.

    (ii) What is the expected number of idle machines not being served?

3. Given an average arrival rate of 20 per hour; is it better for a customer to get on service, at a single channel with mean service that of 22 customers or at one of the two channels in paralle with mean service rate of 11 customers for each of the two channels? Assume that both the queues are of M/M/S type.

                                              [JNTU (Mech.) 2001]

Answer : Better to get service at two channels.

5. A telephone exchange has two long distance operators. The telephone company finds that during the peak load, long distance calls arrive in a Poisson fashion at an average rate of 15 per hour. The length of service on these calls is approximately exponentially distributed with mean length 5 min.

    (a) What is the probability that a subscriber will have to wait for his long distance call during the peak hours of the day?

    (b) If subscribers will wait and are served in turn, what is the expected waiting time?

Answer : (a) 0.48   (b) 32 min.

6. A company currently has two tool cribs, each having a single clerk, in its manufacturing area. One tool crib handles only the tools for the heavy machinery. While the second one handles all other tools. It is observed that for each tool crib the arrivals follow a Poisson distribution with a mean of 20 per hour and the service time distribution is negative exponential with a mean of 2 minutes. The tool manager feels that, if tool cribs are combined in such a way that either clerk can handle any kind of tool as demand arises, would be more efficient and the waiting problem could be reduced to some extent. It is believed that the mean arrival rate at two tool cribs will be 40 per hour, while the servive time will remain unchanged.

Compare in status of queue and the proposal with respect to the total expected number of machines at the tool crib(s), the expected waiting time including service time for each machine and probability that he has to wait for more than five minutes.

Answer : For separate tool cribs : 2 arrivals, 6 min.

                    For combined tool cribs : 2.4 arrivals, 3.6 min.

7. In a machine maintenance a mechanic repairs four machines. The mean time between service requirement is 5 hours for each machine and forms an exponential distribution, The mean repair machine down time costs Rs. 25 per hour and the machine costs Rs. 55 per day of 8 hours.

    (i) Find the expected number of operating machines.

    (ii) Determine the expected down time cost per day.

    (iii) Would it be economical to engage two machines each repairing only two machines?

8. A petrol pump station has two pumps the service times follow the exponential distributions with a mean of 4 min and cars arrive in a Poisson distribution at a rate of 10 cars per hour. Find the probability that a car has to wait. What proportion of time the pump remain idle?

Answer : 0.167; 67%

9. A two-channel waiting line with Poisson arrival has a mean arrival rate of 50 per hour and exponential service with a mean service rate of 75 per hour for each channel. Find

    (a) the probability of an empty system

    (b) the probability that an arrival in a system will have to wait.

Answer : 0.83, 0.167

10. A telephone company is planning to install telephone booths in a new air port. It has established the policy that a person should not have to wait more than 10% of the times he tries to use a phone. The demand for use is estimated to be poisson with an average 30 per hour. The average phone call has an exponential distribution with a mean time of 5 min. How many phone booths should be installed.

Answer : 6 phone booths

Review Questions

1. Write short notes on waiting line situations.   **[CSE - M 97]**

2. What are waiting line costs ? Explain them with the aid of graph.
                                          [Mech - 98/S, MBA - BRAOU - 98]

3. Write short note on queuing models   **[Mech 98/S, MBA - OU - 98]**

4. Write short note on basic structure of queuing model**[Mech. 99 M, Mech - 2000 M]**

5. Discuss queuing system   **[Mech 99/S, MBA - IGNOU - 97]**

6. Briefly explain the important characteristics of a queuing system [Mech 2000 - OR]

7. Briefly explain the representation of queue models based on Kendall-Lee notations. [Mech - 97/P]

8. Discuss the operating characteristics of basic queue model [CSE - 96/C]

9. Discuss different queue disciplines used in waiting lines models. [Mech - 98/C]

10. Discuss the arrival and service patterns used in Q-models [MBA - OU - 98]

11. Explain various queue configurations [MBA - IGNOU - 96]

12. What do you understand by (M/M/1) : (∞/FCFS). Explain the terms [MBA IGNOU - 97, 99, 2001]

13. Discuss the customer behaviours that are often found in queue systems. [MBA - IGNOU - 98]

14. Discuss important features of a waiting line model [Mech - 96/P, MBA - OU - 98/S]

15. Explain the applications of queue models in business situations. [Mech - 96/C, MBA - IGNOU 97, BRAOU - 95]

Objective Type Questions

1. A customer buying tickets in a black market at a cinema hall is said to be
   (a) balker   (b) reneger
   (c) jockeyer   (d) dissatisfied

2. An expediting in production shop in example for _____ queue discipline
   (a) FIFO   (b) LIFO
   (c) SIRO   (d) pre-emptive

3. The dead bodies coming to a grave yard is an example of
   (a) pure death process   (b) pure birth process
   (c) birth and death process   (d) not a queue process

4. In (M/M/S) : (N/FIFO), which of the following is wrongly stated
   (a) poisson arrival   (b) exponential service
   (c) single server   (d) limited service

5. Which of the following is not considered as the negative behaviour of customer according to queue disciplines
   (a) reneging   (b) jockeying
   (c) balking   (d) boarding

6. A steady state can exist in a queue if

    (a) $\lambda > \mu$ (b) $\lambda < \mu$

    (c) $\lambda \leq \mu$ (d) $\lambda \geq \mu$

7. The system of loading and unloading of goods usually follows
    (a) FIFO   (b) LIFO
    (c) SIRO   (d) pre-emptive

8. If the operating characteristics of a queue are dependent on time, then it is said to be
    (a) steady state   (b) transient state
    (c) busy period   (d) explosive state

9. A person who goes out of queue by losing patience to wait is said to be
    (a) reneging   (b) jockeying
    (c) balking   (d) boarding

10. An office filing system follows

    (a) FIFO   (b) LIFO

    (c) SIRO   (d) non-pre-emptive

11. The queue discipline in stack of plates is

    (a) FIFO   (b) LIFO

    (c) SIRO   (d) non-pre-emptive

12. The queue discipline in selecting the contestants for a game such as "Kaun Banega Corodpathi (KBC)" is

    (a) FIFO   (b) LIFO

    (c) SIRO   (d) pre-emptive

13. SIRO discipline is found generally in

    (a) office filing   (b) trains arriving to platform

    (c) lottery   (d) loading and unloading

14. A Poisson arrival, exponential service by single server to limited queue selected randomly is represented as

    (a) $(M/E/S) : (\infty/SIRO)$ (b) $(M/M/S) : (N/SIRO)$

    (c) $(M/M/1) : (N/SIRO)$ (d) $(M/M/1) : (\infty/SIRO)$

15. For a simple queue (M/M/1), $\rho = \dfrac{\lambda}{\mu}$ is called

    (a) traffic intensity   (b) fraction of busy period
    (c) utilization factor   (d) any of the above

16. The unit of traffic intensity is _____
    (a) Poisson
    (b) Markov
    (c) Erlang
    (d) Kendall

17. In a (M/M/1) : (∞/FCFS) model, the length of the system (L_s) is given by

    (a) $\dfrac{\rho^2}{1-\rho}$

    (b) $\dfrac{\lambda^2}{(\mu-\lambda)}$

    (c) $\dfrac{\rho}{(1-\rho)}$

    (d) $\dfrac{\lambda^2}{\mu(\lambda-\mu)}$

18. In (M/M/1) : (∞/FIFO) model, $\dfrac{1}{(\mu-\lambda)}$ represents
    (a) length of system (L_s)
    (b) length of queue (L_q)
    (c) waiting time in queue (W_q)
    (d) waiting time in system (W_s)

19. Which of the following model is same as that of (M/M/1) : (∞/FIFO)
    (a) (M/M/1) : (N/FIFO)
    (b) (M/M/S) : (∞/FIFO)
    (c) (M/M/1) : (∞/SIRO)
    (d) none

20. The characteristics of queue model are independent of
    (a) service pattern
    (b) number of service points
    (c) limit of queue
    (d) queue discipline

21. Which of the following is wrong relation in a simple queue model
    (a) $L_q = \lambda W_q$
    (b) $W_s = W_q + \mu$
    (c) $L_s = L_q + \rho$
    (d) $\lambda = \mu\rho$

22. By seeing the large queue size, a customer may pass out without entering the queue thinking that he may not get service. This is called
    (a) walk - out
    (b) balking
    (c) jockeying
    (d) reneging

23. A doctor rushing to an emergency case leaving his regular service is said to be
    (a) pre-emptive$Q$ discipline
    (b) non-pre-emptive discipline
    (c) reneging
    (d) balking

24. A service system in which customer is stationary but server is moving is found with
    (a) buffet meals
    (b) person attending breakdown of heavy machines
    (c) vehicles at petrol bunk
    (d) out-patients at doctor's clinic

25. The waiting time in system of a simple queue model is equal to

    (a) $W_q + \mu$

    (b) $\dfrac{L_q}{\lambda} + \dfrac{1}{\mu}$

    (c) $\dfrac{\mu}{\mu - \lambda}$

    (d) $\dfrac{1}{\mu - \lambda}$

26. Mean arrival rate at a telephone booth is Poissonal with 6 per hr and average phone call time is distributed exponentially at 3 min. Then the probability that a person arriving has to wait is

    (a) $\dfrac{1}{2}$

    (b) 0.3

    (c) 0.6

    (d) 0.1

27. On an average 30 goods trains arrive at a station Poissonally loading time is distributed exponentially at 36 min. Then average number of trains in queue will be _____

    (a) 2

    (b) 3

    (c) 4

    (d) none of the above

28. A T.V repairer spends about 30 min on each job exponentially distributed, and is busy for 5 hrs in a 8 hr day. The mean arrival rate is _____

    (a) 10 per 8 hr day

    (b) $\dfrac{5}{2}$ per hr

    (c) 10 per hour

    (d) 6 per hour

29. In a Four teller bank counter, service rate is exponential with 3 per hr while customers arrive at Poisson rate of 10 per hr. Then utilization factor is

    (a) $\dfrac{3}{10}$

    (b) $\dfrac{3}{40}$

    (c) $\dfrac{10}{3}$

    (d) $\dfrac{10}{12}$

30. Given $L_q = 3.24$ customers $\lambda = 10/\text{hr}$; $\mu = 3/\text{hr}$; $W_s = $ _____

    (a) 32.4 min

    (b) 35.4 min

    (c) 32.73 min

    (d) 39.42 min

Fill in the Blanks

1. Customers losing patience out of inordinate delay in service walk out of queue system, are said to be _____

2. Pre paid taxi at an airport follow _____ queue discipline

3. The steady state can exist in queue system if λ _____ μ

4. The storage or memory of files in a computer (RAM) in accordance with _____ type of queue discipline

5. The unit of utilization factor is _____

6. The state of queue system whose characteristics are time dependent is _____ state

7. A Poisson arrival, exponential service, 2 server system with infinite input at first come first served is represented in Kendall-Lee notations as _____

8. In a 20-bed hospital, a doctor makes rounds and treats in the exponential distribution according to the severity of the desease. If impatients are expected in Poisson fashion, the queue model is represented in Kendall -Lee notations as _____

9. A medical representative when went to an out-patient clinic of a doctors, he was asked to wait till the current service in the doctor's cabin is finished, and then was given appointment. This is _____ type of queue discipline.

10. The dresses and cloths kept in a suitcase follow _____ queue discipline

11. For [(M/M/1) : (∞/FIFO)] model, the length of queue (L_q) in terms of busy period (ρ) is _____

12. The characteristics of queue model are independent of _____

13. Value of $\dfrac{\rho}{(1-\rho)^2}$ of [(M/M/1) : (∞/FIFO)] model states _____ of queue length

14. The cost of waiting per unit time _____ with increase of service level.

15. Given traffic intensity = 0.75, the server will be idle for _____ hours in a 8 - hr day

16. A lottery system follows _____ queue discipline

17. For (M/M/C) : (∞/FIFO) model, with the parameters as arrival rate (λ) and service rate (μ), the traffic intensity (ρ) = _____

18. An Erlangian service distribution for Poisson arrival of single server in a limited size queue selected at random is represented in Kendall-Lee notation as _____

19. The cost of operating a service factility per unit time _____ with the increase of level of service

20. Aeroplanes landing on a run way at an aerodrom follow _____ queue discipline

Answers

Objective Type Questions :				
1. (c)	2. (d)	3. (b)	4. (c)	5. (d)
6. (c)	7. (b)	8. (b)	9. (a)	10. (b)
11. (b)	12. (c)	13. (c)	14. (c)	15. (d)
16. (c)	17. (c)	18. (c)	19. (c)	20. (d)
21. (b)	22. (b)	23. (a)	24. (b)	25. (b)
26. (b)	27. (b)	28. (a)	29. (d)	30. (d)

Fill in the Blanks :	
1. reneging	2. FCFS or FIFO
3. < (less than)	4. SIRO
5. Erlang	6. transient
7. (M/M/2) : (∞/FCFS)	8. (M/M/1) : (N/priority)
9. non-per-emptive	10. LIFO
11. $\dfrac{\rho^2}{1-\rho}$	12. queue discipline
13. variance or fluctuation	14. decreases
15. 2	16. SIRO
17. $\dfrac{\lambda}{\mu\, c}$	18. (M/E$_k$/1) : (N/SIRO)
19. increases	20. FIFO or FCFS

9 Chapter

Sequencing

CHAPTER AT A GLANCE

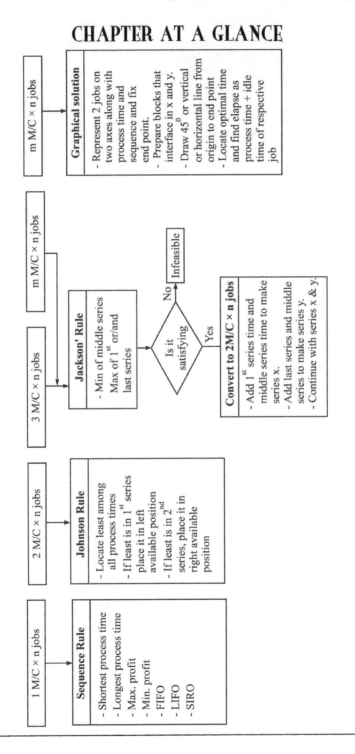

9.0 *Introduction*

It is common problem in the production process to find the sequence of the jobs that will result in least idle time for the better utilisation of equipment. However, it is not applicable for those operations where the technological sequence is already fixed.

For example, one has to wear socks before wearing shoe. Here, the sequence is fixed where as the sequence of jobs in which they have to be processed (the order of jobs is not fixed) is important when some jobs are awaiting the operations with different timings (the technological sequence of operation is assumed to be fixed). For example there are five books to be printed and bound. Here, the operational sequence i.e., printing and then binding is fixed but the sequence in which the books are to be taken for operations is not fixed. The right sequence depending on the time required for each operation will minimise the idle time of printer and binder. This chapter deals with such techniques to find the right sequence that minimises the idle time of operations.

It will be most convenient to study the job sequencing in the following models.

1. n jobs X 1 machine (or man or department)

2. n jobs X 2 machines

3. n jobs X 3 machines

4. n jobs X m machines

5. 2 jobs X n machines

These are explained here below.

9.1 *Sequencing of n Jobs X 1 Machine*

When different letters are to be typed by a typist, he has to decide the order in which he has to type. When a student has some questions to answer in an exam, he would first choose the order in which he has to present to make it effective. When certain jobs are to be done in a production shop on one machine only, then it is the turn of the engineer to decide the order of these jobs to process. If only one operator is there to paint some cars in a automobile repairing shop, he decides the sequence of the cars to be painted. In all such situations one has to follow certain discipline or rule. The following are the rules from which a person can select for his job sequencing when only machine is available. (The term 'machine' here is used in wide sense to mean anything like men, material, department, equipment or any relevant one as the case may be).

1. **First In First Out (FIFO) or First Come First Serve (FCFS) :** When the jobs do not require any preferential treatment, this rule is considered. Complaints at telecom department, electricity departments, trains on single track or platform, customers at a retailer or a telephone booth etc., will follow this rule.

2. **Last In First Out (LIFO) :** When a machine is dismantled for repair or overhaul, the parts are put back in LIFO system. Pipeline laying in water works, electrical wiring maintenance system will follow this model. The office filling system, loading and unloading of trucks, dressing / undressing shirt and a coat, wearing/removing socks and shoe are some more common examples that occur in daily life to follow LIFO.

3. **Earliest Due Date (EDD):** Most of the times the production departments are asked for the probable time of completion of the job and based on these promises the production engineer plans his shop production.

4. **Shortest Process Time (SPT) :** This is another policy to apply on the engineering jobs with the concept that the jobs those take less time to perform are taken first. A student while writing his exam would prefer to write the question that takes lesser time to that takes longer duration.

5. **Longest Process Time (LPT) :** In contrast to the above, the job takes longer time will be taken up first and the jobs which take smaller time will be taken up last.

6. **Pre-emptive Priority Rule :** When a job is very urgent, it will be taken up on priority basis by attending immediately stopping all other jobs. Under this rule the highest priority job is allowed to enter into the service immediately even if another job with lower priority is already in service. Doctors use this discipline when emergency cases arrive to their clinics.

7. **Non Pre-emptive Priority Rule :** In this case, highest priority goes ahead in sequence but service is started immediately after completion of the current service. For example, a doctor gives priority to a medical representative, but he gives the appointment after the finishing the current job, but will not stop the current service unlike in an emergency case.

8. **Priority Service :** Priority is given to certain jobs by the virtue of their importance or recommendation of top officials or expediting jobs etc.

9. **Select In Random Order (SIRO) :** Under this rule the jobs are selected at random for operation irrespective their arrival, urgency, due date etc.

10. **Minimum Cost Rule :** The jobs are selected in the ascending order of their costs.

11. **Maximum Profit Rule :** The jobs are selected in the descending order of the profits.

In the above the first five are static in nature and the rest are dynamic

9.2 *Sequencing of n Jobs X2 Machines : Johnson's Rule*

S.M. Johnson suggested a sequencing rule for a situation where there is a group of n jobs to be processed through two successive work centers. The rule ensures minimum completion time for the group of n jobs by minimising the total idle times of the work centers.

9.2.1 *Assumptions for n Jobs X2 Machines Sequencing*

The following assumptions are usually made while dealing with sequencing problems.

1. No passing rule is observed strictly i.e., the same order of jobs is maintained over each machine. In otherwords, a job can not be processed on 2nd machine unless it is processed on 1st machine.

2. Only one operation is carriedout on a machine at a time.

3. Processing times are known and deterministic.

4. Processing time of a job on a machine is independent of other jobs.

5. The time involved in moving jobs from one machine to another is negligible

6. Each operation, once started, must be completed.

7. An operation must be completed before its succeding operation can start.

8. One machine of each type is available.

9. A job is processed as soon as possible, but only in the order specified.

10. All the men/machines work with consistent efficiency

11. Setting time is either included in process time or neglected

12. No reworking is allowed.

9.2.2 *The Rule*

The rule is

1. *Identify the job with lowest processing time among all the jobs (2n) on both the machines.*

2. *If this shortest processing time belongs to the first machine (or work centre), then this job is placed first in sequence, if the shortest processing time belongs to the second work centre, this job is put last in the sequence.*

3. *For the remaining jobs, repeat step 2, until all the jobs are placed in the sequence. Following example illustrates Johnson's rule.*

ILLUSTRATION 1 ————————————————————————

> *Venkatesh automobile works has six cars for repair. The repair consists of two step procedure viz. Dent removing and pointing. The time estimates are as follows:*
>
> | Car Number | 1 | 2 | 3 | 4 | 5 | 6 |
> |---|---|---|---|---|---|---|
> | Time estimate (dent removing) | 16 | 10 | 11 | 13 | 8 | 18 |
> | Time estimate (painting) | 15 | 9 | 15 | 11 | 12 | 14 |
>
> *Find the correct sequence of operations and prepare Gantt chart and time schedule chart so as to minimise the idle time.* [JNTU (Mech.) 92/P]

Solution :

Sequencing :

Step 1 : The most minimum time among all 12 jobs (6 dent removing jobs and 6 painting jobs), is taken by car number 5 (say C_5) for dent removing i.e., 8 hours.

Step 2 : As the least time 8 for C_5 falls under first operation, this is to be sequenced at extreme left.

C_5					

Now this C_5 is deleted for further sequencing.

Now the problem is

Cars	1	2	3	4	5	6
Dent.	16	10	11	13	⑧	18
Paint	15	9	15	11	1̸2̸	14

Step 3 : Step - 2 is to be repeated till the sequencing is completed.

As the next shortest processing time i.e., 9 for C_2 belongs to the second work centre (i.e., painting) this is to be sequenced to the extreme right.

C_5					C_2

Cars	1	2	3	4	6
Dent.	16	1̸0̸	11	13	18
Paint	15	9	15	11	14

(The next minimum is 10 for, again C_2, which is already sequenced and deleted from further consideration). Then next minimum is 11, which appears in C_3 for work centre - 1 as well as in C_4 for work centre 2. Therefore C_3 goes to left and C_4 goes to right available positions.

C_5	C_3			C_4	C_2

Cars	1	3	4	6
Dent.	16	⑪	1̸3̸	18
Paint	15	1̸5̸	⑪	14

(Similarly, the next shortest processing times are 12 for C_5 for work centre-2 and 13 for C_4 for work centre-1, but these are already allocated and deleted for the consideration). Thus C_6 occupies next position towards right side of the left out positions as this has got the next least process time i.e., 14 hours at work centre-2.

C_5	C_3		C_6	C_4	C_2

Cars	1	6
Dent.	16	1̸8̸
Paint	15	⑭

Obviously, C_1 comes in the left out position. Thus the sequence is . . .

C_5	C_3	C_1	C_6	C_4	C_2

Schedule Chart :

Car No.	First Work Centre (dent removing)			Second Work Centre (Painting)			Idle Time	Remarks
	Time In	Job Time	Time Out	Time in	Job Time	Time Out		
C_5	0	8	8	8	12	20	8*	Painter can start at 8th hr. only
C_3	8	11	19	20	15	35	0	C_3 waits for 1 hour.
C_1	19	16	35	35	15	50	0	No waiting for job and painter
C_6	35	18	53	53	14	67	3	Painter waits for 3 hours.
C_4	53	13	66	67	11	78	0	C_4 waits for one hour.
C_2	66	10	76**	78	9	87	0	C_2 waits of 2 hours.
Total elapsed time			76			87	Total idle time = 11	Painter may be asked to come 8* hours late to the duty, and denting removing operator 11 hrs early.

* Some authors treat this as slack time but not the idle time, as painter may be asked to come late to the duty 8 hours hence to the dentor. If this duration is not considered as the idle time of waiting time for painter, the idle time is only 3 hours.

** Similarly the first operator, i.e., dentor can be asked to go early and need not stay till 87 hours and hence the difference 11 hours is not treated as idle time.

Thus the delivery schedule may given as C_1 on 50th hr, C_2 on 87th hr, C_3 on 35th hr, C_4 on 78th hr, C_5 on 20th hr and C_6 on 67th hr so as to minimise the idle times for the machines or the repair men.

Gantt Chart :

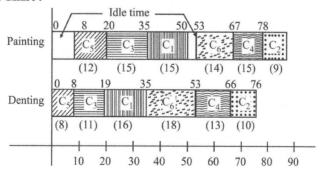

FIGURE 7.1 : GANTT CHART

Remarks:

1. If there is same minimum processing time for two jobs, one at work centre-1 and other at work centre-2 then, the one with shortest time at work centre-1 is kept at most left and the other at most right.

2. If there is a tie in shortest processing time for any two jobs in first work centre, then the sequence is to be determined by checking their minimum time at the second work centre. The job having minimum time is to be allocated most right and the other is kept just before it.

3. If the tie is for two jobs at second work centre, check their minimum time at first work centre and the job having minimum time is to be given left position and the other just after it.

4. If there is a tie in minimum processing times of two jobs with both the machines or work centers then the sequencing is done by selecting arbitrarily. In this case the problem will lead to multiple or alternate optimal solutions (more than one optimal solution).

5. If there is a tie i.e., with same minimum process time for a particular job in both the series of machines, then also, we get an alternate optimal solution, as it can take either left or right available position.

6. If the profits are given, then it is maximisation case for which we have to choose highest value (instead of minimum value) among all and apply Johnson's rules with all assumptions same.

The above remarks can be observed in the sequencing problems illustrated below.

Special Cases In Sequencing : *Case (i): Tie in Different Series :*

When least values appear at more than one place in different series. (No effect).

ILLUSTRATION 2 ────────────────────────────────

Find the sequence for the following data so as to minimise the idle time in hours.

	Jobs			
		A	**B**	**C**	**D**
Machines	**I**	**5**	**6**	**8**	**4**
	II	**4**	**7**	**9**	**10**

What will be the earliest delivery time that B can be promised.

[JNTU (Mech.) 89/S, (CSE) 92/S]

Solution :

According to Johnson's rule, with usual assumptions, we sequence four jobs as follows.

The least time among all the timings is 4 and it appears at two places. One is in the first series (MI) which is to be kept in extreme left position and second is in second (MII) series that comes extreme right.

D			A

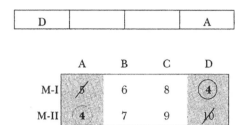

	A	B	C	D
M-I	5̸	6	8	④
M-II	④	7	9	1̸0̸

Then we are left with the following

	B	C
M-I	⑥	8
M-II	7	9

Least of the above timings is 6 that appears in M-I series, therefore *B* occupies next left extreme and obviously the left out position is occupied by *C*.

Then sequence is

D	B	C	A

Time Chart :

Sequence	M-I			M-II			Idle time
	In	Process	Out	In	Process	Out	
D	0	4	4	4	10	14	4
B	4	6	10	14	7	21	-
C	10	8	18	21	9	30	-
A	18	5	23	30	4	34	-

Total elapse = 34, idle time = 4 for M - II

Job *B* can be delivered at the end of 21st hour of starting the operations.

Case (ii) : Tie in Same Series :

i.e., when lowest values appear at more than one place in the same series.

ILLUSTRATION 3 ——————————————————————————

An old shopping complex is intended to be renovated. This is done in two phases namely repairing the patches etc., and then white washing. The time estimates in hours for each of these floors are given below:

	Floors				
	1	2	3	4	5
Repair	20	28	25	20	22
White Wash	24	26	15	27	15

Find the sequence that minimises the idle time of repairing and white washing jobs. Also find the total elapse. Assume no passing and both come and go at the same time and find the idle time of each jobs. [OU (Stat) 84]

Solution :

Let us apply Johnsons rule. On observation we find most minimum time at two places i.e., for third and fifth floors of white washing series (second series), i.e., both are competing for last position. Thus last two positions should be given to these, but the question is that in which order these two are to be placed. This tie is resolved by checking their first series time estimates. These are 25 and 22 respectively. Minimum of these two is 22 and we know that this should be placed to left if it belongs to first series. Therefore fifth floor occupies left of the last two positions and third floor occupies the last position as shown below.

			⑤	③

Floors	1	2	3	4	5
Repair	20	28	25	20	22
White wash	24	26	⑮	27	⑮

Next minimum among 1, 2 and 4 floors is 20, that appears twice i.e., for first and fourth floors in first series making a tie for first position. Therefore they occupy first two positions and the order can be fixed by seeing their second series.

In second series first and fourth floors have 24 and 27 hours respectively, of which 24 is least. As this belongs to second series, this should go to right and thus fourth floor is placed first.

④	①		⑤	③

Floors	1	2	4
Repair	⑳	28	⑳
White wash	24	26	27

Obviously, the second floor is placed in the left out position i.e., third slot. Thus the optimal sequence is

④	①	②	⑤	③

Elapse Time Calculation :

Sequence	Repair			White Wash			Idle Time	
	Start	Process	End	Start	Process	End	Repair	White wash
4	0	20	20	20	27	47	-	20
1	20	20	40	47	24	71	-	-
2	40	28	68	71	26	97	-	-
5	68	22	90	97	15	112	-	-
3	90	25	115	115	15	130	15	3
							15	23

As both repairing and white washing are starting and ending at same time, the repairer will be idle for 15 hours and white washer will be idle for 23 hours.

Total idle time = 15 + 23 = 38 hours.

Total elapsed time = 130 hours.

Case (iii) & (iv) : Multiple or Alternate Optimal Solution : Tie in both series.

If there is a tie with minimum value in first and second series simultaneously or if there are identical values in two jobs. [Multiple optimal solutions]

ILLUSTRATION 4

A Desk Top Publishing (DTP) operator has 5 jobs to process for which he has to first type a copy of DTP and then printing. The estimated timings (in hours) are as follows. Find the optimal sequence in minimise the idle time.

			Jobs		
	A	*B*	*C*	*D*	*E*
DTP	*5*	*6*	*7*	*9*	*6*
Printing	*5*	*8*	*10*	*11*	*8*

What will be expected delivery time for job C. If there is urgency for C what sequence is preferred.   [JNTU 95/CCC]

Solution :

Among all the jobs the most minimum is 5 that appears for *A* in both series i.e., DTP (first series) and printing (second series).

Since it appears in first series, it should occupy first place (say solution - 1) and as it is found in second series it can go to last place also (say solution - 2). Thus we find two solutions as shown below :

Solution 1 :

A				

Solution 2 :

				A

    The next minimum value is read as 6 that appears in first series at *B* and *E* places. Therefore, we check their corresponding second series times, but these are also same. Thus in the first two positions these two can occupy in any order (*BE* or *EB*) resulting in two more alternate solutions.

Solution 1(a) :

A	B	E		

Solution 1(b) :

A	E	B		

Solution 2(a) :

B	E			A

Solution 2(b) :

E	B			A

    Next minimum is 7 with *C* in first series and so placed left of remaining vacancy positions and *D* takes the left out position.

Solution 1(a) :

A	B	E	C	D

Solution 1(b) :

A	E	B	C	D

Solution 2(a) :

B	E	C	D	A

Solution 2(b) :

E	B	C	D	A

Time Charts :

For Solution 1(a) :

Sequence	DTP			Printing			Idle Time	
	In	**Process**	**Out**	**In**	**Process**	**Out**	**DTP**	**Printing**
A	0	5	5	5	5	10	-	5
B	5	6	11	11	8	19	-	1
E	11	6	17	19	8	27	-	-
C	17	7	24	27	10	37	-	-
D	24	9	33	37	11	48	15*	-
						Total Idle Time	15	6

For Solution 2 (b) :

Sequence	DTP			Printing			Idle time	
	In	Process	Out	In	Process	Out	DTP	Printing
E	0	6	6	6	8	14	-	6
B	6	6	12	14	8	22	-	-
C	12	7	19	22	10	32	-	-
D	19	9	28	32	11	43	-	-
A	28	5	33	43	5	48	15*	-
						Total Idle Time	15	6

Total elapse (or completion time of all jobs = 48 hours.

* 15 hours for DTP is not considered as idle time, if DTP operator can be asked to go after completion of his all jobs, i.e., after 33 hours.

The expected delivery time of C is 32 hours or 37 hours. If there is urgency the sequence E-B-C-D-A may be followed.

Case (v) : Maximisation Case :

ILLUSTRATION 5 ———————————————

Mr. Shyam receives Rs. 10/- per hour of processing the jobs. Every job has to undergo two operations, machining and polishing. The time estimates for four jobs to be processed in this week are given below. What should be the sequence if he intentionally delays his jobs by maximising the idle time of operators so as to maximise his returns. What is the amount that he can gain or lose, as compared to that had to be processed with minimum delay.

Jobs	P	Q	R	S
Machining in (hours)	*10*	*45*	*20*	*25*
Publishing in (hours)	*35*	*15*	*40*	*30*

Solution :

The above problem has two cases, viz case (i) maximisation (ii) minimisation.

Case (i) : Maximisation i.e., when Mr. Shyam intentionally delays the process. To find the sequence that maximises the idle time, we choose highest among all the timings of the jobs of both operations and if this falls in first series, place it left otherwise right.

Highest among all is 45 for Q that should occupy extreme left since it is in first series.

Q			

Next highest is 40 for R in second series therefore goes last.

Q			R

Next highest is 35 for P in second series and so goes right side. And the left out position is occupied by S. The sequence is

Q	S	P	R

Schedule is as Follows :

Sequence	Machining			Polishing			Idle time	
	In	Process	Out	In	Process	Out	Machining	Polishing
Q	0	45	45	45	15	60	–	45
S	45	25	70	70	30	100	–	10
P	70	10	80	100	35	135	–	–
R	80	20	100	135	40	175	75*	–
						Total	75 Idel time	55

Total elapse is 175 hours.

Case (ii) : If he wants to minimise.

We choose least for sequencing by Johnsons rule.

The least is 10 for P in first series, so goes to extreme left and next least is 15 for Q in second series, thus goes right. Next least is for R with 20 in first series to occupy second position and the rest is for S.

P	R	S	Q	**

Now the time chart for this sequence is as follows :

Sequence	Machining			Polishing			Idle time	
	In	Process	Out	In	Process	Out	Machining	Polishing
P	0	10	10	10	35	45	–	10
R	10	20	30	45	40	85	–	–
S	30	25	55	85	30	115	–	–
Q	55	45	100	115	15	130	30*	–
						Total	30 Idle time	10

Total elapse = 130

Difference in the total elapse = 175 - 130 = 45 hours.

Job	Total elapse of each job		Returns @ Rs 10 per hr		Difference
	Maximise case	Minimise case	Maximise case	Minimise case	Rs
P	135	45	1350	450	900
Q	60	130	600	1300	-700
R	175	85	1750	850	900
S	100	115	1000	1150	-150
Total	470	375	4700	3750	950

Remarks :

* Assumed that machining operator can go after completing all his jobs and hence no idle time/waiting time till polishing is completed.

** The sequence of maximisation is not the reverse order of minimisation case.

Practice Problems

1. Find the sequence that minimizes the total elapsed time (in hours) required to complete on the following two machines.

Task	A	B	C	D	E	F	G
Machine I	2	5	4	9	6	2	7
Machine II	6	8	7	4	3	9	3

[SVU - (M.Sc.) 83, AU-BE (Elect) 83, JNTU (CSIT) 2000/S
JNTU (Mech. & ECE) 2001/CCC]

Answer : Sequence in FACDEG

2. Find the sequence that minimises the total elapsed time (in days) required to repair and test the following seven jobs.

Activity	1	2	3	4	5	6	7
Repairing time in days	3	5	9	9	6	3	3
Testing time in days	4	5	1	6	3	3	4

[JNTU (Mech.) 88, OU - 86]

Answer : Sequences 3-1-7-6-2-4-5 or 3-1-7-2-4-6 or

3-7-1-6-2-4-5 or 3-7-1-2-4-6-5

3. There are 7 jobs each of which has to pass through machines M_1 and M_2 in order, $M_1 \rightarrow M_2$. Processing times (in hours) are given below. Determine the sequence that will minimise the elapsed time and prepare the time schedule chart and Gantt chart.

Job	1	2	3	4	5	6	7
M1	3	15	12	6	10	11	9
M2	8	10	10	6	12	1	3

[JNTU (Mech.) 98]

Answer : Optimal Sequence is : $1 - 4 - 5 - 3 - 2 - 7 - 6$ or $1 - 5 - 3 - 2 - 4 - 7 - 6$

	M1		M2	
JOB	Time in	Time out	Time in	Time out
1	0	3	3	11
4	3	9	11	17
5	9	19	19	31
3	19	34	34	44
2	34	46	46	56
7	46	55	56	59
6	55	66	66	67

Total time elapse= 67 hours.

Idle time on $M1 = 1$ hour.

Idle time on $M2 = 17$ hours.

4. In a pipe fabrication shop the times required for cutting and bending operation for 6 different sizes of pipes are given. Determine the order in which those pipes should be processed in order to minimise the total time for processing all the pipes.

Pipe	1	2	3	4	5	6
Cutting Time	3	12	5	2	9	11
Welding Time	8	10	9	6	3	1

[JNTU (Mech.) 96/P/S]

Answer : Optimal Sequence is : $4 - 1 - 3 - 2 - 5 - 6$

JOB	M1		M2	
	Time in	Time out	Time in	Time out
4	0	2	2	8
1	2	5	8	16
3	5	10	16	25
2	10	22	25	35
5	22	31	35	38
6	31	42	42	43

Total time elapse = 43 hours.

Idle time on $M1$ = 1 hour.

Idle time on $M2$ = 6 hours.

5. The repairing activity on eight machines take the time as shown below. Sequence them to minimise the idle time of repairmen and schedule in a chart.

Machine	A	B	C	D	E	F	G	H
Repair team - 1	5	4	22	16	15	11	9	4
Repair team - 2	6	10	12	8	20	7	2	21

[JNTU (CSE) 96/S]

Answer : $H \rightarrow B \rightarrow A \rightarrow E \rightarrow C \rightarrow D \rightarrow F \rightarrow G$; min. Total time = 90 hrs.

6. The time required in days for reparing and overhauling six machines is shown in the following table. Also, given that overhauling is to be done only after repairing job. Presently, the following schedule is choosen.

Machine sequence	M_1	M_2	M_3	M_4	M_5	M_6
Repair time	6	6	4	6	5	8
Overhauling time	4	2	10	6	3	6

(i) Determine optimal sequence

(ii) If downtime of any of the six machines costs Rs. 800 per day, idle time of repairmen costs Rs. 250 per day and idle time for overhauling costs Rs. 430 per day, which of the two schedules, the present one and the one found in (i), will be more economical. What are their respective cots.

[JNTU (CSE) -96]

Answer : (i) $M_3 \rightarrow M_6 \rightarrow M_4 \rightarrow M_1 \rightarrow M_5 - M_2$; (ii) Present costs = Rs. 64, 630, Optimal cost = Rs. 60,310.

7. Find the optimal sequence and min. eplased time for the following data.

Job	1	2	3	4	5
Machine P	5	1	9	3	10
Machine Q	2	6	7	8	4

[JNTU (Mech - ECE) 96/C]

Answer : $2 - 4 - 3 - 5 - 1$; 30 hrs, Idle time on $A = 2$ hrs, $B = 3$ hrs

8. Ram & Co plans to fill six positions for which it conducts two types of tests. The experts who came to test the candidates have approximated their probable time and given in minutes of time as follows. Also it is decided to test aptitude first followed by interview. Find the sequence that can minimise the waiting time of the experts.

Position	I	II	III	IV	V	VI
Aptitude test	140	180	150	200	7	100
Interview	70	120	110	80	100	90

Answer :   II – III – V – VI – IV – I ; Min time = 440 min.

9. The Hi-tech Publishers have to publish 8 text books for which it spends the time in days for DTP preperation and printing are estimated as follows. find the best sequence to take up these books to minimise total publishing time and also find the idle time of DTP operator and printing operator.

Books	B_1	B_2	B_3	B_4	B_5	B_6	B_7	B_8
DTP	14	26	17	11	9	26	18	15
Printing	21	15	16	21	22	12	13	25

Answer : $B_5 \to B_4 \to B_1 \to B_8 \to B_3 \to B_2 \to B_7 \to B_6$

Optimal publishing time = 154 days,

Idle time for DTP operator = 18 hrs for printing = 9 hrs.

10. Find the sequence that minimises the total elapsed time (in hours) required to complete following tasks on two machines :

Task	1	2	3	4	5	6	7	8	9
Machine A	2	5	4	9	6	8	7	5	4
Machine B	6	8	7	4	3	9	3	8	11

[JNTU (Mech. & ECE) 98/CCC, CSE/ECE 2001]

Answer : $1 \to 9 \to 3 \to 2 \to 8 \to 5 \to 4 \to 5 \to 7$ or

$1 \to 9 \to 3 \to 8 \to 2 \to 6 \to 4 \to 5 \to 7$

Minimum time is 61 hours, Idle time on $M_A = 11$ hrs,

on $M_B = 2$ hrs.

11. A company has six works on hand, coded A to F. All the works have to go through two machines M_1 and M_{II}. The time required for works on each machine, in hours, is given below :

Work	G	H	I	J	K	L
Machine I	1	4	6	3	5	2
Machine II	3	6	8	8	1	5

Draw a sequence table scheduling the six works on the two machines.

[JNTU (Mech.) 96/P/S]

Answer : $G \rightarrow L \rightarrow J \rightarrow H \rightarrow I \rightarrow K$

Minimum time is 32 hours.

12. A company has 3 works on hand. Each of these must be passed through two departments, the sequential order for which is :

Department A : Press shop   *Department B :* Finishing

The table below lists the number of days required by each job in each department

	Work I	Work II	Work III
Department A :	8	6	5
Department B :	8	3	4

Find the sequence in which the 3 works should be processed so as to take minimum time to finish all the 3 works.

Answer : Work I → work III → work II ; Minimum time is 23 days.

13. Six jobs go first over machine I and then over machine II. The order of the completion of job has no significance. The following table gives the machine times in hours for six jobs as the two machines.

Job	A	B	C	D	E	F
Time on machine I	5	9	4	7	8	6
Time on machine II	7	4	8	3	9	5

Find the sequence of jobs that minimise the total elapsed time to complete the jobs. Find the minimum time by using Gantt's chart or by any other method.

[JNTU (CSE) 92/S]

Answer : $C \rightarrow A \rightarrow E \rightarrow F \rightarrow B \rightarrow D$; Minimum time is 42 hours.

14. Five jobs are to be processed on two machines. Machining times are given in the following table :

Machines/Jobs	J_1	J_2	J_3	J_4	J_5
M_1	10	5	15	22	3
M_2	6	8	12	4	15

Schedule the jobs on machines. **[JNTU B.Tech. (Mech.) 2001]**

Answer : $J_5 \rightarrow J_2 \rightarrow J_3 \rightarrow J_1 \rightarrow J_4$; Elapse = 59 hrs.

15. Solve the following $2M/C \times n$ job sequencing.

Job	1	2	3	4	5
M - 1	3	7	4	5	7
M - 2	6	2	7	3	4

[JNTU (CSE) 96]

Answer : 1 – 3 – 5 – 4 – 2; Min. time = 28 hrs;
idle time 2 hrs on M_1 and 6 hrs on M_2

16. A book binder has to print and bind six books. the binding times required in minutes are 80, 100, 90, 60, 30 and 10 while that for printing are 30, 120, 50, 20, 90 and 110 min. respectively. Find optimum sequence, minimum elapsed time and idle time for each operator.

Answer : Sequence $4 \rightarrow 1 \rightarrow 3 \rightarrow 2 \rightarrow 5 \rightarrow 6$, Min. Elapse = 430 min.

Idle time for printer = 10 min, binder = 40 min.

17. Solve the sequencing problem for six jobs whose process times in hrs on machine A, Machine B are (4, 6), (8, 3), (3, 7), (6, 2), (7, 8) and (5, 4).

Answer : 3 – 1 – 5 – 6 – 2 – 4; Total Elapse = 35 hrs

18. Determine the sequence and minimum elapse for the following process times in hrs.

Job	1	2	3	4	5
M/C P	10	2	18	6	20
M/C Q	4	12	14	16	8

[JNTU (Mech.) 93/P/S]

Answer : 2 – 4 – 3 – 5– 1; Total elapse = 60 hrs.

19. In factory, there are six jobs to perform each of which should go through two machines A and B, in the order A, B. The processing timings (in hours) for the jobs are given below. You are required to determine the sequence for performing the jobs that would minimize the total elapsed time T. What is the value of T?

Job	Machine A	Machine B
J_1	1	5
J_2	3	6
J_3	8	3
J_4	5	2
J_5	6	2
J_6	3	10

[JNTU Mech/Prod/Chem 2001/S]

9.3 *Sequencing n Jobs X 3 Machines : Jackson's Rule*

For sequencing and scheduling n jobs on three machines, Jackson proposed a simple rule by which the $n \times 3$ is converted to $n \times 2$ model for preparing the sequence. And then schedule may be prepared as usually for the three machines individually either by time chart or Gantt chart.

For converting the $n \times 3$ to $n \times 2$ the given problem has to satisfy one or both of the following conditions.

Condition 1 :

The minimum process time among all the jobs to be performed on first machine is greater than or equal to the maximum process time among the jobs to be performed on second machine. i.e., Min $(M_1) \geq$ Max (M_2)

Condition 2 :

The minimum process time among all the jobs to be performed on third machine is greater than or equal to the maximum process time among the jobs to be performed on second machine. Min $(M_3) \geq$ Max (M_2)

(In other words the maximum of second machine is less than or equal to the minimum of first or/and third machine timings). Max $(M_2) \leq$ Min $(M_1$ or/and $M_3)$

If the problem satisfies one or both of the above conditions, the $3 \times n$ sequencing problem is converted to $2 \times n$ sequencing problem by adding the job timings of first machine with the corresponding job timings of second machine to make as first center and similarly adding job timings of second machine with their respective job timings of third machine to make it as second work center. Now, the sequencing is made for the two hypothetical work centers with these new timings according to Johnson's rule.

i.e., $WC_1 = (M_{1j} + M_{2j})$ and $WC_2 = (M_{2j} + M_{3j})$

After finding the sequence, the Gantt chart and time chart calculations are done by considering three machines with the timings given in the original problem.

A numerical example is shown here below to illustrate

ILLUSTRATION 6

A company uses its maintenance crew in three teams for their preventive maintenance of their heavy vehicles. First team looks after the replacement of worn out parts, the second oiling and resetting and the third checking and tests running. The estimated time for maintenance of each of these vehicles is given in hours in the following table and passing is not allowed. Find the sequence and schedule them so as to minimise the total elapsed time and idle time.

Team/Vehicle No.	1	2	3	4	5	6	7
Replacement team	3	8	7	4	9	8	7
Resetting team	4	3	2	5	1	4	3
Inspection team	6	7	5	11	5	6	12

[JNTU-(Mech.) 2001/C/S]

Solution :

Seven vehicles are given to be processed through three repair teams say R_1, R_2, R_3. In order R_1, R_2, R_3. Now, the minimum time in R_1 series is 3, and that in R_3 is 5 while the maximum in R_2 is 5. Since minimum $R_3 \geq$ maximum R_2 (condition 2) is satisfied. Hence the problem can be converted to that of 7 jobs X 2 work centers.

Thus $WC1 = R_1 + R_2$ and $WC2 = R_2 + R_3$

The problem is re-written as

WC/V.No	1	2	3	4	5	6	7
WC1	3 + 4 = 7	8 + 3 = 11	7 + 2 = 9	4 + 5 = 9	9 + 1 = 10	8 + 4 = 12	7 + 3 = 10
WC2	4 + 6 = 10	3 + 7 = 10	2 + 5 = 7	5 + 11 = 16	1 + 5 = 6	4 + 6 = 10	3 + 12 = 15

Using Johnson's optimal sequence algorithm, the correct sequence is*

V_1	V_4	V_7	V_2	V_6	V_3	V_5

(* Students are advised to write sequence step by step as shown in the previous illustrations.)

For calculating the total time elapsed and the idle times, we draw the time chart as follows

Team Vehicle No.	Replacement Team Timings (R_1)			Resetting Team Timings (R_2)			Inspection Team Timings (R_3)			Idle Time / Waiting Time		
	In	Repair	Out	In	Repair	Out	In	Repair	Out	R_1	R_2	R_3
V_1	0	3	3	3	4	7	7	6	13	-	3	7
V_4	3	4	7	7	5	12	13	11	24	-	-	-
V_7	7	7	14	14	3	17	24	12	36	-	2	-
V_2	14	8	22	22	3	25	36	7	43	-	5	-
V_6	22	8	30	30	4	34	43	6	49	-	5	-
V_3	30	7	37	37	2	39	49	5	54	-	3	-
V_5	37	9	46	46	1	47	54	5	59	13	7+12	-
Total		46	46		22	47		52	59	13 + 37 + 7 = 57		

The table indicates that the total elapsed time is 59 hours. Idle time of replacement team is 13 hours at the end of project. Resetting team gets 37 hours of idle time of which are 3 hours at the starting and 12 hours at the ending and 22 hours in middle of the project. The inspection team will have to wait 7 hour at the beginning of the project.

Gantt Chart :

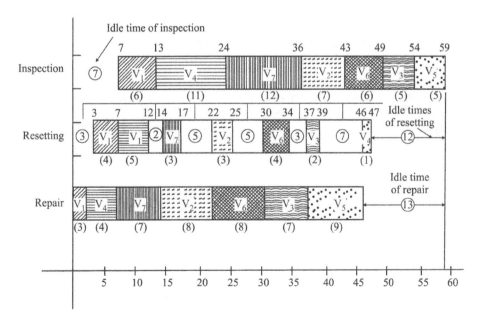

FIGURE 9.2 : GANTT CHART

ILLUSTRATION 7

Find the sequence that minimizes total machining time to complete the following data:

Tasks	A	B	C	D	E	F
Time on Machine I	4	9	8	5	10	9
Time on Machine II	5	4	3	6	2	5
Time on Machine III	7	8	6	12	6	7

[JNTU (CSE) 97]

Solution :

Conditions : Min. of $M_1 \geq$ Max. of M_2 or Min of $M_3 \geq$ Max. of M_2

Here $6 = 6$, Second condition of Jackson's rule of sequencing algorithm is satisified. Hence we can convert the above 3 machines $X\,n$ jobs case to 2 machines $X\,n$ jobs case.

Rearranging the Table :

Tasks	A	B	C	D	E	F
$M_1 + M_2$	9	13	11	11	12	14
$M_2 + M_3$	12	12	9	18	8	12

Note : *Identify the least processing time of all the values. If this falls in 1st machine, allocate the task at left and if it falls in 2nd machine allocate to right. Repeat the process till all the tasks are scheduled. (Students are advised to show all steps)*

Optimal Sequence is : $A \rightarrow D \rightarrow B \rightarrow F \rightarrow C \rightarrow E$

Machines Job	M_1			M_2			M_3			Idle Time		
	Time in	Process Time	Time out	Time in	Process Time	Time out	Time in	Process Time	Time out	M_1	M_2	M_3
A	0	4	4	4	5	9	9	7	16	–	4	9
D	4	5	9	9	6	15	16	12	28	–	–	–
B	9	9	18	18	4	22	28	8	36	–	3	–
F	18	9	27	27	5	32	36	7	43	–	5	–
C	27	8	35	35	3	38	43	6	49	–	3	–
E	35	10	45	45	2	47	49	6	55	10	7 + 8	–
						Total				55 10	30	9

Total elapsed time = 55 hrs ;   Idle time on m/c 1 = 10 hrs.

Idle time on m/c 2 = 30 hrs. and

Idle time on m/c 3 = 9 hrs.

ILLUSTRATION 8

A company has 6 jobs which go through 3 machines X, Y and Z in the order XYZ.
The processing time in minutes for each job on each machine is as follows:

Machines		Job					
		1	*2*	*3*	*4*	*5*	*6*
	X	*18*	*12*	*29*	*36*	*43*	*37*
	Y	*7*	*12*	*11*	*2*	*6*	*12*
	Z	*19*	*12*	*23*	*47*	*28*	*36*

Find the optimal sequence, total elapsed time and idle times for each machine.

[JNTU (Mech.) 99]

Solution :

Since $x_{min} \geq y_{max}$, we have

		1	2	3	4	5	6
$\therefore$	$X + Y$	25	24	40	38	49	49
	$Y + Z$	26	24	34	49	34	48

The correct sequence is

$$J_2 \rightarrow J_1 \rightarrow J_4 \rightarrow J_6 \rightarrow J_3 \rightarrow J_5 \text{ or } J_1 \rightarrow J_4 \rightarrow J_6 \rightarrow J_3 \rightarrow J_5 \rightarrow J_2$$

	Machine X				Machine Y				Machine Z			
Job	Initial Time	Process Time	Final Time	Idle Time	Initial Time	Process Time	Final Time	Idle Time	Initial Time	Process Time	Final Time	Idle Time
2	0	12	12	–	12	12	24	12	24	12	36	24
1	12	18	30	–	30	7	37	6	37	19	56	1
4	30	36	66	–	66	2	68	29	68	47	115	12
6	66	37	103	–	103	12	115	35	115	36	151	0
3	103	29	132	–	132	11	143	17	151	23	174	0
5	132	43	175	34	175	6	181	32 + 28	181	28	209	7

(i) Total clapped time = 209 minutes

(ii) Idle time on machine $X = (209 - 175) = 34$ minutes

(iii) Idle time on machine $Y = 12 + 6 + 29 + 35 + 17 + 32 + (209 - 181)$
     = 159 minutes

(iv) Idle time on machine $Z = 24 + 1 + 12 + 7 = 44$ minutes.

ILLUSTRATION 9 ――――――――――――――――――――――――――――――――――

There are 5 projects to be made, each of which must under go three stages viz., information feeding, analysing and printing. A programmer has decided to operate three computers namely Zebronics, Wipro and Acer for these operations. The printer is connected to Zebronics only.
Processing times are given below:

Project No.	1	2	3	4	5
Acer	8	5	4	6	5
Zebronics	10	13	11	10	12
Wipro	6	2	9	7	4

(i) Determine the total elapsed time and idle time of each computer.
(ii) If it can be processed by the sub-contract services by three outside parties, NIIT, CMC and STG to process on Zebronics, Wipro and Acer respectively, schedule the parties optimally, under the condition that the parties may be called on any day but the contract once started should be continued till the last job of the respective party is completed and the payment should be made for the waiting idle times also.
(iii) What will be the amounts paid to each party if it costs Rs. 10/- for each working hour and Rs. 5/- for each waiting hour.
(iv) If any project that waits will cost Re.1/- per hours, what will be the cost incurred on waiting of projects. Assume all projects are taken up now and waiting time is counted from the time it is taken up.

Solution :

As per the given data, the printing is the last operation and is connected to Zebronics only. Therefore process times on Zebronics should be taken last i.e., as machine - 3, and the other two can be taken in any order if they satisfy Jackson's conditions.

Now let us examine which order of machines is applicable here to convert this n jobs $X3$ machines case to n jobs $X2$ machines case.

Project	1	2	3	4	5
Acer	8	5	4	6	5
Wipro	6	2	9	7	4
Zebronics	10	13	11	8	12

Case (i) : Order of machines as Acer → Wipro → Zebronics

Condition Checking :

Min. of Acer times (A_i)	: 5 —	(Machine - 1)
Min. of Zebronics times (Z_1)	: 8 —	(Machine - 3)
Max. of Wipro times (W_i)	: 9 —	(Machine - 2)

Here Min. of Zebronics times $(Z_r) \not\geq$ Max of Wipro (W_i) and Min. of Acer $(A_i) \not\geq$ Max. of Wipro (W_i). i.e., 5, 8 < 9.

Both the Jackson's conditions are not satisfied in this case and hence we cannot proceed with this order.

However, let us check the condition for case (ii)

Case (ii) : Order of machine as Wipro $\rightarrow$ Acer $\rightarrow$ Zebronics.

Condition Checking :

Min. of Wipro (W_i) : 2 — (Machine - 1)

Min. of Zebronics (Z_i) : 8 — (Machine - 3)

Max. of Acer (A_i) : 8 — (Machine - 2)

Here Jackson's condition (ii) is satisfied as

Min. of Zebronics $(Z_i) \geq$ Max. of Acer (A_i)

i.e., 8 = 8

Therefore we go by the order of the machines as

Wipro $\rightarrow$ Acer $\rightarrow$ Zebronics

Project	1	2	3	4	5
Wipro	6	2	9	7	4
Acer	8	5	4	6	5
Zebronics	10	13	11	8	12

Since, the second condition of Jackson's rule is satisfied we can convert the above n jobs X 3 M/C model to n jobs X 2 M/C model to find the optimal sequence. This is done by taking two hypothetical machines say X and Y where the process times are given by

$$X_i = W_i + A_i \quad \text{and} \quad Y_i = A_i + Z_j$$

Thus the problem reduces to

Project	1	2	3	4	5
Machine X	14	7	13	13	9
Machine Y	18	18	15	14	17

Now, the least among all the ten timings is 7 that appears in first series for project - 2. Therefore this is placed extreme left and project 2 is deleted for the next iteration.

②				

Project	1	3	4	5
Machine X	14	13	13	9
Machine Y	18	15	14	17

Now, least is 9 in first series for project - 5, and so occupies left side i.e., second place, and project - 5 is deleted for next iteration.

②	⑤			

Project	1	3	4
Machine X	14	13	13
Machine Y	18	15	14

Next least timing (13) is in tie between project - 3 and 4, both in first series for next two positions. Therefore, we see second series. The least (14) appears for project - 4, therefore it goes right side of the third and fourth position. The sequence will be

②	⑤	③	④	

Obviously project - 1 occupies the last position.

Thus the optimal sequence is $2 \rightarrow 5 \rightarrow 3 \rightarrow 4 \rightarrow 1$.

The scheduling chart for the above sequence is calculated as follows.

Sequence	WIPRO			ACER			ZEBRONICS			IDLE TIMES			Waiting Time For Projects			
	In	Proc	Out	In	Proc	Out	In	Proc	Out	W	A	Z	At W	At A	At Z	Total
Proj. - 2	0	2	2	2	5	7	7	13	20	-	*-	*-	NIL	NIL	NIL	
Proj. - 5	2	4	6	7	5	12	20	12	32	-	-	-	2	1	8	11
Proj. - 3	6	9	15	15	4	19	32	11	43	-	3	-	6	-	13	19
Proj. - 4	15	7	22	22	6	28	43	8	51	-	3	-	15	-	15	30
Proj. - 1	22	6	28	28	8	36	51	10	61	*-	*-	-	22	-	15	37
Elapse time	28	28		28	36-2 = 34			54	61-7 =54	-	6	-	45	1	51	97

* As we can call the parties at any time, it is better to call CMC first to work on Wipro from 0 to 28 hours and STG to work from second hour to 36th hours on Acer and NIIT to start at 7th hour to finish at 61st hour on Zebronics.

The cost calculations are

Party	Computer	Cost of working	Cost of waiting	Total cost
CMC	WIPRO (0 to 28)	$28 \times 10 = 280$	NIL	280
STG	ACER (2 to 36)	$28 \times 10 = 280$	$6 \times 5 = 30$	310
NIIT	ZEBRONICS (7 to 61)	$54 \times 10 = 540$	NIL	540
* 28 hrs actual working and 6 hrs ; Total 34 for ACER			Total cost (Rs)	1130

Waiting cost of projects : $97 \times 1 = $ Rs. 97/-

9.4 *Sequencing n Jobs X m Machines*

Sequencing the jobs may not be always in the form of $n X 2$ or $n X 3$. In fact in real practice, we often come across with the n jobs to sequence on m machines or men or work centers. However it is not very difficult to sequence in such case. The method we apply here is again based on the Jackson's conditions as explained below.

Let there be n jobs, of which each is to be processed through m work centers/machines say $W_1, W_2, W_3, \ldots W_m$.

Now, the iterative procedure is as follows :

Step 1 : Find minimum process time of W_1 and W_m series, and maximum process time of all the middle series.

Step 2 : Check whether

Condition 1 : Min of W_1 series $\geq$ max of middle series.

Condition 2 : Min of W_m series $\geq$ max of middle series.

Step 3 : If both conditions of step 2 are not satisfied, the method fails. If at least one of the conditions is true, then go to next step.

Step 4 : Convert m work center problem to two-work center problem by assuming two fictitious machines say A and B, where the process times on A and B are

$A_i = $ the algebraic sum of process times of W_1 and all corresponding timings of the mediocre series and $B_i = $ the algebraic sum of process times of W_m and all corresponding timings of the mediocre series.

i.e., $A_i = W_{1j} + W_{2j} + W_{3j} + \ldots . W_{(m-1)j}$

and $B_i = W_{2j} + w_{3j} + W_4j + \ldots . W_{mj}$

(However in general practice we can find the sequence by taking the timings of W_1 and W_m if the sums of the mediocre series are fixed positive constants).

ILLUSTRATION 10

Five machine parts are to be reassembled after yearly maintenance in the order P, Q, R, S and T on four machines A, B, C and D. Find the optimal sequence for the machines when passing is not allowed of which the repair time (in hours) is given below. Also find the total elapsed time, and individual idle times for the repairmen on each machine and machine part. (Consider that the machine will be handed over to the production department immediately after reassembling all the parts).

Machine/Part	P	Q	R	S	T
A	7	5	2	3	9
B	6	6	4	5	10
C	5	4	5	6	8
D	8	3	3	2	6

Solution :

Here the machines are to be sequenced

Checking Sequencing Algorithm Rules :

Minimum repair time of P series [Min. (t_{Pj})] = (t_{PC}) = 5

Minimum repair time of T series [Min. (t_{Tj})] = (t_{TD}) = 6

And maximum repair time of Q, R, S series [Max $(t_{Qj}\ t_{Rj}\ t_{Sj})$] = {6, 5, 6}

Since the condition Min. (t_{Tj}) ≥ Max (t_{Qj}, t_{Rj}, t_{Sj}) is satisfied the given problem can be converted into a four jobs and two machines (say X and Y) problem as

$$t_{Xj} = (t_{Pj} + t_{Qj} + t_{Rj} + t_{Sj})$$

$$t_{Yj} = (t_{Qj} + t_{Rj} + t_{Sj} + t_{Tj})$$

Machine/Job	A	B	C	D
Machine X	17	21	20	16
Machine Y	19	25	23	14

Now, using the optimal sequence algorithm, the following optimal sequence can be obtained.

A	C	B	D

The total elapsed time corresponding to the above sequence can be computed as shown in the following table, using the individual processing times as given in the original problem.

Machine/Part	P	Q	R	S	T
A	0-7	7-12	12-14	14-17	17-26
C	7-12	12-16	16-21	21-27	27-35
B	12-18	18-24	24-28	28-33	35-45
D	18-26	26-29	29-32	33-35	45-51

From the above table, it can be understood that the minimum total elapsed time is 51 hours.

The idle times for the man who reassembles the machine parts are as follows :

For $P = 51 - 26 = 25$ hours.

For $Q = 7 + (18 - 16) + (26 - 24) + (51 - 29) = 33$ hours.

For $R = 12 + (16 - 14) + (24 - 21) + (29 - 28) + (51 - 32) = 37$ hours.

For $S = 14 + (21 - 17) + (28 - 27) + (51 - 35) = 35$ hours.

For $T = 17 + (27 - 26) = 18$ hours.

The waiting times for the machines are as follows :

For A = No waiting time and will be handed over after 26 hours.

For $B = 12 + (35 - 33) = 14$ waiting hours and will be handed over after 45 hours.

For $C = 7$ waiting hours and will be handed over after 35 hours.

For $D = 18 + (33 - 32) + (45 - 35) = 29$ waiting hours and will be handed over after 51 hours.

Practice Problems

1. A mechanic has to assemble 5 machines and each assembly takes three stages for which he employed three persons. The sequence of the stages is fixed as stage I, stage II and stage III. Determine the optimal sequence for the machines to be assembled to minimise the time elapsed from the start of first machine to the end of last machine. The time required, in days for each machine at each stage is given in the following matrix. Also find the time that the mechanic could promise to deliver each machine. If the mechanic can delink a person from the present process if his stage work is completed or not started, what will be the net idle time that the process demands?

Machine Code	Stage I	Stage II	Stage III
M0235988	6	8	14
K0242115	16	10	18
V6566573	14	2	10
G3042721	10	4	12
S6567763	8	6	20

Answer : Optimal Sequence is : $M - G - S - K - V$

Machine	Stage - I		Stage - II		Stage - III		Delivery day
	Time in	Time out	Time in	Time out	Time in	Time out	From now
M	0	6	6	14	14	28	28
G	6	16	16	20	28	40	40
S	16	24	24	30	40	60	60
K	24	40	40	50	60	78	78
V	40	54	54	56	78	88	88

Total elapsed time is 88 hours and idle time for stage - I - 34 hours (the mechanic can delink all the last 34 hours), for stage - II - 58 hours (in which $6 + 32 = 38$ hours can be delinked) and for stage III - 14 hours (in which all the 14 can be delinked). Net idle time due to process is 20 (stage II operator) if delinked

2. There are 5 spare parts are to be made each of which must go through machines $M1$, $M2$, and $M3$ in order $M1$, $M3$, $M2$. Processing times are given below.

   (i) Determine the optimal sequence, total elapsed time and idle time of each machine.

   (ii) If it can be processed by the sub-contract services by three outside parties, $P1$, $P2$ and $P3$ to process on $M1$, $M2$ and $M3$ respectively, schedule the parties optimally, under the conditions that the parties may be called on any day but the contract once started should be continued till the last job of the respective party is completed and the payment should be made for the process delays also.

   (iii) What will be the amounts paid to each party if it costs Rs. 10/- for each working hour and Rs. 5/- for each waiting hour?

Spare Part	1	2	3	4	5
$M1$	8	5	4	6	5
$M2$	10	13	11	10	12
$M3$	6	2	9	7	4

[JNTU (Mech.) 96/CCC, (ECE) -96]

Answer : (i) Optimal sequence is : $2 - 5 - 3 - 4 - 1$

Job	M1		M2		M3	
	Time in	Time out	Time in	Time out	Time in	Time out
2	0	5	5	7	7	20

5	5	10	10	14	20	32
3	10	14	14	23	32	43
4	14	20	23	30	43	53
1	20	28	30	36	53	63

Total elapsed time = 63 hours.

Idle time on $M1$ = 35 hours, on $M2$ = (27 + 5) = 32 hrs and on $M3$ = 7 hrs.

(ii) $P1$ starts at 0th hour and ends at 28th hour (28 hours), $P2$ starts at 5th hour and ends at 36th hour (31 hours) and $P3$ starts at 7th hour and ends at 63th hour (56 hours).

(iii) Payments for $P1$ - Rs. 280/- (all are working hours), for $P2$ - Rs, 295/- (28 working hours and 3 waiting hours). and for $P3$ - Rs. 560/- (all working hours).

3. Find the sequence that minimises the total elapsed time in hours to complete the following jobs on 3 machines. Prepare Gantt chart.

	A	B	C	D	E
$M1$	3	8	7	5	2
$M2$	3	4	2	1	5
$M3$	5	8	10	7	6

[JNTU (Mech.) 87]

Answer : Optimal Sequence is : $A - D - E - C - B$ or $D - A - E - C - B$

JOB	M1		M2		M3	
	Time in	Time out	Time in	Time out	Time in	Time out
A	0	3	3	6	6	11
D	3	8	8	9	11	18
E	8	10	10	15	18	24
C	10	17	17	19	24	34
B	17	25	25	29	34	42

Total elapsed time = 42;　　　Idle time on M1 = 17 hours.

Idle time on M2 = 27 hours.

Idle time on M3 = 6 hours.

4. Find the sequence that minimises total machining time. The Machine sequence is PRQ

Tasks	A	B	C	D	E	F
Time on M/C - P	4	9	8	5	10	9
Time on M/C - Q	7	8	6	12	6	7
Time on M/C - R	5	4	3	6	2	5

[JNTU (Mech. & ECE) 97/CCC/S]

Answer: Optimal Sequence is : $A - D - B - F - C - E$

JOB	P		R		Q	
	Time in	Time out	Time in	Time out	Time in	Time out
A	0	4	4	9	9	16
D	4	9	9	15	16	28
B	9	18	18	22	28	36
F	18	27	27	32	36	43
C	27	35	35	38	43	49
E	35	45	45	47	49	55

Total elapsed time = 55 hours. ;   Idle time on M/C P = 10 hours.
Idle time on M/C R = 30 hours.   Idle time on M/C Q = 9 hours.

5. Find the sequence that minimises total elapsed time if the order of machines is $M_1 \to M_3 \to M_2$.

Tasks	A	B	C	D	E	F	G
Time on M - 1	3	8	7	4	9	8	7
Time on M - 2	6	7	5	11	5	6	12
Time on M - 3	4	3	2	5	1	4	3

[JNTU (Mech. & ECE) 2001/C/S]

Answer : Optimal Sequence is : $A - D - G - B - F - C - E$

JOB	M1		M3		M2	
	Time in	Time out	Time in	Time out	Time in	Time out
A	0	3	3	7	7	13
D	3	7	7	12	13	24
G	7	14	14	17	24	36
B	14	22	22	25	36	43
F	22	30	30	34	43	49
C	30	37	37	39	49	54
E	37	46	46	47	54	59

Total elapsed time = 59 hours.;   Idle time on M1 = 13 hours.
Idle time on M2 = 7 hours.   Idle time on M3 = 37 hours.

6. We have six jobs, each of which must go through machines P, Q and R in the order PQR. Processing time (in hours) are given in the following table.

Job	1	2	3	4	5	6
Machine P (P_i)	8	3	7	2	5	1
Machine Q (Q_i)	3	4	5	2	1	6
Machine R (R_i)	8	7	6	9	10	9

Determine a squence for the five jobs that will minimise the elapsed time T.

[JNTU - CSE 97]

Answer : Optimal sequences : (i) $4 - 5 - 6 - 2 - 1 - 3$ (ii) $4 - 5 - 6 - 2 - 3 - 1$

Idle time is 27 hr. for machine P, 32 hr. for machine Q, and 4 hr. for machine R. Total ellapsed time = 53.

7. Find the sequence that minimises the total elapsed time required to complete the following tasks. Each job is processed in the order I - II - III.

Job	A	B	C	D	E	F	G
Machine I	12	6	5	11	5	7	6
Machine II	7	8	9	4	7	8	3
Machine III	3	4	1	5	2	3	4

Answer : Optimal sequence : $B - F - C - A - E - D - G$ [JNTU - Mech 97/P/S]

Elapsed time : 59 hours

Idle time is 7 hours for machine I, 13 hours for machines B and 37 hours for machine III.

8. There are five jobs, each of which must go through machines A, B and C in the order ABC. Processing times (in hours) are given in the following table.

Job	1	2	3	4	5
Machine A	8	10	6	7	11
Machine B	5	6	2	3	4
Machine C	4	9	8	6	5

Answer : $3 - 2 - 4 - 1 - 5$; minimum time is 51 hours   [JNTU - CSIT 2000]

9. Determine the optimal sequence and idle time of each mahine that will minimise the total elapsed time.

Job No.	1	2	3	4	5
Machine A	5	7	6	9	5
Machine B	2	1	4	5	3
Machine C	3	7	5	6	7

[JNTU - CSE - 2000]

Answer : $2 - 5 - 4 - 3 - 1$ or $5 - 4 - 3 - 2 - 1$ or $5 - 2 - 4 - 3 - 1$

Elapsed time = 4- hours; Idle time is 8 hrs on machine A, 25 hrs on Machine B and 12 hrs on Machine C.

10. Give the following data

(a)

Job 1	1	2	3	4	5	6
Machine A	12	10	9	14	7	9
Machine B	7	6	6	5	4	4
Machine C	6	5	6	4	2	4

(b) Order of the processing of each job : ACB

(c) Sequence suggested : Jobs 5, 3, 6, 2, 1, 4

   (i) Find the total time elapsed for the sequence suggested?

   (ii) Is the given sequence optimal?

   (iii) If your answer to (ii) is 'No', find the optimal sequence and the total elapsed time associated with it.   **[JNTU CSE 2001/S]**

Answer :   (i) Total elapse = 70 hrs, Idle times on machine A = 9; Machine B = 38; Machine C = 43

   (ii) No.

   (iii) Optimal sequene is $1 - 3 - 2 - 4 - 6 - 5$
   Idle time on Machine A = 6, on Machine B = 35 and on Machine C = 40 hours

11. Find an optimal sequence for the following sequencing problem of four works and five machines when passing is not allowed. Processing time (in hours) is given below;

Work	A	B	C	D
Machine M_1	6	5	4	7
Machine M_2	4	5	3	2
Machine M_3	1	3	4	2
Machine M_4	2	4	5	1
Machine M_5	8	9	7	5

Also find the total elapsed time.

Answer : Optimal sequence : $A - C - B - D$;
   Minimum total elapsed time is 43 hrs.

9.5 *Sequencing of 2 Jobs X n Machines*

Suppose there are two jobs say I and II to be processed on 'n' machines say M_1, M_2 ... M_n in two different orders. The technological sequence of the operations on the two jobs is already fixed, but the exact or expected time of processing time is not known. Now, the objective of the problem is to determine the optimal sequence of processing that results in minimum idle time. In such cases, a graphical solution is most suitable by taking the two jobs on the two co-ordinate axes. This method is summarised in the following steps.

Algorithm :

Step 1 : Draw the perpendicular axes, such that x-axis represents Job - 1 and y-axis represents Job - 2. Thus we can say that when process moves horizontally along x-axis, Job - I is under process while Job -2 remains idle. And if we move along y-axis (i.e., vertically), Job - 2 will be under process while Job - 1 is idle. If we move along $45°$ to either axis, both jobs are simultaneously processed.

Step 2 : Mark the technological sequences and the process times on the axes as given.

Step 3 : Construct various blocks starting from origin, to represent area of the machine that will be common to both jobs in first quadrant. These are the areas at which both jobs can not be performed simultaneously. Mark the end point i.e., farthest corner of last block.

Step 4 : Draw the straight line starting from origin in $45°$ angle to either of axis. When the line is obstructed by a block, go along the side of block either horizontally or vertically as can be moved. Continue the line till end point is reached.

Step 5 : The diagonal line (at $45°$) represents that both jobs are simultaneously processed, vertical line represents that Job-1 is idle while Job-2 is under process and horizontal line represents that Job-2 is idle while Job-1 is under process. Thus, choose the path in which diagonal movement is maximum. This is said to be optimal path.

Step - 6 : Calculate the total horizontal distance (idle time of Job-2) and total vertical distance (idle time of Job-1) on optimal path. These distance are added to the process time represented by the co-ordinates of end of point, as follows.

Total Elapsed Time of Job-1 :

= value of x-co-ordinate of end point + total vertical distance on optimal path

= process time of Job 1 + idle time of Job-1.

Total Elapsed Time for Job-2 :

= value of y co-ordinate of end point + total horizontal distance on optimal path.

= process time of job 2 + idle time of job 2.

The above procedure is illustrated through an example given below :

ILLUSTRATION 11

There are two jobs to be processed through five machines A, B, C, D and E. The prescribed technological order is

   Job -1: $A \rightarrow B \rightarrow C \rightarrow D \rightarrow E$

   Job -2: $B \rightarrow C \rightarrow A \rightarrow D \rightarrow E$

The process times in hours are given below.

Machine	A	B	C	D	E
Job - 1	3	4	2	6	2
Job - 2	3	5	4	2	6

Find out the optimal sequencing of the jobs on machines and the minimum time required to process these jobs.

Solution :

The above problem is in 2 jobs X n machine model for which we use graphical method. Let us take Job-1 on x-axis and Job -2 on y-axis and represent the process time in their technological sequence.

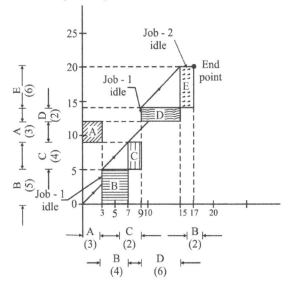

FIGURE 9.3 :

Now, we prepare the blocks that are common to both Job 1 and 2. Then we find a diagonal line at 45° starting from origin to reach the end point. The block B obstructs after moving to (3, 3) units and therefore we move vertically for 2 units and then go by 45° line. At (9, 11) we can get two paths. One of them is to go 3 units vertically from (9, 11) to (9, 14) and again at 45° from (9, 14) to (15, 20), then go 2 units horizontally [(15, 20) to (17, 20)] to reach the end point. This gives 2 units horizontal and (2 + 3) units vertical distances.

The second path is along $(9, 11) \rightarrow (15, 12) \rightarrow (17, 14) \rightarrow (17, 20)$. This path gives a total of 8 $(2 + 6)$ units of vertical and 5 units of horizontal distance. These idle times are more than those in first path, therefore first path is optimal.

The calculation of elapsed time

For Job-1 is

Process time + idle time

= x-coordinate of end point + total vertical distance on optimal path.

= $17 + (2 + 3) = 22$ hours.

For Job 2 :

Process time + idle time

= y-coordinate of end point + total horizontal distance on optimal path.

= $20 + 2 = 22$ hours.

Remark :

If we go by second path, the elapse would have been $17 + 8 = 25$ for Job - 1 and $20 + 5 = 25$ for Job - 2 which are not optimal.

Practice Problems

1. Use the graphical method to find the minimum elapsed time sequence of 2 works and 5 machines when we are given the following information.

Work 1	Sequence	A	B	C	D	E
	Time (hours)	2	3	4	6	2
Work 2	Sequence	C	A	D	E	B
	Time (hours)	4	5	3	2	6

[JNTU (Mech.) 99/P/S]

Answer : Idle time is 3 hours for work 1 and zero hours for work 2.

Elapsed time for work 1 is $17 + 3 = 20$ hours, for work 2 it is $(20 + 0) = 20$

2. Two jobs are to be processed on four machines A, B, C and D technological order for these jobs on machines is as follows ;

Job A	1	2	3	4
Job B	4	2	1	3

Processing times are given in the following table

Machines

Jobs	1	2	3	4
Job A	4	6	7	3
Job B	4	7	5	8

Find the optimal sequence of jobs on each of the machine.

[JNTU (Mech.) 97/P/S]

Answer : Idle time is 4 hours for job A and zero for Job B.

Elapsed time for Job A is 20 + 4 = 24, for job B it is (24 + 0) = 24.

3. A machine shop has six machines A, B, C, D, E and F. Two jobs must be processed through each of machines. The times one machines and the necessary sequence of the jobs through the shop are given below.

Order	1	2	3	4	5	6
Job - I	$A – 20$	$C – 10$	$D – 10$	$B – 30$	$E – 25$	$F – 16$
Job - II	$A – 10$	$C – 30$	$B – 15$	$D – 10$	$F – 15$	$E – 20$

[AU (B.E.) 86]

Answer : Minimum total time = 150 hrs.

4. A machine shop has four machines A, B, C and D. Two jobs must be processed through each of these machines. The time and necessary sequence of jobs through the shop are given below.

Job 1	Sequence	A	B	C	D
	Time (hrs)	2	4	5	1
Job 2	Sequence	D	B	A	C
	Time (hrs)	6	4	2	3

Use graphic method to obtain the total minimum elapsed time.

[JNTU (Mech.) 95/CCC]

Answer : 15 hrs

5. Solve the following 2 jobs $\times$ n machine sequencing problem. Also determine which jobs is to be taken first on each machine.

Job 1	Sequence	A	B	C	D	E
	Time (hrs)	3	4	2	6	2
Job 2	Sequence	B	C	A	D	E
	Time (hrs)	5	4	3	2	6

[JNTU (CSE) 97/S, (Mech.) - 98]

Answer : Total time = 22 hrs, idle time for Job 1 = (2 + 3) hrs, for
Job - 2 = 2 hrs. A $(J1)$, B $(J2)$, C $(J2)$, D $(J2)$, E $(J1)$

Flow chart for Sequencing Problems :

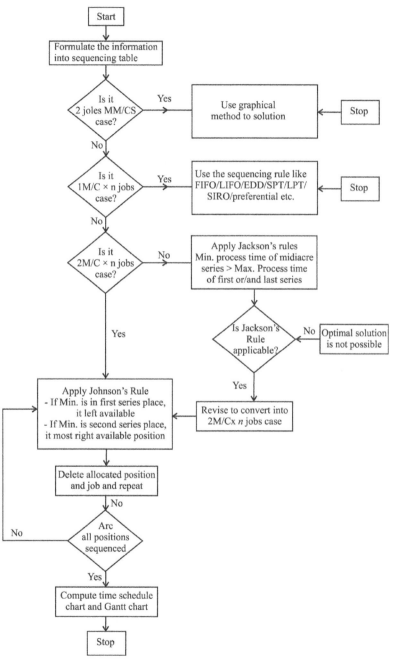

FIGURE 9.4 :

Review Questions

1. What are the assumptions made in sequencing problems? **[JNTU (Mech.) 98/S]**

2. Describe the procedure of optimal solution for processing n jobs through two machines. **[JNTU (Mech.) 98/S]**

3. Describe the procedure of optimal solution for processing n jobs through three machines and hence for n jobs through m machines.

                   [JNTU (CSE) 96]

4. Explain the rules of sequencing 'n' jobs when there is only one machine to process.

5. Describe various sequencing models. **[JNTU (Mech.) 97/P]**

6. Explain the working rules of n jobs X 2 machines and n jobs X 3 machines with suitable examples **[JNTU (Mech.) 98/P]**

7. Explain the situations leading to multiple optimal solutions in sequencing problems. **[JNTU (Mech.) 97/CCC]**

8. Explain the possibility and working rules of a maximisation case in sequencing.

9. Explain the Jackson's conditions in working with n jobs X 3 machine sequencing problems.

10. Explain the relevance and importance of Gantt chart in sequencing.

11. Write a short note on the contributions of S.M. Johnson & Jackson towards Operations Research. **[JNTU (Mech.) 94]**

12. Explain the algorithm for sequencing of 2 jobs to process on 'n' machines.
          [JNTU - (Mech.) 95, Mech/Mechatronics/Chem. 2001/S]

13. Explain "No passing rule" with reference to sequencing problems.
                   [OU (Maths) 85]

14. Give justification of Johnson's rule for sequencing n jobs X 2 machines.
                   [OU (Maths) 87]

15. Write the sequencing algorithm for 5 jobs to process on 3 machines.

16. Give justification of graphical method used to sequence 2 jobs X M machines.
                   [JNTU (Mech.) 98/S]

17. Explain the elements that characterise the sequencing problem.

18. Give different examples of sequencing problem of various models from your daily life. **[JNTU Mech/Prod - 99, CSE/ECE 2001]**

19. Discuss the following terms in the context of sequencing problems.
    (a) No. of machines   (b) Waiting time of job
    (c) Idle time of machine   (d) No passing rule

20. Critically appreciate the significance of job sequencing in a job/batch production. How this importance would vary in mass/flow production.

Objective Type Questions

1. The time required for printing of four project reports P,Q,R and S is 4,7,9 and 6 hrs while its data entry requires 6,3,2 and 5 hrs respectively. The sequence that minimises the total elapse is _____

   (a) PSQR   (b) PQRS   (c) PSRQ   (d) RQSP

2. Printing of five books P,Q,R,S and T takes 9,3,4,6 and 7 hrs while for binding they take 6,5,8,5 and 2 hrs respectively. The correct sequence that maximises the total elapse is _____

   (a) QRPST   (b) TSPRQ   (c) PTSQR   (d) RQSTP

3. In which of the following we get multiple optimal solutions_____

   (a) Same least time for two jobs in both series of machines of $2^{m/c} \times n$ jobs sequence model.

   (b) If we can connect the zeros as the corners of a rectangular loop, allocated cells alternatively in AP or TP

   (c) If objective function has same slope to that of one of non-redundant constraint of an LPP.

   (d) all the above

4. Which of the following is wrong in assumptions of sequencing 2 machines × n jobs by Johnson's Rule

   (a) no passing is allowed

   (b) processing times are known

   (c) time of moving job from one machine to the other is negligible

   (d) the time of processing is dependent on order of processing

5. According to the assumptions of $2^{m/c} \times n$ job sequencing

    (a) once a job loaded can be removed before completing

    (b) two jobs can be simultaneously processed on a machine

    (c) machines used are different type

    (d) the order of completion of jobs are dependent on sequence of jobs

6. According to Johnson's rule, the smallest processing time if occurs in

    (a) first series, place it in the first available position

    (b) second series, place it in first available position

    (c) first series, place it middle

    (d) second series, place it middle.

7. The sequencing of $3^{m/c} \times n$ jobs should pass Jackson's rule according to which Maximum process time of 2nd series

    (a) $\geq$ minimum process time of 1st or 3rd series

    (b) $\geq$ maximum process time of 1st or 3rd series

    (c) $\leq$ minimum process time of 1st or 3rd series

    (d) $\leq$ maximum process time of 1st or 3rd series.

8. Which of the following does not characterise the sequencing problem

    (a) no passing rule

    (b) idle time of machine

    (c) number of machines and number of jobs

    (d) number of movements while processing the jobs.

9. FCFS rule is most applicable to _____ case of sequencing

    (a) $1^{m/c} \times n$ jobs   (b) $2^{m/c} \times n$ jobs

    (c) $3^{m/c} \times n$ jobs   (d) $n^{m/c} \times 2$ jobs

10. The sequencing rule usually followed at a petrol bank when 'n' vehicle are waiting, is

    (a) FIFO   (b) LIFO

    (c) lowest process time   (d) Highest profit rule

11. If two jobs J_1 and J_2 have same minimum process time in M_1 series but process time J_1 is less than that of J_2 in M_2 series then J_1 occupies_____

    (a) first available place from left

    (b) second available place from left

    (c) first available place from right

    (d) second available place from right

12. If two jobs J_1 and J_2 are in tie at M_2 series but for M_1, process time for J_1 is less than that of J_2, then among the available positions, J_2 occupies

    (a) last position   (b) last but one position

    (c) first position   (d) second position

13. If two jobs J_1 and J_2 have same process times in both series we prefer

    (a) J_1 (b) J_2

    (c) J_1 and J_2 (d) J_1 or J_2

14. Multiple optimal solutions are obtained if two jobs have same process times in

    (a) first m/c series   (b) second m/c series

    (c) both series   (d) any one series

15. If a job is has minimum process time in first m/c series as well as in second m/c series, the job is placed

    (a) available first position   (b) available last position

    (c) any one (first or last) position (d) both first and last position.

16. Jobs A to E have process times as 12,4,20,14 and 22 on 1st M/C and 6,14, 16, 18 and 10 on 2nd M/C, the optimal sequence is _____

    (a) CEABD   (b) AECDB

    (c) BCDEA   (d) BDCEA

17. In the above problem, if machines have idle times 6 and 14 hrs on M_1 & M_2 respectively. Then total elapse = _____ hrs.

    (a) 66   (b) 78

    (c) 58   (d) 86

18. Consider the following process times (in hrs)

Jobs	P	Q	R	S	T
M_1	5	1	9	3	10
M_2	2	6	7	8	4

The optimal sequence is _____

(a) PTRSQ   (b) QSRTP   (c) QRSTP   (d) PTSRQ

19. In the above problem idle times for M_1 and M_2 are 2 and 3 hrs respectively. The total elapsed time = _____ hrs

(a) 28   (b) 33   (c) 30   (d) 31

20. In 3 m/c × 5 jobs case, the least of processing times on machines A, B and C are 5, 1 and 3 while highest of process times are 9, 5 and 7 respectively. The Jackson's rule is applicable if order of the machines is _____

(a) $B - A - C$ (b) $A - B - C$

(c) $C - B - A$ (d) any order

21. In a maximisation case of sequencing $2M/C \times n$ jobs the job is placed at available left first position if it has _____ process time in _____ machine series

(a) least, first   (b) highest, first

(c) least, second   (d) highest second

22. The fundamental assumption for Johnson's method of sequencing is _____

(a) no passing rule

(b) passing, rule

(c) similar machines must be used for first and second process

(d) non zero process time

23. If a job has zero process time for any machine the job must

(a) posses first position only

(b) possess last position only

(c) possess extreme position

(d) be deleted from the sequencing

24. No passing rule means

    (a) a job once loaded on a m/c should not be removed until it is completed

    (b) a job can not be processed on second m/c unless it is processed on first machine

    (c) a machine should not be started unless the other is ready to start

    (d) no job should be processed unless all other are kept ready to start.

25. A sequencing problem may be infeasible or non optimal or not applicable in case of

    (a) 1 $M/C \times n$ jobs   (b) 2 $M/C \times n$ jobs

    (c) 3 $M/C \times n$ jobs   (d) 2 jobs $\times n$ M/C

26. If passing is allowed, the problem can be solved by

    (a) Johnson's (2 $M/C \times n$ jobs) rule

    (b) Jackson's (3 $M/C \times n$ jobs) rule

    (c) hungarian floods technique

    (d) none of the above

27. The technological order of machines to be operated is fixed in

    (a) 1 $M/C \times n$ jobs   (b) 2 $M/C \times n$ jobs

    (c) 3 $M/C \times n$ jobs   (d) n $M/c \times$ jobs

28. In a 2 jobs $\times n$ M/C sequencing, a line at 45° represents

    (a) job 1 is idle   (b) job 2 is idle

    (c) both jobs are idle   (d) no job is idle

29. In 2 jobs $\times$ n machine case, the elapsed time for job 1 (on X axis) is reported as

    (a) process time of job 1 + idle time of job 2

    (b) process time of job 2 + idle time of job 1

    (c) process time of job 1+ total vertical distance

    (d) process time of job 1 + total horizontal distance

30. Given process time for 6 jobs on three machines P, Q and R as follows.

Job l	1	2	3	4	5	6
M/C P	12	10	9	14	7	9
M/C Q	7	6	6	5	4	4
M/C R	6	5	6	4	2	4

The optimal sequence can obtained if the order of M/C to process is _____

(a) PQR   (b) PRQ

(c) QPR   (d) RPQ

Fill in the Blanks

1. If there is some least process time for two jobs in both series, then the sequencing problem will have _____

2. The basic assumption for Johnson's rule for 2 $M/C \times n$ jobs sequencing is that _____ is not allowed

3. According to Johnson's rule if smallest processing time appears in 2nd M/C series, then its position in the sequence is _____

4. The sequence of 3 $M/C \times n$ jobs is converted to 2 $M/C \times n$ jobs if _____

5. FIFO, LIFO, SPT, LPT etc. are the rules of _____ sequencing

6. If in two jobs, A has same process time as that of B in 1st machine series but greater in 2nd machine series, then the most left available position will be occupied by _____

7. The most min. process time for (J_1, M_2) is equal to that of (J_2, M_2) and process times for (J_1, M_1) greater than that of (J_2, M_1) in a 2 $M/C \times 5$ jobs sequencing. Then position of J_2 is _____

8. The position of J_1 in the above question is _____

9. If J_1 and J_2 have equal process times in both M_1 and M_2 series, then we choose _____

10. If min. process time appears in both M_1 series and M_2 series for the same job, then the job is placed _____ of available positions

11. Jobs P, Q, R and S have process times 10, 11, 8 and 6 on machine M_1 and 9, 7, 5, 6 on machine M_2 then the sequence is _____ or _____

12. In a maximisation case of sequencing $2 M/C \times n$ jobs, a job is placed in left (first) available position if it has _____ process time (or profit) in first M/C series

13. If process time for job is zero in 1st series, then this job should occupy _____ position

14. The idle time of second machine decreases if the job with _____ process time is taken first on first machine

15. No possing means _____

16. If passing is allowed, the sequencing problem becomes a case in _____ problem.

17. In 2 jobs $\times n$ M/C sequencing _____ is fixed

18. In a 2 jobs $\times n$ M/C case, a horizontal line represents the idle time of _____ assuming J_1 and J_2 an X and Y axes respectively.

19. In the graphical solution for 2 jobs $\times n$ machines, when both jobs are processed, then line moves at _____

20. For binding five books P, Q, R, S and T take 9, 3, 4, 6 an 7 hrs while for printing they take 6, 5, 8, 5 and 12 hrs respectively. The optimal sequence is _____

Answers

Objective Type Questions :				
1. (d)	2. (c)	3. (d)	4. (d)	5. (c)
6. (a)	7. (c)	8. (d)	9. (a)	10. (a)
11. (b)	12. (a)	13. (d)	14. (c)	15. (c)
16. (d)	17. (b)	18. (b)	19. (c)	20. (b)
21. (b)	22. (a)	23. (c)	24. (b)	25. (c)
26. (c)	27. (d)	28. (d)	29. (c)	30. (b)

Fill in the Blanks :	
1. Alternate or multiple optimal solution	2. Passing
3. least available or extreme right	4. min. process time of 1st or/and 3rd M/C series $\geq$ max. process times of 2nd M/C
5. 1 $M/C \times n$ jobs	6. job B
7. fourth	8. fifth
9. arbitrarily/either J_1 or J_2	10. first or least/left or right
11. SPQR or PQSR	12. largest or highest
13. first position	14. lowest or minimum
15. A job can not be processed on 2nd M/C until it is processed on 1st machine	16. assignment
17. Technological order of M/C	18. Job 2
19. $45°$	20. SPTRQ

10 Chapter

Replacement Analysis

CHAPTER OUTLINE

CHAPTER AT A GLANCE

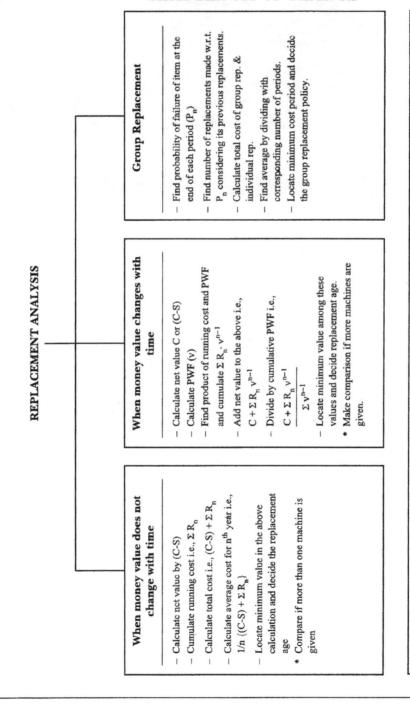

REPLACEMENT ANALYSIS

When money value does not change with time

- Calculate net value by (C-S)
- Cumulate running cost i.e., ΣR_n
- Calculate total cost i.e., $(C-S) + \Sigma R_n$
- Calculate average cost for n^{th} year i.e., $1/n \{(C-S) + \Sigma R_n\}$
- Locate minimum value in the above calculation and decide the replacement age
- * Compare if more than one machine is given

When money value changes with time

- Calculate net value C or (C-S)
- Calculate PWF (v)
- Find product of running cost and PWF and cumulate $\Sigma R_n \cdot v^{n-1}$
- Add net value to the above i.e., $C + \Sigma R_n v^{n-1}$
- Divide by cumulative PWF i.e., $\dfrac{C + \Sigma R_n v^{n-1}}{\Sigma v^{n-1}}$
- Locate minimum value among these values and decide replacement age.
- * Make comparison if more machines are given.

Group Replacement

- Find probability of failure of item at the end of each period (P_n)
- Find number of replacements made w.r.t. P_n considering its previous replacements.
- Calculate total cost of group rep. & individual rep.
- Find average by dividing with corresponding number of periods.
- Locate minimum cost period and decide the group replacement policy.

C = Cost of machine • S = Salvage value • R_n = Running cost in n^{th} year

N = Number of years • At C = Average (Annual) total cost • v^{n-1} Present Worth Factor (PWF)

10.0 *Introduction*

No machine is immortal and immune completely to any failures. No matter how safely you run, how closely you follow the instructions of the manufacturer or supplier, how best you maintain to its standards and specification. Perhaps we can only try to prevent or prolong the occurrence of failure if we know the probable reason for its occurrence. Even in such cases, some times we can temporarily stop the occurrence of the failure and some other times we can reduce its impact or volume of failure. However, this is possible if we have the complete knowledge of the failures that may occur on the given equipment and their causes, effects or costs and the remedial measures. The awareness of equipment failures often makes the engineer so confident that after the rectification of the failure he will be able to assure the production manager about its running condition.

10.1 *Failure Patterns*

10.1.1 *Definition*

Failure is defined in many ways. Of them, the popular and appropriate are given below.

Failure is defined as inability of a machine or equipment to perform the intended or specified job under specified conditions.

A failure is defined as an event that changes a product or machine or equipment from an operational condition to a non-operational condition.

Failure can be defined as "Non-conformance to some defined performance criterion".

10.1.2 *Behaviour of Machines*

Almost all machines or equipment whether electrical or mechanical or electronic, assumed to behave in the same manner. These machines are normally expected to have one of the three types of behaviours with reference to the failures that occur. These three behaviours are as follows :

1. The rate of failures is decreasing.
2. The rate of failures is constant.
3. The rate of failures is increasing.

Amazingly, every machine is found to have all the three behaviours siginificantly and distinctly in certain period of its lifetime. Therefore, it has gained so much of significance in replacement analysis studies.

Machine Life Cycle (MLC) - Bath Tub Curve :

Machine Life Cycle is classified into in three phases, which is analogous to the three phases of Human Life Cycle as shown below.

1. Infancy Phase : Early failures or infant failures.

2. Youth Phase : Random failures or rare event failures.

3. Old Age Phase : Wear out failures or old age failures

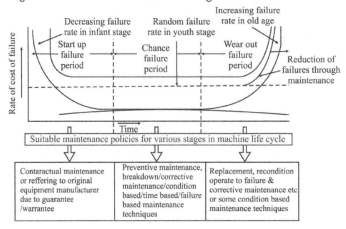

FIGURE 10.1 : BATH TUB CURVE

1. Early Failures or Failures at Infant Stage of Machine :

The machine immediately after it is brought from the Original Equipment Manufacturer (OEM), may not work with full efficiency due to various reasons such as the initial friction between the moving or rotating parts, not adjusting to the environment, lack of skill and knowledge of the person operating on it etc. To control these failures one has to know about the machine thoroughly and follow the instructions strictly. Inspite of following the instructions given by OEM strictly, the output may be low or slow or delayed. Yet, it is necessary in the interest of good running and long life of the machine. For example, when you buy a scooter, the OEM gives you instructions that you have to run the vehicle at not more than 40 KMPH upto first 1500 km.This obviously restricts and slows down the job and may cause the inconvenience to the user. But if this is not followed it may lead to a catastrophic failure by affecting the piston movement in the cylinder or high fuel consumption which will be uneconomical. Such problems are very common to any new machine. However, the period of these infant failures may vary from one machine to the other. (See figure and table). During this failure period, in the event of machine is usually referred to OEM (of the machine is covered under guarantee.warantee) or any contractual maintenance.

2. Random or Rare - Event Failures :

These failures occur in young stage of the machine. After passing over the infant stage, the machine will be running with its full efficiency and the user will enjoy its full fruit in this period only. The effective usage and correct maintenance can enhance this portion of its life. User's care may come down in this stage due to the facts that the user might have got boredom or monotony or some sort of negligence as he observes the machine will be working with full efficiency though much care is not taken. This act in fact may or may not result as failure immediately, but its impact will be there on long run by affecting its life and wear out of the parts etc. However, the immediate failures are known as random or rare-event failures. The reliability of machine will be very high in this stage. Usually preventive maintenance and breakdown maintenance are adapted in this stage to low cost failures while RCM (Reliability Centered Maintenance) and CBM (Condition Based Maintenance) techniques are employed on the machines that are high cost or in case the failure may lead to high cost or damage. (See figure and table)

3. Old Age of Wear Out Failures :

Most of the efforts put by maintenance are attributed to this stage and costs could go high at this stage if this stage is not detected. Calender time is not only the scale for detection of this stage. For instance a machine used sparingly and a machine used continuously will not reach to the old age stage at the same time, though they are bought at same time. Hence one should notice that the operational period, conditions maintained while usage, care and efforts put on it to increase its life during its infant and young stages etc., are a few factors governing the old age failures wever, these failures are mainly because of the worn out parts of the machines. As and when this stage is noticed in the machine, a plant engineer may have to choose one of the following alternative strategies. (See table and figure)

- Replacement of the machine with the new one.

- Reconditioning of the machine.

- Updating with the new technological features.

- Operate to failure and corrective maintenance (as long as its average annual maintenance cost is less than or equal to the interest on the cost of the new machine) or selling in seconds.

- Scrapping.

It is important to note here that the age of a machine (infant/young/middle/old) is not just decided by the calendar time but by its running time and other factors such as environment, usage etc.

The above phases are summarised in the following tabular form.

TABLE :

Phase	Type of Failure	Failure Rate	Cause of Failure	Cost of Failures	Suitable Maintenance Policy
Infant Phase	Early failures or infant failures	Decreasing	Faulty Design, Erratic operation, Environment problems, Installation errors, Initial friction	Medium to high	(Under warantee / guarantee) refer to OEM
					Contractual maintenance (No warantee)
Youth Phase	Random or Chance or Rare-Event failures	Constant	Operation errors, Heavy workloads, Over run	Low to medium	Break-down maintenance, Preventive maintenance
				Medium to high	Reliability centered maintenance, condition based maintenance
Olde Age Phase	Wear-out or Age failures	Increasing	Wear, Tear, Creep, Fatigue, Weakened parts	Low	Operate to fail and corrective maintenance
				High	Reconditioning or replacement policies or scrapping

10.1.3 *Failures Based on the Volume of Failure*

1. **Small Failures :** These are the failures, which can be rectified in few minutes. These failuers are very common on any machine. These will not have any considerable impact on the machine performance and on the operator.

2. **Minor Failuers :** These are the failures, which can be rectified in a few hours or with a little effort. The failures of this kind will have a little to considerable effect on the productive work and could hardly damage the machine or men or environment

3. **Major Failures :** These failures take few days to rectify. They also may require large work force, knowledge and skill to rectify. These failures may cause considerable damage to the machine, minor injuries to men and affects the regular work.

4. **Catastrophic Failures :** These are the costliest failures. The occurrence of such failures may cause the damage to the machine, men and some times the environment also. The rectification or recovery from the losses of this failure may take a few weeks to months even.

10.1.4 *Types of Failures Based on the Mode of Failure*

1. **Sudden Failures :** These types of failures occur in items after giving some period of desired service rather than deterioration while in service. This period of giving desired service is not constant but follows some frequency distribution, which may be progressive, retrogressive or random in nature.

   (a) *Progressive Failure :* If the probability of failure in the beginning of an item is less and gradually increases in its life, then such failure is called progressive failure. For example, light bulbs and tubes fail progressively.

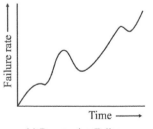 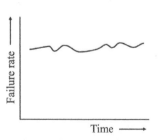

(a) Progressive Failure   (b) Retrogressive Failure   (c) Random Failure Pattern

FIGURE 10.2 : FAILURE PATTERNS OF SUDDEN FAILURES

(b) Retrogressive Failure : If probability of failure in the beginning of the life of an item is more but as time passes the chances of its failure become less then such failure is said to be retrogressive.

(c) Random Failure : In this type of failure, the constant probability of failure is associated with items that fail from random causes such as physical shocks, not related to age. For example, vacuum tubes in air-burn equipment have been found to fail at a rate of the age of tube.

2. ***Gradual Failures :*** Gradual failure is progressive in nature. i.e. as the life increases, its operational efficiency also deteriorates resulting in increased running (maintenance and operating) costs. They also cause decrease in its productivity and decrease in the resale or salvage value. Mechanical items like pistons, rings, bearings etc., and automobile tyres fall under this category.

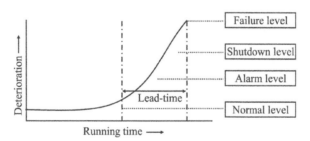

FIGURE 10.3 : GRADUAL FAILURE PATTERN

10.2 *Failure Costs*

Cost Involved in Machine Failure Analysis :

While analysing the machine failure we are concerned with the following costs.

1. Purchase cost of machine or equipment capital investment.

2. Depreciation or salvage value or scrap value.

3. Running costs including Maintenance, Repair and Operating (MRO) costs.

4. Fialure costs and damage costs.

In the above four, the first three are inevitable, but failure costs can be prevented by better maintenance policies. On critical examination, we can notice that the first three costs depend on age of the machine. These costs vary with the running age as follows.

1. Purchase costs of machine or equipment is considered to be independent of age of machine. However the interest on the investment is assumed to be lost.

2. Resale value or salvage value or scrap value of the machine decreases with the running age of the machine. As the machine grows old, its resale value comes down. This decrease depends on actual or expected condition of the machine.

3. Running cost or operating cost or maintenance cost of the machine is due to its minor failures or preventive maintenance or operating costs etc. These will increase on the machine as the age grows. This is assumed to be increasing due to the wear and tear on the moving parts of the machine.

The above costs are shown graphically on a hypothetical machine here below.

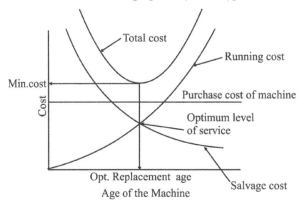

FIGURE 10.4 :   COST INVOLVED IN REPLACEMENT ANALYSIS

Summing, the above three costs, we can notice that the average total cost decreases for certain period and then increases. The age when the graph shows its minimum costs will be optimum age of replacing the machine.

Necessity or Significance of Replacement :

The replacement of parts or entire machine will become significant and necessary in the following cases.

1. When average costs of repairs or maintenance or operating goes higher than the costs of machine. Or in other words, the cost of maintenance will increase to such an extent that the average annual repair or maintenance cost is greater than or equal to costs of new machine.

2. When the machine completely fails to work very frequently by which the production schedules are interrupted.

3. Machine runs with less efficiency and therefore not economical.

4. Modified or new designs in the market may give an edge of advantage such as reduced cost of production or ease and comfort of operation or more functions are available in new design etc.

5. If it is expected that the existing model may become absolete or resale value may drastically come down.

6. If the item is non-repairable type or use & throw type.

10.3 *Types of Replacement Problems*

The replacement of machine is considered as the following three types in this book.

Type - I : *Replacement policy for items when money value remains unchanged with time.*

Type - II : *Replacement policy for items when money value changes with time.*

Type - III : *Group replacement policy.*

These are explained in detail with examples and illustrations in the sections to follow.

10.4 *Type - I : Replacement Policy when Money Value does not Change with Time*

Let us now find the optimal policy for the case of replacement when money value does not change with time.

Let C = capital or purchase cost of new item.

S = scrap or salvage or resale value of the item at the end of t years.

$R(t)$ = running cost for the yeat t.

n = replacment age of the equipment.

Here two cases arise.

Case (i) : *When time t is a continuous variable :*

If the equipment is used for 't' years, then the total cost incurred over this period is given by

T_C = capital (or purchase) cost - scrap value at the end of t years + running cost for t years.

$$= C - S + \int_0^n R(t)\, dt$$

Theretore the average cost per unit time incurred over the period of n years is

$$ATC_n = \frac{1}{n}\left\{ C - S + \int_0^n R(t)\,dt \right\} \qquad \dots\ (1)$$

To obtain optimal value of n for which ATC_n is minimum, differentiate ATC_n with respect to n and set the first derivative equal to zero. i.e. minimum of ATC_n.

$$\frac{d}{dn}\left[ATC_n \right] = -\frac{1}{n^2}[C - S] + \frac{R(n)}{n} - \frac{1}{n^2}\int_0^n R(t)\,dt = 0$$

or $$R(n) = \frac{1}{n}\left\{ C - S + \int_0^n R(t)\,dt \right\}.\ n \neq 0$$

$$R(n) = ATC_n \qquad \dots\ (2)$$

Hence the following replacement policy can be derived with the help of equation (2)

Policy : *Replace the equipment when the average annual cost for n years becomes equal to the current/annual running cost.*

i.e. $$R(n) = \frac{1}{n}\left\{ C - S + \int_0^n R(t)\,dt \right\}$$

Case (ii) : *When time t is a descrete variable :*

The average cost incurred over the period n is given by

$$ATC_n = \frac{1}{n}\left\{ C - S + \sum_{t=0}^n R(t) \right\} \qquad \dots\ (3)$$

If $C\text{–}S$ and $\displaystyle\sum_{t=0}^n R(t)$ are assumed to be monotonically decreasing and increasing respectively, then there will exist a value of n for which ATC_n is minimum. Thus we shall have inequalities

$$ATC_{n-1} > ATC_n < ATC_{n+1}$$
$$ATC_{n-1} - ATC_n > 0$$

and $$ATC_{n+1} - ATC_n > 0$$

Rewriting equation (3) for period $n + 1$, we get

$$ATC_{n+1} = \frac{1}{n+1} \left\{ C - S + \sum_{t=1}^{n+1} R(t) \right\}$$

$$= \frac{1}{n+1} \left\{ C - S + \sum_{t=1}^{n} R(t) + R(n+1) \right\}$$

$$= \frac{n}{n+1} \cdot \frac{\left\{ C - S + \sum_{t=1}^{n} R(t) \right\}}{n} + \frac{R(n+1)}{n+1}$$

$$= \frac{n}{n+1} \cdot ATC_n + \frac{R(n+1)}{n+1}$$

Therefore
$$ATC_{n+1} - ATC_n = \frac{n}{n+1} ATC_n + \frac{R(n+1)}{n+1} - ATC_n$$

$$= \frac{R(n+1)}{n+1} + ATC_n \left[\frac{n}{n+1} - 1 \right]$$

$$= \frac{R(n+1)}{n+1} - \frac{ATC_n}{n+1}$$

Since $ATC_{n+1} - ATC_n > 0$ we get

$$\frac{R(n+1)}{(n+1)} - \frac{ATC_n}{n+1} > 0$$

$$R(n+1) - ATC_n > 0$$

$$R(n+1) > ATC_n \qquad \qquad \cdots \ (4)$$

Similarly $ATC_{n-1} - ATC_n > 0$ implies that $R(n) < ATC_{n-1}$.

This provides the following replacement policy.

Policy 1 : *If the next year, running cost, $R(n+1)$ is more than average cost of nth year, ATC_n then it is economical to replace at the end of n years*

i.e. $\quad R(n+1) > \dfrac{1}{n} \left\{ C - S + \displaystyle\sum_{t=0}^{n} R(t) \right\}$

Policy 2 : *If the present year's running cost is less than the previous year's average cost, ATC_{n-1} then do not replace.*

$$\text{i.e.} \quad R\,(n) < \frac{1}{n-1}\left\{C - S + \sum_{t=0}^{n-1} R\,(t)\right\}$$

The Procedure :

Thus the procedure for obtaining the decision when to replace the equipment in the case of money value not changing with time can be outlined as follows.

Step 1 : Draw the table with columns as shown below and enter the values of column 1, 2, 3 and 4 as given in the problem.

Year of service (n)	Cost of equipment (C)	Salvage value (S)	Running Cost R_n	Net value $(C-S)$	Cumulative running cost $\sum R_n$	Total Cost $(TC) =$ $(C-S) + \sum R_n$	Average total cost $= ATC =$ $\frac{1}{n}\left[(C-S) + \sum R_n\right]$
(1)	(2)	(3)	(4)	(5) = (2) - (3)	(6) = $\sum$ (4)	(7) = (5) + (6)	8 = (7)/(1)

Step 2 : Calculate net value by difference of cost and salvage value $(C-S)$ i.e., 2nd column value minus 3rd column value and enter in column 5

Step 3 : Calculate cumulative running cost i.e., $\sum R_n$ cumulation of column 4 and enter in column 6.

Step 4 : Calculate total cost $TC = (C-S) + \sum R_n$ i.e., sum of 5th column value and 6th column value to enter in col. 7.

Step 5 : Calculate average total cost for n years, i.e., divide 7th column value by 1st column value i.e., $ATC = \frac{1}{n}\left\{(C-S) + \sum R_n\right\}$ and enter in col. 8.

Step 6 : Observe the values in column 8 and identify the minimum value. The year corresponding to this minimum value is the age of the equipment to be replaced.

This is illustrated through the numerical example given below.

ILLUSTRATION 1

A firm is thinking of replacing a particular machine whose cost price is Rs. 12,200. The scrap price of the machine is only Rs. 200. The maintenance costs are found to be as follows:

Year	1	2	3	4	5	6	7	8
Maintenance cost	220	500	800	1200	1800	2500	3200	4000

Determine when the firm should get the machine replaced. [JNTU B. Tech. (CSE) 97/S]

Solution :

The calculations of average running cost per year during the life of the machine are shown in the following table :

Cost price machine (C) = 12,200; S = 200; $C - S$ = 12,000

Year n	Cost C	Salvage value (Rs) S	Running cost (Rs) $R(n)$	Depreciation Cost (Rs) $C–S$	Commulative Running Cost (Rs) $\sum R(n)$	Total Cost (Rs) TC	Average Cost (Rs) Per Year At C_n
(1)	(2)	(3)	(4)	(5) = (2) – (3)	(6) = Σ (4)	(7)=(5)+(6)	(8)=(7)/(1)
1	12200	200	220	12000	220	12220	12220
2	12200	200	500	12000	720	12720	6360
3	12200	200	800	12000	1520	13520	4506.67
4	12200	200	1200	12000	2720	14720	3680
5	12200	200	1800	12000	4520	16520	3304
6	12200	200	2500	12000	7020	19020	3170 ←
7	12200	200	3200	12000	10220	22220	3174.29
8	12200	200	4000	12000	14220	26220	3277.5

From the above table it may be noted that the average cost per year, ATC_n is *minimum in the sixth year (Rs. 3170)*. And this average cost is increasing from 7th year onwards. Hence the machine should be replaced *after 6 years.*

ILLUSTRATION - 2

A plant Manager is considering replacement policy for a new machine. He estimates the following costs in Rupees.

Year	1	2	3	4	5	6
Replacement cost at beginning of year	100	110	125	140	160	190
Salvage value at end of year	60	50	40	25	10	0
Operating Costs	25	30	40	50	65	80

Find an optimal replacement policy and corresponding minimum cost.

[JNTU (Mech.) 98/S]

Solution :

The calculations for replacement of the machine are shown in the following table :

Year	Replacement Cost	Resale value (S)	Net Value	Operating Cost	Commulative Running Cost ΣR_n	Total Cost $C - S + \Sigma R_n$	Average Annual cost ATC_n
(1)	(2)	(3)	(4)= (2) – (3)	(5)	(6) = $\Sigma(5)$	(7)= (4) + (6)	(8)=(7)/(1)
1	100	60	40	25	25	65	65
2	110	50	60	30	55	115	57.5 ←
3	125	40	85	40	95	180	60
4	140	25	115	50	145	260	65
5	160	10	150	65	210	360	72
6	190	0	190	80	290	480	80

From the above table, it may be noticed that the average cost per year, ATC_n is minimum in the 2nd year, i.e., ATC_2 is Rs. 57.5 which is less than ATC_1 (Rs. 65) and ATC_3 (Rs. 130/-). Hence the machine should be replaced at the end of second year.

ILLUSTRATION 3

A fleet owner finds from his past records that the cost per year of running a vehicle whose purchase price is Rs. 50000 are as under:

Year	1	2	3	4	5	6	7
Running cost Rs.	*5000*	*6000*	*7000*	*9000*	*21500*	*16000*	*18000*
Resale Value Rs.	*30000*	*15000*	*7500*	*3750*	*2000*	*2000*	*2000*

There after running costs increase by Rs. 2000, but resale value remains constant at Rs. 2000. At what age is a replacement due?

[JNTU B.Tech. (Mech.) 97/CCC, (ECE) 97/CCC]

Solution :

The required calculations are shown in the following table :

Year (1)	C (2)	S (3)	C–S (4) =(2)–(3)	R(n) (5)	cum R(n) (6) = Σ (5)	TC (7)=(4)+(6)	ATC $8 = \dfrac{(6)}{(1)}$
1	50,000	30,000	20,000	5000	5000	25,000	25,000
2	50,000	15,000	35,000	6000	11,000	46,000	23,000

3	50,000	7500	42,500	7000	18,000	60,500	20166.6
4	50,000	3750	46,250	9000	27,000	73250	**18312.5**
5	50,000	2000	48,000	21500	48,500	96,500	19300
6	50,000	2000	48,000	16000	64,500	112500	18750
7	50,000	2000	48,000	18000	82,500	130500	**18642.8**
8	50,000	2000	48,000	20000	102500	150500	18812.5
9	50,000	2000	48,000	22000	124500	170500	18944.4

The machine should be replaced at the end of 4th year or at the end of 7th year if it is not replaced at the end of 4th year.

ILLUSTRATION 4

Machine A cost Rs. 45,000 and the operating costs are estimated of Rs. 1000 for the first year, increasing by Rs. 10,000 per year in the second and subsequent years. Machine B costs Rs. 50,000 and operation costs are Rs. 2000 for the first year, increasing by Rs. 4000 in the second and subsequent years. If we now have a machine of type A, should we replace it with B? If so when? Assume that both machine have no resale value and future costs are not discounted.

[JNTU (Mech.) 95, (CSE) 2000/S]

Solution :

The calculations of average costs running per year during the life of machine A and B are shown in tables give below.

TABLE : CALCULATIONS OF AVERAGE RUNNING COST FOR MACHINE A.

Year of Service (n)	Running Cost in Rs. $R(n)$	Cumulative Running Cost in Rs. $\sum R(n)$	Depreciation Cost (Rs.) $C-S$	Total Cost (Rs.) TC	Average Cost (Rs) Rs. ATC_n
(1)	(2)	(3)	(4)	(5) = (3) + (4)	(6) = (5) / (1)
1	1000	1000	45,000	46,000	46,000
2	11,000	12,000	45,000	57,000	28,500
3	21,000	33,000	45,000	78,000	**26,000**
4	31,000	64,000	45,000	1,09,000	27,200
5	41,000	1,05,000	45,000	1,50,000	30,000
6	51,000	1,56,000	45,000	2,01,000	33,500

From the above table it may be noted that the average running cost per year is *lowest in the third year, i.e., Rs. 26,000.* Hence, machine A should be replaced after every three years of service.

TABLE : CALCULATIONS OF AVERAGE RUNNING COST FOR MACHINE B

Year of service n	Running cost (Rs.) $R(n)$	Cumulative running cost (Rs.) $\sum R(n)$	Depreciation cost (Rs.) $C-S$	Total cost (Rs.) TC	Average cost (Rs.) ATC_n
(1)	(2)	(3)	(4)	(5) = (3) + (4)	(6) = (5) / (1)
1	2,000	2,000	50,000	52,000	52,000
2	6,000	8,000	50,000	58,000	29,000
3	10,000	18,000	50,000	68,000	22,667
4	14,000	32,000	50,000	82,000	20,500
5	18,000	50,000	50,000	1,00,000	20,000
6	22,000	72,000	50,000	1,22,000	20,333

From the above table it may be noted that the average running cost per year is *lowest in the fifth year, i.e., Rs. 20,000*. This cost is less than the lowest average running cost (Rs. 26,000) per year for machine A. Hence machine A should be replaced by machine B.

Now to find the time of replacement of machine A by machine B, the total cost of machine A in the successive years is computed as follows :

Year	1	2	3	4
Total cost incurred (Rs)	46,000	57,000-46,000 = 11,000	78,000-57,000 = 21,000	1,09,000-78,000 = 31,000

Machine A should be replaced by machine B at the time (age) when its running cost for the next year exceeds the lowest average running cost (Rs. 20,000) per year of machine B.

Calculations show that the running cost (Rs. 21,000) of machine A in the third year is more than lowest in average cost (Rs. 20.000) per of machine B. Hence machine A should be replaced by machine B *after two years*.

Practice Problems

1. The cost of a machine is Rs. 6100 and its scrap value is only Rs. 100. The maintenance costs are found from experience to be

Year	1	2	3	4	5	6	7	8
Maintenance cost Rs.	100	250	400	600	900	1200	1600	2,000

When should the machine be replaced ?   **[JNTU (Mech.) 93]**

Answer : At the end of sixth year Rs. 1583.33

2. A truck - owner finds from his past experience that the maintenance costs are Rs. 200 for the first year and then increase by Rs. 2,000 every year. The cost of truck type A is Rs. 9,000. Determine the best age at which to replace the truck. If the optimum replacement is followed what will be the average yearly cost of owning and operating the truck? Truck type B costs Rs. 20,000. Annual operating costs are Rs. 400 for the first year and then increase by Rs. 800 every year. The truck owner has now the truck type A which is one year old. Should it be replaced by B type, and if so, when?

Answer : After every 3rd year, truck A is to be replaced (Rs. 5200). After every 6th years truck B (Rs. 5733.3) is replaced. Comparing both, cost of truck B is lower than that of Truck A from 3rd year onwards, hence after 3rd year replace Truck A by Truck B.

3.   (a) Machine A costs Rs. 9000. Annual operating costs are Rs. 200 for the first year, and then increase by Rs. 2,000 every year. Determine the best age at which to replace the machine. If the optimum replacement policy is followed, what will be the average yearly cost of owning and operating the machine?

  (b) Machine B costs Rs. 10,000. Annual operating costs are Rs. 400 for the first year, and then increase by Rs. 800 every year. You now have a machine of type A which is one year old. Should you replace it with B, if so, when?

[ICWA, June 1985]

Answer : (a) Total cost in the 3rd year is least i.e., Rs. 5200 hence replace at the end of 3rd year.

  (b) Average annual cost is least at the end of 5th year. The total cost Rs. 2200 for 1 year old machine A remains less than minimum average cost 400/- on machine B until second year, therefore replace machine A before 3rd year.

4. A firm is considering replacement of a machine whose cost price is Rs. 12,200 and the scrap value is Rs. 200. The maintenance costs are found from experience to be as follows :

Year	1	2	3	4	5	6	7	8
Maintenance cost (Rs)	200	500	800	1200	1800	2500	3200	4000

[JNTU (CSE) 96/S, (Prod./Mech,Mechatronics/Chem.) 2000]

Answer : Replace at the end of the 6th year

5. The following table gives the running costs per year and resale price of a certain equipment whose purchase price is Rs. 5000.

Year	1	2	3	4	5	6	7	8
Running costs (Rs)	1500	1600	1800	2100	2500	2900	3400	4000
Resale value (Rs)	3500	2500	1700	1200	800	500	500	500

At what year is the replacement due? **[JNTU (CSE) 94/S, ICWA (DCC) 86, 85, C.A (May) 82]**

Answer : At the end of the 4th year ATC_3 (2733) > ATC_4 (2700) < ATC_5 (2740)

6. A truck owner finds from his past records that the maintenance costs per year of a truck whose purchase price is Rs. 8000, are given below :

Year	1	2	3	4	5	6	7	8
Maintenance cost (Rs)	1000	1300	1700	2200	2900	3800	4800	6000
Resale Price (Rs)	4000	2000	1200	600	500	400	400	400

Determine at what time it is profitable to replace the truck.

[JNTU (Mech.) 93]

Answer : Replace at the end of the 5th year

7. A fleet owner finds from his past records that the costs per year of running a vehicle whose purchase price is Rs. 5000 are as under :

Year	1	2	3	4	5	6	7
Running costs (Rs)	500	600	700	900	2150	1600	1800
Resale value (Rs)	3000	1500	750	375	200	200	200

Thereafter, running cost increases by Rs. 200 but resale value remains constant at Rs. 200. At what age is a replacement due?   **[JNTU (CSE) 96]**

Answer : At the end of 5th year

8. A truck owner from his past experience estimated that the maintenance cost per year of a truck whose purchase price is Rs. 1,50,000 and the resale value of truck will be as follows.

Year	1	2	3	4	5	6	7	8
Maintenance cost (Rs)	10,000	50,000	20,000	25,000	30,000	40,000	45,000	50,000
Resale value (Rs)	1,30,000	1,20,00	1,15,000	1,05,000	90,000	75,000	60,000	50,000

Determine at which time it is profitable to replace.   **[JNTU (CSE) 95]**

Answer : Though the minimum value of the average cost occurs at the end of first year, it is impractical to replace the truck every year, so next time the minimum average cost occurs at the end of 4th year. Hence it is better to replace the truck after every 4 years.

Note : *Here the condition after every 4 years, policy maintenance cost increases with time violating if in the 3rd year.*

9. A machine costs Rs. 10,000 operating costs are Rs. 50 per year for the five years. In the sixth and succeeding years operating costs increase by Rs. 100 per year. Find optimum length of time to hold the machine before replacing it.

[JNTU (Mech.) FDH 95]

Answer : End of 15 years.

10. (a) Machine *A* costs Rs. 3600. Annual operating costs are Rs. 40 for first year and then increases by Rs. 360 every year. Assuming machine *A* has no resale values, determine the best replacement age.

   (b) Another machine *B* which is similar to machine *A* costs Rs. 4000. Annual running costs are Rs. 200 for first year and then increase by Rs. 200 every year. It has resale value oᶠ Rs. 1500, Rs. 1000 and Rs. 500. If replaced at the end of first, second and third year respectively. It has no resale value during fourth year onwards. Which machine would you prefer to purchase. Further costs are not discounted. **[JᴺTU (CSIT) 98/S]**

Answer : Replace machine *A* by machine *B* at the end of 4th year.

11. A truck with first cost of Rs. 80,000 has depreciation and service pattern as shown below.

Year	1	2	3	4	5	6
Depreciation	28,000	20,000	14,000	5,000	4,000	4,000
Service cost	18,000	21,000	2 ,000	29,000	34,000	40,000

Assume no interest charges. Find age for replacement.

[JNTU (Mech.) 98/CCC]

Answer : After six years

12. A machine owner finds from his past records that the costs per year of maintaining a machine whose purchase price is Rs. 6,000 are as given below.

Year	1	2	3	4	5	6	7	8
Maintenance cost (Rs)	1000	1200	1400	1800	2300	2800	3400	4000
Resale price	3000	1500	750	375	200	200	200	200

Determine at what age is a replacement due?

[JNTU (Mech.) 97/CCC, (ECE) 97/C, (CSIT) 2000]

Answer : The machine should be replaced at the end of fifth year.

13. The machine owner has three small machines of purchase price Rs. 6000 each and cost per year of maintaining each machine is same. Two of these machines are two years old and the third is one year old. He is considering a new machine of purchase price Rs. 8000 with 50% more capacity than one of two old ones. The estimates of maintaining cost and resale price for new machine are as given below ;

Year	1	2	3	4	5	6	7	8
Maintenance cost (Rs)	1200	1500	1800	2400	3100	4000	5000	6100
Resale price (Rs)	4000	2000	1000	500	300	300	300	300

Assuming that the loss of flexibility due to fewer machines is of no importance, and he continues to have sufficient work for three of the old machines. What should his policy be?   **[JNTU (CSE) 94]**

Answer : All the three small machines should be replaced after two years.

16. Fleet cars have increased their costs as they continue in service due to increased direct operating costs (gas and oil) and increased maintenance (repairs, tyres, batteries etc.) The initial cost is Rs. 3,500 and the trade-in value drops as time passes until it reaches a constant value of Rs. 500.

Given the cost of operating, maintaining and the trade-in value, determine the proper length of service before cars should be repalced.

Year of service	1	2	3	4	5
Year end trade-in value	1900	1050	600	500	500
Annual operating cost	1500	1800	2100	2400	2700
Annual maintaining cost	300	400	600	800	1000

[JNTU (Mech. & ECE) 98/S/CCC]

Answer : The cars should be replaced at the end of the third year.

10.5 *Replacement Policy for Items When Money Value Changes With Time*

Money value changes with time. Suppose you keep Rs. 100 in a bank which gives you an interest at the rate of 10% p.a. After one year you will recieve Rs. 110. Thus we can say that "if we have Rs. 100 today, it is worth having Rs. 110 after one year". Similarly, a rupee possessed today is worth 1.1 after one year @10%.

In the other way, if Rs. 1.10 after one year is Re. 1.00 today, Re. 1.00 after one year will be 1/1.1 or $(1.1)^{-1}$ today.

In a similar fashion one rupee today will be $(1.1)^2$ two years hence because it is 1.1 after one year and 1.1×1.1 in the next i.e., second year. Thus a rupee two years hence is worth $(1.1)^{-2}$ today. Thus the present worth of one rupee, n years hence will be $(1.1)^{-n}$ or $(1 + 0.10)^n$ today, when money value is changing @10%.

Now if the money value is supposed to be changing at the rate of r percent, the present worth of one rupee, n years hence will be $(1 + r)^{-n}$ or $\dfrac{1}{(1 + r)^n}$

Present Worth Factor (PWF) : *It is the factor that converts a rupee 'n' years hence changing (money grows but value decreases) at the rate of 'r' percent is worth today.* This is denoted by $v = (1 + r)^{-n}$. This is also called "discount rate" or "depreciation rate" or "present value factor".

Now let C = Initial cost of the equipment

and   if R_1 = Operating cost in first year, (present worth is also R_1)

if R_2 = Operating cost in second year present worth of

$$R_2 = (1 + r)^{-1} R_2 \text{ or } v R_2$$

if R_3 = Operating cost in third year, and present worth of

$$R_3 = (1 + r)^{-2} R_3 \text{ or } v^2 R_3 \text{ and so on}$$

if R_n = Operating cost in n^{th} year present worth of

$$R_n = v^{n-1} R_n$$

Thus the present value of all future discounted cost in 'n' year assuming scrap value of the equipment be zero, is

$$P_n = C + R_1 + v R_2 + v^2 R_3 + \ldots + v^{n-1} R_n$$

Thus P_n is the amount of money required to pay all future costs of purchasing the equipment and operating it assuming that it is to be replaced after 'n' years.

Now if we assume that the manufacturer invests the amount P_n by borrowing at the rate of interest 'r' and repays it in 'n' years with a fixed annual instalment 'x' on diminishing balance,

Then $P_n = x + v x + v^2 x + \ldots + v^{n-1} x$

$$= [1 + v + v^2 + \ldots + v^{n-1}] x$$

$$= \left[\frac{1 - v^n}{1 - v} \right] x$$

$\therefore$ $x = \left[\dfrac{1 - v}{1 - v^n} \right] P_n$

Hence the best period to replace the machine is the period n which minimises $\dfrac{1 - v}{1 - v^n} \cdot P_n$. But $1 - v = A$ positive constant quantity and so we can write $F_n = \dfrac{P_n}{1 - v^n}$ and find out the value of n the period at which to replace the machine that minimises F_n. Since n can assume only discrete values $[1, 2, 3, \ldots]$ we can use the method of finite differences to calculate its optimal values.

F_n will be minimum if $\Delta F_{n-1} < 0 < \Delta F_n$ .... (b)

Now $\quad \Delta F_n = F_{n+1} - F_n$ .... (c)

$$= \frac{P_{n+1}}{1 - v^{n+1}} - \frac{P_n}{1 - v^n} = \frac{(1 - v^n) P_{n+1} - (1 - v^{n+1}) P_n}{(1 - v^{n+1})(1 - v^n)}$$

$$= \frac{1}{(1 - v^{n+1})(1 - v^n)} \left[(P_{n+1} - P_n) + (v^{n+1} P_n - v^n P_{n+1}) \right] \quad \text{.... (d)}$$

We have $\quad P_{n+1} = (C + R_1 + v R_2 + \ldots + v^{n-1} R_n) + v^n R_{n+1}$

$$= P_n + v^n . R_{n+1} \text{ or } v^n R_{n+1} = P_{n+1} - P_n$$

Hence $\Delta F_n = \dfrac{1}{(1 - v^{n+1})(1 - v^n)} \left[v^n . R_{n+1} + v^{n+1} P_n - v^n \left[P_n + v^n R_{n+1} \right] \right]$

$$= \frac{1}{(1 - v^{n+1})(1 - v^n)} \left[v^n R_{n+1} (1 - v^n) - v^n P_n (1 - v) \right]$$

$$= \frac{v^n (1 - v)}{(1 - v^{n+1})(1 - v^n)} \left[\frac{1 - v^n}{1 - v} . R_{n+1} - P_n \right] \quad \text{.... (e)}$$

$$= \text{A positive constant} \times \left[\frac{1 - v^n}{1 - v} . R_{n+1} - P_n \right]$$

Hence F_n has always the same sign as $\left[\dfrac{1 - v^n}{1 - v} R_{n+1} - P_n \right]$

∴   From equation (b), n will be optimal if

$$\left[1 - \frac{v^{n+1}}{1 - v} R_n - P_{n-1} \right] < 0 < \left[\frac{1 - v^n}{1 - v} . R_{n+1} - P_n \right] \quad \text{.... (f)}$$

From above equation we have $\dfrac{1 - v^n}{1 - v} R_{n+1} - P_n > 0$

or $\quad R_{n+1} > P_n \left[\dfrac{1 - v}{1 - v^n} \right] \quad$ or $\quad R_{n+1} > \dfrac{P_n}{[(1 - v^n)/(1 - v)]}$

or $\quad R_{n+1} > \dfrac{C + R_1 + v R_2 + v^2 R_3 + \ldots + v^{n-1} R_n}{1 + v + v^2 + \ldots + v^{n-1}}$

or $\quad R_{n+1} > \dfrac{C + \displaystyle\sum_{r=1}^{n} R_r v^{r-1}}{\displaystyle\sum_{r=1}^{n} v^{r-1}} \quad \text{.... (g)}$

The right hand expression is the weighted average (denoted by w_r) of all costs up to and including period $(n - 1)$. The weights $1, v, v^2 .. v^{n-1}$ are the discount factors applied to the costs for each period.

From equation (f), the left hand side of expression can be expressed as

$$R_{n+1} < \dfrac{C + \displaystyle\sum_{r=1}^{n} R_r\, v^{r-1}}{\displaystyle\sum_{r=1}^{n} v^{r-1}} \qquad \dots \text{(h)}$$

From equation (g) and (h) we conclude that

1. The machine should be replaced if the next period cost is greater than the weighted average of previous costs.

2. The machines should not be replaced if the next period's cost is less than the weighted average of previous costs.

Note : *When money value is not considered,* $v = 1$

$$R_{n+1} > \dfrac{C + R_1 + R_2 + .. R_n}{1 + 1 + \dots \text{ to } n \text{ terms}} \quad \text{or} \quad R_{n+1} > \dfrac{P_n}{n} \text{ where } \dfrac{P_n}{n}$$

average yearly cost with no resale value.

This is identical to the previous case when money value was ignored. In real practice replacement policy is greatly influenced by complicated tax laws prevailing. Discussion in this regard is not included in the scope of this book. In actual dealings the influence of tax has got to be taken into consideration.

Steps to Find the Policy when Money Value Changes with Time :

Step 1 : Note the values of capital cost of machine, salvage value, rate of depreciation *PWF* etc.

Step 2 : Construct the tabular form as given below, and enter the first 2 columns as per the given data.

Year	Running Cost (R_n)	PWF v^{n-1}	$R_n\, v^{n-1}$	$C + \Sigma\, R_n\, v^{n-1}$	$\Sigma\, v^{n-1}$	Average Annual Total Cost $T\, C_n$
(1)	(2)	(3)	(4) = (2) − (3)	(5) = C+Σ (4)	(6) = Σ (3)	(7) = (5)/(6)

Step 3 : Calculate the present worth factor for each year by the formula $\dfrac{1}{(1+r)^{n-1}}$ and put in col. 3.

Step 4 : Calculate $R_n\, v^{n-1}$ i.e., the product of 2nd and 3rd col. values, and enter them in col. 4.

Step 5 : Col. 5 is summation of cummulative running cost and capital cost including (salvage if any) i.e., $C + \Sigma R_n v^{n-1}$.

Step 6 : Find cummulation of col. 3. and enter in col. 6 $\left[\Sigma v^{n-1}\right]$

Step 7 : Calculate the value of col. (5) divided by col. (6) i.e., $\left(C + \Sigma R_n v^{n-1}\right) / \Sigma v^{n-1}$ and enter in col. 7.

Step 8 : The lowest value in col. 7. corresponds to the year in which the machine is to be replaced.

Note : *Students are advised to check the problem thoroughly and note whether depreciation is given or depreciated amount is given or discount rate is given etc.*

ILLUSTRATION 5

The initial price of equipment is Rs. 5000. The running cost varies as shown below.

Year	1	2	3	4	5	6	7
Running cost (Rs.)	400	500	700	1000	1300	1700	2100

Taking a discount rate of 0.09. Find out the optimum replacement interval.

Solution :

Year = n	Maintenance cost = R_r	Discount factor = v^{r-1}	Discount cost $R_r v^{r-1}$	$C + \sum_{r=1}^{n} R_r v^{r-1}$	$\sum_{r=1}^{n} v^{r-1}$	ATC
(1)	(2)	(3)	(4) = (2) − (3)	(5) = $C + \Sigma$(4)	(6) = Σ (3)	(7) = $\dfrac{(5)}{(6)}$
1	400	1.000	400	5400	1.000	5400
2	500	0.900	450	5850	1.900	3079
3	700	0.810	567	6417	2.710	2368
4	1000	0.729	729	7146	3.439	2083
5	1300	0.656	853	7999	4.095	1953
6	1700	0.599	1016	9015	4.694	**1921**
7	2100	0.531	1115	10117	5.225	1936

From the above table, we observe that $ATC_5 > ATC_6 < ATC_7$

i.e., 1953 > 1921 < 1936 *the optimum replacement period is 6 years.*

ILLUSTRATION 6

A manufacturer is offered 2 machines A and B. A is priced Rs. 5000 and running costs are estimated at Rs. 800 for each of the first five years increasing by Rs. 200 per year in the sixth and subsequent years. Machine B which has the same capacity as A with Rs, 2500 but will have running cost of Rs. 1200 per year for 6 years, increasing by Rs. 200 per year thereafter. If money is worth 10% per year, which machine should be purchased?

(Assume that the machines will eventually be sold for scrap at a negligible price)

[JNTU Mech/Mechatronics/Chem/Prod. 2001/S]

Solution :

TABLE FOR A

$$r = 10\% \qquad C_A = 5000$$

Year	R_r	v^{n-1}	$R_r v^{n-1}$	$C + \sum_{r=1}^{n} R_r v^{n-1}$	$\sum_{r=1}^{n} v^{n-1}$	$\dfrac{5}{6}$
(1)	(2)	(3)	(4) = (2) × (1)	(5) = C + Σ (4)	(6)= Σ (3)	(7) = (S) (6)
1	800	1.0000	800	5800	1.0000	5800.00
2	800	0.9091	727	6527	1.9091	3418.88
3	800	0.8264	661	7188	2.7355	2627.67
4	800	0.7513	601	7789	3.4868	2233.85
5	800	0.6830	546	8335	4.1698	1998.89
6	1000	0.6209	621	8956	4.7907	1869.45
7	1200	0.5645	677	9633	5.3552	1798.81
8	1400	0.5132	718	10351	5.8684	1763.85
→ 9	1600	0.4665	746	11097	6.3349	1751.72
10	1800	0.4241	763	11860	6.7590	1754.70

TABLE FOR B

$$R = 10\% \qquad C_B = 2500$$

1	1200	1.0000	1200.00	3700.00	1.0000	3700.00
2	1200	0.9091	1090.91	4790.91	1.9091	2509.51
3	1200	0.8264	991.98	5782.59	2.7353	2113.91
4	1200	0.7513	901.56	6684.15	3.4868	1916.99
5	1200	0.6830	819.60	7503.75	4.1698	1799.55
6	1200	0.6209	745.08	8248.83	4.7909	1721.84
7	1400	0.5645	790.30	9039.13	5.3552	1687.92
→ 8	1600	0.5132	821.12	9860.25	5.8684	1680.23
9	1800	0.4665	839.70	10699.95	6.3349	1689.25
10	2000	0.4241	848.20	11548.15	6.7590	1708.56

From table for A

$$ATC_8 \, (1763.85) > ATC_9 \, (1751.72) < ATC_{10} \, (1754.70)$$

∴   Replacement period for A = 9 year.

From table for B

$$ATC_7 (1687.92) > ATC_8 (1680.23) < ATC_9 (1689.25)$$

∴   Replacement period for B = 8 year.

The fixed annual payment for $-A = x_A = \dfrac{1-v}{1-v^2} \cdot P\,(9)$

$$= \frac{1-0.9091}{1-(0.9091)9} \times 11097 = \text{Rs. } 1752$$

$$x_B = \frac{1-0.9091}{1-(0.9091)8} \times 9860.25 = \text{Rs. } 1680 < x_A.$$

Hence purchase machine B instead of A.

Alternately weighted average cost in 9 year for A = 1781.72 and that for B in 8 year = 1680.23 which is the lowest.

∴   Purchase B.

Practice Problems

1. A company has the option to buy one of the mini computer : MINICOMP and CHIPCOMP. MINICOMP costs Rs. 5 lakh and running and maintenance costs are Rs. 60,000 for each of the first five years increasing by Rs. 20,000 per year in the sixth and subsequent year. CHIPCOMP has the same capacity as MINICOMP, but costs only Rs. 2,50,000. However its running and maintenance costs are Rs. 1,20,000 per year for first five year and increase by 20000 per year thereafter. If the money is worth 10% per years, which computer should be purchased? What are the optimal replacement periods for each of the computer? Assume that there is no salvage value for either computer. Explain your analysis.

   [JNTU (CSE) 96/S]

Answer : Better replacement period for CHIPCOMP computer is after six years.

2. Assume that the present value of one rupee to be spent in a years time is Rs. 0.90 and C = Rs. 3000 capital cost of equipment and the running cost are given in the table as follows.

Year	1	2	3	4	5	6	7
Running cost (Rs)	500	600	800	1000	1300	1600	2000

When should the machine be replaced?

Answer : After 5 years

3. An Engineering company is offered two types of material handling equipments *A* and *B*. *A* is priced at Rs. 60,000 including costs of installation, operation and maintenance are estimated to be Rs. 10000. for each of the first 5 years, increasing every year by Rs. 3000 per year in the sixth and subsequent year. Equipment *B* with rated capacity same as *A*, requires an initial investment of Rs. 30000 but in term of operation and maintenance costs are estimated to be Rs. 13,000 per year for the first six years, increasing every year by Rs. 4000 per year from the seventh year onwards. The company expects a return of 10% on all its investments. Neglecting the scrap value of the equipment at the end of its economic life, determine which equipment the company should buy.

[I.I.T.E Grad. 1981, JNTU (Mech. & FDH) 95]

Answer : Purchase *B*; Replace after 7 years

4. A manufacturer is offered two machine *A* and *B*. *A* is priced at Rs. 50000 and running costs are estimated at Rs. 8000 for each of the first five years increasing by Rs. 2000 per year in the sixth and subsequent year. Machine *B* which has the same capacity as *A*, costs Rs. 25000 but will have running costs of Rs. 12000 per year for six years, increasing by Rs. 2000 per year thereafter.

If money is worth 10 percent per year, which machine should be purchased? (Assume that the machines will eventually be sold for scrap at a negligible price.

[JNTU (Mech. & FDH) 96]

Answer : Machine *B*; Replace after 8 year]

5. A truck is priced for Rs. 60000 and running costs are estimated at Rs. 6000 for each of first 4 years increasing by Rs. 2000 per year in the fifth and subsequent years. If money is worth 10%, when should the truck be replaced. Assume scrap value = 0.   [JNTU (Mech.) 96/C/S, (ECE) 96/S/CCC]

Answer : Replace after 9 year

6. A machine costs Rs. 10,000 and operating costs are Rs. 500 per year for first five years. Operation cost increases by Rs. 100 per year in the sixth and subsequent year. Assuming 10% discount rate of money per year find optimum length of time to hold the machine before it is replaced. State clearly assumptions made.   [JNTU (Mech.) 93, Madras B.E. (Mech) 1984]

Answer : Nineteen years

7. A truck has been purchased at a cost of Rs. 1,60,000. The value of the truck is depreciated in the first 3 years by Rs. 20000 each year and Rs. 16000 per year thereafter. Its maintenance and operating cost for the first three years are Rs. 16000, 18000 and 20000 in that order and then increase by Rs. 4000 every year. Assuming an interest rate of 10%. Find the economic life of the truck.

[JNTU (Mech.) 2001/CCC, JNTU (ECE) CCC 2001/43301]

Answer : 7 years

8. A truck owner estimates that the running costs and the salvage value of trucks for various years will be as illustrated below. If the purchase price of a truck is Rs. 80000 estimate the optimum replacement age for the truck. Take money's value as 15% per annum.

Year	1	2	3	4	5	6	7	8
Running Cost (Rs)	6000	7000	9000	12000	15000	20000	25000	30000
Resale Price (Rs)	60000	40000	30000	30000	25000	20000	20000	20000

[JNTU (CSE) 98]

Answer : By the end of 5th year

9. If you wish to have a return of 10% per annum on your investment. Which of the following plans would you prefer.

	Plan A	Plan B
First cost	2,00,000	2,50,000
Scrap value after 15 years	1,50,000	1,80,000
Excess of annual revenue over manual disbursement	25,000	30,000

Answer :   Plan A

10. Singareni Collieries Company Ltd (SCCL) has received offers for two types of load hauling dumpers A and B. A has a pay load of 25 tonnes and is priced at Rs. 4,00,000 while B also with pay load of 25 tonnes, is priced at Rs. 3,60,000. The operating costs over the estimated life of 5 years for both the types are as follows.

Year	1	2	3	4	5
Type A (Rs)	8000	9000	10000	11000	12000
Type B (Rs)	14000	16000	18000	20000	22000

Which type of dumper is to be preferred.

Answer : Type B

11. A manual stamper currently valued at Rs. 1000 is expected to last two years and costs Rs. 4000 per year to operate. An automatic stamper which can be purchased for Rs. 3000 will last for 4 years and can be operated at an annual cost of Rs. 3000. If money carries a rate of interest of 10% per annum, determine which stamper should be purchased.   [Bangalore B.E. (Mech) 1982]

Answer : Purchase an automatic stamper

12. There are two machines A and B. A is replaced after every 3 year where as B is replaced after every 8 years. The yearly cost of the machines are given below :

Year	1	2	3	4	5	6
A	1000	200	400	1000	200	400
B	1700	100	200	300	400	500

Assuming the value of money to be 10% per year, determine which machine should be purchased. **[JNTU (CSE) 95/S]**

Answer : Considering six years period for both the machines, machine A appears to be cheaper than B. Hence machine A should be purchased.

13. The cost of a new car is Rs. 10,000. Compare the optimum moment of replacement assuming the following cost information :

Assuming that repairs are made at the end of each year only if the car is to be retained and are not necessary if the car is to be sold for its salvage value. Also assume that the rate of discount is 10%.

Age of car	Repair costs / year in Rs.	Salvage value at the end of the year in Rs.
1	5000	8000
2	10000	6400
3	10000	5120

[JNTU (CSE) 93/S]

Answer : The car must be replaced at the end of first year.

14. Let $v = 0.9$ and initial price is Rs. 5000. Running cost varies as follows :

Year	1	2	3	4	5	6	7
Running cost in (Rs)	400	500	700	1000	1300	1700	2100

What would be the optimum replacement interval?

Answer : Six years

15. A pipeline is due for repairs. It will cost Rs. 10000 and last for 3 years. Alternatively a new pipeline can be laid at a cost of Rs. 30000 and last for 10 years. Assuming cost of capital to be 10% and ignoring salvage value, which alternative should be chosen? **[C.A. (May) 1981]**

Answer : Existing pipe line should be continued

16. The cost pattern of two machines M_1 and M_2 when money value is not considered is

	Cost at the beginning of the year in (Rs)	
Year	M_1	M_2
1	900	1400
2	600	100
3	700	700

Find the cost pattern for each machine when money is worth 10% per year and hence find the machine which is less costly.

Answer : Machine B should be purchased

17. An individual is planning to purchase a car will cost Rs. 1,20,000. The resale value of the car at the end of the year is 85% of the previous year value. Maintenance and operation costs during the first year are Rs. 20,000 and they increase by 15% every year. The minimum resale value of car can be Rs. 40,000.

    (i) When should the car be replaced to minimise average annual cost (ignore initial)?

    (ii) If interest of 12% is assumed, when should the car be replaced?

Answer : At the end of 1st year

18. The cost of a new machine is Rs. 5000 the maintenance cost of nth year is given by $R_n = 500 (n - 1)$, $n = 1, 2 \ldots$ suppose that the discount rate per year is 0.5. After how many years it will be economical to replace the machine by a new one?   **[JNTU (CSE) 2001]**

Answer : 5 years

19. The annual costs of two machines X and Y are given below. Find their cost patterns if money value is 10% per year, what conclusion do you derive about which machine is economical?

Year	1	2	3
Machine X (Rs)	1900	1100	1400
Machine Y (Rs)	2700	300	1400

Answer : Machine X is more economical.

20. The initial cost of an item is 15000 and maintenance and running costs in rupees for different years are given below :

Year	1	2	3	4	5	6	7
Running cost	2500	3000	4000	5000	6500	8000	10000

What is the replacement policy to be adopted if the capital is worth 10% and there is no salvage value.

Answer : Optimal replacement period = 5 years (Rs. 7609).

21. A special lathe machine costs Rs. 1 lakh, annual operation and machine cost is Rs. 1,000 and increases at a rate of Rs. 750 for 2nd and 3rd year and at a rate of Rs. 1250 for the remaining life. The life of equipment is 10 years. Salvage value at the end of the first year is Rs. 80,000 and falls at a rate of 10,000 every year for the first 4 years. Afterwards it has no solvage value. What should be economic life of lathe if the rate of return on capital invested is 25%.

Answer : At the end of 2nd year (Rs. 18000) [JNTU - CSE 97]

10.6 *Group Replacement*

Group Replacement of Items That Fail Completely :

This policy is concerned with the items that either work perfectly, or work partially, or inefficiently or fail completely. This situation generally happens when the system consists of a large number of identical low cost items that are increasingly liable to failure with age. In such cases, the replacement of individual items would incur a set of costs, which is independent of the number replaced. However it may be advantageous to replace all the items at a time at a fixed interval. This policy is known as group replacement policy and is very attractive, particularly when

1. The value of any individual item is so small.

2. The cost of keeping records of individual ages is high that cannot be justified.

3. The purchase of such identical items in bulk can be had at discounted rate.

4. Average individual replacement would be costlier than the average group replacement.

5. If sufficient number of standby machines are available.

6. New designs of the equipment considerably increase the production rate.

In all the above cases the two types of replacement policies considered are :

1. **Individual Replacement :** Under this policy, an item is replaced immediately after its failure.

2. **Group Replacement :** Under this policy a decision will be taken so as to replace all the items irrespective of the fact that the items have failed - not failed, provided if any item fails, before the optimal time it may be replaced individually.

Algorithm for Deciding Group Replacement Policy :

Step 1 : Find the probability of failure of items at the end of each period.

Step 2 : Find the number of replacements made at the end of each period with reference to probability of failures at the end of each period considering its previous replacements.

Step 3 : Calculate cost of individual replacement at the end of each period.

Step 4 : Calculate cost of group replacement at the end of each period.

Step 5 : Calculate total cost of group replacement including individual replacements by adding step 3 and step 4.

Step 6 : Calculate the average cost per period by dividing the result in step 5 with period number.

Step 7 : Identify the least among the average cost per period as the period of group replacement policy.

ILLUSTRATION · 7

1000 bulbs are in use and it costs Rs. 10 to replace an individual bulb which has burnt out. If all bulbs were replaced simultaneously it would cost Rs. 4 per bulb. It is proposed to replace all bulb's at fixed intervals of time, whether or not they have burnt out and to continue replacing burnt out bulbs as and when they fail.

The failure rates have been observed for certain type of light bulbs are as follows:

Week	1	2	3	4	5
Percent failing by the end of week	10	25	50	80	100

At what intervals all the bulbs should be replaced? At what group replacement price per bulb would a policy of strictly individual replacement become preferable to the adopted policy.   [JNTU CSE 2001]

■ Solution :

Step 1 : To find out the probability of failure of items at the end of each week.

The probability of failure of light bulbs in first week

$$= P_1 = \frac{10}{100} = 0.10$$

The probability of failure of light bulbs in second week.

$$= P_2 = \frac{(25-10)}{100} = 0.15$$

The probability of failure of light bulbs in third week

$$= P_3 = \frac{(50-25)}{100} = 0.25$$

The probability of failure of light bulbs in fourth week

$$= P_4 = \frac{(80-50)}{100} = 0.30$$

The probability of failure of light bulbs in fifth week

$$= P_5 = \frac{(100-80)}{100} = 0.20$$

Sum of all probabilities is 1.

i.e., $\qquad P_1 + P_2 + P_3 + P_4 + P_5 = 1$

$\therefore \qquad$ All further probabilities P_6, P_7, P_8 and so on will be zero.

Step 2 : Calculation of number of replacements made considering previous replacements.

Let N_i be the number of replacements made at the end of ith week, if all 1000 bulbs are new initially.

Thus

$$N_0 = N_0 = 1000$$

$$N_1 = N_0\,P_1 = 1000 \times 0.1 = 100$$

$$N_2 = N_0\,P_2 + N_1\,P_1 = 1000 \times 0.15 + 100 \times 0.10 = 160$$

$$N_3 = N_0\,P_3 + N_1\,P_2 + N_2\,P_1 = 1000 \times 0.25 + 100 \times 0.15 + 160 \times 0.10 = 281$$

$$N_4 = N_0\,P_4 + N_1\,P_3 + N_2\,P_2 + N_3\,P_1 \qquad = 377$$

$$N_5 = N_0\,P_5 + N_1\,P_4 + N_2\,P_3 + N_3\,P_2 + N_4\,P_1 = 350$$

$$N_6 = 0 + N_1\,P_5 + N_2\,P_4 + N_3\,P_3 + N_4\,P_2 + N_5\,P_1 = 230$$

$$N_7 = 0 + 0 + N_2\,P_5 + N_3\,P_4 + N_4\,P_3 + N_5\,P_2 + N_6\,P_1 = 286$$

From above results it is clear that number of bulbs burnt out increases upto fourth week and decrease upto sixth week and again start increasing. The whole system comes to a steady state where the proportion of bulbs failing in each week is the reciprocal of their average life.

As the mean age of bulbs.

$$= 1 \times P_1 + 2 \times P_2 + 3 \times P_3 + 4 \times P_4 + 5 \times P_5$$

$$= 1 \times 0.1 + 2 \times 0.15 + 3 \times 0.25 + 4 \times 0.30 + 5 \times 0.20 = \textbf{3.35 week.}$$

$\therefore$ number of failures in each week in steady state become $= \dfrac{1000}{3.35} = 299$

$\therefore$ cost of replacing bulbs individually only on failure $= 10 \times 299 =$ Rs. **2990.**

Step 3 : Calculating the cost of individual replacement at the end of each period cost of individual replacement at

end of first week $= 100 \times 10 = 1000$

end of second week $= 160 \times 10 = 1600$

end of third week $= 281 \times 10 = 2810$

end of fourth week $= 377 \times 10 = 3770$

Step 4 : Calculating the cost of group replacement at the end of each period.

end of first week $= 1000 \times 4 = 4000$

end of second week $= 4000 + 1000 = 5000$

end of third week $= 5000 + 1600 = 6600$

end of fourth week $= 6600 + 2810 = 4410$

Step 5 : Calculating total cost of group replacement including individual replacement i.e., adding values of step 3 and step 4.

end of first week $= 4000 + 1000 = 5000$

end of second week $= 5000 + 1600 = 6600$

end of third week $= 6600 + 2810 = 9410$

end of fourth week $= 9410 + 3770 = 13180$

Step 6 : Calculate average cost per week

end of first week $= \dfrac{5000}{1} = 5000$

end of second week $= \dfrac{6600}{2} = 3300$

end of third week $= \dfrac{9410}{3} = 3136.67$

end of fourth week $= \dfrac{13180}{4} = 3295$

Step 7 : To identify the least among average cost per period.

It is identified as ***third week. i.e., 3136.67,*** so it would be optimal to ***replace all the bulbs after every 3 weeks,*** other wise the average cost will be increasing.

LLUSTRATION · 8 ───

A factory has a large number of bulbs all of which must be in working condition. The mortality of bulbs is given in the following table :

Week	Proportion of Bulbs Failing During the Week
1	0.1
2	0.15
3	0.25
4	0.35
5	0.12
6	0.03

If a bulb fails in service, it costs 3.50 to replace but if all bulbs are replaced at a time it costs Rs. 1.20 each. Find the optimum group replacement policy. (Assume 1000 bulbs as available in the beginning). [JNTU (Mech.) 99]

Solution :

$$N_o = 1000$$

$$N_1 = N_o P_1 = 1000 \times 0.1 = 100 \text{ bulbs}$$

$$N_2 = N_1 P_1 + N_o P_2$$

$$= 100 \times 0.10 + 1000 \times 0.15$$

$$= 160 \text{ bulbs}$$

$$N_3 = N_2 P_1 + N_1 P_2 + N_o P_3$$

$$= 160 \times 0.01 + 100 \times 0.15 + 1000 \times 0.25$$

$$= 281 \text{ bulbs}$$

$$N_4 = N_3 P_1 + N_2 P_2 + N_1 P_3 + N_o P_4$$

$$= 281 \times 0.1 + 160 \times 0.15 + 100 \times 0.25 + 1000 \times 0.35$$

$$= 427$$

$$N_5 = N_4 P_1 + N_3 P_2 + N_2 P_3 + N_4 P_4 + N_o P_5$$

$$= 427 \times 0.1 + 281 \times 0.15 + 160 \times 0.25 + 100 \times 0.35 + 1000 \times 0.12$$

$$= 279.$$

$$N_6 = N_5 P_1 + N_4 P_2 + N_3 P_3 + N_2 P_4 + N_1 P_5 + N_o P_6$$

$$= 279 \times 0.1 + 427 \times 0.15 + 281 \times 0.25 + 160 \times 0.35 + 100 \times 0.12 + 1000 \times 0.03.$$

$$= 260 \text{ bulbs.}$$

End of the week	Total no. of bulbs failed	Cummulative no. of failure	Cost of replacement (Rs. 3.50)	Cost of group replacement (Rs. 1.20)	Total cost	Average cost per week
1	100	100	350	1200	1550	1550
2	160	260	910	1200	2110	1055
3	281	541	1893.5	1200	3093.5	1031.1
4	427	968	3388.5	1200	4588	1147
5	279	1247	4364.5	1200	5564.5	1113
6	260	1507	5274.5	1200	6474.5	1079.08

∴ *bulbs should be replaced by the end of every 3rd week.*

Practice Problems

1. Find the cost per period of individual replacement policy of an installation of 300 lighting bulbs, given following :

    (i) Cost of replacing individual bulbs is Rs. 3.

    (ii) Conditional probability of failure is given below :

Week No.	0	1	2	3	4
Conditional probability of failure	0	$\dfrac{1}{10}$	$\dfrac{1}{3}$	$\dfrac{2}{3}$	1

JNTU (CSE) 95]

Answer : Expected life of each light bulb is 3 week and the average cost of individual replacement of 300 light bulbs is Rs. 206.

2. An electric company which generates and distributes electricity conducted a study on the life of poles. The appropriate life data are given in the following table :

Year after installation	1	2	3	4	5	6	7	8	9	10
Percentage poles failing	1	2	3	5	7	12	20	30	16	4

(i) If the company now installs 5000 poles and follows a policy of replacing poles only when they fail, how many poles are expected to be replaced each day during the next ten year.

(ii) If the cost of replacing individually is Rs. 160 per pole and if we have a common group replacement policy, it costs Rs. 30 per pole. Find out the optimal period to group replacement.   **[JNTU (ECE) 95/CCC]**

Answer : (i) 5533 poles (ii) All poles after 6th year

3. The following failure rates have been observed for a certain type of light bulbs.

End of week	1	2	3	4	5	6	7	8
Probability of failure to date	0.05	0.13	0.25	0.43	0.68	0.88	0.96	1.00

The cost of replacing an individual bulb is as 2.25, the decision is made to replace all bulbs simultaneously at fixed intervals and also to replace bulbs as they fail in service. If the cost of group replacement is 60 paise per bulb and the total number of bulbs is 1000. What is best interval between group replacement? **[OU (Mech.) 88]**

Answer : Replacement at the end of third week.

4. Suppose that a special purpose type of light bulb never lasts longer than two weeks. There is chance of 0.3 that a bulb will fail at the end of first week. There are 100 new bulbs initially. The cost per bulbs for individual replacement is Rs. 1.25 and the cost per bulb for a group replacement is Re. 0.50.

Is it cheapest to replace all bulbs;

    (i) Individually.   (ii) Every week.

    (iii) Every second week.   (iv) Every third week.   **[OU - Mech 89]**

Answer : Individual replacement.

6. The probability P_n of failure just before age n are shown below. If individual replacement costs Rs. 1.25 and group replacement costs Re. 0.50 per item, find the optimal group replacement policy.

$n =$	1	2	3	4	5	6	7	8	9	10	11
$P_n =$	0.01	0.03	0.05	0.07	0.10	0.15	0.2	0.15	0.11	0.08	0.05

[JNTU (Mech.) 87]]

Answer : Replacement after every 5 weeks.

6. It has been suggested by a data processing firm that they adopt a policy of periodically replacing all the 1000 tubes in a certain piece of equipment. A given type of tube is known to have a mortality distribution (probability of failure) shown in the following table.

Tube failure / week	1	2	3	4	5
Probability of failure	0.3	0.1	0.1	0.2	0.3

The cost of replacing the tubes on an individual basis is estimated to be Re. 1.00 per tube. The cost of group replacement policy average Rs. 0.30 per tube. Compare the cost of preventive replacement with that of remedial replacement. **[JNTU (Mech.) 91, OU (EEE) 87]**

Answer : It is optimal to have a group replacement after every 4 week along with the individual replacement.

7. A decorative series lamp set circuit contains 10,000 bulbs, when any bulb fails it is replaced and the cost of replacing a bulb individually is Re. 1 only. If all the bulbs are replaced at the same time the cost per bulb would be reduced to 35 paise. The percent surviving say $s(t)$ at the end of month t, are given as

t	1	2	3	4	5	6
$s(t)$	97	90	70	30	15	0

Determine the optimal replacement policy.   **[JNTU (ECE) 93/S CCC]**

Answer : After 3rd month

8. A computer has a large number of electronic tube, that are subject to mortality as given below :

Period	Age of failure (hours)	Probability of failure
1	0-100	0.10
2	101-2	0.26
3	201-300	0.35
4	301-400	0.22
5	401-300	0.07

If the tubes are group replaced, the cost of replacement is Rs. 15 per tube. Group replacement can be done at fixed intervals in the night shift when the computer is not normally used. Replacement of individuals tubes which fails in services costs Rs. 60 per tube. How frequently should the tubes be replaced?

[JNTU (CSE) 92/S]

Answer : Minimum cost for group replacement is Rs. 18600 for an interval of 200 hours.

9. There are 1000 bulbs in the system survival rate is given below :

Week	0	1	2	3	4
Bulb in operation at the end of week	1000	850	500	200	100

The group replacement of 1000 bulbs costs Rs. 100 and individual replacement is Re. 0.50 per bulb. Suggest suitable replacement policy.   **[JNTU (Mech.) 90]**

Answer : One week group replacement policy ; Average cost = Rs. 175 per week; Cost of individual replacement / week = Rs. 196.

10. The following mortality rates have been observed for a certain type of light bulbs.

Week	1	2	3	4	5
Percent failure by end of week	10	35	58	80	100

There are 1000 bulbs in use and it cost Rs. 2 to replace an individual bulb which has burnt out. If all bulbs were replaced simultaneously it would cost 0.50 paise per bulb. It is proposed to replace all bulbs at fixed intervals whether or not they have burnt out and to continue replacing burnt out bulbs as they fail. At what interval should all bulbs be replaced.   **[Bangalore B.E. (Mech) 1984**
JNTU (FDH, Mech.) 92]

11. In a machine shop, a particular cutting tool cost Rs. 6 to replace. If a tool breaks on a job, the production disruption and associates with amount to Rs. 30. The past life of a tool is given as follows :

Job Number	1	2	3	4	5	6	7
Proportion of broken tools	0.01	0.03	0.04	0.13	0.25	0.55	0.95

[I.A.S 1989]

After how many jobs should the shop replace a tool before it breaks down.

Answer : Group replacement after a week.

12. The following mortality rates have been found for a certain type of coal cutter motor.

Weeks	10	20	30	40	50
Total %failure upto end of 10 weeks	5	15	35	65	100

If the motors are replaced over the week, the total cost is Rs. 200. If they fail during the week the total cost is Rs. 100 per failure. Is it better to replace the motors before failure and if so when?

Answer : Motors should be replaced every 20 weeks.

13. The mortality rates as obtained for an electronic component are noted in the table below :

Month	1	2	3	4	5	6
Percentage fails at the end of the month	8	22	45	70	85	100

There are 1500 items in operation. It costs Rs. 20 to replace an individual item and Re. 0.50 per item if all items are replaced simultaneously. It is decided to replace all items at fixed intervals and to continue replacing individual item as and when they fail. At what intervals should all items be replaced?

[JNTU (Mech.) 91]

14. The following failure rates have been observed for certain types of light blubs

Week	1	2	3	4	5
Percent failure by end of week	10	25	58	80	100

There are 1000 bulbs in use, and it cost Rs. 2 to replace an individual bulb which is burnt out. If all bulbs were replaced simultaneously it would cost 50 paise per bulb. It is proposed to replace all bulbs at fixed intervals, whether they have burnt out or not, and to continue replacing bulbs on and when they fail. At what intervals, all the bulbs should be replaced? **[JNTU - CSE/ECE 2001]**

Review Questions

1. What are the costs involved in failure and replacement analysis of equipment. Explain.

2. Discuss the effect of running age of machine on failures and machine behaviour with the aid of a graph (Bath Tub Curve). Where do you fit machine replacement in the graph.

3. Explain Machine Life Cycle, with reference to costs.

4. How does an equipment behave in the following stages? What type of maintenance do suggest?
   (a) Early stage when just bought and installed.
   (b) When it is producing at its highest rate.
   (c) When its expected life is almost completed.

5. When do you recommend replacement of a machine with a new one?

6. Discuss the effect of age on machine resale value and running cost with an aid of a graph.

7. Discuss the replacement policy when money value is not changing with time.

8. Write a note on replacement policy when money value is changing with time.

9. What do you understand by group replacement? When do you opt this policy?

10. Discuss the step by step procedure to decide the interval of group replacement.

11. Write a short note on group replacement with examples of its application.

12. Distinguish between the individual replacement and group replacement policies.

13. What policy do you recommend for the items that fail completely. Justify your answer with examples.

14. Discuss the use of probabilities in replacement policies.

15. Distinguish between repairable and non-repairable system with reference to replacement policies.

16. Discuss the importance of group replacement in the case of non-repairable systems.

17. Discuss different types of failures that occur on a machine.

18. Distinguish between gradual failures and sudden failures and their effects.

19. Discuss the following failures and suggest the maintenance policies to be adopted.
   (a) Early failures or infant failures.
   (b) Random failures or chance failures.
   (c) Wear-out failures or old-age failures.

20. Discuss the effect of money value on machine replacement.

Objective Type Objectives

1. When a machine is in infant stage, the major maintenance policy considered is
   (a) corrective maintenance   (b) preventive maintenance
   (c) contractual maintenance with OEM (d) condition monitoring

2. When money value changes with time at 10% then PWF for first year is

   (a) 0.909   (b) 1   (c) 0.826   (d) 0.9

3. Which of the following maintenance policy is not used in old age stage of a machine
   (a) operate to failure and corrective maintenance
   (b) renovation/reconditioning
   (c) replacement
   (d) scheduled preventive maintenance

4. When money value is changing with time @ 20%, the discount factor for 2nd year =

   (a) 1   (b) 0.833   (c) zero   (d) 0.6944

5. If the probability of failure of a machine is gradually decreasing, the failure mode is said to be
   (a) progressive   (b) retrogressive
   (c) regressive   (d) recursive

6. Which of the following replacement policy is probability model
   (a) when money value does not change with time where time is continuous variable
   (b) when money value does not change with time where time is a discrete variable
   (c) when money value changes with time
   (d) group replacement policy

7. Which of the following replacement policy is considered to be dynamic
   (a) when money value does not change with time where time is continuous variable
   (b) when money value does not change with time where time is descrete variable
   (c) when money value changes with time
   (d) none of the above

8. A machine is to be replaced if the average running cost
   (a) is not equal to current running cost
   (b) of current period is greater than that of next period
   (c) till current period is greater than that of next period
   (d) of current period is less than that of next period.

9. Which of the following is correct assumption for replacement policy when money value does not change with time

   (a) no initial cost   (b) scrap value is zero

   (c) scrap value is constant   (d) no maintenance cost

10. In replacement analysis the maintenance cost is the function of

    (a) present value   (b) time

    (c) maintenance policy   (d) resale value

11. Find the oddman out

    (a) present worth factor (PWF)   (b) discounted rate (DR)

    (c) depreciation value (DV)   (d) mortality tables

12. Which of the following is not used in group replacement policy decisions

    (a) failure probability   (b) loss due to failure

    (c) cost of individual replacement   (d) present worth factor

13. Reliability of an item is

    (a) failure probability   (b) 1 - failure probability

    (c) 1/ failure probability   (d) warranty period

14. Group replacement policy is applicable for

    (a) repairable items   (b) items that fail partially

    (c) items that fail completely   (d) dissimilar items

15. The maintenance cost are assumed to be mostly dependent on

    (a) calendar age   (b) running age

    (c) user's age   (d) manufacturer's age

16. Group replacement is not preferred for

    (a) large number of identical items

    (b) low cost items where record keeping may be costly or difficult

    (c) items that fail completely

    (d) repairable items

17. Which cost of the following is irrelevant to replacement analysis

    (a) capital (purchase) cost of machine

    (b) operating cost of machine

    (c) production cost of the machine

    (d) maintenance cost

18. Group replacement policy is most suitable for

    (a) blood group   (b) street lights

    (c) truck   (d) any of the above

19. Replacement problem arises when

    (a) old item becomes too expensive to operate or maintain

    (b) old item completely fails to work

    (c) a better design or efficient model is available that can increase rate of production

    (d) all the above

20. The type of failure that usually occurs in old age machine is

    (a) early failures   (b) random failure

    (c) wear-out failures   (d) chance failures

21. Decreasing failure rate is usually observed in _____ stage of the machine

    (a) infant   (b) youth

    (c) old age   (d) any

22. Chance failures that occur on a machine are commonly found on a graph of time (X-axis) Vs failure rate (Y axis) as

    (a) hyperbolic

    (b) parabolic

    (c) straight line nearly parallel to X axis

    (d) straight line nearly parallel to Y axis

23. The curve used to interpret machine life cycle is

    (a) learning curve   (b) bath tub curve

    (c) time series   (d) ogive curve

24. If machine becomes old, the failure rate expected will be _____

    (a) decreasing   (b) constant

    (c) increasing   (d) may increase or decrease

25. The production rate on a machine is expected to be highest or peak in _____ stage

    (a) infant   (b) youth   (c) old age   (d) any

26. A replacement decision is common in _____ stage

    (a) youth   (b) infant   (c) old age   (d) none

27. The preventive maintenance is expected to enhance _____ of the machine

    (a) infancy   (b) youth   (c) old age   (d) no stage

28. The replacement policy that is imposed on an item irrespective of its failure (i.e. it is replaced even though it has not failed) is _____

    (a) individual replacement   (b) group replacement

    (c) if money value is constant   (d) if money value varies with time

29. The replacement model that uses failure probability is _____

    (a) individual replacement
    (b) group replacement
    (c) replacement when money value is constant
    (d) replacement when money value is time dependent

30. Replacement stage is said to be reached if _____ is increasing

    (a) failure rate   (b) failure cost

    (c) failure probability   (d) any of these

Fill in the Blanks

1. When money value changes with time @ 20%, then present worth factor (PWF) during first year is taken as _____

2. When money value changes with time @ r then discount factor for n^{th} year = _____

3. There are two machines whose costs are calculated as follows.

   Machine - 1 : Discounted @ 10% twice a year

   Machine - 2 : Discounted @ 20% per annum.

   which machine is preferable _____

4. In retrogressive failures, the failure probability _____ with time

5. The probabilistic model among replacement policies is _____

6. The replacement policy when money value does not change with time can be placed under _____ models of OR classification

7. A machine is to be replaced if the average running cost of current period is _____ to that of _____ period

8. Scrap value is assumed to be _____ for replacement policy when money value does not change with time

9. The maintenance cost is function of _____

10. If failure probability of machine is 'f', then its reliability is _____

11. The replacement policy applicable for the items that completely fail is _____

12. The suitable replacement policy applicable for low cost, identical, large number of items is _____

13. The wear-out failures usually occur in _____ stage o machine

14. The highest and most efficient production rate is found in _____ stage of the machine life cycle

15. The replacement decision is a major concern for machine in _____ stage of machine life cycle

16. The shape of graphical representation of machine life cycle is _____

17. The shape of early failure rate in a machine life cycle graph is _____

18. The preventive maintenance is expected to increase the life of machine in _____ stage of the machine

19. The discount factor at the end of 1st year when money value changes @ 10% is _____

20. Contractual maintenance with original equipment manufacturer (OEM) is preferred in _____ stage of the machine

Answers

Objective Type Questions :				
1. (c)	2. (b)	3. (d)	4. (b)	5. (b)
6. (d)	7. (c)	8. (d)	9. (c)	10. (b)
11. (d)	12. (d)	13. (b)	14. (c)	15. (b)
16. (d)	17. (c)	18. (b)	19. (d)	20. (c)
21. (a)	22. (c)	23. (b)	24. (c)	25. (b)
26. (c)	27. (b)	28. (b)	29. (b)	30. (d)

Fill in the Blanks :	
1. 1	2. $(1 + r)^{-n+1}$
3. machine - 1	4. decreases
5. group replacement	6. static
7. less than or equal to ($\leq$), next	8. constant
9. running time or running age of machine	10. $1 - f$,
11. group replacement	12. group replacement
13. old age	14. youth
15. old age	16. bath tub
17. decreasing hyperbolically	18. youth
19. 0.909	20. Infant or early

11
Chapter

Game Theory

CHAPTER AT A GLANCE

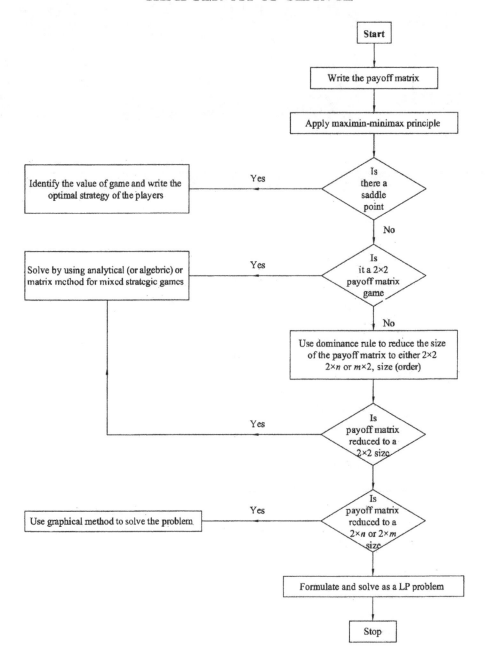

11.0 *Introduction*

In the competitive world, it is essential for an executive to study or at least guess the activities or actions of his competitor. Moreover, he has to plan his course of actions or reactions or counter actions when his competitor uses certain technique. Such war or game is a regular feature in the market and the competitors have to make their decisions in choosing their alternatives among the predicted outcomes so as to maximise the profits or minimising the loss.

The game theory has taken its importance in the business management in 1944, when Von Neumann published "Theory and practice of Games and Economic Behaviour".

11.1 *Theory of Games : The Terminology Used*

A competitive situation is called game, in this context.

11.1.1 *Properties of a Game*

1. There are finite number of competitors called *'players'*.

2. Each player has a finite number of possible courses of action called *'strategies'*.

3. All the strategies and their effects are known to the players but player does not know which strategy is to be choosen.

4. A game is played when each player chooses one of his strategies. The strategies are assumed to be made simultaneously with an outcome such that no player knows his opponents strategy until he decides his own strategy.

5. The game is a combination of the strategies and in certain units (usually in terms of money) which determines the gain (shown in positive figures) or loss (shown in negative figures).

6. The figures (either gain or loss) shown as the outcomes of strategies in a matrix form is called "*pay-off matrix*"

7. The player playing the game always tries to choose the best course of action which results in optimal pay off, called "*optimal strategy*".

8. The expected pay off when all the players of the game follow their optimal strategies is known as "*Value of the game*". The main objective of a problem of games is to find the value of the game.

9. The game is said to be *'fair game'* if the value of game is zero, otherwise it is known as *'unfair'*.

11.1.2 *Definitions of Terms Used*

1. **Startegy :** A startegy for a player has been defined as a set of rules or alternative courses of action available to him in advance, by which player decides the course of action that he should adopt.Startegy may be of two types :

   (a) *Pure Strategy :* If the players select the same strategy each time, then it is referred to as pure-strategy. In this case each player known exactly what the other is going to do i.e., there is a deterministic situation and the objective of the players is to maximize gains or to minimize losses.

   (b) *Mixed Strategy :* When the players use a combination of strategies and each player always kept guessing as to which course of action is to be selected by the other player at a particular occasion then this is known as mixed-strategy. Thus, there is a probabilistic situation and objective of the player is to maximize expected gains or to minimize losses. Thus, mixed strategy is a selection among pure strategies with fixed probabilities.

2. **Optimum Strategy :** A course of action or play which puts the player in the most preferred position, irrespective of the strategy of his competitors is called an optimum strategy. Any deviation from this strategy results in a decreased pay-off for the player.

3. **Value of the Game :** It is the expected pay-off of play when all the players of the game follow their optimum strategies. The game is called *fair* if the value of the game is zero and unfair if it is non-zero.

4. **Two-Person Zero-Sum Game :** There are two types of two person zero-sum games. In one, the most preferred position is achieved by adopting a single strategy and therefore the game is known as the *pure strategy game*. The second type requires the adoption by both players a combination of different strategies in order to achieve the most preferred position and is, therefore, referred to as the mixed strategy game.

5. **Payoff Matrix :** A two-person zero-sum game is conveniently represented by a matrix as shown below. The matrix, which shows the outcome of the game as the players select their particular strategies, is known as the *payoff matrix*. It is important to assume that each player knows not only his own list of possible courses of action but also of his opponent.

   Let player A have m courses of action $(A_1, A_2, \ldots A_m)$ and player B has n courses of action $(B_1, B_2, \ldots B_n)$. The number n and m need not be equal. The total number of possible outcome is, therefore $(m \times n)$. These outcomes are shown in the following table.

$$
\begin{array}{c}
\textit{A's payoff matrix} \\
\textit{Player B}
\end{array}
\qquad\qquad
\begin{array}{c}
\textit{B's payoff matrix} \\
\textit{Player B}
\end{array}
$$

$$
\text{Player } A \;
\begin{array}{c}
A_1 \\ A_2 \\ \vdots \\ A_m
\end{array}
\begin{bmatrix}
B_1 & B_2 & \cdots & B_n \\
a_{11} & a_{12} & \cdots & a_{1n} \\
a_{21} & a_{22} & \cdots & a_{2n} \\
\vdots & \vdots & \vdots & \vdots \\
a_{m1} & a_{m2} & \cdots & a_{nm}
\end{bmatrix}
\; ; \quad
\text{Player } A \;
\begin{array}{c}
A_1 \\ A_2 \\ \vdots \\ A_m
\end{array}
\begin{bmatrix}
B_1 & B_2 & \cdots & B_n \\
-a_{11} & -a_{12} & \cdots & -a_{1n} \\
-a_{21} & -a_{22} & \cdots & -a_{2n} \\
\vdots & \vdots & \vdots & \vdots \\
-a_{m1} & -a_{m2} & \cdots & -a_{nm}
\end{bmatrix}
$$

11.2 *Types of Games*

Games can be classified into different categories with reference to various aspects as given below.

11.2.1 *With Reference to Number of Players*

With reference to the number of players playing the game, the games are divided into two categories Viz. (1) Two-person games and (2) Multi-person games.

1. **Two Person Games :** If only two players are playing the game then the game is said to be two person game. The rows and column in the pay off matrix represent the strategies of these two players.

   e.g. : (a) Chess players

   (b) Marketing strategies of Coke and Pepsi in India.

2. **Multi-person Game :** It is a game in which more than two persons play.

   e.g. : Competition among tooth paste producers, soap producers.

11.2.2 *With Reference to Nature of Out Comes*

1. Zero-sum game.   2. Non-zero sum game.

1. **Zero-sum Game :** If loss of a player is gain of other player and vice-versa, then the game is called zero-sum game.

2. **Non-zero Sum Game :** In contrast to the above, if a gain of player is not a loss of another player (positive sum game) or loss of a player is not a gain of another player (negative sum game), then such games are non-zero sum games.

   e.g. : Suppose there are two players, say Colgate and Pepsodent. In a colony there are 1000 people who either use Colgate or Pepsodent. Now, if say 20 people have changed their brand from Colgate to Pepsodent due to effective advertisement then the loss of Colgate is gain for Pepsodent. This is a zero-sum game.

   If a group of 30 people has stopped using any toothpaste and use neem-sticks. Here loss of any player is not a gain to other player. This game is non-zero sum (negative sum) game.

11.2.3 *With Reference to Certainty*

1. Deterministic games.   2. Probabilistic games.

1. **Deterministic Games :** If the game yields a solution with one certain single strategy (pure strategy) for each player, the game is deterministic. In this game the strategies are pure and saddle points exist.

2. **Probabilistic Games :** If a player adopts more than one strategy with some probabilities or prepares a blend of two or more strategies, these games are said to be probabilistic. Here. mixed strategies are used as no saddle points exist.

11.2.4 *With Reference to Value of the Game*

    1. Fair game   2. Unfair game

    1. Fair Game : If the value of the game is zero, i.e., neither player wins nor loses (drawn), the game is said to be fair.

    2. Unfair Game : If one of the player wins (other loses), or the value of the game is non zero (may be positive or negative), it is unfair.

    The following is the classification of games shown with a chart.

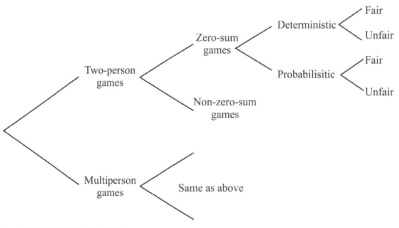

FIGURE 11.1 : CLASSIFICATION OF GAMES

11.3 *Two Person-zero Sum Games or Rectangular Games*

    A game with only two persons with a condition that the loss of one player is gain of the other and vice-versa (so that total sum is zero) is said to be two-person zero-sum game or rectangular game.

    Pay Off Matrix : The qualitative measures of returns of the players in terms of gains or losses when players select their particular strategies can be represented in the form of a matrix in m rows and n columns, called pay off matrix. The m rows are the m courses of action of one player while n columns represent the strategies of the other player. Thus one player say A has m strategies $A_1, A_2, A_3 \ldots A_m$ and player B has n strategies as $B_1, B_2, B_3 \ldots B_n$. When each player chooses one of his strategies say A_i and B_j, then the expected out come x_{ij} is represented as the elements of the matrix. It is a general convention to read the pay off matrix with reference to row player (here A) always. Thus a positive sign of x_{ij} is gain to A or loss to B while the negative sign of x_{ij} (i.e., $- x_{ij}$) is loss of A or gain of B.

Obviously player '*A*' wishes to maximise his gains while player *B* wishes to minimise his loss.

An exemplary pay of matrix of two-person zero-sum game is shown below.

(x_{11} is pay off when *A* uses A_1 strategy and *B* uses B_1 strategy x_{12} is pay off when *A* applies A_1 and *B* applies B_2 strategies and so on).

$$
\begin{array}{c}
\text{Player } B\text{'s strategies}\\
\begin{array}{cccccc}
& B_1 & B_2 & B_3 & \cdots & B_n
\end{array}\\
\begin{array}{c}
A_1\\ A_2\\ A_3\\ \cdot\\ \cdot\\ \cdot\\ A_m
\end{array}
\left[
\begin{array}{cccccc}
x_{11} & x_{12} & x_{13} & \cdots & x_{1n}\\
x_{12} & x_{22} & x_{32} & \cdots & x_{2n}\\
x_{13} & x_{23} & x_{33} & \cdots & x_{3n}\\
\cdot & \cdot & \cdot & \cdots & \cdot\\
\cdot & \cdot & \cdot & \cdots & \cdot\\
\cdot & \cdot & \cdot & \cdots & \cdot\\
x_{m1} & x_{m2} & x_{m3} & \cdots & x_{mn}
\end{array}
\right]
\end{array}
$$

Player *A's* strategies

11.3.1 *Assumptions of the Game*

1. Each player has finite number of possible courses of actions (strategies) with him.

2. The list of strategies of each player need not be same.

3. Player in Rows attempts to maximise his gains while player in columns tries to minimise his losses.

4. The decision of both players are made individually prior to the play with out any communication between them.

5. The decision is supposed to be made simultaneously and also announced simultaneously so that neither player has any advantage over the other due to the direct knowledge of the other player's decision.

6. Both players know their own pay offs as well as the pay off of each other.

7. The positive sign of the pay off indicates the gain to row player or loss to column player and negative sign indicates loss to row player or gain to column player.

11.3.2 *Solving Two Person-zero Sum Games*

Two person zero sum games may be deterministic or probabilistic. The deterministic games will have saddle points and pure strategies exist in such games. In contrast, the probabilistic games will have no saddle points and mixed strategies are taken with the help of probabilities. These are explained in following sections.

Games With Saddle Points - Deterministic Games - with Pure Strategies :

These games can be solved by using one of the following methods.

1. Minimax and Maximin principle.
2. Dominance principle.

Minimax and Maximin Criteria :

In a given pay off matrix of two players, it is evident that the row player will maximise his minimum profits, thus applies maxi-min principle. Similarly, the column player attempts to minimise his maximum loss using his maxi-max rule.

For example for player A (in rows), minimum value in each row represents his least gain when he chooses to play that particular strategy. In other words, it is the minimum gain he can be confident, what so ever the strategy that his opponent uses. Thus in each row, when such minimal values are calculated, among all these least gains, he wishes to select the maximum of these values. This selection of maximum gains (largest value) among the row minimum values is referred to as Maxi-Min principle.

Similarly, for player in columns (say B) who is assumed to be loser, the maximum value of each column denotes the maximum loss to him if he adopts that particular columns strategy. In other words, it is the worst case of B for which he must be prepared. These are written beneath the pay off matrix called column maxima from which he can choose least (to minimise his losses). This is called Mini-max principle. If maximum value of row minima i.e., maximin value of row player is equal to the minimax value of column player (i.e., minimum value of column maxima), then the game is said to have '*Saddle (Equilibrium) Points*', and the corresponding strategies are said to be '*Optimal Strategies*'. The value of the pay off (say V) at this saddle (equilibrium) point is known as '*Value of the Game*'. A game may have more than one saddle point but the value will be same.

Steps To Find Saddle Points by Applying Minimax-Maximin Criteria :

Step 1 : Write pay off matrix.

Step 2 : Select the minimum value in each row of the pay off matrix under the head row minimum written at right end of each row put a circle ($\bigcirc$) over these elements in the pay off matrix.

Step 3 : Select largest among the row minima found from step-2 and write beneath the row minima.

Step 4 : Select largest value (maximum) of each column and put a box or rectangle ($\square$) over these values in the pay off matrix. Write these values of column maximum beneath the pay off matrix.

Step 5 : Select minimum value of column maxima (found in step-4) and write this at the right end.

Step 6 : If Max $(R_{\min})$ = Min $(C_{\max})$ i.e., value in step - 3 equals value in step-5 then saddle point exits. Thus saddle point is found at the element on which both circle and box are enclosed. Note this pay off element as the value of the game.

This is illustrated with the following numerical example.

ILLUSTRATION 1 ————————————————————————————

Players Adithi and Sahithi haue a competition to reach college early. Each has two strategies viz going by car or going by a two-wheeler. If both go by cars, Sahithi reaches 5 minutes early but if both go by two-wheelers, Adithi can reach 4 minutes early. If Adithi goes by car and Sahithi by a two wheeler, Adithi reaches 2 minutes early while Adithi by two wheeler and Sahithi by car keeps Sahithi 7 minutes early. If both can choose their strategies independently and simultaneously and know these estimated pay offs each other, find who will win. What will be outcome of the game and their optimal choices.

Solution :

Step 1 : *To Write Pay Off Matrix :* Let us first formulate the information into the pay off matrix.

Assume Adithi is the row player and Sahithi is column player. With this assumption, we can say that gains to Adithi are positive and that to Sahithi are negative.

The pay offs for various alternatives are as follows.

Adithi by car when Sahithi by car is *-5* (since Sahithi wins by 5 min).

Adithi by car when Sahithi by two wheeler is *2* (since Adithi wins by 2 min).

Adithi by two wheeler when Sahithi by car is *-7* (since Sahithi wins by 7 min)

Adithi by two wheeler while Sahithi by two wheeler is *4* (since Adithi wins by 4 min).

Now, these pay offs are arranged in a matrix form as follows :

		Sahithi	
		Car	Two Wheeler
Adithi	Car	– 5	2
	Two Wheeler	– 7	4

Step 2 : Find row minima and encircle these values.

		Sahithi		R_{min}
		Car	Two Wheeler	
Adithi	Car	(– 5)	2	– 5
	Two Wheeler	(– 7)	4	– 7

Step 3 : Find maximum of row minima

Step 4 : Find column maxima and enrectangle these elements.

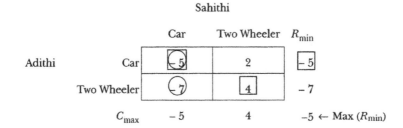

Step 5 : Find Minimum of column maxima

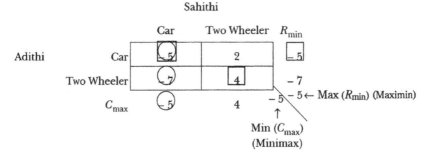

As Maximin = Minimax or Max (R_{min}) = Min (C_{max}) the saddle point exists at first row-first column element which is both encircled and enrectangled.

The Optimal Strategies :

                For Adithi   —   TO GO BY CAR

                For Sahithi —   TO GO BY CAR

Value of the game = -5

Which means Sahithi will win by 5 minutes or Adithi will lose by 5 minutes.

[It is a deterministic (pure strategies) unfair game]

ILLUSTRATION 2

> *Solve the following game*
>
> | | | B_1 | B_2 | B_3 | B_4 |
> |---|---|---|---|---|---|
> | | A_1 | -5 | -2 | 0 | 7 |
> | A | A_2 | 5 | -6 | -4 | 8 |
> | | A_3 | 4 | 0 | 2 | 3 |

Solution :

Finding R_{min} and then maximim among R_{min}

		B_1	B_2	B_3	B_4	R_{min}
A	A_1	-5	-2	0	7	-5
	A_2	5	-6	-4	8	-6
	A_3	4	0	2	3	0

0

Find C_{max} and then min of C_{max}

		B_1	B_2	B_3	B_4	R_{min}
A	A_1	-5	-2	0	7	-5
	A_2	5	-6	-4	8	-6
	A_3	4	0	2	3	0 $\leftarrow$ Max (R_{min})
C_{max}		5	0	2	8	0

$\uparrow$
Min (C_{max})

As Max (R_{min}) = Min (C_{max})

Saddle point exists at A_3, B_2

$\therefore$ A uses pure (deterministic) strategy A_3

       B uses pure (deterministric) strategy B_2

       and value of the game $(v) = 0$

i.e.,   Game is drawn (i.e., neither A nor B wins)

       (This is a fair game as $v = 0$)

11.4 *Dominance Principle*

The dominance principle is used to reduce the order of the pay off matrix. Thus if saddle points exist, the matrix can be reduced to a matrix with one element which is equal to value of the game.

Similarly, we can say that the dominance principle is used to delete rows and/ or columns of the pay off matrix which are inferior (i.e., less attractive) to at least one of the remaining rows and / or columns (strategies) in terms the pay offs. The deleted rows or columns would never be used by both the players for determining their optimum strategy.

The dominance principle is composed of the following rules.

Rule 1 : Row Dominance :

For a row player, in a pay off matrix, if every element in a particular row is less than or equal to the corresponding element of another row, then the former row is said to be inferior or dominated by the latter. Therefore the row player will never employ the former row (strategy) for whatever strategy that is used by his opponent. Hence this row can be deleted from the pay off matrix for further iteration.

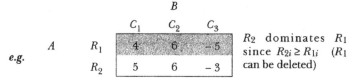

In the above pay off every element in R_1 is less than or equal to its corresponding element in R_2. $(4<5, 6=6$ and $-5<-3)$. Therefore R_1 can be deleted since row player (A) will never use R_1 for whatever strategy B uses among C_1, C_2, and C_3.

Note : The rule can not be applied even if one element does not obey rule.

Rule - 2 : Column Dominance :

For a column player, in a pay off matrix if every element in a column is greater than or equal to corresponding element of another column, the former column strategy never yields any better result than the latter and therefore the former column will never be used whatsoever may be the opponent's strategy. Hence the former column can be deleted from the pay off matrix for further iteration.

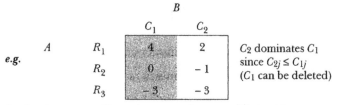

In the above pay off matrix, every element of C_1 is either greater than or equal its corresponding element of C_2. This suggests that, first column is inferior to the second in any case and hence deleted.

Rule 3 : Modified Row Dominance :

When no single row (pure) strategy has dominance over the other, then the comparison may be made between a row and the average of a group of rows. If every element is less than or equal to average of corresponding elements of a group, then the former row can be removed. If a row dominates over the average of group of rows, then the group may be discarded.

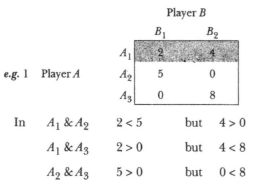

$e.g.$ 1   Player A

In $A_1 \& A_2$ $2 < 5$ but $4 > 0$

$A_1 \& A_3$ $2 > 0$ but $4 < 8$

$A_2 \& A_3$ $5 > 0$ but $0 < 8$

Thus there is no pure row dominance.

But by comparing the A_1 average values of A_2 and A_3 we have,

From $A_2 \& A_3$ $\dfrac{5+0}{2} > 2$ of A_1 and $\dfrac{0+8}{2} = 4$ of A_1

Therefore the row A_1 is inferior to average of A_2 and A_3. Hence A_1 can be deleted thus we reduce matrix to $\begin{bmatrix} 5 & 0 \\ 0 & 8 \end{bmatrix}$

                    B

                B_1 B_2

        A A_1 8   3

$e.g.$ 2 A_2 15 -1

        A_3 1   4

In the above pay off, between

$A_1 \& A_2$ $8 < 15$ but $3 > -1$

$A_1 \& A_3$ $8 > 1$ but $3 < 4$

$A_2 \& A_3$ $15 > 1$ but $-1 < 4$

Thus there is no pure dominance.

But comparing average of $A_2 \& A_3$ with A_1, we have

From A_2 & A_3 $\dfrac{15+1}{2} = 8 \le 8$ of A_1,

From A_2 & A_3 $\dfrac{-1+4}{2} = \dfrac{3}{2} \le 3$ of A_1,

Therefore the average is dominated by A_1 and hence both A_2 and A_3 are deleted leaving the matrix as $\begin{bmatrix} 1 & 4 \end{bmatrix}$

Rule - 4 : Modified Column Dominance :

Similar to the above rule, when no pure column dominance exists, a convex linear combination i.e., average of certain pure strategies may be compared with another pure strategy applying rule 2, i.e., if every element in a column is greater than or equal to corresponding element of other column the former is omitted.

Player B

		B_1	B_2	B_3	
Player A	A_1	8	15	1	$8 = \dfrac{15+1}{2}$
e.g.	A_2	3	-1	4	$3 > \dfrac{-1+4}{3}$

Clearly, their is neither pure column nor pure row dominance exist in the above pay off. But the average (convex linear combination) of B_2 and B_3 is greater than or equal to corresponding element of first column. Therefore B_1 can be deleted.

Thus we get,

$$\begin{bmatrix} 15 & 1 \\ -1 & 4 \end{bmatrix}$$

These rules used to solve the games are illustrated through following numerical examples.

ILLUSTRATION 3

Let us the take the game of illustration - 1 and solve by using dominance :
After formulating we get

Sahithi

		Car	Two-wheeler	
Adithi	Car	-5	2	$C_{2j} > = C_{1j}$ (*C_2 can be deleted*)
	Two-wheeler	-7	4	

Solution :

Now, from Sahithi's point of view, she never uses her second strategy i.e., two-wheeler since, for what ever Adithi's strategy, she will be the loser. Therefore her car strategy dominates over two-wheeler strategy (every element of first column is less than or equal to corresponding element of second column). Hence Sahithi's two-wheeler strategy is deleted.

Thus we get,

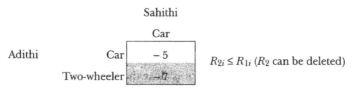

Sahithi

Car

Adithi   Car -5 $R_{2i} \le R_{1i}$ (R_2 can be deleted)

Two-wheeler -7

Now, as Adithi has her better strategy with car (her loss is less here), she will never got for two-wheeler strategy. (Every element in first row is greater than or equal to the corresponding element in the second row). Hence second row is deleted.

Now the game is reduced to a single element matrix i.e., *both using car strategy* with value of the game as **- 5**. i.e., Sahithi wins by 5 units.

ILLUSTRATION 4

An engineering student was frequently absent to the classes in a semester. To safe guard himself, he can choose one of the alternatives given below and the professor also had four strategies. The student has approximated the probable percent of marks in the following pay off matrix against various strategies.

The students strategies are showing reasons as

S_1*: due to ill health* S_2*: to attend sister's marriage are* S_3 *: went on project work* S_4 *: attended inter college celebrations. The professor's strategies are* P_1*: Not giving attendance* P_2*: Giving exam tough* P_3*: Evaluating strictly* P_4*: Complaining to principal.*

The pay of is

	P_1	P_2	P_3	P_4
S_1	55	53	32	62
S_2	40	30	74	50
S_3	57	54	44	53
S_4	54	54	72	56

Use dominance principle so that the student may choose his optimal strategy.

Solution :

In the given pay off, by examining thoroughly we find that every element of P_2 column is less than or equal to the corresponding element of P_1 column, hence professor will never use the P_1 strategy (P_2 dominates P_1), hence P_1 is deleted. The pay off is then reduced to

	P_1	P_2	P_3	P_4
S_1	55	53	32	62
S_2	40	30	74	50
S_3	57	54	44	53
S_4	54	54	72	56

deleted

$P_{1j} \geq P_{2j}$ (P_2 dominates and P_1 is deleted). After P_1 is deleted S_3 can be deleted in the next iteration, Since $S_{3i} \leq S_{4i}$

From the above pay off, it is clear that, the student will never use S_3 because every element of S_4 is greater than or equal to corresponding element of S_3 (S_4 dominates S_3). Hence by using the principle of dominance we delete S_3 and revise the pay of as

	P_2	P_3	P_4
S_1	53	32	62
S_2	30	74	50
S_4	54	72	56

$P_{4j} \geq P_{2j}$ (P_4 is deleted)

Again P_2 dominates P_4, hence P_4 is deleted, we get

	P_2	P_3
S_1	53	32
S_2	30	74
S_4	54	72

$S_{1i} \leq S_{4i}$ (S_1 is deleted)

Now, S_4 dominates S_1, hence S_1 is deleted.

	P_2	P_3
S_2	30	74
S_4	54	72

$P_{3j} \geq P_{2j}$ (P_3 can be deleted)

Now, As P_2 dominates P_3 , P_3 is deleted

	P_2
S_2	30
S_4	54

$S_{2i} \leq S_{4i}$ (S_2 can be deleted)

Now, S_4 dominates S_2, hence S_2 is deleted. Thus student uses S_4 strategy while professor uses P_2 strategy i.e. The professor gives the exam tough when student claims that his absence is due to attending intercollege celebrations and the value of the game is 54.

ILLUSTRATION 5

Let us consider the example in illustration - 2.

$$B$$

		B_1	B_2	B_3	B_4
A	A_1	-5	-2	0	7
	A_2	5	-6	-4	8
	A_3	4	0	2	-3

Solution :

In this game, on observing we can notice the dominance of B_2 over B_3 and B_4. (by Rule - 2)

∴ The game reduces to

$$B$$

		B_1	B_2
A	A_1	-5	-2
	A_2	5	-6
	A_3	4	0

Now, A_3 dominates over A_1 and A_1 is deleted, So, we get

$$B$$

		B_1	B_2
A	A_2	5	-6
	A_3	4	0

Now, as B_2 dominates B_1 and so, we delete B_1

$$B$$

		B_2
A	A_2	-6
	A_3	0

As A_3 is dominating A_2 we have the value of game as '**zero**'.

i.e., Fair game (drawn)

A uses A_3 strategy and B uses B_2 strategy.

Modified Dominance :

ILLUSTRATION 6 ——————————————————————————————

Use dominance principle to reduce the following game to 2×2 game, Is the game stable.

Player B

		B_1	B_2	B_3	B_4
Player A	A_1	6	−10	9	0
	A_2	6	7	8	1
	A_3	8	7	15	1
	A_4	3	4	−1	4

Solution :

We can easily verify that the above game has no saddle points. Therefore it is *not stable*.

Now using row dominance, the third row A_3 dominates over A_1 as well as A_2. Therefore A_1 and A_2 are deleted (shaded in the pay off matrix of the problem given above).

Thus the play reduces to

Player B

		B_1	B_2	B_3	B_4
Player A	A_3	8	7	15	1
	A_4	3	4	− 1	4

Now, we can delete second column (B_2 strategy) as it is dominated by fourth strategy (every element is less than or equal to corresponding element).

Player B

		B_1	B_3	B_4
Player A	A_3	8	15	1
	A_4	3	− 1	4

Now, there is no pure dominance, however we can apply modified dominance with a convex combination of B's B_3 and B_4 strategies dominates his first strategy.

$$\frac{1}{2}(15 + 1) = 8 \le 8$$

$$\frac{1}{2}(-1 + 4) = \frac{3}{2} \le 3$$

Therefore B's B_1, strategy is deleted.

Thus the game reduces to

Player B

		B_3	B_4
Player A	A_3	15	1
	A_4	-1	4

This game can be solved by using either algebraic or any other suitable method, which will be explained in the sections to follow.

11.5 *Games Without Saddle Points: Mixed Strategy*

In certain cases, no pure strategy solutions exist for the game. In other words, saddle points do not exist. In all such game, both players may adopt an optimal blend of the strategies called Mixed strategy to find a saddle (equilibrium) point. The optimal mix for each player may be determined by assigning each strategy a probability of it being chosen. Thus these mixed strategies are probabislitic combinations of available better (or dominant) strategies and these games hence called probabilistic games.

Summarily, when saddle points do no exist in the game, we can use the dominance principle to reduce to 2×2 game or at least $2 \times n$ or $m \times 2$ matrices. If this is also not possible we can formulate it as Linear programming problem.

Thus the probabilistic mixed strategy games without saddle points are commonly solved by any of the following methods :

S.No.	Method	Applicable to
1.	Algebraic method	2×2 games
2.	Graphical method	2×2, $m \times 2$ and $2 \times n$ games
3.	Linear programming method	2×2, $m \times 2$, $2 \times n$ and $m \times n$ games.

These are explained here below :

1. Algebraic Method :

Suppose a general game without saddle points as given below :

Player B

$$\text{player } A \quad \begin{array}{c} A_1 \\ A_2 \end{array} \begin{bmatrix} \begin{array}{cc} B_1 & B_2 \\ a & b \\ c & d \end{array} \end{bmatrix}$$

Since it is assumed that there are no saddle points, A wishes to use a mixed strategy of A_1 and A_2 with the probabilities of p_1 and p_2, and B to use a mixed strategy of B_1 and B_2 with the probabilities of q_1 and q_2 respectively.

According to the assumption that A always tries to maximise his minimum returns and B tries to minimise his maximum losses, they employ their strategies most optimally such that the $B's$ minimax losses will be equal $A's$ maximin gains.

Hence, out come of probability mix (of p_1 and p_2) by A in his A_1 and A_2 strategies when B uses his B_1 strategy i.e., $ap_1 + cp_2$ is the value of the game.

Otherwise, when B uses his B_2 strategy, $A's$ returns will be $ap_1 + dp_2$, is also the value of the game. These both must be equal.

$$\therefore \qquad ap_1 + cp_2 = bp_1 + dp_2$$

$$(a - b)\, p_1 = (d - c)\, p_2$$

$$\frac{p_1}{p_2} = \frac{d - c}{a - b} \qquad\qquad \dots\text{(i)}$$

Similarly, from $B's$ point of view

$$aq_1 + bq_2 = cq_1 + dq_2$$

i.e., $\qquad (a - c)\, q_1 = (d - b)\, q_2$

$$\therefore \qquad \frac{q_1}{q_2} = \frac{(d - b)}{(a - c)} \qquad\qquad \dots\text{(ii)}$$

From the rule of probabilities, we know that $p_1 + p_2 = 1$ and $q_1 + q_2 = 1$

Thus we have

$$p_1 = 1 - p_2 \quad \text{or} \quad p_2 = 1 - p_1 \qquad\qquad \dots\text{(iii)}$$

and $\qquad q_1 = 1 - q_2 \quad \text{or} \quad q_2 = 1 - q_1 \qquad\qquad \dots\text{(iv)}$

Thus we have (on solving the equations)

$$p_1 = \frac{(d - c)}{(d - c) + (a - b)} = \frac{(c - d)}{(c - d) - (a - b)} = \frac{(c - d)}{(b + c) - (a + d)}$$

and $\quad q_1 = \dfrac{(d - b)}{(d - b) + (a - c)} = \dfrac{(b - d)}{(b - d) - (a - c)} = \dfrac{(b - d)}{(b + c) - (a + d)}$

Where $p_2 = 1 - p_1,\ q_2 = 1 - q_1,$

(and $v = ap_1 + cp_2$ or $bp_1 + dp_2$ or $aq_1 + bq_2$ or $cq_1 + dq_2$)

i.e., $\qquad v = \dfrac{ad - bc}{(a + d) - (b + c)}$

This method is illustrated with the following numerical example.

* Illustration - 6 is continued :

Previous to this section, the problem in Illustration - 6 was reduced from 4×4 to 2×2 and left unsolved there.

Now we can use the above formulae and find the solution.

Player B

		B_3	B_4
A_3		15	1
A_4		-1	4

Player A

Assuming the probabilities p_1 & p_2 for A and q_1 & q_2 for B to use A_3, A_4 and B_3, B_4 respecitvely, we have

$$p_1 = \frac{-1-4}{(-1+1)-(15+4)} = \frac{5}{19}, p_2 = \frac{14}{19}$$

$$q_1 = \frac{1-4}{(-1+1)-(15+4)} = \frac{3}{19} ; q_2 = \frac{16}{19}$$

and $\quad V = \frac{(15)(4)-(-1)(1)}{(15+4)-(-1+1)} = \frac{61}{19}$

Thus A uses his A_3, A_4 strategies with probabilities $\left(\frac{5}{19}, \frac{14}{19}\right)$ while B uses his B_3, B_4 strategies with probabilities $\left(\frac{3}{19}, \frac{16}{19}\right)$ and the value of the game $= \frac{61}{19}$ i.e., A wins.

ILLUSTRATION 7 ─────────────────────────────

Children Srija and Himaja play a game who have some 25 paise coins and 50 paise coins. Each draw a coin from their bags with out knowing other's choice. If the sum of coins drawn by both is even Srija wins them, otherwise Himaja wins.

Find the best strategy for each player and also find the value of the game.

Solution :

First let us formulate the game. According to the data given i.e., each has some 25 paise coins and some 50 paise coins in their bags and draw one at a time.

Now, from this we can say that each has two strategies.

Strategy 1 : Drawing 25 paise coins.

Strategy 2 : Drawing 50 paise coins.

Let us choose Srija as row players and Himaja as column player.

Now if Srija draws 25 paise (say strategy S_1) and Himaja draws 25 paise (say, strategy H_1), then the sum is even. Therefore Srija wins them. Her gains (pay off) is 25 paise.

Similarly, if Srija draws 25 paise (strategy S_1) and Himaja draws 50 paise (say strategy H_2) then Himaja wins these coins and her gain (pay off) is 25 paise or otherwise Srija's loss i.e., -25 paise.

Again, if Srija draws 50 paise (strategy S_2) and Himaja draws 25 paise (strategy H_1), then Himaja wins 50 paise or Srija's loss i.e., - 50 paise.

Also, when both draw 50 paise (S_2 and H_2) Srija wins 50 paise.

Thus, the pay off is as follows.

Himaja

		H_1 (25 paise)	H_2 (50 paise)
Srija	S_1 (25 paise)	25	- 25
	S_2 (50 paise)	- 50	50

On examination we can find that there are no saddle points, therefore the players use their mixed strategies, say with the probabilities as p_1 and p_2 for Srija on S_1 and S_2 respectively, and as q_1 and q_2 for Himaja on H_1 and H_2 respectively.

Thus we have,

$$25\,p_1 - 50\,p_2 \qquad = -25\,p_1 + 50\,p_2 \qquad \ldots \ldots (1)$$
$$25\,q_1 - 25\,q_2 = -50\,q_1 + 50\,q_2 \qquad \ldots \ldots (2)$$

Also $\qquad p_1 + p_2 = 1 \qquad \ldots \ldots (3)$

$\qquad\qquad q_1 + q_2 = 1 \qquad \ldots \ldots (4)$

From (1) $\qquad 50\,p_1 = 100$

$$\frac{p_1}{p_2} = \frac{2}{1} \qquad \ldots \ldots (5)$$

From (3) and (5)

We get $\qquad p_1 = \dfrac{2}{3}; \quad p_2 = \dfrac{1}{3}$

Also from (2) we have

$$75\,q_1 = 75\,q_2$$

$$\frac{q_1}{q_2} = 1 \qquad \ldots \ldots (6)$$

From (4) and (6)

We get $\qquad q_1 = \dfrac{1}{2} \quad q_2 = \dfrac{1}{2}$

Thus Srija is expected to use her strategies S_1 and S_2 (i.e., Drawing 25 paise and 50 paise coins respectively) in probabilities of (2/3, 1/3).

While for Himaja it is optimal to use her strategies with (1/2, 1/2) probability of H_1 and H_2, i.e., drawing 25 paise and 50 paise coins respectively.

The value of the game is zero.

And it is fair game ($V = 0$)

$$[\text{Since} \quad V = 25 \times \frac{1}{2} - 25 \times \frac{1}{2} = 0$$

$$V = -50 \times \frac{1}{2} + 50 \times \frac{1}{2} = 0$$

or $\qquad V = 25 \times \dfrac{2}{3} - 50 \times \dfrac{1}{3} = -25 \times \dfrac{2}{3} + 50 \times \dfrac{1}{3} = 0]$

ILLUSTRATION 8

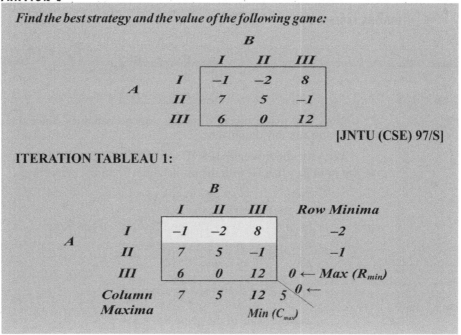

Find the best strategy and the value of the following game:

		B		
		I	*II*	*III*
A	*I*	−1	−2	8
	II	7	5	−1
	III	6	0	12

[JNTU (CSE) 97/S]

ITERATION TABLEAU 1:

		B			Row Minima
		I	*II*	*III*	
A	*I*	−1	−2	8	−2
	II	7	5	−1	−1
	III	6	0	12	0 ← Max (R_{min})
Column Maxima		7	5	12	5

Min (C_{max})

Solution :

By minimax (maximin) principle, we have no saddle points as Max (R_{min}) ≠ Min (C_{max}).

Hence, we check the principle of dominance. It is found that Row AIII is dominating AI, since every element in A(III) is > the corresponding element of A(I).

Therefore A will never use the strategy A(I), hence deleted. Now, the new iterated pay off is

ITERATION TABLEAU 2 :

		B		
		I	II	III
A	II	7	5	− 1
	III	6	0	12

The column B(II) dominates B(I) since every element in B(II) is less than its corresponding element of B(I). Hence B(I) is deleted (B will never use B–I as it yields more losses against any strategy played by A).

ITERATION TABLEAU 3 :

$$B$$

		II	III
A	II	5	-1
	II	0	12

Now, they play a mixed strategies with probabilities. Since there are neither saddle points nor any dominance.

Let A use the strategies II & III with probabilities p_1 and p_2 respectively and B use the strategies II & III with the probabilities q_1 and q_2 respectively.

then $\qquad p_1 + p_2 = 1; q_1 + q_2 = 1$

And from the pay off we know that

$$5p_1 + op_2 = -1p_1 + 12p_2 \text{ and } 5q_1 - q_2 = 0.q_1 + 12q_2$$

$$\therefore \quad \frac{p_1}{p_2} = 2 \text{ and } \frac{q_1}{q_2} = \frac{13}{5}$$

$$\therefore \quad p_1 = \frac{2}{3}; p_2 = \frac{1}{3} \text{ and } q_1 = \frac{13}{18}; q_2 = \frac{5}{18}$$

and the value of the game is $5 \times \dfrac{2}{3} = \dfrac{10}{3}$.

(by substituting in any of the above probability expressions).

Thus, the mixed strategies of A are (II, III) with the probabilities $\left(\dfrac{2}{3} \text{ and } \dfrac{1}{3}\right)$.

the mixed strategies of B are (II and III) with the probabilities (13/18 and 5/18).

The expected value of the game is $\dfrac{10}{3}$ (A wins the game).

ILLUSTRATION 9

A and B play a game in which each has three coins a 5P, a 10P and 20P. Each selects a coin without the knowledge of the others choice. If the sum of the coins is an odd amount, A wins B's coins. If the sum is even B wins A's coins. Find the best strategy for each player and the value of the game.

[JNTU (Mech.) 99/CCC, (ECE) 99/CCC]

Solution :

When A draws 5p, if B draws 5p, then B gains 5p of A, if B draws 10 p, A wins 10 p and if B draws 20 p then also A wins 10p and so on.

Thus, the pay off matrix for player A is formulated as :

			Player B		
			5p	10p	20p
			B_1	B_2	B_3
Player A	5p	A_1	-5	10	20
	10p	A_2	5	-10	-10
	20p	A_3	5	-20	-20

It is clear that this game has no saddle point. Therefore, further we try to reduce the size of the given pay off matrix. Note that every element in column B_3 is more than or equal to every corresponding element of column B_2. Evidently, the choice of strategy B_3 by the player B will always result in more losses as compared to that of selecting the strategy B_2. Thus strategy B_3 is inferior to B_2. Hence, deleting B_3 strategy from the pay off matrix. The reduced pay off matrix is shown below:

Player B

		B_1	B_2
Player A	A_1	−5	10
	A_2	5	− 10
	A_3	5	−20

After the column B_3 is deleted, it may be noted that strategy A_2 of player A is dominated by his A_3 strategy, since the profit due to strategy A_2 is greater than or equal to the profit due to strategy A_3, regardless of which strategy player B selects. Hence strategy A_3 (row 3) can be deleted from further considerations. Thus the reduced pay off matrix is

Player B

		B_1	B_2
Player A	A_1	−5	10
	A_2	5	− 10

Since in the reduced 2×2 matrix, the maximin. value is not equal to the minimax value. Hence there is no saddle point and one cannot determine the point of equillibrium. For this type of game situation, it is possible to obtain a solution by applying the concept of mixed strategies.

Let p_1, p_2 and q_1, q_2 be the probabilities of A and B playing the strategies (A_1, A_2) and (B_1, B_2) respectively and we have $p_1 + p_2 = 1$ and $q_1 + q_2 = 1$.

Now, from the above pay off,

$$-5 p_1 + 5 p_2 = 10 p_1 - 10 p_2$$

and $$-5 q_1 + 10 q_2 = 5 q_1 - 10 q_2$$

Thus we have $\dfrac{p_1}{p_2} = 1$ and $\dfrac{q_1}{q_2} = 2$.

By solving the above equations, we have

$$p_1 = \frac{1}{2}, p_2 = \frac{1}{2}, q_1 = \frac{2}{3} \text{ and } q_2 = \frac{1}{3}$$

And the value of the game is $-5 \times \dfrac{1}{2} + 5 \times \dfrac{1}{2} = 0$ thus the mixed strategy of $A (A_1, A_2)$ is with the probabilities of $\left(\dfrac{1}{2}, \dfrac{1}{2}\right)$ and that of $B (B_1, B_2)$ is $\left(\dfrac{2}{3}, \dfrac{1}{3}\right)$ the value of the game is zero (fair game).

11.6 *Graphical Solutions*

Graphical solutions are helpful for the two-person zero-sum games of order $2 \times n$ or $m \times 2$ i.e., when one of the players have only two dominant strategies to mix while other has many strategies to play. However, graphical solutions used with an assumption that optimal strategies for both the players assign non-zero probabilities to the same number of pure strategies. Therefore, if one player has only two strategies, the other also should have to use same (two) number of strategies. This method is helpful in finding out which two strategies can be used.

Here we have two cases viz $2 \times n$ and $m \times 2$ cases.

Case (i) : **2 × n games :**

In this case, row player has two strategies to play while column player has n strategies. The graphical method is applied to find which two strategies of column player are to be used.

Suppose a pay off given below :

$$
\begin{array}{c}
& & \text{Player } B \\
& & B_1 \quad B_2 \quad B_3 \quad \cdots \quad B_n \\
\text{Player } A \quad
\begin{array}{c} A_1 \\ A_2 \end{array}
&
\begin{bmatrix}
a_{11} & a_{12} & a_{13} & \cdots & a_{1n} \\
a_{21} & a_{22} & a_{23} & \cdots & a_{2n}
\end{bmatrix}
\end{array}
$$

If player A mixes his strategies A_1 and A_2 with probabilities p_1 and $p_2 \geq 0$ where $p_1 + p_2 = 1$, then for each strategy of B, A will have the expected pay off as follows.

B's pure strategy	A's pay off
B_1	$a_{11}p_1 + a_{21}p_2$
B_2	$a_{12}p_1 + a_{22}p_2$
B_3	$a_{13}p_1 + a_{23}p_2$
.	.
.	.
B_n	$a_{1n}p_1 + a_{2n}p_2$

As A is assumed to always try to maximise his minimum gains. The highest point of lower envelop (lower boundary) formed by drawing these as straight lines represents the optimal probability mix. The lines corresponding to this point will yield a 2×2 matrix from which the value of the game, optimal strategies and their probability mix can be found.

Thus the two series are to represented on two parallel axes and these are joined to represent respective pay offs. Then the lower envelop is identified and upon this highest point is located. The coefficients of these lines make a 2×2 matrix for further iteration / calculation.

The above method is illustrated through the following example :

ILLUSTRATION 10 ─────────────────────────────

Solve the following game graphically.

$$\begin{bmatrix} -6 & 0 & 6 & -3/2 \\ 7 & -3 & -8 & 2 \end{bmatrix}$$

[JNTU (Mech.) 2001]

Solution :

Let Row player's (say A) strategies are A_1 and A_2 are used with the probabilities p_1 and p_2, then his expected pay offs when his opponent (column player) uses his pure strategies are shown below :

Column players pure strategies	Row player ($A's$) expected pay off
I	$-6p_1 + 7p_2$
II	$0p_1 - 3p_2$
III	$6p_1 - 8p_2$
IV	$-(3/2)p_1 + 2p_2$

These are graphically represented as follows as two parallel axes of unit distance apart.

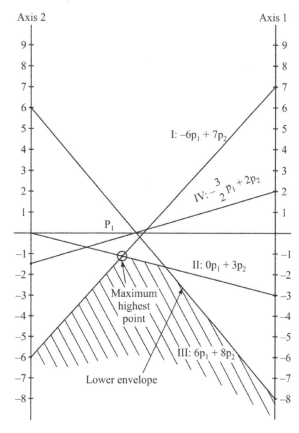

FIGURE 11.2 :

The highest point on lower envelop appear at the intersection of the lines represented by column players I and II strategies.

$$-6 p_1 + 7 p_2 \text{ and } 0 p_1 - 3 p_2$$

∴   The required 2×2 matrix is

$$
\begin{array}{c}
\quad\quad q_1 \quad\quad q_2 \\
\begin{array}{c} p_1 \\ p_2 \end{array}
\begin{bmatrix} -6 & 0 \\ 7 & -3 \end{bmatrix}
\end{array}
$$

We have $-6 p_1 + 7 p_2 = 0 p_1 - 3 p_2 \quad \Rightarrow \quad 6 p_1 = 10 p_2$

$$p_1 + p_2 = 1$$

$$p_1 = \frac{10}{16} = \frac{5}{8}, \qquad p_2 = \frac{6}{16} = \frac{3}{8}$$

and $-6 q_1 + 0 q_2 = 7 q_1 - 3 q_2 \quad \Rightarrow \quad 13 q_1 = 3 q_2$

$$\frac{q_1}{q_2} = \frac{3}{13} \quad \text{and} \quad q_1 + q_2 = 1$$

∴ $q_1 = \frac{3}{16}, \qquad q_2 = \frac{13}{16}$

∴ Row player will use his strategies with the probabilities as $(5/8, 3/8)$ respectively where the column player uses the first two strategies mix at the probabilities of $(3/16, 13/16)$.

The value of the game

$$-6 \times \frac{5}{8} + 7 \times \frac{3}{8} = -\frac{9}{8}$$

or $0 \times \frac{5}{8} - 3 \times \frac{3}{8} = -\frac{9}{8}$

The **column player wins the game** with a gain of **9/8 units** or **row player loses** the game with a **loss of 9/8 units.**

e (ii) : m × 2 games :

In this case a column player has two strategies while row player can choose optimal mix of best two among m strategies. The graphical method is used to find which two strategies of rows player would be the best.

Suppose a pay off of two players A and B is given below

Player B

$$
\begin{array}{c}
\quad\quad\quad B_1 \quad\quad\quad B_2 \\
\begin{array}{c} A_1 \\ A_2 \\ \cdot \\ \cdot \\ \cdot \\ A_m \end{array}
\begin{bmatrix} a_{11} & a_{12} \\ a_{21} & a_{22} \\ \cdot & \cdot \\ \cdot & \cdot \\ \cdot & \cdot \\ a_{m1} & a_{m2} \end{bmatrix}
\end{array}
$$

Player A

When player B mixes his strategies B_1 and B_2 with non-zero probabilities q_1 and q_2 where $q_1 + q_2 = 1$, then for each strategy of A, B's expected pay off is given by

A's pure strategy	B's pay off
A_1	$a_{11} q_1 + a_{21} q_2$
A_2	$a_{12} q_1 + a_{22} q_2$
.	. .
.	. .
.	. .
A_m	$a_{m1} q_1 + a_{m2} q_2$

Now, as B is assumed to be the looser always, he will try to minimise his maximum losses. Thus when these expected pay offs of B, when drawn as straight lines between two parallel axes the lowest point (minimising) of the upper envelop (maximum losses) represents his optimal strategy. The line consisting of this point will represent the best two strategies at its edges on the two parallel axes of unit distance apart. The perpendicular drawn from the minimum point of envelop on to x-axis divides segment on x-axis into two parts equal to the probabilities used by B.

This is illustrated through the following example.

ILLUSTRATION 11

Solve the following game graphically.

$$
\begin{array}{c} & B \\ & \begin{array}{cc} B_1 & B_2 \end{array} \\ \begin{array}{c} A_1 \\ A_2 \\ A_3 \\ A_4 \\ A_5 \\ A_6 \end{array} & \left[\begin{array}{cc} 1 & -3 \\ 3 & 5 \\ -1 & 6 \\ 4 & 1 \\ 2 & 2 \\ -5 & 0 \end{array} \right] \end{array}
$$

Solution :

Let B's strategies B_1 and B_2 be used with the probabilities q_1 and q_2 where $q_1 + q_2 = 1$. Then the pay offs of B for each of the strategies used by A are given below.

A's pure strategy	B's expected pay off
A_1	$q_1 - 3q_2$
A_2	$3q_1 + 5q_2$
A_3	$-q_1 + 6q_2$
A_4	$4q_1 + q_2$
A_5	$2q_1 + 2q_2$
A_6	$-5q_1 + 0q_2$

The above pay offs against A's strategies are drawn as straight lines between two parallel axes of unit distance apart.

Since column player B wishes to minimise his maximum expected pay off (loss), we consider the lowest point (minimum) of intersection G in the upper envelop $EFGH$ (maximum) which represents the minimax of B's expected pay off.

The minimax point G is the point of intersection of the lines A_2 and A_4. Thus the pay off of A is found most optimal with the blend of A_2 and A_4 strategies. Thus the 6×2 matrix reduces to 2×2 matrix with A's A_2 and A_4 and B's B_1 and B_2 strategies which can be evaluated by algebraic method.

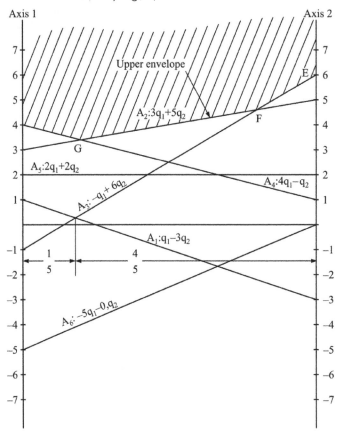

FIGURE 11.3 :

$$B$$

$$A \quad \begin{array}{c} \\ A_2 \\ A_4 \end{array} \begin{array}{cc} B_1 & B_2 \\ \begin{bmatrix} 3 & 5 \\ 4 & 1 \end{bmatrix} \end{array}$$

Now, if p_1 and p_2 are the probabilities of A to use A_2 and A_4 and q_1 and q_2 are the probabilities of B to play B_1 and B_2 then

$$3p_1 + 4p_2 = 5p_1 + p_2 = V \text{ (the value of the game)}$$

and $$3q_1 + 5q_2 = 4q_1 + q_2 = V \text{ (the value of the game)}$$

Where $p_1 + p_1 = 1$ and $q_1 + q_2 = 1$

Solving the above equations we get,

$$2p_1 = 3p_2 \quad \Rightarrow \quad \frac{p_1}{p_2} = \frac{3}{2}$$

and $p_1 + p_2 = 1$

$$p_1 = \frac{3}{5} \quad \text{and} \quad p_2 = \frac{2}{5}.$$

Also $q_1 = 4q_2 \Rightarrow \quad \frac{q_1}{q_2} = \frac{4}{1}$

and $q_1 + q_2 = 1$

$\therefore \quad q_1 = \frac{4}{5}$ and $q_2 = \frac{1}{5}$

Thus A plays his A_2 and A_4 strategies with the probabilities $(3/5, 2/5)$ while B applies his B_1 and B_2 strategies with the probability mix $(4/5, 1/5)$ and the value of the game

$$V = 4 \times \frac{4}{5} + \frac{1}{5} = \frac{17}{5}.$$

Remark : *Observe the strategies A_1, A_5 and A_6 and compare each with A_2. You can find that A_2 is dominating these three since it is up over these in graph*

11.7 *Linear Programming Method*

When a two person zero sum games are in larger size (say $m \times n$) these can be solved by linear programming method. This is the major advantage of this method and hence can be considered as universal method.

To illustrate this method first let us consider a 3×3 matrix given below :

$$
A \quad
\begin{array}{c}
 \\
A_1 \\
A_2 \\
A_3
\end{array}
\overset{\begin{array}{ccc} B_1 & B_2 & B_3 \end{array}}{
\begin{bmatrix}
a_{11} & a_{12} & a_{13} \\
a_{21} & a_{22} & a_{23} \\
a_{31} & a_{32} & a_{33}
\end{bmatrix}}
$$

Now, as per the assumptions, A always attempts to choose the set of strategies with the non-zero probabilities say p_1, p_2 and p_3 (where $p_1 + p_2 + p_3 = 1$) that maximises his minimum expected gains.

Similarly, according to B, he would choose the set of strategies with the non-zero probabilities, say q_1, q_2 and q_3 (where $q_1 + q_2 + q_3 = 1$), that minimises his maximum expected losses.

From the above statements, we can draw the following conclusions.

From A's point of view .

He would mix all his strategies A_1, A_2 and A_3 with non-zero probabilities p_1, p_2 and p_3 in such a way that the sum should be atleast equal to the value of the game (V) for whatever strategy that B may play. In other words, the sum of his expected pay off should be greater than or equal to V.

Therefore,

A's expected gain is $a_{11} p_1 + a_{21} p_2 + a_{31} P_3 \geq V$ when B plays B_1.

Similarly, A's gains when B uses his second strategy (B_2) and third strategy (B_3) respectively are

$$a_{12} p_1 + a_{22} p_2 + a_{32} p_3 \geq V$$

and $a_{13} p_1 + a_{23} p_2 + a_{33} p_3 \geq V$

And also $p_1 + p_2 + p_3 = 1.$

Simplifying the above inequations, by dividing by V through out,

$$a_{11} \left(\frac{p_1}{V}\right) + a_{21} \left(\frac{p_2}{V}\right) + a_{31} \left(\frac{p_3}{V}\right) \geq 1$$

$$a_{12} \left(\frac{p_1}{V}\right) + a_{22} \left(\frac{p_2}{V}\right) + a_{32} \left(\frac{p_3}{V}\right) \geq 1$$

$$a_{13} \left(\frac{p_1}{V}\right) + a_{23} \left(\frac{p_2}{V}\right) + a_{33} \left(\frac{p_3}{V}\right) \geq 1$$

and $\dfrac{p_1}{V}, \dfrac{p_2}{V}, \dfrac{p_3}{V} \geq 0$

Since, the objective of A is to maximise the value of the game V, which is equivalent to minimising the value of $\dfrac{1}{V}$, the objective function of A is

$$\text{Minimise } Z_A = \frac{1}{V} = \frac{p_1}{V} + \frac{p_2}{V} + \frac{p_3}{V}$$

Rewriting the above the assuming

$$\frac{p_1}{V} = x_1, \qquad \frac{p_2}{V} = x_2 \text{ and } \frac{p_3}{V} = x_3, \text{ we get}$$

The LPP formulation as

$$\begin{aligned}
\text{Minimise} \qquad & Z_A = x_1 + x_2 + x_3 \\
\text{Subject to} \quad & a_{11} x_1 + a_{21} x_2 + a_{31} x_3 \geq 1 \\
& a_{12} x_1 + a_{22} x_2 + a_{32} x_3 \geq 1 \\
& a_{13} x_1 + a_{23} x_2 + a_{33} x_3 \geq 1 \\
\text{and} \quad & x_1 \geq 0, x_2 \geq 0, x_3 \geq 0
\end{aligned}$$

Similarly, from B's view, we get the L.P problem as the dual of the above problem.

(Player B minimises his maximum losses).

For player B, it is written as

$$\text{Maximise} \quad Z_B = y_1 + y_2 + y_3$$

$$\text{Subject to} \quad a_{11} y_1 + a_{12} y_2 + a_{13} y_3 \leq 1$$

$$a_{21} y_1 + a_{22} y_2 + a_{23} y_3 \leq 1$$

$$a_{31} y_1 + a_{32} y_2 + a_{33} y_3 \leq 1$$

$$\text{and} \quad y_1, y_2, y_3 \geq 0.$$

$$\left(\text{where } y_1 = \frac{q_1}{V}, y_2 = \frac{q_2}{V} \text{and } y_3 = \frac{q_3}{V} \right)$$

− This problem now can be solved using usual simplex method, and the problem can be extended to any size $m \times n$. Now let us illustrate this method with an example.

ILLUSTRATION 12

Two companies A and B are competing for the same product. Their different strategies are given in the following pay off matrix.

Company A

		A_1	A_2	A_3
Company B	B_1	2	−2	3
	B_2	−3	5	−1

Use linear programming to determine the best strategies for both the players.

[JNTU-CSIT-99]

Solution :

Assume the non-zero probabilities p_1, p_2 for B to use his strategies B_1 and B_2 respectively (where $p_1 + p_2 = 1$) and q_1, q_2, q_3 for A to use A_1, A_2, A_3 respectively (where $q_1 + q_2 + q_3 = 1$)

Now with reference to A, $Max \ Z_A = \frac{1}{V}$ or $\frac{q_1}{V} + \frac{q_2}{V} + \frac{q_3}{V}$

$$2q_1 - 2q_2 + 3q_2 \leq V$$

$$-3q_1 + 5q_2 - q_3 \leq V$$

$$q_1 + q_2 + q_3 = 1$$

$$q_1, q_2, q_3 \geq 0$$

Diving by V through out the inequalities and equation,

$$2\left(\frac{q_1}{V}\right) - 2\left(\frac{q_2}{V}\right) + 3\left(\frac{q_3}{V}\right) \le 1, \quad -3\left(\frac{q_1}{V}\right) + 5\left(\frac{q_3}{V}\right) \le 1$$

and $\quad \dfrac{q_1}{V} + \dfrac{q_2}{V} + \dfrac{q_3}{V} = \dfrac{1}{V}$

Since A wishes to minimise his maximum losses V or maximises $\dfrac{1}{V}$ the objective function of A will be

$$\text{Maximise} \quad Z_A = x_1 + x_2 + x_3$$

$$\text{Subject to} \quad 2x_1 - 2x_2 + 3x_3 \le 1$$

$$-3x_1 + 5x_2 - x_3 \le 1$$

$$\text{and} \quad x_1, x_2, x_3 \ge 0$$

$$\left(x_1 = \frac{q_1}{V}, \ x_2 = \frac{q_2}{V} \text{ and } x_3 = \frac{q_3}{V} \right)$$

Converting the inequalities into equalities by introducing slack variables, we get,

$$\text{Maximise} \quad Z_A = x_1 + x_2 + x_3 + os_1 + os_2$$

$$\text{Subject to} \quad 2x_1 - 2x_2 + 3x_3 + s_1 = 1$$

$$-3x_1 + 5x_2 - x_3 + s_2 = 1$$

$$x_1, x_2, x_3 \ge 0, \qquad s_1, s_2, \ge 0$$

IBFS :

$$x_1 = 0, \ x_2 = 0, \ x_3 = 0, \text{ (non basic)}$$

$$s_1 = 1, \ s_2 = 1 \text{ (basic)}$$

ITERATION TABLEAU 1 :

Entering variable

C_B	BV	C_j	1	1	1	0	0	Min Ratio	Remarks
		SV	x_1	x_2	x_3	S_1	S_2		
0	S_1	1	2	-2	3	1	0	1/2	Key row
0	S_2	1	-3	5	-1	0	1	-ve	(Ignore -ve ratio)
		Z_j	0	0	0	0	0		
Leaving variable		$Z_j - C_j$	-1	-1	-1	0	0		

Key column

[Here there is a tie for entering variables x_1, x_2 and x_3, all three being decision variables, selection is made arbitrarily].

ITERATION TABLEAU 2 :

Entering variable

C_B	BV	C_j	1	1	1	0	0	Min Ratio	Remarks
		SV	x_1	x_2	x_3	S_1	S_2		
1	x_1	1/2	1	1	3/2	1/2	0	-ve	$R_1^N \to \frac{1}{2} R_1^0$
0	S_2	5/2	0	2	7/2	3/2	1	5/4	$R_2^N \to R_2^0 + 3R_1^N$
		Z_j	1	1	3/2	1/2	0		
		$Z_j - C_j$	0	-2	1/2	1/2	0		

Leaving variable

Key column

ITERATION TABLEAU 3 :

C_B	BV	C_j	1	1	1	0	0	Min Ratio	Remarks
		SV	x_1	x_2	x_3	S_1	S_2		
1	x_1	7/4	1	0	13/4	5/4	1/2		$R_1^N \to R_1^0 + R_2^N$
1	x_2	5/4	0	1	7/4	3/4	1/2		$R_2^N \to \frac{1}{2} R_2^N$
		Z_j	1	1	5	2	1	$Z_{max} = \frac{1}{V} = 3$ or $V = \frac{1}{3}$	
		$Z_j - C_j$	0	0	4	2	1		

Since $Z_j - C_j \geq 0$ for all the variables we have attained optimal solution.

The solution is

$$x_1 = \frac{7}{4}, \ x_2 = \frac{5}{4}, \ x_3 = 0$$

and   Maximise $Z_A = \frac{7}{4} + \frac{5}{4} + 0 = \frac{12}{4} = 3$

$\therefore$ $\frac{1}{V} = 3 \Rightarrow$ value of the game $(V) = \frac{1}{3}$

and $q_1 = x_1 V = \frac{7}{4} \times \frac{1}{3} = \frac{7}{12}$

$q_2 = x_2 V = \frac{5}{4} \times \frac{1}{3} = \frac{5}{12}$

$q_3 = x_3 V = 0 = 0$

To find B's probabilities of using B_1 and B_2 we use $Z_j - C_j$ values of final table [by duality theorems].

$$Z_j - C_j \text{ for } s_1 = 2 \text{ corresponds to } y_1 = 2$$

$$\text{and} \quad Z_j - C_j \text{ for } s_2 = 1 \text{ corresponds to } y_2 = 1$$

Where y_1, y_2 are the variables of dual problem of the above (primal) problem,

i.e.,   Minimise $Z_B = y_1 + y_2$

         Subject to $2y_1 - 3y_2 \geq 1,$

$$-2y_1 + 5y_2 \geq 1,$$

$$3y_1 - y_2 \geq 1$$

       and $y_1, y_2 \geq 0$

Where $y_1 = \dfrac{p_1}{V} \text{ and } y_2 = \dfrac{p_2}{V}.$

Thus

$$p_1 = V y_1 = \frac{1}{3} \times = \frac{2}{3}$$

$$p_1 = V y_2 = \frac{1}{3} \times 1 = \frac{1}{3}$$

Thus A uses his A_1, A_2 strategies with probability mix as (7/12, 5/12).

While B uses his B_1, B_2 strategies with probability mix as (2/3, 1/3) and the value of the game = 1/3.

i.e., B wins by 1/3 units.

11.8 *Distinction Between Deterministic and Probabilistic Games*

(Distinction Between Games With Saddle Points and Games Without Saddle Points)

S.No.	Games With Saddle Points (Deterministic)	Games Without Saddle Points (Probabilistic)
1.	These games have saddle points.	Saddle points do not exist in these games.
2.	Pure strategies are used by the players in these games.	Players use mixed strategies in these games.
3.	The game is strictly deterministic.	The game uses probabilities
4.	These games can be solved by Mini-max and Maxi-min criteria or by dominance rules.	Algebraic or linear programming techniques are used to solve these games. However, dominance rules, graphical methods can be used to reduce to 2×2 size and then algebraic method is applied.

11.9 *Limitations of Game Theory*

The game theory was received with a great enthusiasm initially but later has been found with lot of limitations. The most prominent limitations are listed here below.

1. **Unrealistic Assumption :** The assumption that the players know their own as well as other's (opponent's) payoffs in unrealistic. One can only guess of his own and opponent's strategies or pay offs.

2. **Complexity with More Number of Players :** As the number of players increase, the problem becomes so complex that is becomes difficult to solve.

3. **Oligopolic Situations can not be Solved :** Duopolic situations are easy to solve but oligopolic situations can not be solved in many occasions by using Game theory and in general practice, we have mostly, oligoplic situations in business.

4. **Minimax - Maximin in Impractical :** The assumption of maximin - minimax show that the players are risk-averse and have thorough knowledge of strategies, which seems impractical.

5. **Limitation on Collusion :** the game theory can not be applicable in case the players share their business secrets to workout a collusion.

6. **Non-Dynamic or Static Nature :** In markets, generally the changes are very frequent and fast. The game theory gives least scope to workout the strategy immediately unless some ground work to known pay-off is made.

7. **No base for calculating pay off :** There is no correct formula and scientific base for calculating pay offs of game.

11.10 *Flow Chart of Game Theory*

The theory of games and various solution methods applied are summarized through a flow chart given in page 11-2.

Practice Problems

I GAMES WITH SADDLE POINTS ──────────────────────────────────────

1. Solve the game when pay off matrix is given by

Player B

		B_1	B_2	B_3
Player A	A_1	– 3	12	– 1
	A_2	– 6	– 4	– 3
	A_3	– 5	15	– 8

[JNTU (CSE) 97]

Answer : Opt. Strategies (A_1, B_1) with $V = -3$ i.e., B wins by 3 units.

2. Verify your result in the above problem by dominance principle.

3. The pay off matrix in respect of a two person zero-sum game is

A's Strategy	B's Strategy				
	B_1	B_2	B_3	B_4	B_5
A_1	8	10	- 3	- 8	- 12
A_2	3	6	0	6	12
A_3	7	5	- 2	- 8	17
A_4	- 11	12	- 10	10	20
A_5	- 7	0	0	6	2

(a) Write the maximin and minimax strategy.

(b) Is it a strictly determinable game?

(c) What is the value of the game?

(d) Verify your answer using dominance principles.   **[JNTU (CSE) 2000]**

Answer : (a) Minimax of B with $B_3 =$ zero, Maximin of A with $A_2 =$ zero.

         (b) Yes, it is strictly determinable game, since saddle point exists at (A_2, B_3) i.e., zero.

         (c) Value of the game is zero i.e., fair game.

4. Consider the game with the following pay off

player B

$$\begin{array}{c} & \begin{array}{cc} B_1 & B_2 \end{array} \\ \text{player } A \quad \begin{array}{c} A_2 \\ A_4 \end{array} & \left[\begin{array}{cc} 2 & 6 \\ -2 & \lambda \end{array} \right] \end{array}$$

(a) Show that the game is strictly determinable, whatever λ may be.

(b) Determine the value of the game.

Answer : Value of the game is 2 with optimal strategy for player A and B is A_1 and B_1, respectively.

5. A company management and the labour union are negotiating a new three year settlement. Each of these has four strategies.

I   : Hard and aggressive bargaining.

II   : Reasoning and logical.

III   : Legalistic strategy.

IV   : Conciliatory approach.

The costs to the company are given for every pair of strategy choice.

		Company strategies			
		I	II	III	IV
Union strategies	I	20	15	12	35
	II	25	14	8	10
	III	40	2	10	5
	IV	- 5	4	11	0

What strategy will the two sides adopt? Also determine the value of the game. Use minimax - maximin rule and then verify your result with dominance rule.

[OU (MBA) 87]

Answer : Value of the game = 12 for which company will always adopt strategy - III - legalistic strategy while union will always adopt strategy - I hard and aggressive bargaining.

6. Two competitive manufacturers are producing a new toy under license from a patent holder. In order to meet the demand they have the option of running the plant for 8, 16 or 24 hours a day. As the length of production increases so does the cost. One of the manufacturers, say A, has setup the matrix given below, in which he estimates the percentage of the market that he could capture and maintain the different product schedules.

Manufacturer A	Manufacturer B		
	C_1 : 8 hrs	C_2 : 16 hrs	C_3 : 24 hrs
S_1 : 8 hrs	60%	56%	34%
S_2 : 16 hrs	63%	60%	55%
S_3 : 24 hrs	83%	72%	60%

(i) At which level should each produce.

(ii) What percentage of the market will B have.   **[JNTU (FDH, Mech.) 93/S]**

Answer : (i) Optimal strategies : (S_3, C_3) i.e., both should produce 24 hrs a day

(ii) B loses 60% market share which is most optimum, otherwise he losses more than 60%. Thus B will have 40% of market.

7. Two firms are competing for market share for a certain product. Each firm is considering what promotional strategy to employ for coming period. Assume the following pay off matrix describes the increase in market share of the firm A and decrease the market share for firm B. Determine the optimal strategies for each firm.

Firm B

	No promotion	Moderate promotion	Much promotion
No promotion	5	0	-10
Moderate promotion	10	6	2
Much promotion	20	15	10

Firm A (labels for rows: No promotion, Moderate promotion, Much promotion)

(i) Which firm would be the winner in terms of market share?

(ii) Would the solution strategies necessarily maximise profits for either of the firms.

(iii) What might the two firms do to maximise their profits?

[JNTU (Mech. & ECE) 94/S]

Answer : Optimal strategy for both A and B is "Much promotion" with value of game = 10.

(i) Firm A is the winner.

(ii) Yes, because it is a zero sum game.

(iii) They have to go for much promotion.

8. Solve the game using dominance principle and verify with saddle points of minimax - maximin.

Firm B

	B_1	B_2	B_3	B_4	B_5
A_1	3	-1	4	6	7
A_2	-1	8	2	4	12
A_3	16	8	6	14	12
A_4	1	11	-4	2	1

Firm A

[IGNOU (Assignment, MBA) 1996]

Answer : Optimal strategy is (A_3, B_3) with value of game = 6.

9. Solve the following game by minimax - maximin and then verify by dominance rules.

9	3	1	8	0
6	5	4	6	7
2	4	3	3	8
5	6	2	2	1

[IGNOU (MBA) 97]

Answer : 4 in second row, third column.

10. Solve by dominance and verify by finding saddle points.

-2	0	0	-5	3
3	2	1	2	2
-4	-3	0	-2	6
5	3	-4	2	-6

[Dr. BRAOU (Assignment) 97]

Answer : Value = 1 in second row and third column.

11. For what value of λ, the game with following pay off is strictly determinable.

λ	6	2
-1	λ	-7
-2	4	λ

[Dr. BRAOU (Assignment) 97]

Answer : $-7 \geq \lambda \geq 6$.

12. Solve the following games by minimax - maximin

-1	0	0	2	3
3	2	2	2	2
-4	-3	0	2	6
5	3	-4	2	6

[JNTU (Mech., FDH) 96]

Answer : Value with the game = 2 when row player uses his 2nd strategy while column player can use either 3rd or 4th strategies. Two saddle points.

13. Solve the following games by dominance.

3	-5	0	6
-4	-2	1	2
5	4	2	3

[JNTU (Mech., FDH) 93]

Answer : Value = 2 for 3rd row and 3rd column.

14. Use minimax - maximin and verify by any other method to solve the following game.

	I	II	III
I	6	8	6
II	4	12	2

Answer : $V = 6$ with (I, II)

15. Solve the game and check your answer by graph.

–1	2	–2
6	4	–6

Answer : (I, III), $V = -2$

16. Determine the range of value of p and q that will make the pay off element a_{22} a saddle point for the game whose payoff matrix (a_{ij}) is given below :

Player B

2	4	5
10	7	q
4	p	8

Player A

[JNTU Mech/Prod/Chem/ Mechatronics 2001/S]

17. Use dominance principles to solve the following.

5	1	10
50	1	1
50	0.1	10

Answer : (A_2, B_2), $V = 1$

[JNTU Mech. 2002/C/P]

II GAMES WITHOUT SADDLE POINTS

1. Solve the following game without saddle points.

Player **B**

2	5
7	3

Player A

[JNTU (CSE) 2001]

Answer : A wins with $\dfrac{29}{7}$, A uses probl. $\left(\dfrac{4}{7}, \dfrac{3}{7}\right)$ and B uses with prob $\left(\dfrac{2}{7}, \dfrac{5}{7}\right)$

2. Use concept of dominance to reduce the size of the matrix of given problem to 2×3 matrix and solve the game.

Player B

1	8	3
6	4	5
0	1	2

Player A

[JNTU (CSIT) 97/S]

3. Consider the game

B

5	50	50
1	1	0.1
10	1	10

A

Verify that the strategies (1/6, 0, 5/6) for player A and (49/54, 5/54,0) for B are optimal and find the value of the game. [JNTU (CSE) 2003/S (Set - 1)]

4. Use dominance principles to solve the following.

2	2	3	-2
4	3	2	6

[OU (MSc.) 89]

5. By modified dominance method solve the following

4	4	2	-4	-6
8	6	8	-4	0
10	2	4	10	12

[OU (MSc.) 92]

Answer : Row player (II, III) : $\left(\dfrac{4}{9}, \dfrac{5}{9}\right)$; col player (II, IV) : $\left(\dfrac{7}{9}, \dfrac{2}{9}\right)$; $V = \dfrac{34}{9}$

6. By L.P.P solve the following game

3	-1	-3
-3	3	-1
-4	-31	3

Answer : $A\left(\dfrac{20}{45}, \dfrac{11}{45}, \dfrac{14}{45}\right)$, $B\left(\dfrac{14}{45}, \dfrac{11}{45}, \dfrac{20}{45}\right)$; $V = -\dfrac{29}{45}$

7. By L.P.P solve the following game

-1	2	1
1	-2	2
3	-4	-3

Answer : $A\left(\dfrac{17}{46}, \dfrac{20}{46}, \dfrac{9}{46}\right)$, $B\left(\dfrac{7}{23}, \dfrac{6}{23}, \dfrac{10}{23}\right)$ $V = \dfrac{15}{23}$

8. By L.P.P solve the following game

7	1	7
9	-1	1
5	7	6

Answer : $A\left(\dfrac{1}{10}, \dfrac{1}{10}, \dfrac{8}{10}\right)$; $B\left(\dfrac{19}{30}, \dfrac{7}{30}, \dfrac{4}{30}\right)$, $V = \dfrac{28}{5}$

9. By L.P.P solve the following game

		Q		
		Q_1	Q_2	Q_3
	P_1	9	1	4
P	P_2	0	61	3
	P_3	5	2	8

Answer : $(P_1, P_2, P_3) : \left(\dfrac{3}{8}, \dfrac{13}{24}, \dfrac{1}{12}\right)$, $(Q_1, Q_2, Q_3) : \left(\dfrac{7}{24}, \dfrac{5}{9}, \dfrac{11}{72}\right)$; $V = \dfrac{91}{24}$

10. Solve the game whose pay off matrix,

3	2	4	0
2	4	3	4
4	2	4	0
0	4	0	8

Answer : $V = \dfrac{8}{3}, \left(0, 0, \dfrac{2}{3}, \dfrac{1}{3}\right) \left(0, 0, \dfrac{2}{3}, \dfrac{1}{3}\right)$

11. Explain the notion of dominance and use it to solve the game.

0	0	0	0	0	0
4	2	0	2	1	1
4	3	1	3	2	2
4	3	7	−5	1	2
4	3	4	−1	2	2
4	3	3	−2	2	2

Answer : Maximising player $\left\{0, 0, \dfrac{6}{7}, \dfrac{1}{7}, 0, 0\right\}$

Minimising player $\left\{10, 0, \dfrac{4}{7}, \dfrac{3}{7}, 0, 0\right\}$ and $V = \left(\dfrac{13}{7}\right)$.

12. Solve the following two person zero-sum game.

Player B

		B_1	B_2	B_3
	A_1	5	7	11
Player A	A_2	2	−1	8
	A_3	18	−6	10

Find the optimal strategies for the two players and the value of the game to each of the player. **[JNTU (Mech.) 98/S]**

Answer : Player A uses $(A_1, A_3) : (12/13, 1/13)$

Player B uses $(B_1, B_2) : (1/2, 1/2)$ value of game = 55/26 i.e., A wins.

13. Solve the following games graphically.

2	−1	5	−2	6
−2	4	−3	1	0

Answer : $V = \left(-\dfrac{2}{7}\right)\left(\dfrac{3}{7}, \dfrac{4}{7}\right)$ and $\left\{\dfrac{3}{7}, 0, 0, \dfrac{4}{7}, 0\right\}$

14. Solve the following games graphically.

1	6
4	5
5	3

Answer : $V = \dfrac{13}{3}, A\left(0, \dfrac{2}{3}, \dfrac{1}{3}\right), B\left(\dfrac{2}{3}, \dfrac{1}{3}\right)$

15. Solve the following games graphically.

(a)

2	7
3	5
11	2

(b)

1	3	11
8	5	2

Answer : $V = \dfrac{73}{14}, A\left(\dfrac{9}{14}, 0, \dfrac{5}{14}\right), B\left(\dfrac{5}{14}, \dfrac{9}{14}\right)$

16. Solve the following games graphically.

6	2	7
1	9	3

Answer : $V = \dfrac{13}{14}, \left(\dfrac{2}{3}, \dfrac{1}{3}\right)\left(\dfrac{7}{12}, \dfrac{5}{12}\right)$

17. Solve the following games graphically.

6	8	6
4	12	2

Answer : $V = 6$; (I, III)

18. Solve the following games graphically.

–1	2	–2
6	4	–6

Is the game (i) fair (ii) strictly determinable?

Answer : Yes, Strictly determinable. $V = -2$ of 1st row, 3rd column

19. Solve the following games graphically.

3	2	3	6
5	3	2	–1

Answer : $V = \dfrac{5}{2}, \left(\dfrac{1}{2}, \dfrac{1}{2}\right)\left(0, \dfrac{1}{2}, \dfrac{1}{2}, 0\right)$ or $\left[0, \dfrac{7}{8}, 0, \dfrac{1}{8}\right]$.

20. Solve the following games graphically.

2	5
3	1
0	3

Answer : $A\left(\dfrac{2}{5}, \dfrac{3}{5}, 0\right)$, $B = \left(\dfrac{4}{5}, \dfrac{1}{5}\right) V = \dfrac{13}{15}$

21. Determine the best strategy for player A in the following game :

Player B

		I	II
	I	3	−5
	II	1	−1
Player A	III	2	−3
	IV	−1	3
	V	0	1

Answer : In case of graphical solution all the lines pass through the minimax point, A's optimal strategies are $\left\{\dfrac{1}{3}, 0, 0, \dfrac{2}{3}, 0\right\}$, $\left\{\dfrac{1}{9}, 0, 0, 0, \dfrac{8}{9}\right\}$, $\left\{0, \dfrac{2}{3}, 0, \dfrac{1}{3}, 0\right\}$, $\left\{0, \dfrac{1}{3}, 0, 0, \dfrac{2}{3}\right\}$, $\left\{0, 0, \dfrac{4}{9}, \dfrac{5}{9}, 0\right\}$, $\left\{0, 0, \dfrac{1}{6}, 0, \dfrac{5}{6}\right\}$ value of the game is $\dfrac{1}{3}$.

22. Solve the following game graphically.

1	3	−1	4	2	−5
−3	5	6	1	2	0

[JNTU (Mech.) 94/CCC]

Answer : $\left\{\dfrac{4}{5}, \dfrac{1}{5}\right\}$, $\left(0, \dfrac{3}{5}, 0, \dfrac{2}{5}, 0, 0\right)$; $V = \dfrac{17}{5}$.

23. Solve the followng game graphically

	B_1	B_2	B_3	B_4	B_5
A_1	2	-4	6	-3	5
A_2	-3	4	-4	1	0

Answer : $A\left\{\dfrac{4}{9}, \dfrac{5}{9}\right\}$, $B\left\{\dfrac{4}{9}, 0, 0, \dfrac{5}{9}, 0\right\}$, $V = -\dfrac{7}{9}$

24. Obtain the optimal strategies for both players and the value of the game for two-person zero-sum game whose payoffs matrix is given as follows :

Player B

	B_1	B_2	B_3
A_1	6	2	7
A_2	1	9	3

Player A

Answer : $A\ (A_1, A_2)\ ; \left(\dfrac{3}{11}, \dfrac{8}{11}\right);\ B\ (B_1, B_2) : \left(\dfrac{2}{11}, \dfrac{9}{11}\right);\ V = \dfrac{49}{11}$

25. Use graphical method to solve the game

Player B

	B_1	B_2
A_1	–6	7
A_2	4	–5
A_3	–1	–2
A_4	–2	5
A_5	7	–6

Player A

[ICWA, June 1985]

26. A soft drink company calculated the market share of two products against its major competitor having three products and found out the impact of additional advertisement in any one of its products against the order.

Competitor B

	B_1	B_2	B_3
A_1	6	7	15
A_2	20	12	10

Company A

What is the best strategy for the company as well as the competitor? What is the pay off obtained by the company and the competitor in the long run? Use graphical method to obtain the solution. **[Delhi. Univ., MBA April 1989]**

27. Consider the following payoff matrix for two firms. Find the best strategies for both the firms.

Firm II

	No advertising	Medium advertising	Large advertising
No advertising	60	50	40
Medium advertising	70	70	50
Large advertising	80	60	75

(m I labels the rows)

Answer :   Firm I : $\left\{0, \dfrac{3}{7}, \dfrac{4}{7}\right\}$ Firm. II : $\left\{0, \dfrac{5}{7}, \dfrac{2}{7}\right\}$, $V = \dfrac{50}{3}$

28. Two children Divya and Savya play the following game, named "scissors, paper, stone". Both players simultaneously call one of the three "scissors, paper and stone" scissors beat paper (as paper can be cut by scissors), paper beats stone (as stone can be wrapped in paper) and stone beats scrissors. If the two players name the same item, then there is a tie. If there is one point for win, zero for tie and -1 for loss, form the pay off matrix of the game and solve it.

Answer : The game is symmetric. Hence $V = 0$ optimal strategies for both players are $\left\{\dfrac{1}{3}, \dfrac{1}{3}, \dfrac{1}{3}\right\}$.

Hint :

Savya

	Sc	P	St
Sc	0	1	-1
P	-1	0	1
St	1	-1	0

(Divya labels the rows)

29. Solve the following games by LPP.

1	−1	3
3	5	−3
6	2	−2

[JNTU (Mech., FDH) 94]

Answer : $V = 1$, for maximising player $\left\{\dfrac{2}{3}, \dfrac{1}{3}, 0\right\}$ for minimising player $\left\{0, \dfrac{1}{2}, \dfrac{1}{2}\right\}$.

30. Two firms A and B are competing for the same product. Their different strategies are given in the following pay off matrix.

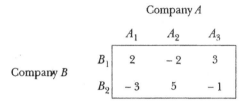

Company A

		A_1	A_2	A_3
Company B	B_1	2	-2	3
	B_2	-3	5	-1

Use linear programming to determine the best strategies for both the players.

[IGNOU (MCA, Assignment) 96]

Answer : $A\ (A_1, A_2) : \left(\dfrac{7}{12}, \dfrac{.5}{12}\right)$,

$B\ (B_1, B_2): \left(\dfrac{2}{3}, \dfrac{1}{3}\right); V = \dfrac{1}{3}$

31. Raju and Ravi plays game as follows they simultaneously and independently write one of the three numbers 1, 2 and 3. If the sum of the numbers written is even, Ravi pays to Raju this sum in rupees. If it is odd, Raju pays the sum to Ravi in rupees. Form the payoff matrix of player Raju and solve the game to find out the value of the game and probabilities of mixed strategies of Raju and Ravi. [DOECCA 98]

Answer : $\begin{bmatrix} 2 & -3 & 4 \\ -3 & 4 & -5 \\ 4 & -5 & 6 \end{bmatrix}$

Raju $\left(\dfrac{1}{4}, \dfrac{1}{2}, \dfrac{1}{4}\right)$, Ravi $\left(\dfrac{1}{4}, \dfrac{1}{2}, \dfrac{1}{4}\right)$, $V = 0$

32. For the following pay off matrix, find the value of the game and the strategies of players A and B using linear programming.

	A		
B	-1	2	-2
	6	4	-6

[JNTU (Mech., FDH) 92]

Answer : (B_1, A_3), $v = -2$; Game is deterministic and unfair

33. Find the best strategy and the value of the following game, using LPP method

$$
\begin{array}{cc}
 & B \\
 & \begin{array}{ccc} \text{I} & \text{II} & \text{III} \end{array} \\
A \quad \begin{array}{c} \text{I} \\ \text{II} \\ \text{III} \end{array} & \left[\begin{array}{ccc} -1 & -2 & 8 \\ 7 & 5 & -1 \\ 6 & 0 & 12 \end{array} \right]
\end{array}
$$

[JNTU (CSE) 97/S]

Answer : A uses (II, III) with prob. (2/3, 1/3)

B uses (II, III) with prob. (13/18, 5/18) value of the game $= 10/3$
i.e., A wins.

34. A and B play a game in which each has three coins a 5 paise, a 10 paise and a 20 paise. Each selects a coin without the knowledge of the others choice. If the sum of the coins is an odd amount, A wins B's coins. If the sum is even B wins A's coins. Find the best strategy for each player and the value of the game. Use LPP method to solve the game.   **[JNTU (Mech.) 99, (ECE) 99]**

Answer :   Player A (A_1, A_2) : (1/2. 1/2),

Player B (B_1, B_2) : (2/3, 1/3) value of the game is zero i.e., fair game.

Hint :

$$
\text{pay-off's} A \quad \begin{array}{c} A_1 \\ A_2 \\ A_3 \end{array} \begin{array}{c} \begin{array}{ccc} B_1 & B_2 & B_3 \end{array} \\ \left[\begin{array}{ccc} -5 & 10 & 20 \\ 5 & -10 & -10 \\ 5 & -20 & -20 \end{array} \right] \end{array}
$$

35. Firm X is fighting for its life against the determination of firm Y to drive it out of the industry. Firm X has the choice of increasing price, leaving it unchanged or lowering it. Firm Y has the same three options. Firm X's gross sales in the event of each of the pairs of choices are shown below.

	Firm Y's pricing strategies		
Firm X's pricing strageties	Increase price	Do not change	Reduce price
Increase price	90	80	110
Do not change	110	100	90
Reduce price	120	70	80

Assuming firm X as the maximising one, formulate and solve the problem as a linear programming problem.   **[Osmania MBA, Nov. 1990]**

Review Questions

1. Explain (i) strategy (ii) pay off matrix (iii) saddle points.

2. What is optimal strategy? Discuss rectangular games without saddle points.

3. Write short notes on solution of games with saddle points.

4. Write short notes on

    (i) Applications of game theory in business systems.

    (ii) Distinguish between games with saddle points and without saddle points.
    [JNTU Mech/Mechatronics/Chem/Prod. 2001/S

5. Write short notes on minimax criteria. **[JNTU Mech. 1998/S**

6. Distinguish between zero-sum and non-zero sum game.

7. Define a rectangular game. Briefout its solution methods.

8. Distinguish between deterministic and probabilistic games. What approaches do you apply to solve each of them.

9. Explain the approaches to solve rectangular games through a flow chart.

10. How is concept of dominance used in simplifying rectangular games.

11. Explain the graphical method of solving $2 \times n$ games and $m \times 2$ games.

12. How do you apply game theory in the following fields.
    (a) marketing (b) personnel department.

13. Explain the relevance of game theory to managerial problems.

14. Explain how you can apply linear programming to game theory.

15. Give a critical appreciation of game theory with reference to duality concept in LPP.

16. Write short notes on
    (a) Two-person zero-sum games.
    (b) Pure and mixed strategies.

17. Distinguish the following

    (a) Two-person Vs. multi-person games.

    (b) Zero-sum Vs. non-zero-sum games.

    (c) Fair Vs. unfair games.

    (d) Deterministic Vs. probabilistic games.

18. Distinguish the terms minimax and maxi-min with reference to rectangular games.   [JNTU Mech/Prod/Chem. 2001/S]

19. Discuss briefly different types of rectangular games and their solution methods.

20. Give the algorithm of finding saddle points in a rectangular game.

21. List out the assumptions made in the theory of game.

                                                        [JNTU CSE/ECE 2001]

Objective Type Questions

1. If the value of the game is zero, then that game/strategy is known as

   (a) pure strategy   (b) mixed strategy

   (c) fair game   (d) pure game

2. When minimax and maximin criteria match, then

   (a) saddle points exist   (b) fair game is resulted

   (c) mixed strategies are adopted   (d) game will be unfair

3. A predetermined plan of action based on which the games are played that does not change during the game is said to be

   (a) fair strategy   (b) pure strategy

   (c) mixed strategy   (d) value of the game

4. Which of the following is true in the case of row dominance in a game theory

   (a) least of the row $\geq$ highest of another row

   (b) least of the row $\leq$ highest of another row

   (c) every element of a row $\geq$ corresponding element of another row

   (d) every element of a row $\leq$ corresponding element of another row

5. If the gain of a player is loss of another player then the game is called

   (a) fair game   (b) unfair game

   (c) zero sum game   (d) non zero sum game

6. Which of the following is never called a two person game.

   (a) chess   (b) carrons

   (c) Horse race   (d) cricket

7. In a two person-zero sum game, the following assumption is wrong.

   (a) row player is always a loser

   (b) column player always minimises losses

   (c) there are only two persons

   (d) if one loses, the other gains

8. Games with saddle points are _____ in nature

   (a) deterministic   (b) probabilistic

   (c) stochastic   (d) normative

9. If a two person zero sum game is converted to LP problem

   (a) no. of variables are two only

   (b) no. of constraints are two only

   (c) if row player represents primal, column player represents dual

   (d) it will not have objective function

10. Which of the following wrong

    (a) games without saddle points are probabilistic

    (b) games with saddle points will have pure strategies

    (c) games without saddle points use mixed strategies

    (d) games with saddle points can not be solved by dominance rule

11. If saddle point does not exist,

    (a) it is deterministic game

    (b) mixed strategies are used

    (c) maximin - minimax criteria are used

    (d) it is a fair game

12. A game is said to be fair if

    (a) saddle point exists   (b) pure strategies are used

    (c) value of the game is zero   (d) it is a zero sum game

13. Which of the following method can not be used for the games without saddle points

    (a) algebraic method   (b) linear programming

    (c) graphical solution   (d) minimax - maximin criteria

14. If there are more than two players, the game is

    (a) non-zero sum game   (b) multiplayer game

    (c) zero sum game   (d) open game

15. If loss in sales of pepsi is gain in sales of coke, then the game between pepsi and coke is a

    (a) fair game   (b) pure game

    (c) zero - sum game   (d) one - one game

16. In a colony there are 100 families which either use colgate or pepsodent tooth paste. Due to raise in price of colgate some of them discontinued to use any tooth paste. This situation can be fit in

    (a) zero sum game   (b) positive sum game

    (c) negative sum game   (d) fair game

17. The value of a game whose pay off is a 2×2 unit matrix is _____

    (a) zero   (b) unity

    (c) 0.5   (d) none of these

18. Which of the following is a fair game

    (a) $\begin{bmatrix} 1 & 0 \\ 0 & 1 \end{bmatrix}$ (b) $\begin{bmatrix} -1 & 0 \\ 0 & 1 \end{bmatrix}$

    (c) $\begin{bmatrix} -1 & 0 \\ 0 & -1 \end{bmatrix}$ (d) $\begin{bmatrix} 1 & -1 \\ -1 & 1 \end{bmatrix}$

19. The value of the game whose pay off is $\begin{bmatrix} 1 & 2 \\ -2 & 3 \end{bmatrix}$

    (a) 1   (b) 2

    (c) – 2   (d) 3

20. Value of the game whose pay off is given by $\begin{bmatrix} 2 & -4 & 6 \\ 0 & -3 & -2 \\ 3 & -5 & 4 \end{bmatrix}$ will be

    (a) zero (b) – 3

    (c) 3 (d) – 5

21. The pay off $\begin{bmatrix} 2 & 6 \\ -2 & \lambda \end{bmatrix}$ is strictly determinable, then the value of λ is _____

    (a) zero (b) positive value

    (d) negative value (d) any value

22. For the pay off $\begin{bmatrix} -5 & 2 \\ -7 & -3 \end{bmatrix}$ the row player always uses

    (a) first strategy

    (b) second strategy

    (c) mixed strategies of both first and second

    (d) no strategy

23. In the above problem what will be better strategy for a column player

    (a) first strategy

    (b) second strategy

    (c) mixed strategy with $\left(\dfrac{1}{2}, \dfrac{1}{2}\right)$

    (d) mixed strategy with $\left(\dfrac{1}{5}, \dfrac{1}{2}\right)$

24. In the problem given in Q.No. 22, the value of the game is _____

    (a) – 5 (b) 2

    (c) – 7 (d) – 3

25. In the pay off of problem given in Q.No.22, the winner is

    (a) row player (b) column player

    (c) neither of the two (d) none of the above

26. Srija and Sanja play a game with two types of coins 5 ps and 10 ps. Each draw one coin randomly and if the sum is even Srija wins the coins, otherwise Sanja. The value of the game is _____

    (a) 1/2   (b) 5

    (c) zero   (d) – 5

27. In the above problem, the strategy used by Srija is _____

    (a) 5 ps always

    (b) 10 ps always

    (c) both strategies with probability $\left(\dfrac{1}{2}, \dfrac{1}{2}\right)$

    (d) mixed strategies of $\left(\dfrac{1}{4}, \dfrac{3}{4}\right)$

28. In a game between cricket teams India and Pakistan the toss is supposed to play major role. Both captains take one coin and toss. If both get same, India wins otherwise Pakistan. If we take '1' for winning and -1 for losing, the value of the game is _____

    (a) 1   (b) -1

    (c) zero   (d) 1/2

29. Identify the dominance in $\begin{bmatrix} 1 & 7 & 3 \\ 5 & 6 & 4 \\ 7 & 2 & 0 \end{bmatrix}$

    (a) R_1 dominates R_2 (b) R_2 dominates R_3

    (c) C_3 dominates C_2 (d) C_2 dominates C_3

30. Which of the following game is not fair

    (a) $\begin{bmatrix} 0 & 0 \\ 0 & 0 \end{bmatrix}$ (b) $\begin{bmatrix} 1 & -1 \\ -1 & 1 \end{bmatrix}$

    (c) $\begin{bmatrix} -5 & +5 \\ +10 & -10 \end{bmatrix}$ (d) $\begin{bmatrix} 1 & 0 \\ 0 & 1 \end{bmatrix}$

Fill in the Blanks

1. A game is said to be _____ if the value of the game is zero.

2. If a game uses pure strategies, it is called _____

3. A graphical solution can be used for games if the order of the pay off is _____ or _____

4. If gain of a player is loss of the other then the game is called_____

5. Deterministic games use _____ strategies

6. _____ strategies are used in the games without saddle points

7. If every element in R_1 is less than or equal to corresponding element in R_2, then _____ is said to be dominating.

8. According to basic assumptions of game theory, row player always wishes to _____ and column player wishes to _____

9. The value of the game whose pay off is a 2×2 identity matrix is _____

10. When a game is represented by LPP, if row player represents primal, column player represents _____

11. A game with more than two players is called_____ player game

12. If loss of a player is not a gain to the other, then the game is called _____ game

13. The game whose pay off is null matrix is _____ game

14. The best strategy for row player in the game with pay off $\begin{bmatrix} 1 & -1 \\ -1 & 1 \end{bmatrix}$ is ____

15. The value of the game with pay of $\begin{bmatrix} 1 & -1 \\ -1 & 1 \end{bmatrix}$ is _____

16. In the pay off $\begin{bmatrix} 5 & 6 \\ -2 & 2 \end{bmatrix}$ the saddle point is at _____ and is equal to _____

17. If minimax value is same as maximin value the game said to have _____

18. The corresponding strategy of each player at equilibrium point is _____ strategy

19. The value of the game whose pay off is $\begin{bmatrix} 2 & 0 \\ 0 & 4 \end{bmatrix}$ is _____

20. The list of all possible actions that will take for every pay off that might result is called _____

Answers

Objective Type Questions :				
1. (c)	2. (a)	3. (b)	4. (d)	5. (c)
6. (b)	7. (a)	8. (a)	9. (c)	10. (d)
11. (b)	12. (c)	13. (d)	14. (b)	15. (c)
16. (c)	17. (c)	18. (b)	19. (a)	20. (b)
21. (d)	22. (a)	23. (a)	24. (a)	25. (b)
26. (c)	27. (c)	28. (c)	29. (c)	30. (d)

Fill in the Blanks :	
1. fair	2. deterministic
3. $2 \times n$ or $n \times 2$	4. zero sum game
5. pure	6. mixed
7. R_2	8. Maximise gains, minimise losses
9. $\dfrac{1}{2}$	10. dual
11. multi	12. non zero sum
13. fair	14. mixed strategy of both rows with (1/2, 1/2)
15. zero	16. (R_1, C_1), 5
17. saddle point	18. optimal
19. $\dfrac{4}{3}$	20. strategy

12
Chapter

Dynamic Programming

CHAPTER AT A GLANCE

DYNAMIC PROGRAMMING

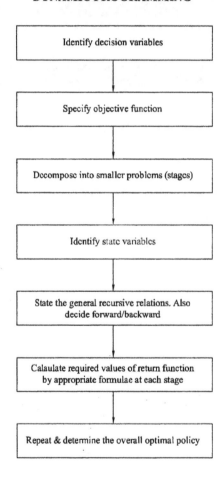

Identify decision variables

Specify objective function

Decompose into smaller problems (stages)

Identify state variables

State the general recursive relations. Also decide forward/backward

Calaulate required values of return function by appropriate formulae at each stage

Repeat & determine the overall optimal policy

12.0 *Introduction*

Dynamic programming is an optimization technique of multistage decision process. The word "dynamic" means to the situations of change in several stages, such as every week, every day or every month etc., and 'programming' is the term used in the mathematical sense of selecting an optimum allocation of resource.

Rechard Bellman propounded this technique of finding optimal solution by developing recursive relation. In dynamic programming, a large problem is split into smaller sub problems each of which involve in a few variables. These sub problems are sequentially optimized so as to make the total problem optimal.

An important point to note here is that – the problem of successive stages is treated separately even though by the very nature of the problem, these stages are dependent. The answer to this question is based on "Bellman's principle of optimality", which is stated as *"An optimal policy (set of decisions) has the property that what ever the initial state and decisions are, the remaining decisions must constitute an optimal policy with regard to the state resulting from the first decision"*.

Discrete and continuous, deterministic as well as probablistic models can be solved by this method. Thus dynamic programming method is very useful in obtaining optimal solutions of various problems such as inventory, replacement, allocation, linear programming, reliability calculations etc. A single constraint problem is relatively simple, but in the problem of more than two constraints more complexities may appear.

12.1 *The Approach & Algorithm*

The three basic features that are present in the structure of the Dynamic Probramming Problem (DPP) problem are as follows.

1. The DP problem can be decomposed into sequence of smaller sub–problems called stages of the original problem. At every stage there will be number of decision alternatives and a decision is made by most suitable one out of given list. Stages usually represent different time periods associated with planning period of problem, places, people, or any other suitable parameter

   e.g. In replacement problem, each year is a stage, in the sales man allocation problem, each territory may be a stage.

2. The states are represented in various conditions of decision process at a stage. The variables that specify the condition of decision process at a particular stage are called states. These states can provide information for analysing the possible effect that the current decision could have upon future action.

   e.g. A specific city is referred to as state variable in any stage of the shortest root problem.

3. At each stage, a decision is made which can influence the state of the system at the next stage to arrive the optimal solution at the current stage. This decision described in the form of an algebraic equation to explain its worth or benefit, is referred to as Return function.

The solution of a DPP is based on Bellman's principle of optimality (recursive optimization technique) which states:

An optimal policy has the property that, whatever the initial state and initial decision, the remaining decisions must constitute an optimal policy with regard to the state resulting from first decision.

With reference to the above principle, the solution procedure goes by solving sequentially adding a series of sub problems until the overall optimum is obtained. *If the problem is sequenced from last stage to first (i.e., initial stage), the method is said to be **backward recursion** while it is said to be **forward recursion** if it is sequenced from first (i.e., initial) to the last.*

General Algorithm :

Step 1 : Identify the decision variables of the problem and specify objective function to be optimised under the constraints, if any.

Step 2 : Decompose the given problem into number of smaller problems (or stages). Identify the state variables at each stage and write down the transformation function as a function of state variable and decision variables at the next stage.

Step 3 : State the general recursive relationship for computing the optimal policy. Decide whether to apply the forward or backward recursion to solve the problem.

Step 4 : Construct appropriate tables to show the required values of the return function at each stage as shown below.

Possible decision (d_n) entering states (S_n)	$\dfrac{F_n(s_n, d_n)}{d_n}$	Optimal Return $F_n^*(S_n)$	Optimal Decision (d_n^*)

Step 5 : Repeat and Determine the overall optimal policy or decision and its value at each stage. There may be more than one such optimal policy also.

12.2 *Salient Features & Their Definitions of DPP*

On summarising the above topics we get the following salient features in DPP.

1. **Recursive Function :** A dynamic programming problem is solved by an objective function which is recurring in nature. This recursive approach can be applied in two directions.

    (a) *Forward Recursive Approach :* In this approach the main problem is atacked by dividing it from intial stage to final stage.

    (b) *Back Ward Recursive Approach :* In this approach the problem is dealt by sub dividing the problem from last to first and then solving from final to intial stages.

2. **Stage :** A large problem is decomposed into smaller and sequential sub-problems. Each sub-problems is referred to as 'a stage' where a decision is called for. Thus each stage can be considered to have a starting and ending. the ending of one stage will be the beginning of its next stage.

Suppose an engineering students has to acquire good marks in his degree after studying for four years. Each year is considered to be a stage. When a student completes his first year successsfully, he is said to have ended 1st stage (and started his second stage) having $1/4^{th}$ of his large problem is completed. Thus he moves to optimise his returns in second stage i.e.,2nd year with similar objective function, but this includes the results of 1st year, and after completion of his second stage he acquires $2/4$ of problem solved. Similarly, he goes to 3rd stage, finishing $3/4^{th}$ that should include the results of 1st and 2nd year. Finally, after completing his fourth year, he completes his total problem that includes results of his previous three stages.

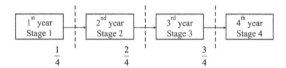

FIGURE 12.1 : STAGES OF A DPP

3. **State :** Any stage just before starting or just after completion will have certain status called ' the state' and the variable that links up two stages is called state variable.

4. **Return Function :** The decision of a stage described in the form of an algebraic equation with state variables to explain the worth or benefit (or cost is known as Return function or transformation equation.

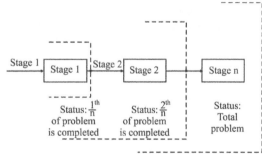

FIGURE 12.2 : STAGES AND STATES IN A DPP

12.3 *Characteristics of DPP*

The basic the characteristics of the DPP are given below.

1. The problem can be sub-divided into sub-problems referred to as stages with policy decision at each stage.

2. A stage is a device to sequence the decisions. This means that the sequential summation of these decisions obtained from optimal solution of sub-problems will be the part of the optimal solution of the total problem.

3. Every stage consists a number of states associated with it. The states are the different possible conditions in which system may find itself at that stage of the problem.

4. Decision at each stage converts the current stage into state associated with the next stage.

5. The state of the system at a stage is described by a set of variables called state variables.

6. When the current state is known, an optimal policy for the remaining stages is independent of the policy of the previous stages.

7. To identify the optimum policy for each state of the system, a recursive equation is formulated with n stages remaining, given the optimal policy for each state with $(n - 1)$ stages left.

8. Using recursive equation approach each time the solution moves either backwards or forwards by stage obtaining the optimum policy of each state for that particular stage, till it attains the optimum policy beginning at final or initial stage respectively.

ILLUSTRATION 1

A medical representative located at city has to travel to city 10. He knows the distance of alternative routes from city 1 to city 10 and has drawn the network map as shown in the following figures along with the distance between the cities. Find shortest possible route. Also find the shortest routes from any city to city 10.

[JNTU (Mech.) 96]

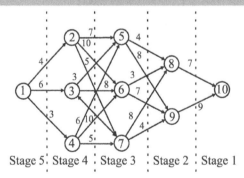

Stage 5 : Stage 4 : Stage 3 : Stage 2 : Stage 1

FIGURE 12.3 :

Solution :

The above problem can be divided into five stages as shown with dotted lines in the figure, and backward recursion is chosen with the objective to find shortest route. It is solved in the tabular form method as given below.

At stage 1, if the medical representative is at city 10, it is the destination itself, he need not travel any more.

Stage 2 : At stage 2, i.e., if he is at city 8 or city 9, the shortest path calculation is shown below.

Decisions / State Variable (S_1)		Destinations i.e., city 10		Minimum Distance (Optimal)	Optimal Decision
		Stage Distance	Cummulative Distance		
Entering State	8	7	7	7	8 to 10
	9	9	9	9	9 to 10

Stage 3 : Now, using these optimal results of stage 2, we proceed to stage 3.

Decisions / State Variable (S_2)		Destinations				Min. (Optimal) Distance	Optimal Stage Decision	Cummul. Optimal Decision
		City 8 (7)		City 9 (9)				
		Stage Distance	Cummul. Distance	Stage Distance	Cummul. Distance			
Entering Variables	5	4	11	8	17	11	5-8	5-8-10
	6	3	10	7	16	10	6-8	6-8-10
	7	8	15	4	13	13	7-9	7-9-10

Stage 4 : With the above optimal decision we compute stage-4 decisions.

(Figures in parantheses are the optimal distances derived from previous table).

Decisions / State Variable (S_3)		Destinations						Min. (Optimal) Distance	Optimal Stage Decision	Cummu. Optimal Decision
		City 5 (11)		City 6 (10)		City 7 (13)				
		Stage Dist.	Cummu. Dist.	Stage Dist.	Cummu. Dist.	Stage Dist.	Cummu. Dist.			
Entering States	2	7	18	10	20	5	18	18	2-5 or 2-7	2-5-8-10 or 2-7-9-10
	3	3	14	8	18	4	17	14	3-5	3-5-8-10
	4	6	17	10	20	5	18	17	4-5	4-5-8-10

Stage 5 : With the optimal (18, 14 and 17) distance decisions we now proceed to calculate final decision table of stage 5.

Decisions / State Variable (S_4)	Destinations						Min. (Optimal) Distance	Optimal Stage Decision	Cummu. Optimal Decision
	City 2 (18)		City 3 (14)		City 4 (17)				
	Stage Dist.	Cummu. Dist.	Stage Dist.	Cummu. Dist.	Stage Dist.	Cummu. Dist.			
1	4	22	6	20	3	20	20	1-3 or 1-4	1-3-5-8-10 or 1-4-5-8-10

Thus two optimal solutions are obtained as follows.

Solution set 1 $\begin{cases} \text{sequence (city)} \textcircled{1} \rightarrow \textcircled{3} \rightarrow \textcircled{5} \rightarrow \textcircled{8} \rightarrow 10 \\ \text{Distance } (km) \qquad\quad 6 \qquad 3 \qquad 4 \qquad 7 \end{cases} = 20$

Solution set 2 $\begin{cases} \text{sequence (city)} \textcircled{1} \rightarrow \textcircled{4} \rightarrow \textcircled{5} \rightarrow \textcircled{8} \rightarrow 10 \\ \text{Distance } (km) \qquad\quad 3 \qquad 6 \qquad 4 \qquad 7 \end{cases} = 20$

We can thus state that medical representative may have to travel, min. of 20 *km* from city 1 to city 10. If he is at other stations, the following optimal decisions may be made

Stations	Decision	Min. Distance to travel	Decision taken at
1	1-3-5-8-10 or 1-4-5-8-10	20 *km*	Stage - 5
2	2-5-8-10 or 2-7-9-10	18 *km*	Stage - 4
3	3-5-8-10	14 *km*	Stage - 4
4	4-5-8-10	17 *km*	Stage - 4
5	5-8-10	11 *km*	Stage - 3
6	6-8-10	10 *km*	Stage - 3
7	7-9-10	13 *km*	Stage - 3
8	8-10	7 *km*	Stage - 2
9	9-10	9 *km*	Stage - 2

ILLUSTRATION 2

Suppose the above problem is to maximise the distance assuming that he gets more allowance if travels more. Now the decision tables will be as follows.

Stage - 2:

Decisions / State Variable (S_1)	Destinations city 10		Maximum Distance (Optimal)	Optimal Decision
	Stage Distance	Cummulative Distance		
Entering State Variables 8	7	7	7	8 - 10
9	9	9	9	9 - 10

Stage 3 :

Decisions / State Variable (S_2)		Destinations				Max (Optimal) Distance	Overall Optimal Decision
		City 8 (7)		City 9 (9)			
		Stage Distance	Cummul. Distance	Stage Distance	Cummul. Distance		
Entering Variables	5	4	11	8	17	17	5-9
	6	3	10	7	16	16	6-9
	7	8	15	4	13	15	7-8

Stage 4 :

Decisions / State Variable (S_3)		Destinations						Max. (Optimal) Distance	Optimal Stage Decision	Overall Decision
		City 5 (17)		City 6 (16)		City 7 (15)				
		Stage Dist.	Cummu. Dist.	Stage Dist.	Cummu. Dist.	Stage Dist.	Cummu. Dist.			
Entering States	2	7	24	10	26	5	20	26	2-6	2-6-9-10
	3	3	20	8	24	4	19	24	3-6	3-6-9-10
	4	6	23	10	26	5	20	26	4-6	4-6-9-10

Stage 5 :

Decisions / State Variable (S_4)	Destinations						Max. (Optimal) Distance	Optimal Stage Decision	Overall Decision
	City 2 (22)		City 3 (24)		City 4 (26)				
	Stage Dist.	Cummu. Dist.	Stage Dist.	Cummu. Dist.	Stage Dist.	Cummu. Dist.			
1	4	30	6	30	3	29	30	1-2 or 1-3	1-2-6-9-10 or 1-3-6-9-10

Here also we get two solutions as given below :

$$\text{Solution set 1} \begin{cases} \text{sequence (city) } ①\rightarrow②\rightarrow⑥\rightarrow⑨\rightarrow⑩ \\ \text{Distance } (km) \quad\quad 4 \quad\; 10 \quad\; 7 \quad\; 9 \end{cases} = 30$$

$$\text{Solution set 2} \begin{cases} \text{sequence (city) } ①\rightarrow③\rightarrow⑥\rightarrow⑨\rightarrow⑩ \\ \text{Distance } (km) \quad\quad 6 \quad\; 8 \quad\; 7 \quad\; 9 \end{cases} = 30$$

(Other stage decisions can be read at the respective tables).

Total max. distance, the medical representative can travel upto 30 *km* without repitition.

ILLUSTRAION 3

> *Find number of each of three items to be included in a package so that value of package will be maximum. Total weight of package must not exceed 5 Kgs.*
>
> | Item | Weight in Kgs | Value in Rs. |
> |------|---------------|--------------|
> | 1 | 1 | 30 |
> | 2 | 3 | 80 |
> | 3 | 2 | 65 |
>
> [JNTU (CSE) 98]

Solution :

The given problem can be formulated as follows :

Let x_1, x_2 and x_3 be the number of items of each type

maximise $f(x) = 30 x_1 + 80 x_2 + 65 x_3$ (The total value is to be maximised)

subject to $\quad x_1 + 3x_2 + 2x_3 \leq 5$

$\quad x_1, x_2, x_3 \geq 0$

We have to determine how many units of three types of items are to be loaded. So it is a three stage dynamic programming problem (DPP). Let x_j $(j = 1, 2, 3)$ denote three decisions. Let $f_j(x_j)$ denote the value of the optimal allocation for the three types of items.

If $f_j(S, x_j)$ be the value associated with the optimum solution $f_j^*(s)$, $(j = 1, 2, \ldots n)$ then we have

$$f_1^*(s) = \max_{0 \leq x_1 \leq s} f_1(s, x_j)$$

and $\quad f_j^*(s) = \max_{0 \leq x_1 \leq s} \left\{ p_j(x_j) + f_{j-1}^*(s - x_j) \right\},$

$$j = 1, 2, 3$$

where $p_j(x_j)$ denotes the expected value obtained from allocation of x_j units of weight to the item j.

Now, for stage - I problem (i.e., for one item loading).

$$f_1^* = \max_{x_1} \left\{ 30 x_1 \right\}$$

where the largest value of $x_1 = \left[\dfrac{W}{w_1} \right] = 5$

(Here W is total weight and w_1 is weight of each of item 1.)

Hence we have 6 alternatives i.e., 0, 1, 2, 3, 4 and 5 for x_1

We have the following computations.

x_1	Value of 30 x_1						Optimum solution	
S	0	1	2	3	4	5	$f_1^*(S)$	x_1^*
0	0						0	0
1	0	30					30	1
2	0	30	60				60	2
3	0	30	60	90			90	3
4	0	30	60	90	120		120	4
5	0	30	60	90	120	150	150	5

For 2–stage problem, the largest value of x_2 is $\lceil 5/3 \rceil = 1$ and thus we have two alternatives i.e., 0 and 1 for x_2

$$f_2^*(s) = \max_{x_2} \{80 x_2 + f_1^* (S - 3 x_2)\}.$$

The computations in the 2–stage problem are as follows.

x_2	Value of $80 x_2 + f_1^*(s - 3x_2)$		Optimum solution	
S	0	1	$f_2^*(s)$	x_2^*
0	0 + 0 = 0		0	0
1	0 + 30 = 30		30	0
2	0 + 60 = 60		60	0
3	0 + 90 = 90	80 + 0 = 80	90	0
4	0 + 120 = 120	80 + 30 = 110	120	0
5	0 + 150 = 150	80 + 60 = 140	150	0

For 3–stage problem the largest value of x_3 is $\lceil 5/2 \rceil = 2$ and so we have 0, 1 and 2 as the alternatives for x_3

And, $$f_3(s) = \max_{x_3} \{65 x_3 + f_2^* (S - 2x_3)\}$$

The computations of 3–stage problem are as follows.

x_2	Value of $65 x_3 + f_3^* (S - 3x_3)$			Optimum solution	
S	0	1	2	$f_f^*(s)$	x_3^*
0	0 + 0 = 0			0	0
1	0 + 30 = 30			30	0
2	0 + 60 = 60	65 + 0 = 65		65	1
3	0 + 90 = 90	65 + 30 = 95		95	1
4	0 + 120 = 120	65 + 60 = 125	130 + 0 = 130	130	2
5	0 + 150 = 150	65 + 90 = 155	130+30=160	160	2

For the given total weight W = 5, the optimum solution is ,

$$x_1^* = 1 \quad, \quad x_2^* = 0 \quad, \quad x_3^* = 2$$

and $f_3^*(s) = 160$ is the maximum value of load. In other way,

Item No.	Qty. to be Loaded	Unit Value (Rs.)	Total Value (Rs.)
1	1	30	30
2	0	80	0
3	2	65	130
Grand total	3	Rupees	160

ILLUSTRATION 4

Solve the following using tabular method of solution. The owner of a chain of four grocery stores has purchased six crates of fresh fruits. The estimated probability distributions of potential sales of the fresh fruits before spoilage differ among the four stores. The following table gives the estimated total expected fruits at each store, when it is allocated to various number of crates:

Number of Crates	Stores			
	1	*2*	*3*	*4*
0	*0*	*0*	*0*	*0*
1	*4*	*2*	*6*	*2*
2	*6*	*4*	*8*	*3*
3	*7*	*6*	*8*	*4*
4	*7*	*8*	*8*	*4*
5	*7*	*9*	*8*	*4*
6	*7*	*10*	*8*	*4*

For administrative reasons, the owner does not wish to split crates between stores. However, he is willing to distribute zero crates to any of his stores. Find the allocation of six crates to four stores so as to maximise the expected profit.

[JNTU (EEE) 98/OT]

Solution :

Let the four stores be considered as four stages in a dynamic programming formulation. The decision variables x_j $(j = 1, 2, 3, 4)$ denote the number of crates allocated as the jth stage from the previous one.

Now, let $p_j(x_j)$ be the expected profit from allocation of x_j crates to store j. Then the problem can be formulated as an L.P.P as follows:

Maximize $Z = p_1(x_1) + p_2(x_2) + p_3(x_3) + p_4(x_4)$

Subject to the constraints

$$x_1 + x_2 + x_3 + x_4 = 6,$$
$$x_1, x_2, x_3, x_4 \geq 0$$

Let there be S crates available for j remaining stores and x_j be the initial allocation. Define $f_j(x_j)$ as the value of the optimal allocation for stores 1 to 4 both inclusive.

Thus for stage $j = 1$, $f_1(S, x_1) = \{p_1(x_1)\}$

If $f_j(S, x_j)$ be the profit associated with the optimum solution $f_j^*(S)$ $(j = 1, 2, 3, 4)$ then

$$f_1^*(S) = \underset{0 \le x_1 \le S}{Max} \ p_1(x_1)$$

Thus the recursive relation is

$$f_j(S, x_j) = p_j(x_j) + f_{j+1}^*(S - x_j)$$
$$\text{for} \quad j = 1, 2, 3, 4$$
$$\text{and} \quad f_j^*(S) = \max_{0 \le x_j \le S} \ \{p_j(x_j) + f_j + 1(S - x_j)\}$$

The solution to this problem starts with $f_4^*(S)$ and is completed when $f_1^*(s)$ is obtained. (Backward recursion)

The computations for stage - I problem (i.e., for $j = 4$) are as follows :

S	f_4	x_4^*
0	0	0
1	2	1
2	3	2
3	4	4
4	4	3,4
5	4	3,4,5
6	4	3,4,5,6

For $j = 3$, we have a two–stage problem. The computations are shown below :

$$f_3(S, x_3) = P_3(x_3) + f_4^*(S - x_3)$$

x_3 / S	$f_3(S, x_3)$							Optimal Solution	
	0	1	2	3	4	5	6	$f_3(s)$	x_3^*
0	0 + 0							0	0
1	0 + 2	6 + 0						6	1
2	0 + 3	6 + 2	8 + 0					8	1,2
3	0 + 4	6 + 3	8 + 2	8 + 0				10	2
4	0 + 4	6 + 4	8 + 3	8 + 2	8 + 0			11	2
5	0 + 4	6 + 4	8 + 4	8 + 3	8 + 2	8 + 0		12	2
6	0 + 4	6 + 4	8 + 4	8 + 4	8 + 3	8 + 2	8 + 0	12	2, 3

For $j = 2$, we have three stage problem. So we have

$$f_2(S, x_2) = P_2(x_2) + f_3^*(S - x_2)$$

S \ x_2	0	1	2	3	4	5	6	$f_2^*(s)$	x_2^*
0	0 + 0							0	0
1	0 + 6	2 + 0						6	1
2	0+ 8	2 + 6	4 + 0					8	0, 1
3	0+10	2 + 8	4 + 6	6 + 0				10	0, 1, 2
4	0+11	2 +10	4 + 8	6 + 6	8 + 0			12	1, 2, 3
5	0 +12	2 +11	4 +10	6 + 8	8 + 6	9 + 0		14	2, 3, 4
6	0 +12	2 +12	4 +11	6 +10	8 + 8	9 + 6	10+0	16	3, 4

For $j = 4$, we have the required four–stage problem :

$$f_1(S, x_1) = P_1(x_1) + f_2^*(S - x_1)$$

S \ x_1			$f_1(S, x_1)$					Optimal Solution	
	0	1	2	3	4	5	6	$f_1^*(s)$	x_1^*
6	0+16	4+14	6+12	7+10	7+8	7+6	7+0	18	1, 2

From above computations, it is observed that the maximum profit of Rs. 18 can be obtained by choosing the following eight alternative solutions :

Store 1 or x_1^*	Store 2 or x_2^*	Store 3 or x_3^*	Store 4 or x_4^*
1	2	2	1
1	3	1	1
1	3	2	0
1	4	1	0
2	1	2	1
2	2	1	1
2	2	2	0
2	3	1	0

ILLUSTRATION 5

Solve the following L.L.P by dynamic programming method:
Maximize $Z = 2x_1 + 5x_2$
subject to $2x_1 + x_2 \leq 430$, $2x_2 \leq 460$
        $x_1, x_2 \geq 0$

[JNTU (CSE) 97/S]

Solution :

$$\text{Max} \quad Z = 2x_1 + 5x_2 \text{ subject to } 2x_1 + x_2 \leq 430.$$

$$2x_2 \leq 460 \,; x_1, x_2 \geq 0$$

The L.P problem can be considered as a two–stage, two–state problem because there are two decision variables* and two constraints*. Starting with second stage backward recursion, the procedure is as follows :

The optimal value of $f_2(c_1, c_2)$ at the second stage is given by

$$f_2(c_1, c_2) = \underset{0 \leq x_2 \leq c}{Max} \quad \{5x_2\}$$

where $c_1 = 430$, $c_2 = 460$. The feasible value of x_2 is a non–negative value which satisfies all the given constraints $x_2 \leq c_1 (= 430)$ and $2x_2 \leq c_2 (=460)$. Thus the maximum value 'c' that x_2 can assume is; $c = Min \left\{ 430, \dfrac{460}{2} \right\} = 230.$

Therefore $f_2(c_1, c_2) = \underset{0 \leq x_2 \leq c}{Max} \quad \{5x_2\}$

$$= 5 \min \{ 430 - 2x_1 \,; 230\}$$

and $x_2^* = Min \{430 - 2x_1, 230\} = 230$

Proceeding backwards to stage 1 ($j = 1$), the recursive relation for optimizaiton can be expressed as :

$$f_1(c_1, c_2) = \underset{0 \leq x_1 \leq c}{Max} \quad \{2x_1 + f_2^*(c_1 - 2x_1), 230)\}$$

$$= \underset{0 \leq x_1 \leq 212}{Max} \quad [2x_1 + 5 \min \{430 - 2x_1, 230\}]$$

Where maximization variable x_1, satisfying the conditions

$2x_1 \leq c_1 (= 430)$ and $ox_1 \leq c_2 (= 460)$ is minimum of

$$c = Min \left\{ \frac{430}{2}, \frac{460}{0} \right\} = 215.$$

Since the minimum (i.e., 230) of $(430-2x_1, 230)$ is obtained at $x_1 = 100$ for $0 \leq x_1 \leq 215$, we get

$$f_1^*(c_1, c_2) = Max [2x_1 + 5 Min \{430-2x_1; 230\}]$$

$$= Max [2 \times 100 + 5 \times 230]$$

$$= 1350$$

and $\quad x_1^* = 100$

Hence the optimum solution to the given L.P. problem is $x_1 = 100$; $x_2 = 230$ and $Max. Z = 1350$.

---

* The number of decision variables of LPP represent the stages of DPP and the number of constraints of LPP correspond to the states of DPP

ILLUSTRATION 6 ——————————————————————

> *A firm has accepted to supply 75 bulbs at the end of the first month and 115 bulbs at the end of the second month. The cost of manufacturing any bulb in any month is given by $45x + 0.2x^2$ rupees where x is the number of bulbs manufactured in the month. However the company can manufacture more number of bulbs and carryout to the next month, and, an inventory carrying charges of Rs.8/- per bulb is to be charged. Find the number of bulbs to be manufactured in each month to maintain the total cost to minimum. (Use calculus method of solution). (no initial inventory).*

Solution:

Let us first formulate the *NLP* problem.

Let x_1 = No. of bulbs to be manufactured in first month

 x_2 = No. of bulbs to be manufactured in second month

Objective Function:

Now, manufacturing cost in first month = $45x_1 + 0.2x_1^2$

and   manufacturing cost in second month = $45x_2 + 0.2x_2^2$

and there is no initial inventory, hence no carrying cost in first month.

But, the carrying cost in second month, if x_1 exceeds 75,

$$= 8\,(x_1 - 75)$$

Hence total cost   = $45x_1 + 0.2x_1^2 + 45x_2 + 0.2x_2^2 + 8\,(x_1 - 75)$

This total cost is to be minimised. Therefore objective function is

Minimise $Z = 0.2\,x_1^2 + 0.2\,x_2^2 + 53\,x_1 + 45\,x_2 - 600$

Constraint Set: According to the agreement of supply

 $x_1 \geq 75$ [Requirement constraint on agreed supply for first month]

 $x_1 + x_2 = 190$ [Requirement constraint on agreed supply for first two months]

Conditions of Variables: Since the production of negative number of bulbs is meaningless, we have non-negative conditions for both variable and also can not be fractions

$$x_1 \quad \geq 0, \quad x_2 \geq 0$$

Summary: The problem is summarised below

Min $Z = (45x_1 + 0.2x_1^2) + [(45x_2 + 0.2x_2^2) + 8\,(x_1 - 75)]$

subject to $x_1 \geq 75$

$$x_1 + x_2 = 190 \quad \text{and} \quad x_1 \geq 0, \quad x_2 \geq 0 \qquad \text{and integers}$$

Solution by Calculus Method:

To solve the above problem, we start from second month and go by backward recursion. If I_2 is the inventory at the beginning of the second month, the optimum number of bulbs to be manufactured in the second month is given by

$$x_2^* = 115 - I_2 \qquad\qquad \ldots\ldots \text{(i)}$$

and the cost incurred in the second month by

$$r_2\,(x_2^*, I_2) = 8\,I_2 + 45\,x_2^* + 0.2\,x_2^{*\,2} \qquad\qquad \ldots\ldots \text{(ii)}$$

By using equation (i), R_2 can be expressed as

$$R_2\,(I_2) = 8\,I_2 + 45\,(115 - I_2) + 0.2\,(115 - I_2)^2$$

$$= 8\,I_2 + 5175 - 45\,I_2 + 0.2\,(13225 + I_2^2 - 230\,I_2)$$

$$= 0.2\,I_2^2 - 83\,I_2 + 7820$$

Since the inventory at the beginning of the first month is zero, the cost involved in the first month is given by

$$R_1\,(x_1) = 45\,x_1 + 0.2\,x_2^2$$

Thus the total cost involved is given by

$$f_2\,(I_2, x_1) = (45\,x_1 + 0.2\,x_1^2) + (0.2\,I_2^2 - 83\,I_2 + 7820) \qquad \dots \text{(iv)}$$

But the inventory at the beginning of second month is related to x_1 as

$$I_2 = x_1 - 75 \qquad \dots \text{(v)}$$

Equations (iv) and (v) lead to

$$f = f_2\,(I_2)$$

$$= (45\,x_1 + 0.2\,x_1^2) + \left[0.2\,(x_1 - 75)^2 - 83\,(x_1 - 75) + 7820\right]$$

$$= 0.2\,x_1^2 + 0.2\,(x_1^2 - 150\,x_1 + 5625) + 45\,x_1 - 83\,x_1 + 6225 + 7820$$

i.e., $\qquad f = f(x_1) = 0.4\,x_1^2 - 68\,x_1 + 15170$

Since f is a function of x_1 only, the optimum value of x_1 can be obtained by calculus methods as

$$\frac{df}{dx_1} = \frac{d}{dx_1}\left[0.4\,x_1^2 - 68\,x_1 + 15170\right] = 0$$

$$= 0.8\,x_1 - 68 = 0 \qquad \text{or} \qquad x_1^* = \frac{68}{0.8} = 85$$

As $\qquad \dfrac{d^2 f}{dx_1^2} = 0.8 > 0,$

the value of x_1^* corresponds to the minimum value of f

Thus the optimum solution is given by

$$x_1^* = 85; \qquad x_2^* = 105$$

and $\qquad f_{min}^* = f(x_1^*) = 0.4\,(85)^2 - 68\,(85) + 15170$

$$= \text{Rs. } 12,280/\text{-}$$

[Manufacturing cost is Rs. 5270/- in the first month and Rs. 6390 in the second month while the carrying cost on excess (10 bulbs) manufactured in the first month is Rs. 80/-. Total Rs. 12,280/-]

ILLUSTRATION 7

Use Bellman's principle of optimality to find the optimum solution to the following problem:

Minimize $\qquad Z = y_1^2 + y_2^2 + y_3^3,$ *subject to the constraints*

$$y_1 + y_2 + y_3 \geq 15 \qquad\qquad y_1, y_2, y_3 \geq 0$$

[JNTU (EEE) 98/OT, (CSE) 2000 JNTU Prod/Mech/Chem/Mechatronics - 2001/S]

Solution :

Since the decision variables are y_1, y_2 and y_3 the given problem is a 3 stage problem defined as follows :

$$S_3 = y_1 + y_2 + y_3 \geq 15$$
$$S_2 = y_1 + y_2 = S_3 - y_3 \text{ and}$$
$$S_1 = y_1 = S_2 - y_2$$

Therefore the functional (recurrence) relation is

$$f_1 (S_1) = \min_{0 \leq y_1 \leq S_1} y_1^2 = (S_2 - y_2)^2$$

$$f_2 (S_2) = \min_{0 \leq y_2 \leq S_2} \left\{ y_1^2 + y_2^2 \right\} = \min_{0 \leq y_2 \leq S_2} \left\{ y_2^2 + f_1 (S_2) \right\}$$

and $\quad f_3 (S_3) = \min_{0 \leq y_3 \leq S_3} \left\{ y_1^2 + y_2^2 + y_3^2 \right\} = \min_{0 \leq y_3 \leq S_3} \left\{ y_3^2 + f_2 (S_2) \right\}$

$$f_2 (S_2) = \min_{0 \leq y_2 \leq S_2} \left\{ y_2^2 + (S_2 - y_2)^2 \right\} = \left(\frac{1}{2} S_2 \right)^2 + \left(S_2 - \frac{1}{2} S_2 \right)^2$$

$$= \frac{1}{2} S_2^2$$

According to maxima -minima principles, the minimum value of function $y_2^2 + (S_2 - y_2)^2$ can be obtained by equating the first derivative to zero

$$\frac{d}{dy_2} \left[y_2^2 + (S_2 - y_2)^2 \right] = 0$$

i.e. $\quad 2y_2 - 2 (S_2 - y_2) = 0$

$\Rightarrow \quad 4y_2 = 2 S_2 \qquad \therefore \quad y_2 = \frac{1}{2} S_2$

$$f_3 (S_3) = \min_{0 \leq y_3 \leq S_3} \left\{ y_3^2 + f_2 (S_2) \right\} = \min_{0 \leq y_3 \leq S_3} \left\{ y_3^2 + \frac{1}{2} (S_3 - y_3)^2 \right\}$$

or $\quad f_3 (15) = \min_{0 \leq y_3 \leq S_3} \left\{ y_3^2 + \frac{1}{2} (15 - y_3)^2 \right\}$

since, $S_3 (y_1 + y_2 + y_3) \geq 15$

Since the minimum value of the function $y_3^2 + \frac{1}{2} (15 - y_3)^2$ occurs at $y_3 = 5$, we have

$$f_3 (15) = \left\{ 5^2 + \frac{1}{2} (15 - 5)^2 \right\} = 75$$

Thus, $\qquad S_3 = 15$ implies that $y_3^* = 5$;

$\qquad S_2 = S_3 - y_3 = 15 - 5 = 10$ implies that

$\qquad y_2 = \frac{1}{2} S_2 = 5$

$\qquad S_1 = S_2 - y_2 = 10 - 5 - 5$ implies that

$\qquad y_1^* = S_1 = 5$

Hence the optimal policy is $y_1 = 5$, $y_2 = 5$ and $y_3 = 5$ with $f_3^* (15) = 75$.

ILLUSTRATION 8 ————————————————————————————————

Using dynamic programming approach, solve reliability problem with following data.

Mi	i = 1		i = 2		i = 3	
	R	C	R	C	R	C
1	0.6	2	0.8	3	0.7	1
2	0.7	4	0.8	5	0.8	2
3	0.9	5	0.9	6	0.9	3

The total capital available is 10 (in units of thousand rupees)

[JNTU (Mech.) CCC-2001/S, (ECE) CCC-2001/S]

Solution :

Let R be the total reliability of a system of 'n' components arranged in series and let there be 'k' parallel units per component ($j = 1, 2, 3,$)

Now, we formulate the problem as

Minimise $R = R_1 ; R_2 ; R_3$

Subject to constraints $C_1 + C_2 + C_3 \leq C$ where 'C' is total capital available i.e., 10

Now, to develop the recursive relationship.

Let us define the following.

x_j : The capital allocated to the stage j where $j = 1, 2, 3$.

k_j : Number of parallel units assigned onto main component (j)

$f_j (x_j)$: Reliability of components (stages) 1 to j given that $0 \leq x_j \leq C$

The recursive equations are :

$$f_1 (x_1) = \max_{0 < C_1 \leq x_j} \{R_1 (C_1)\} ; k_1 = 1, 2, 3$$

and $\quad f_j (x_j) = \max_{0 < C_j \leq x_j} \{R_j (C_j) ; f_{j+1} (x_j - C_j)\}; j = 1, 2, 3$ and $k_j = 1, 2, 3$

Starting with stage-1, since main component-1 must include atleast one (parallel) unit. we find that x_1 must at least equal $c_j (1) = 2$. By the same reasoning x_1 cannot exceed $10 - (3 + 1) = 6$; otherwise remaining capital will not be sufficient to provide main components 2 and 3 with atleast one (parallel) unit each. Following the same reasoning we get $x_2 = 5, 6, 7, 8$ or 9 and $x_3 = 6, 7, 8, 9$ or 10.

Stage 1 : $f_1(x_1) = max.\ R_1(C_1)$

x_1	$k_1 = 1$	$k_1 = 2$	$k_1 = 3$	Optimal Solution	
	$R_1 = 0.6\ ;\ C_1 = 2$	$R_1 = 0.7,\ C_1 = 4$	$R_1 = 0.9\ ;\ C_1 = 5$	$f_1(x_1)$	k_1^*
2	0.6	—	—	0.6	1
3	0.6	—	—	0.6	1
4	0.6	0.7	—	0.7	2
5	0.6	0.7	0.9	0.9	3
6	0.6	0.7	0.9	0.9	3

Stage 2 : $f_2(x_2) = R_2(C_2)\,f_1(x_2 - C_2)$

x_2	$k_1 = 1$	$k_1 = 2$	$k_1 = 3$	Optimal Solution	
	$R_2 = 0.8\ ;\ C_2 = 3$	$R_2 = 0.8,\ C_2 = 5$	$R_2 = 0.9\ ;\ C_2 = 6$	$f_2(x_2)$	k_2^*
5	$0.8 \times 0.6 = 0.48$	—	—	0.48	1
6	$0.8 \times 0.6 = 0.48$	—	—	0.48	1
7	$0.8 \times 0.7 = 0.56$	$0.8 \times 0.6 = 0.48$	—	0.56	1
8	$0.8 \times 0.9 = 0.72$	$0.8 \times 0.6 = 0.48$	$0.9 \times 0.6 = 0.54$	0.72	1
9	$0.8 \times 0.9 = 0.72$	$0.8 \times 0.7 = 0.56$	$0.9 \times 0.6 = 0.54$	0.72	1

Stage 3 : $f_3(x_3) = R_3(C_3)\,f_2(X_3 - C_3)$

x_3	$k_1 = 1$	$k_1 = 2$	$k_1 = 3$	Optimal Solution	
	$R_3 = 0.7\ ;\ C_3 = 1$	$R_3 = 0.8,\ C_3 = 2$	$R_3 = 0.9\ ;\ C_3 = 3$	$f_3^*(x_3)$	k_3^*
6	$0.7 \times 0.48 = 0.336$	—	—	0.336	1
7	$0.7 \times 0.48 = 0.336$	$0.8 \times 0.48 = 0.384$	—	0.384	2
8	$0.7 \times 0.56 = 0.392$	$0.8 \times 0.48 = 0.384$	$0.9 \times 0.48 = 0.432$	0.432	3
9	$0.7 \times 0.72 = 0.504$	$0.8 \times 0.56 = 0.448$	$0.9 \times 0.48 = 0.432$	0.504	1
10	$0.7 \times 0.72 = 0.504$	$0.8 \times 0.72 = 0.576$	$0.9 \times 0.56 = 0.504$	0.576	2

Practice Problems

Solve the following LPP by Dynamic programming.

1. Max $Z = 8x_1 + 7x_2$. Subject to $2x_1 + x_2 \le 8$, $5x_1 + 2x_2 \le 5$, $x_1, x_2 \ge 0$

Answer : $x_1 = 0$, $x_2 = 7.5$ Max. $Z = 52.5$

2. Max $Z = 3x_1 + 7x_2$, Subject to $x_1 + 4x_2 \le 8$, $x_2 \le 2$, $x_1, x_2 \ge 0$

Answer : $x_1 = 8$, $x_2 = 0$, $Z_{\max} = 24$

3. Max $Z = 50 x_1 + 100 x_2$, Subject to $2x_1 + 3x_2 \le 48$, $x_1 + 3x_2 \le 42$,
   $x_1 + x_2 \le 21$, $x_1, x_2 \ge 0$

Answer : $x_1 = 6$, $x_2 = 12$, $Z_{\max} = 60$

4. Min $Z = x_1 + 3x_2 + 4x_3$, Subject to $2x_1 + 4x_2 + 3x_3 \ge 60$, $3x_1 + x_2 + 3x_3 \ge 90$
   $x_1, x_2, x_3 \ge 0$

5. Max. $Z = 2x_1 + 5x_2$, Subject to $2x_1 + x_2 \le 43$, $2x_2 \le 46$, $x_1, x_2 \ge 0$

Answer : $x_1 = 10$, $x_2 = 23$, $Z_{\max} = 135$

6. Suggest the best combination of advertising mdeia and advertising frequencies to a shaving blade manufacturing company interested in selecting the advertising, mdeia for product with help of following estimated data.

Frequency / Month	Expected Radio	Sales in (Rs.) Newspaper	Television
1	150	125	220
2	250	200	275
3	300	225	370
4	310	250	340

The cost of advertisting in newspaper, radio and television are Rs. 450, Rs. 1000, Rs. 1800 respectively for one time appreance. The budget allocated for advertising is Rs. 5000/- month.

7. Find shortest path from vertex A to B along arcs joining the verious vertices lying between A and B. The lengths of paths are as shown here.

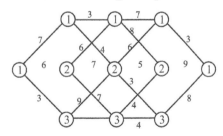

FIGURE 12.4 :

8. Find the shortest path from vertex A to B given. The lengths of paths are shown in the following network.

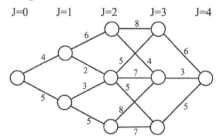

FIGURE 12.5 :

9. Using dynamic programming find the shortest path from city 1 to city 10 in the following network. The distance between two cities is given on arrow connecting. them.

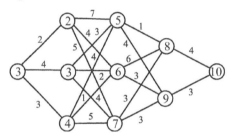

FIGURE 12.6 :

10. The routes of an air line, which connects 16 cities (A ,B, ... P) are shown in fig. below. Journey from one city to another is possible only along the lines (routes). Shown with the associated costs indicated on the path segments. If a persons wants to travel from city A to city P with minimum cost, without any back tracking, determine the optimal path (route) using dynamic programming.

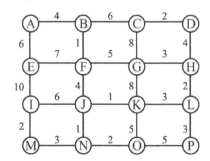

FIGURE 12.7 :

11. A minimum cost pipe line is to be laid between points (towns) A and E. the pipe line is required to pass through one node out of B_1, B_2 and B_3 one out of C_1, C_2 and C_3 and one out of D_1, D_2 and D_3 (see fig. below).

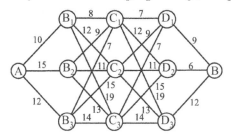

FIGURE 12.8 :

12. Find the minimum path by solving the dynamic programming problem given below.

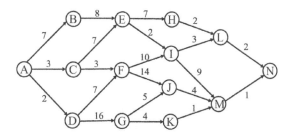

FIGURE 12.9 :

13. Use dynamic programming to find the value of. Max $Z = y_1; y_2; y_3$, Subject to constraint : $y_1 + y_2 + y_3 = 5$ and $y_1, y_2, y_3 \geq 0$

Hints : $f_1(x_1) = \max \ (z_1)$ and $f_j(x_j) \ \max_{0 \leq z_j \leq x_j} \ \{z_j f_{j-1}(x_j - z_j)\}, j = 1, 2, 3.$

Answer : $\left(\dfrac{5}{3}, \dfrac{5}{3}, \dfrac{5}{3}\right)$ with $f_3(5) = \left(\dfrac{5}{3}\right)^3$

14. Use dynamic programming to Max $Z = 2x_1 + 3x_2$ Subject to constraints

$$x_1 - x_2 \leq 1, x_1 + x_2 \leq 3, x_1 + x_2 \geq 0 \text{ and } x_1, x_2, x_3 \geq 0$$

Answer : $x_1 = 0$, $x_2 = 3$, Max $Z = 9$

15. Use DPP method to Min $Z = x_2 + 3x_2 + 4x_3$, Subject to $2x_1 + 4x_2 + 3x_3 \geq 60$,

$3x_1 + 2x_2 + x_3 \geq 60$

$2x_1 + x_2 + 3x_3 \geq 90$ and $x_1, x_2, x_3 \geq 0$

Review Questions

1. Explain the salient features of DPP. [JNTU (Mech.) 96/CC, ECE/CC - 96, (Mech) 94]

2. Write a short note on Dynamic Programming for Optimization.
   [JNTU (CSE) 97, (Mech) 98/S, Mech 99]

3. Define Bellman's principle of optimality and its application to DPP.
   [JNTU (EEE) 97/S, Mech/Prod/Chem/Mechatronics - 2001/S]

4. Write short notes on
   (a) stage   (b) state   (c) recursive approach   [JNTU (CSE) 96/S]

5. Explain tabular method of solving DPP   [JNTU (EEE) 98, (Mech.) 95]

6. Write a short note on the characteristics of DPP   [JNTU - Mech. 94/S]

7. Explain step by step procedure for developing an optimal decision policy in Dynamic Programming Problem with an example.   [JNTU (ECE) 97]

Objective Type Questions

1. In dynamic programming, which of the following does not change at every stage

   (a) objective function   (b) recursive function
   (c) results   (d) state variables

2. An optimal policy (set of decisions) has the property that whatever the initial state and decision are, the remaining decisions must constitute an optimal policy with regard to the _____ resulting from the first decision

   (a) state   (b) function   (c) recursion   (d) stage

3. The sub problem of a main problem in a DPP is said to be

   (a) state   (b) function   (c) recursion   (d) stage

4. An LPP solved by using DPP method assumes the variables as _____

   (a) states   (b) stages
   (c) recursions   (d) return functions

5. Which of the following is considered as recursive optimisation

   (a) linear programming problem   (b) dynamic programming problem
   (c) assignment problem   (d) transportation problem

6. The optimality principle (to solve DPP) is given by

   (a) Dantzig   (b) Richard Brown
   (c) Bellman   (d) Kimball

7. The conditions prevailing in decision process at a sub-problem of a main problem are said to be
   (a) stages   (b) functions
   (c) states   (d) recursions

8. A decisions that is made has its own benefits represented in an algebraic equation is called _____

   (a) stage   (b) state variable

   (c) return function   (d) decision variable

9. Return function of DPP depends on

   (a) state variables   (b) stage decision

   (c) decision policy   (d) all the above

10. The equation that converts all the stages of the DPP is known as

    (a) state function   (b) stage function

    (c) transition function   (d) return function

11. The final stage decision is first solved in _____ recursion

    (a) forward   (b) backward

    (c) upward   (d) downward

12. A DPP will have _____

    (a) no objective function

    (b) only one objective functions

    (c) objective function equal to that of no. of stages

    (d) infinite objective functions

13. A three variable, 2 constraint LPP if solved in dynamic programming approach will have _____ stages

    (a) 2   (b) 3   (c) 6   (d) 5

14. If an LPP has 'n' decision variables in 'm' constraints, then it can be converted into DPP with

    (a) m stages, m states   (b) m stages, n states

    (c) n stages, m states   (d) n stages, n states

15. Which of the following is not true in the case of DPP approach

    (a) main problem is split into sub problems

    (b) recursive relation ship is used

    (c) values of return function remain same at all stages

    (d) there may be more than one optimal policy

16. Find odd man out with reference to dynamic programming

    (a) objective function   (b) bijective function

    (c) recursive function   (d) transformation function

17. Characteristics LPP and DPP approaches differ in

    (a) objective function   (b) constraints

    (c) variables   (d) conditions of variables

18. DPP can not be applied to

    (a) production scheduling problems   (b) inventory problems

    (c) net work analysis   (d) behavioural problems

19. According to optimality principle, if an optimal solution is obtained to a system _____ of it must be optimal
    (a) some portion   (b) final stage
    (c) any portion   (d) at least one stage

20. When input and output are given as fixed which recursion works well ?
    (a) forward   (b) backward
    (c) both recursions are equally good   (d) no recursion can work well

Fill in the Blanks

1. An optimal policy has the property that whatever the _____ and _____ are, the remaining decision must constitute and optimal policy with regard to the stage resulting from the _____

2. A DPP is decomposed into smaller sub problems called _____

3. The conditions of a sub problem are known as _____

4. The algebraic equation that represents the benefits of a decision in DPP is called _____

5. The optimality principle is given by _____

6. The equation that interconnects all the stages of DPP is known as _____

7. An LPP of 'm' decision variables in 'n' constraints can be converted to DPP with _____ stages and _____ stages

8. First stage decision is solved last in _____ recursive approach

9. According optimality principle, if an optimal solution is obtained to a system, it must be optimal to _____ of the system.

10. A 3 - variable, 2 constraint LPP can be solved in DPP approach of _____ states and _____ stages.

Answers

Objective Type Questions :				
1. (b)	2. (a)	3. (d)	4. (b)	5. (b)
6. (c)	7. (c)	8. (c)	9. (d)	10. (c)
11. (b)	12. (c)	13. (b)	14. (c)	15. (c)
16. (b)	17. (a)	18. (d)	19. (c)	20. (c)

Fill in the Blanks :	
1. Initial state, decisions, first decision	2. stages
3. states	4. return function
5. Richard Bellman	6. Transition function or state transformation function
7. m, n	8. backward
9. any stage or portion	10. 2, 3

Printed and bound by CPI Group (UK) Ltd, Croydon, CR0 4YY

17/10/2024

01775694-0016